NEURODEGENERATION AND PRION DISEASE

NEURODEGENERATION AND PRION DISEASE

Edited by

DAVID R. BROWN
University of Bath, UK

Springer

Library of Congress Cataloging-in-Publication Data

Neurodegeneration and prion disease / edited by David R. Brown
p. ; cm.
Includes bibliographical references and index.
ISBN 0-387-23922-7 (alk. paper)
1. Prion Diseases. 2. Nervous system—Degeneration. I. Brown, David R., PhD.
[DNLM: 1. Prion Diseases—physiopathology. 2. Cell Death—physiology. 3. Nerve Degeneration—etiology. 4. Neurons—physiology. 5. Prion Diseases—complications. 6. Prions—pathogenicity. WL 300 N4938152 2005]
QR201.P737N48 2005
616.8′3—dc22

2004062570

Printed in the United States of America.

9 8 7 6 5 4 3 2 1 SPIN 11054078

springeronline.com

To:

Lorna Jessica Hellena Brown
and Hadassah Margaret Irmgard Brown

As you grow and to other things pass on
Your path winding through Spring and storm
And from treacherous heaven away you turn
You remain beloved daughters of this black swan.

David R. Brown, MSc. Ph.D.

David Brown has worked in the field of prion disease or TSEs for over ten years. He was born in Australia but spent part of his early childhood in the United Kingdom. After returning to Australia he completes his schooling in Sydney and attended Sydney University. There he completed a Bachelor of Science degree in biochemistry, a Master of Science degree in neurobiology and a Doctor of Philosophy degree also in Neuroscience. His initial research interests included neuronal growth factors and topographic innervation of toad muscle. Following three years of postdoctoral research, Dr. Brown left Australia in 1994. Since then he has worked in the Albert Einstein College of Medicine in New York, the Department of Neuropathology in Göttingen and the Department of Biochemistry at the University of Cambridge. His interest in prion began during his four years researching in Germany. After returning to work in the United Kingdom to work at Cambridge, Dr. Brown established his own independent research group that quickly gained international recognition. In parallel with research focusing on the function of the prion protein and mechanisms of cell death in neurodegeneration, Dr. Brown's research has also investigated basic aspects of cellular neurobiology including the nature of the interactions between neurones and glial cells. David Brown is currently a Reader in Biochemistry at the University of Bath and his research continues to reap international recognition and acclaim. He is also a member of the Spongiform Encephalopathy Advisory Committee that advises the UK government on issues to do with BSE and variant CJD.

Contents

List of Contributors

Subhabrata Basu
Institute of Pathology, Case Western Reserve University, 2085, Adelbert Road, Cleveland, Ohio, USA.

Katarina Bedecs
Department of Biochemistry and Biophysics, Stockholm University, Svante Arrhenius väg 12, S-10691 Stockholm, Sweden.

Sharmila Bose
Department of Molecular Cellular & Developmental Biology, The University of Michigan, 830 North University Avenue, Ann Arbor, Michigan, USA.

David R. Brown
Department of Biology and Biochemistry, University of Bath, Bath, BA2 7AY, UK.

Herbert Budka
Institute of Neurology, Medical University of Vienna, AKH 4J, Währinger Gürtel 18-20, A-1097 Vienna, Austria.

Orso Bugiani
Istituto Nazionale Neurologico "Carlo Besta", Division of Neuropathology and Biochemistry and Molecular Pharmacology, Via Celoria 11, 20133 Milan, Italy

Roberto Cappai
Department of Pathology and The Centre for Neuroscience, The University of Melbourne, Melbourne, Victoria 3010. Australia.

Richard I. Carp
New York State Institute for Basic Research in Developmental Disabilities, 1050 Forest Hill Road, Staten Island, NY 10314, USA.

John C. Curtis
Biomedical Sciences, Hugh Robson Building, George Square, Edinburgh, EH8 9XD, UK.

Colm Cunningham
CNS Inflammation Group, School of
Biological Sciences, Biomedical Sciences Building, University of Southampton, Southampton SO16 7PX, UK.

Tullio Florio
Section of Pharmacology, Dept. Oncology Biology and Genetics, University of Genova, Genova, Italy and Pharmacology and Neuroscience, National Institute for Cancer Research (IST) Genova, Italy.

Gianluigi Forloni
Istituto di Ricerche Farmacologiche "Mario Negri", Department of Neuroscience, Via Eritrea 62, 20157 Milan, Italy

Bernardino Ghetti
Indiana University School of Medicine, Department of Pathology and Laboratory Medicine, 635 Barnhill Drive MSA138, Indianapolis, IN 46202, USA

Sabine Gilch
Institute of Virology Prion Research Group, Technical University of Munich, Biedersteinerstrasse 29, 80802 Munich, Germany.

Yaping Gu
Institute of Pathology, Case Western Reserve University, 2085, Adelbert Road, Cleveland, Ohio, USA.

Ramanujan S. Hegde
Cell Biology and Metabolism Branch, NICHD, National Institutes of Health, 18 Library Drive, Bldg. 18T, Room 101, Bethesda, MD 20892, USA.

Claudio Hetz
Instituto de Ciencias Biomedicas, Universidad de Chile, Santiago, Chile.

Andrew F. Hill
Department of Biochemistry and Molecular Biology, and Department of Pathology and The Centre for Neuroscience, The University of Melbourne, Melbourne, Victoria 3010, Australia.

James W. Ironside
National Creutzfeldt-Jakob Disease Surveillance Unit,
Western General Hospital, Edinburgh, EH4 2XU, UK.

Alex R. Johnston
Biomedical Sciences, Hugh Robson Building, George Square,
Edinburgh, EH8 9XD, UK.

Boe-Hyun Kim
Ilsong Institute of Life Science, and [b]Department of Microbiology,
College of Medicine, Hallym University, Ilsong Building,
Kwanyang-dong 1605-4, Dongan-gu, Anyang 431-060,
South Korea.

Jae-Il Kim
New York State Institute for Basic Research in Developmental
Disabilities, 1050 Forest Hill Road, Staten Island, NY 10314, USA.

Yong-Sun Kim
Ilsong Institute of Life Science, and Department of Microbiology,
College of Medicine, Hallym University, Ilsong Building,
Kwanyang-dong 1605-4, Dongan-gu, Anyang 431-060, South Korea.

Oscar Kuruvilla
Institute of Pathology, Case Western Reserve University, 2085,
Adelbert Road, Cleveland, Ohio, USA.

Pawel P. Liberski
Laboratory of Electron Microscopy and Neuropathology,
Department of Molecular Pathology and Neuropathology, Medical
University, Lodz, Poland.

Xiu Luo
Institute of Pathology, Case Western Reserve University, 2085,
Adelbert Road, Cleveland, Ohio, USA.

Nikki K. MacLeod
Biomedical Sciences, Hugh Robson Building, George Square,
Edinburgh, EH8 9XD, UK.

Rodrigo Morales
Department of Neurology, University of Texas Medical Branch,
Galveston, Texas, USA.

Michela Morbin
Istituto Nazionale Neurologico "Carlo Besta", Division of
Neuropathology and Neurology 5, Via Celoria 11, 20133 Milan, Italy

Richa Mishra
Institute of Pathology, Case Western Reserve University, 2085, Adelbert Road, Cleveland, Ohio, USA.

Max Nunziante
Institute of Virology Prion Research Group, Technical University of Munich, Biedersteinerstrasse 29, 80802 Munich, Germany.

Pedro Piccardo
Center for Biologics Evaluation and Research, Food and Drug Administration, Rockville, MD 20852, USA and Indiana University School of Medicine, Department of Pathology and Laboratory Medicine, 635 Barnhill Drive MSA138, Indianapolis, IN 46202, USA

Neena S. Rane
Cell Biology and Metabolism Branch, NICHD, National Institutes of Health, 18 Library Drive, Bldg. 18T, Room 101, Bethesda, MD 20892, USA.

Suehiro Sakaguchi
Department of Molecular Microbiology and Immunology, Nagasaki University Graduate School of Biomedical Sciences, Sakamoto 1-12-4, Nagasaki 852-8523 and PRESTO, Japan Science and Technology Agency, 4-1-8 Honcho Kawaguchi, Saitama, Japan.

Mario Salmona
Istituto di Ricerche Farmacologiche "Mario Negri", Department of Biochemistry and Molecular Pharmacology, Via Eritrea 62, 20157 Milan, Italy

Hermann M. Schätzl
Institute of Virology Prion Research Group, Technical University of Munich, Biedersteinerstrasse 29, 80802 Munich, Germany.

Gennaro Schettini
Section of Pharmacology, Dept. Oncology Biology and Genetics, University of Genova, Genova, Italy and Pharmacology and Neuroscience, National Institute for Cancer Research (IST) Genova, Italy.

Neena Singh
Institute of Pathology, Case Western Reserve University, 2085, Adelbert Road, Cleveland, Ohio, USA.

Claudio Soto
Department of Neurology, University of Texas Medical Branch, Galveston, Texas, USA.

Fabrizio Tagliavini
Istituto Nazionale Neurologico "Carlo Besta", Division of Neuropathology and Neurology 5, Via Celoria 11, 20133 Milan, Italy

Stefano Thellung
Section of Pharmacology, Dept. Oncology Biology and Genetics, University of Genova, Genova, Italy.

Ursula Unterberger
Institute of Neurology, Medical University of Vienna, AKH 4J, Währinger Gürtel 18-20, A-1097 Vienna, Austria.

Laura Vella
Department of Biochemistry and Molecular Biology, and Department of Pathology, The University of Melbourne, Melbourne, Victoria 3010, Australia.

Till Voigtländer
Institute of Neurology, Medical University of Vienna, AKH 4J, Währinger Gürtel 18-20, A-1097 Vienna, Austria.

Chapter 1

INTRODUCTION

David R. Brown

Department of Biology and Biochemistry, University of Bath, Bath BA2 7AY, UK

In 1982 Stanley Prusiner and colleagues purified an abnormal protein from the brains of mice experimentally infected with a rare sheep disease called scrapie[1]. This protein was called the prion protein. Earlier work had suggested that this diseases and others, loosely collected together as transmissible spongiform encephalopathies (TSEs), were not transmitted by conventional infectious agents. Prusiner suggested that this new protein was the infectious agent in these diseases[2]. Such a contentious suggestion lead to ferocious debate. Many researchers still maintained that there was no such thing as an infectious protein. Despite this, by 1990 most people accepted that the cause of the TSEs was the abnormal isoform of the prion protein his research group had identified. The most convincing evidence for this had come from the work of Charles Weissmann, whose prion protein knockout mice could not be infected because they lacked expression of the protein that was now forever linked to these disease[3,4]. Since then it has become more widely accepted for these diseases to be termed prion diseases. In 1997 when Stanley Prusiner won the Nobel Prize for his work on prion diseases[5]. Even then, there was still an element of resistance in the scientific community. It was considered that, in order the transmissible agent to truly be a protein only, the protein would have to be generated from a recombinant source.

In 2004 that evidence emerged. Recombinant protein injected into mice led to a prion disease that could then be transmitted to other mice[6]. Naturally, scepticism still continues about this novel theory. Those who work in the prion field know that this is simply part of the game. Intense scepticism of any findings on any aspect of research in prion diseases makes progress in the field very slow. Part of the problem is probably due to a fundamental misunderstanding of the nature of prion diseases. Although the diseases can be transmitted experimentally, prion diseases are not contagious diseases as are bacterial or viral disease.

Some forms of prion diseases are inherited, such as Gerstmann-Sträussler-Scheinker syndrome (GSS)[7–8]. GSS is linked to point mutations in the prion protein gene (*prnp*). Despite inheriting what amounts to a dominant lethal genetic mutation, GSS patients usually reach their fifth decade of life before symtoms of the disease emerge. This clearly links onset of these diseases to something inherant in the normal aging process. Other forms of transmission have occurred because of human intervention. Iatrogenic CJD results from use of tissues or hormone derivatives of tissue from people who had the more conventional disease, sporadic CJD. Kuru, a disease of the native of New Guinea occurred because of the ritual practice of eating brains from older relatives[9]. The new disease, variant CJD had been linked to the eating of food contaminated with the BSE agent. BSE (bovine spongiform encephalopathy) and variant CJD share many similarities and the two diseases clearly have a similar origin[10–11]. It is widely accepted that BSE caused vCJD but there is also considerable doubt that vCJD arose from the eating of BSE contaminated meat. BSE was also mostly the result of human intervention. The feeding of rendered animal remains back to dairy cattle result in tens of thousands of cases of BSE carrying cows. Yet, now that this practice is banned and BSE number have dropped dramatically, there remains a significant number of BSE cases. The cause of these cases of BSE remains unknown. This is similar to the majority of cases of human prion disease. The major form of human prion diseases is sporadic CJD. This disease cannot be linked to any form of infection. Similarly, the disease of sheep called scrapie can also not be linked to any specific infection event. Scrapie is the first described prion disease with reports dating back to the 15th century.

As panic over the BSE epidemic subsides and the predicted exponential increase in variant CJD cases has not happened, more rational thought has entered into the arena to assess the possible cause of the major forms of prion diseases. The two logical explanations that have been put forward are the following: The sporadic forms of prion diseases could arise through a freak event in the normal ageing process. As mentioned above, GSS does not manifest until late in life. This implies that the kinetics of prion formation are very slow and take upwards of 30 years to result in abnormal prion protein forming in the brain. Alternatively some change that occurs as we grow older may be needed to trigger protein conversion to make the normal cellular form of the prion protein to flip conformation and generated PrP^{Sc}. Once a significant amount of this protein is formed it is able to catalyse its own conversion and spontaneous deposition of large amounts of PrP^{Sc} results in the pathological changes leading to CJD[12]. The possible change in the ageing brain is the gradual decay in the balance between oxidative damage and antioxidant defence. The second hypothesis about the cause of sporadic prion diseases is that exposure to an agent in the environment may trigger protein conversion. Evidence for this comes from the existence of localised hot spots for different prion diseases[13]. The disease of deer, chronic wasting disease, is very heavily localised to small areas with US states such as Colorado. In Iceland, some farms have recurrent scrapie problems while other remain consistently scrapie free. What in the environment could have such an effect remains to be verified. However, the prime candidate has been manganese[14]. Manganese accumulates in the brains of patients and animals with prion diseases[15–16]. However, it remains to be shown whether manganese is causative, a factor that enhances the incidence of an ever present disease or is simply coincidental.

The implication of all the forgoing is that prion diseases are fundamentally a result of a normal brain protein becoming conformationally altered. An event or a series of similar events result in the stabilisation in a conformational switch in the isoform of the protein generated by the brain[12]. The trigger of this can be one of three possibilities. The first is the introduction into the cell of preformed PrP^{Sc} aggregates. This is then able to catalyse conformation alteration of prion protein generated by that cell. The second possibility is that the normal cellular isoform of the prion protein (PrP^{c}) encounters a different agent which then catalyses conversion. This could be interaction with a metal that does not normally bind to the protein such as manganese[17]. Lastly, conversion to the abnormal isoform occurs naturally but with a low probability. This implies that, the kinetic equilibrium does not favour PrP^{Sc} formation but that in time a small amount will form that is the sufficient to catalyse further conversion by the first mechanism. An alternative version of this last hypothesis is that PrP^{Sc} is formed in the brain all the time but mechanism are in place which rapidly clear it away before it can autocatalyse further PrP^{Sc} formation. Disease develops when this corrective process falters, possibly as a result of ageing.

Understanding the common threads in all these theories and the clear link between disease progress and expression of the prion protein and the conformation it assumes makes discussion of any "contagion" causing these disease seem absurd. Panic among both the lay and the scientific communities about the inherent infectiousness of prion disease is purely an hysterical response to misinformation or wanton ignorance of the fundamental truth of the cause of these disease. A normal brain glycoprotein becomes converted to a protease resistant isoform, by whatever mechanism, and initiates a series of pathological changes in the brain resulting in death. This implies that to understand these diseases we must know what causes the patient's own prion protein to change conformation and understand how production of this abnormal conformation relates to the pathological changes that cause the death of the patient.

Until recently study of the prion protein has focussed on PrP^{Sc} and the normal form of the protein has been overlooked. However, the gene for PrP was first described in 1986[18], the year that BSE first emerged. The protein is a glycoprotein anchored to the outside of the cell by a glycosylphophatidylinositol (GPI) anchor[19]. This means that the protein is attached to the membrane by a sugar group. Although there has been some speculation about there being a transmembrane form of the protein, this has largely been dismissed[20–22]. Similar reports of dimeric forms of the protein and subcellular localisation of the protein to the cytosol or nucleus are isolated and unconfirmed[23–25]. The protein is highly expressed by neurones and is concentrated in the synapse[26]. The expression of the protein is not specific to neurones and low level expression can be detected in many cell types. The age of the cellular prion protein began in 1995 with the first suggestions that a fragment of the protein could bind copper[27]. This largely was ignored until 1997 when a number of colleagues and I provided the first accepted evidence that the protein binds copper *in vivo*[28]. Since then there has been overwhelming support for the idea that PrP^c is a metalloprotein.

The function of the protein has been the subject of a number of investigations. Despite numerous different approaches the emerging consensus is that lack of expression of PrP^c causes cells to respond poorly to stress[29]. These changes can range from altered electrophysiological parameters[30], altered sleep patterns[31], modified cell adhesion characteristics[32] and disturbed cell signalling pathways[33]. More substantial evidence points to PrP^c being some form of antioxidant. My own research has suggested that PrP^c is a molecule with the ability to clear away superoxide radicals that would otherwise damage cell components[34]. This would make a PrP^c superoxide dismutase. Alternative research has shown that PrP^c can alter copper uptake into cells[35] and that binding of copper to PrP^c is important to the mechanism by which it is internalised from the cell surface[36]. These theories are not contradictory, as sequesting copper is in itself an antioxidant effect. Copper has the potential to generating molecules that can cause oxidative stress. Therefore the leading theory as to the function of PrP^c is that it is an antioxidant.

Conversion of PrP^c to PrP^{Sc} results in the loss of function of the protein without the loss of its expression. In prion protein knockout mice, lack of expression of PrP^c could be compensated for by rapid changes in expression of other proteins that could perform similar functions. As a potential antioxidant, such rapid compensation is very likely considering the wide range of antioxidants that the body can mobilise and the high induciblity of many cellular antioxidants. One implication is that loss of PrP^c function could expose neurones to assaults that cause initiate cell death. Therefore understanding the function of this protein and how to compensate for it could be one possible way to counteract the cell death that occurs in prion disease.

Prion diseases are neurodegenerative conditions. They result in a very rapid loss of neurones in specific areas of the brain. This neuronal loss occurs late in the disease and corresponds to onset of neurological and behavioural symptoms. In experimentally induced prion disease, there is a long incubation time between challenge with the prion disease agent and the onset of symptoms. In the case of BSE this can be years. Following onset of symptoms death from complications follows very rapidly. In humans with CJD this can be a matter of months. If cell death could be stopped then possibly, the CJD patient could recover. Clearly, knowing what causes this cell death is central to understanding these disease. Surprisingly, until recent years, there has been little research on the mechanism of cell death in prion disease. Reviews on the subject of "neurodegeneration and prion disease" often failed to mention mechanisms of cell death in any detail.

The first models of the mechanism of neurodegeneration emerged from cell cultures studies in the early 1990s. These models showed that PrP^{Sc} or a peptide derivative could kills neurones by an apoptotic mechanism[37–38]. The first finding was that was of any significance was that neuronal cell death requires the expression of PrP^c by the target cell[39]. My research was the first to show that neurones from prion protein knockout mice were resistant to toxic prions. This was later confirmed in animal models[40–41].

Advancement in the field of neurodegeneration and prion diseases has resulted in the research described in the chapters of this book. This compilation therefore reflects the great strides that have been made in recent years. Many individual and complementary approaches have been taken, providing a wealth of information that has the potential to one day provide us with a possible way forward in finding preventative treatments to halt the advance of neurodegeneration. Prion diseases are rare but so are reliable models of most human neurodegenerative diseases. In this regard prion diseases are the exception as experiment infection of mice provides us with an accurate and essential tool for research. The implication of this is that study of prion diseases might provide insights into neurodegeneration that are relevant to other diseases like Alzheimer's disease where animal models don't exist.

The main model used by researchers to study prion diseases is the mouse model using mouse passaged scrapie. Some researchers also use hamsters but the availability of transgenic mice makes mice a more attractive choice. Studies with such mice have lead to a whole range of interesting research. The sheep disease scrapie can be divided into a series of strains or different forms. These strains retain a range of characteristics when used to infect mice of a similar genetic background. These characteristics include the length of time the animal takes to fall ill (incubation time), localisation within the brain of pathology and extent and localisation PrP^{Sc} deposition as well as the ratio between the amounts of the three glycoforms (di-, mono-, or non-gycosylated) of PrP^{Sc} deteccted[42]. Challenge with the scrapie agent is usually performed either by force feeding mice scrapie agent laced food (oral challenge) or direct injection of the agent into the brain. Oral challenge is a less successful route of infection but it has provided insight into the mechanism of oral transmission of prion disease.

In terms of the study of neurodegeneration, the mouse model has proved a difficult one to provide mechanism of action. This is because it is difficult to separate the cause of neuronal death from the necessity to introduce PrP^{Sc} into the brain from an external source. Apoptotic cell death occurs in the brain and this is proceeded by the activation of microglia and occurs in parallel with increased astrogliosis[43–46]. Very early changes to neurones can be detected such as loss of dendritic spines[47]. Use of conditional prion protein knockout mice has shown that stopping expression of PrP^c during the disease progress, after considerable PrP^{Sc} has been formed, results in sessation of cell death and recovery of the animal[41]. This really just confirms what was first identified in 1994 using cell culture models[39]. Namely, that PrP^c expression is necessary by the target cell and without it toxic prions cannot kill neurones.

Five chapters in this edition deal with rodent models. The first is the insightful examination of changes to behavior in mice carrying scrapie by Colm Cunningham. The second by Nikki Macleod and colleagues describes how electrophysiological techniques have been used to investigate changes, both as a result of transgenic manipulation, and by scrapie infection. Next, Suehiro Sakaguchi discusses how transgenic mice were used to investigate prion diseases. Herbert Budka and colleagues discuss the use of mouse models to investigate pathological and biochemical changes associated with neurodegeneration in prion disease. Finally, Yong-Sun Kim and colleagues discuss the possible role of oxidative stress and mitochondrial dysfunction in the cause of cell death in prion disease based on their own research using scrapie infected rodents.

Another way of studying neurodegeneration without resorting to models is to examine the pathological changes associated with neurodegeneration. This requires a very thorough study of those changes in both human and animal disease. A very thorough study of the pathology of prion diseases is present in a chapter by Pawel Liberski and James Ironside. In another study, Bernardino Ghetti and colleagues, discusses the pathology of inherited forms of prion diseases.

The remaining chapters concentrate on *in vitro* models. Katarina Bedecs describes the use of scrapie infected cell lines and the way in which they provide insights into changes in the metabolism of cells constituatively generating PrP^{Sc}. Models of the toxic actions of PrP^{Sc} and related peptides are discussed by a number authors. This reflects that this system for studying prion diseases has been in use for over ten years and is one of the easiest to set up and provides results in a matter of weeks rather than years. Studies with scrapie infected mice can take very long because the incubation time is usually around four months. Studies with cattle or sheep take even longer as the incubation period for the disease can be between two and five years. I have provided a chapter summarising the vast body of work produce by my own group since I first began using cell culture models in 1993. Gennaro Schettini, Neena Singh, Roberto Cappai and their colleagues provide the next three chapters also based on work with*in vitro* toxicity models.

Other factors can also contributed to cell survival and the onset of prion disease. The recent work of Claudio Soto investigating the role of endomplasmic reticulum (ER) stress in creating a cellular environment that would favour cell death is presented in a chapter from his group. Although most PrP^c is anchored to the surface by a glycosylphosphatidylinositol anchor, it has been suggested that some amount of the protein could be incorporated into the cell membrane. A putative stop transfer element was described many years ago and at the same time it was suggested that the hydrophobic domain of the protein could potential be a transmembrane domain[48]. Ramanujan Hegde presents his interesting work that proposes that transmembrane forms of PrP could be involved in neurodegeneration in prion diseases. In particular this mechanism has been suggested to be relevant for inherited forms of these diseases.

In the final chapter Hermann Schätzl and colleagues present their elegant and compelling work examining the mechanism of internalisation of the prion protein and the possible effect of prion mutations might have on cell survival.

The broad range of the chapters indicates the great expanse of imaginative directions taken to shine some light into darkness that surrounds this enigmatic field. These various approaches high light the complexity of the subject and the need for continued, international and comprehensive support for the research of the many laboratories dedicated to find a way forward in determining the nature of neurodegeneration in prion disease. Perhaps, there is a long way to go until we develop effective treatments for these disorders, but the science of neurodegeneration and prion diseases has well and truly solidified into a vibrant and compelling field of study.

References

1. D. C. Bolton, M. P. McKinley and S. B. Prusiner, Identification of a protein that purifies with the scrapie prion. *Science*, **218**, 1309–1311 (1982).

2. S. B. Prusiner, Novel proteinaceous infectious particles cause scrapie. *Science* **216**, 136–144 (1982).

3. H. Büeler, Fischer, Y. Lang, H. Bluethmann, H.-P. Lipp, S. J. DeArmond, S. B. Prusiner, M. Aguet and C. Weissmann, Normal development and behaviour of mice lacking the neuronal cell-surface PrP protein. *Nature*, **356**, 577–582, (1992).

4. H . Büeler, A. Aguzzi, A. Sailer, R. A. Greiner., P. Autenried, M. Aguet and C. Weissmann, Mice devoid of PrP are resistant to scrapie. *Cell* **73**, 1339–47 (1993).

5. S. B. Prusiner, Prions. *Proc. Natl. Acad. Sci. USA* **95**, 13363–13383 (1998).

6. G. Legname, I. V. Baskakov, H. O. Nguyen, D. Riesner, F. E. Cohen, S. J. DeArmond and S. B. Prusiner, Synthetic mammalian prions. *Science*. **305**, 673–676 (2004).

7. B. Ghetti, P. Piccardo, B. Frangione, O. Bugiani, G. Giaccone, K.Young, F. Prelli, M. R. Farlow, S. R. Dlouhy and F. Tagliavini, Prion protein amyloidosis. *Brain Pathol.* **6**, 127–145 (1996).

8. D. R. Brown, Molecular advances in understanding inherited prion diseases. *Mol. Neurobiol.* **25**, 287–302 (2002).

9. D. C. Gajdusek and C. J. Gibbs Jr. Transmission of two subacute spongiform encephalopathies of man (Kuru and Creutzfeldt-Jakob disease) to new world monkeys. *Nature* **230**, 588–591 (1971).

10. J. Collinge, Human prion diseases and bovine spongiform encephalopathy (BSE). *Hum. Mol. Genet*. **6** 1699–1705 (1997).

11. M. E. Bruce, R. G. Will, J. W. Ironside, I. McConnell, D. Drummond, A. Suttie, L. McCardle, A. Chree, J. Hope, C. Birkett, S. Cousens, H. Fraser and C. Bostock,

C. J. Transmissions to mice indicate that 'new variant' CJD is caused by the BSE agent. *Nature*, **389**, 498–501 (1997).

12. F. E. Cohen and S. B. Prusiner, Pathologic conformations of prion proteins. *Annu. Rev. Biochem.*, **67**, 793–819 (1998).

13. M. Purdey, Ecosystems supporting clusters of sporadic TSEs demonstrate excesses of the radical generating divalent cation manganese and deficiencies of antioxidant co-factors; Cu, Se, Fe, Zn. Does a foreign cation substitution at PrP's Cu domain initiate TSE? *Med. Hypoth.* **54**, 278–306 (2000).

14. D. R. Brown, BSE did not cause variant CJD: an alternative cause related to post-industrial environmental contamination, *Medical Hypotheses*, **57**, 555–560 (2001).

15. B.-S. Wong, S. G.Chen, M. Colucci, Z. Xie, T. Pan, T. Liu, R. Li, P. Gambetti, M.-S. Sy and D. R. Brown, Aberrant metal binding by prion protein in human prion disease. *J. Neurochem.* **78**, 1400–1408 (2001).

16. A. M. Thackray, R. Knight, S. J. Haswell, R. Bujdoso, and D. R. Brown, Metal imbalance and compromised antioxidant function are early changes in prion disease. *Biochem J*, **362**, 253–258 (2002).

17. D. R. Brown, F. Hafiz, L. L. Glasssmith, B.-S. Wong, I. M. Jones, C. Clive, and S. J. Haswell, Consequences of manganese replacement of copper for prion protein function and proteinase resistance. *EMBO J.* **19**, 1180–1186 (2000).

18. K. Basler, B. Oesch, M . Scott, D. Westaway, M. Walchli, D. F. Groth, M. P. McKinley, S. B. Prusiner and C. Weissmann, Scrapie and cellular PrP isoforms are encoded by the same chromosomal gene. *Cell* **46**, 417–28 (1986).

19. N. Stahl, D. R. Borchelt and S. B. Prusiner, Differential release of cellular and scrapie prion protein from cellular membranes of phosphatidylinositol specific phospholipase C. *Biochemistry* **29**, 5405–5412 (1990).

20. R. S. Hegde, J. A. Mastrianni, M. R. Scott, K. D. Defea, P. Tremblay, M. Torchia, S. J. DeArmond, S. B. Prusiner, and V. R. Lingappa, A transmembrane form of the prion protein in neurodegenerative disease. *Science* **279**, 827–834 (1998).

21. R. S. Stewart and D. A. Harris, Most pathogenic mutations do not alter the membrane topology of the prion protein. *J. Biol. Chem.* **276**, 2212–2220 (2000).

22. A. Holme, M. Daniels, J. Sassoon, and D. R. Brown, A novel method of generating neuronal cell lines from gene-knockout mice to study prion protein membrane orientation. *Eur. J. Neurosci.* **18**, 571–579 (2003).

23. R. K. Meyer, A. Lustig, B. Oesch, R. Fatzer, A. Zurbriggen, and M. Vandevelde, A monomer-dimer equilibrium of a cellular prion protein (PrPC) not observed with recombinant PrP. *J Biol Chem.* **275**, 38081–38087 (2000).

24. X. Roucou, Q. Guo, Y. Zhang, C. G. Goodyer, A. C. LeBlanc, Cytosolic prion protein is not toxic and protects against Bax-mediated cell death in human primary neurons. *J. Biol. Chem.* **278**, 40877–40881 (2003).

25. A. Mange, C. Crozet, S. Lehmann, F. Beranger, Scrapie-like prion protein is translocated to the nuclei of infected cells independently of proteasome inhibition and interacts with chromatin. *J. Cell Sci.* **117**, 2411–2416 (2004).

26. N. Salès, K. Rodolfo, R. Hässig, B. Faucheux, L. Di Giamberdino, and K. L. Moya, Cellular prion protein localization in rodent and primate brain. *Eur. J. Neurosci.* **10**, 2464–2471 (1998).

27. M. P. Hornshaw, J. R. McDermott and J. M. Candy, Copper binding to the N-terminal tandem repeat regions of mammalian and avian prion protein. *Biochem Biophys Res Commun* **207**, 621–629 (1995).

28. Brown D. R., Qin K., Herms J. W., Madlung A., Manson J., Strome R., Fraser P. E., Kruck T., von Bohlen A., Schulz-Schaeffer W., Giese A., Westaway D. and Kretzschmar H. (1997) The cellular prion protein binds copper in vivo. *Nature* **390**, 684–687.

29. D. R. Brown, R. St. J. Nicholas, and L. Canevari, Lack of prion protein expression results in a neuronal phenotype sensitive to stress. *J. Neurosci. Res.* **67**, 211–224 (2002).

30. J. Collinge, M. A. Whittington, K. C. Sidle, C. J. Smith, M. S. Palmer, A. R. Clarke and J. G. Jefferys, Prion protein is necessary for normal synaptic function. *Nature* **370**, 295–297 (1994).

31. R. Huber, T. Deboe and I. Tobler Sleep deprivation in prion protein deficient mice sleep deprivation in prion protein deficient mice and control mice: genotype dependent regional rebound. *Neuroreport.* **13**, 1–4 (2002).

32. E. Graner, A. F. Mercadante, S. M. Zanata, V. R. Martins, D. G. Jay, R. R. Brentani, Laminin-induced PC-12 cell differentiation is inhibited following laser inactivation of cellular prion protein. *FEBS Lett.* **482**, 257–260 (2000).

33. L. B. Chiarini, A. R. Freitas, S. M. Zanata, R. R. Brentani, V. R. Martins, R. Linden, Cellular prion protein transduces neuroprotective signals. *EMBO J.* **21**, 3317–26 (2002).

34. D. R. Brown, B.-S. Wong, F. Hafiz, C. Clive, S. J. Haswell and I. M. Jones, Normal prion protein has an activity like that of superoxide dismutase. *Biochem J* **344**, 1–5 (1999).

35. D. R. Brown, Prion protein expression aids cellular uptake and veratridine-induced release of copper. *J. Neurosci. Res.* **58**, 717–725 (1999).

36. P. C. Pauly and D. A. Harris, Copper stimulates endocytosis of the prion protein. *J. Biol. Chem.* **273**, 33107–33110 (1998).

37. W. E. Müller, H. Ushijima, H. C. Schroder, J. M. Forrest, W. F. Schatton, P. G. Rytik, M. Heffner-Lauc, Cytoprotective effect of NMDA receptor antagonists on prion protein (PrionSc)-induced toxicity in rat cortical cell cultures. *Eur. J. Pharmacol.* **246**, 261–267 (1993).

38. G. Forloni, N. Angeretti, R. Chiesa, E. Monzani, M. Salmona, O. Bugiani and F. Tagliavini, Neurotoxicity of a prion protein fragment. *Nature* **362**, 543–546 (1993).

39. D. R. Brown, J. Herms, and H. A. Kretzschmar, Mouse cortical cells lacking cellular PrP survive in culture with a neurotoxic PrP fragment. *Neuroreport* **5**, 2057–2060 (1994).

40. S. Brandner, S. Isenmann, A. Raeber, M. Fischer, A. Sailer, Y. Kobayashi, S. Marino, C. Weissmann and A. Aguzzi, Normal host prion protein necessary for scrapie-induced neurotoxicity. *Nature*, **379**, 339–343 (1996).

41. G. Mallucci, A. Dickinson, J. Linehan, P. C. Klohn, S. Brandner and J. Collinge, Depleting neuronal PrP in prion infection prevents disease and reverses spongiosis. *Science*. **302**, 871–874 (2003).

42. M. E. Bruce and H. Fraser Scrapie strain variation and its implications. *Curr. Top. Microbiol. Immunol.,* **172**, 125–138 (1991).

43. A. E. Williams, L. J. Lawson, V. H. Perry, and H. Faser, Characterization of microglial response in murine scrapie. *Neuropathol. Appl. Neurobiol.,* **20**, 47–55 (1994).

44. A. Giese, M. H. Groschup, B. Hess, and H. A. Kretzschmar, Neuronal cell death in scrapie-infected mice is due to apoptosis. *Brain Pathol.* **5**, 213–221 (1995).

45. S. Betmouni, V. H. Perry, and J. L. Gordon, Evidence for an early inflammatory response in the central nervous system of mice with scrapie. *Neuroscience*, **74**, 1–5 (1996).

46. A. Giese, D. R. Brown, M. H. Groschup, C. Feldmann, I. Haist and H. A. Kretzschmar, (1998) Role of microglia in neuronal cell death in prion disease. *Brain Pathol*. **8**, 449–457.

47. P. V. Belichenko, D. Brown, M. Jeffrey and J. R. Fraser, Dendritic and synaptic alterations of hippocampal pyramidal neurones in scrapie-infected mice. Neuropathol *Appl Neurobiol.* **26**, 143–149 (2000).

48. C. S. Yost, C. D. Lopez, S. B. Prusiner, R. M. Myers, and V. R. Lingappa, Non-hydrophobic extracytoplasmic determinant of stop transfer in the prion protein. *Nature* **343**, 669–672 (1990).

Chapter 2

NEUROPATHOLOGY OF TRANSMISSIBLE SPONGIFORM ENCEPHALOPATHIES (PRION DISEASES)

Pawel P. Liberski[1] and James W. Ironside[2]

[1]*Laboratory of Electron Microscopy and Neuropathology, Department of Molecular Pathology and Neuropathology, Medical University, Lodz, Poland*
[2]*National Creutzfeldt-Jakob Disease Surveillance Unit, Western General Hospital, Edinburgh, EH4 2XU, UK*

2.1. Introduction

The transmissible spongiform encephalopathies (TSEs) or prion diseases are a group of neurodegenerative disorders which include kuru[1], Creutzfeldt-Jakob disease (CJD)[2], variant Creutzfeldt-Jakob disease (vCJD), Gerstmann-Sträussler-Scheinker (GSS) disease[3], and fatal familial insomnia[4] in man, natural scrapie in sheep, goats and mufflons, transmissible mink encephalopathy in ranch-reared mink[5], chronic wasting disease of mule deer and elk in the USA[6] and Canada, bovine spongiform encephalopathy (BSE) or "mad cow disease"[7] and its analogues in several exotic species of antelopes and wild felids in zoological gardens and feline spongiform encephalopathy in domestic cats.

These disorders are caused by a still not completely understood pathogen variously referred to as a "prion" (predominantly) or a slow, unconventional or atypical virus, or "virino" (rarely). Despite wide acceptance for the prion theory, these names still reflect different views about the true molecular structure of the pathogen and, by the same token, our ignorance of its nature. Those who prefer to view this pathogen as composed "predominantly or exclusively" of a pathologically folded protein

(PrP^{Sc}; Sc from scrapie or PrP^{d}; d from disease), use the term "prion"; hence the term "prion diseases".

The "virino" hypothesis suggests that the pathogen is a molecular chimera composed of a still-to-be-discovered nucleic acid and a shell-protein which is host-encoded (perhaps PrP^{d}). The virus hypothesis simply suggests that the pathogen is a yet-to-be-identified unconventional virus. The "unified theory" of Weissmann[8], not unlike the virino theory, suggests that the agent is a molecular chimera of the misfolded protein that confers infectivity and an unidentified oligonucleotide that specifies strain characteristics.

2.2. Nomenclature

The nomenclature of PrP species is confusing. PrP^{c} is a normal cellular isoform. PrP^{Sc} (PrP^{res} or PrP^{d}, from disease) is a pathological misfolded protein. PrP^{Sc} is operationally defined as resistant to proteinase K (PK) and insoluble in denaturing detergent; however, in some diseases, pathological isoform of PrP is not PK resistant[9]. Thus, we prefer to use the neutral term PrP^{d} which denotes that misfolded species of PrP which is disease-associated; PK-resistant or not. PrP 27-30 is a proteolytic cleavage product of PrP^{d} which is sometimes referred to as PrP^{res} (res from resistant) when generated following incomplete proteolytic digestion in Western blotting.

2.3. PrP, its gene, the "prion" hypothesis

PrP^{c} is a highly conserved sialoglycoprotein encoded by a cellular gene mapped to chromosome 20 in man and 2 in mouse[10]. The gene is ubiquitous; it has been cloned in numerous mammalian species included marsupials and there are analogues of this gene in birds, reptiles, amphibians, and recently fish; those in Drosophila and nematodes appeared to be cloning artefacts. PrP 27-30 was first discovered as a protein co-purifying with infectivity in extracts derived from brains infected with the 263K strain of scrapie agent which led to the conclusion that PrP is a part of infectivity.

The "prion" hypothesis, which is deeply rooted in this association between PrP and infectivity, was formulated by Stanley B. Prusiner in 1982[11]. The hypothesis postulated that the scrapie agent was a proteinaceous infectious particle, because infectivity was dependent on protein but resistant to methods known to inactivate nucleic acids. A similar proposal had been presented a decade earlier by many investigators who all developed the earlier suggestion based on irradiation studies, that scrapie agent was devoid of disease-specific nucleic acid[12].

Like many amyloid proteins, PrP 27-30 is a proteolytic cleavage product of a precursor protein, PrP 33-35^{d}. However, PrP 33–35^{d} is not the *primary* product of the cellular gene. It has an amino acid sequence and posttranslational modifications (like glycosylation and the attachment of GPI, glycophospholipid inositol anchor) identical to those of PrP 33–35^{c}, but strikingly different physicochemical features; in particular, PrP^{c} is completely degraded by a limited proteolysis but PrP^{d} is only partially degraded, yielding a core protein (PrP 27–30) which may be visualised by electron microscopy as scrapie-associated fibrils (SAF), better known as prion rods[13]. To become PrP^{d}, PrP^{c} must be first transported to the cell surface and then through the endosomal-lysosomal pathway.

PrP has several interesting features. As already mentioned, PrP is a glycoprotein with two Asn-glycosylation sites; thus, PrP may exist as deglycosylated, monoglycosylated and di-glycosylated isoforms of different electrophoretic mobilities and glycoforms[14]. The various combinations of glycosylation and codon 129 genotype (see later) correlate to some degree with the phenotypic expression of human TSE. In particular, a distinctive glycosylation pattern is uniquely present in both BSE and vCJD[14,15]. Although glycosylation patterns breed true—i.e., they are retained in passage[14]—changes in electrophoretic mobility may occur in the presence of metal ions[14], and more than one pattern may occur in different regions of the same brain, or brain and peripheral organs in the same patient.

PRNP gene in humans consists of two exons and the whole ORF is confined to the second exon[16]. The polymorphism at codon 129 merits special comment. Codon 129 encodes Met in ca 60% and Val in 40% of alleles in the normal Caucasian population. However, in all forms of CJD, there is marked over-representation of homozygotes over heterozygotes. The codon 129 polymorphism may also exert a modifying effect on the phenotypic expression of a given *PRNP* mutation.

The situation in kuru is particularly interesting. The practice of cannibalism underlying the kuru epidemic created a selective force on the prion protein genotype. As in CJD, homozygosity at codon 129 ($129^{Met\ Met}$ or $129^{Val\ Val}$) is overrepresented in kuru. However, Mead *et al.*[17] found that among Fore women over fifty years of age, there is a remarkable overrepresentation of heterozygosity ($129^{Met\ Val}$) at codon 129, which is consistent with the interpretation that $129^{Vam\ Met}$ makes an individual resistant to TSE agents and that such a resistance was selected by cannibalistic rites. Because of this $129^{Met\ Val}$ heterozygote advantage, it has been suggested that the heterozygous genotype at codon 129 has been sustained by a widespread ancient practice of human cannibalism.

2.4. Classifications

From early days, CJD (the name as Jakob-Creutzfeldt disease was coined by Spielmeyer in 1922) has been sub-classified into several forms. For instance, Daniel[18] singled out the classical cortico-striato-spinal (Jakob) type; Heindenhain type (characterized by cortical blindness due to severe involvement of the occipital lobes); diffuse type (dementia with pyramidal and extrapyramidal signs and symptoms) and ataxic type[19]. Siedler and Malamud[20] discriminated cortical, cortico-striatal, cortico-striato-cerebellar, cortico-spinal and cortico-nigral type. In the literature, CJD exists under more than 50 different names and many of these do not represent CJD in a modern sense. The discrimination of all these variants is merely of historical interest but recent molecular studies substantiated the existence of certain defined phenotypes.

PrP^{res} (after limited proteolytic digestion) may exist as 21 kDa (type 1) and 19 kDa (type 2) isoforms which coupled with the status of codon 129 of the *PRNP* gene underlie the existence of 7 molecular variants—MM1, MV1, MM2- cortical, MM2—thalamic, MV2, VV1 and VV2. These variants differ both clinically and neuropathologically[21]. Type MM1corresponds to classical sporadic CJD with changes in the cerebral cortex, striatum, thalamus and the cerebellum; PrP^d accumulates mostly as synaptic deposits. This type comprises approximately 70% of sCJD cases. Second, most common type, VV2 comprises approximately 15% of all sCJD cases. Changes are confined to the limbic system, striatum, the cerebellum, thalamus and hypothalamus and several brain stem nuclei. The involvement of the cerebral cortex depends on the duration of illness; those cases of short duration may exhibit minimal cortical changes, spongiform change demonstrates laminar distribution while PrP^d accumulates as plaque-like, perineuronal and synaptic deposits. MV2 type (approximately 8%) is reminiscent of VV2 type—spongiform change is confined to the subcortical structures while PrP^d expression is mostly plaque-like. In contrast to MV2 type, in VV2 type—"true" (i.e., congophilic and visible in a routine H & E stain) plaques predominates. MM2 type is further sub-classified into MM2-thalamic, which corresponds to FFI and FSI cases and MM2-cortical, similar to MM1 type, from which differs by limited cerebellum involvement and larger (coarse) vacuoles. VV1 is very rare (<1% of all sCJD)—changes are limited to cerebellar cortex and the striatum while other structures, including the cerebellum are barely involved.

A more refined approach was used by Collinge *et al.*[14,22] who exploited the size of PrP^d fragments following limited PK digestion and the relative abundance of mono-, di- and deglycosylated glycoforms. This approach discriminated PrP^d types 1–4 and 6; type 5 exists in vCJD-infected transgenic mice but not in humans. All type 1 cases are homozygous for Met at codon 129 of the *PRNP* gene; type 2 may exist coupled with every status of the codon 129; type 3 is associated with at least one 129^{Val} allele with the exception of a single CJD cases homozygous for Met at codon 129. Type 4, characterized by predominance of diglycosylated glycoform, is unique for vCJD and BSE[15]. These types differ neuropathologically as well as clinically. Type 1 cases demonstrated widespread spongiform change in the cerebral cortex, mild changes in the basal ganglia, cerebellum and brainstem but no spongiform degeneration in the hippocampus. In type 2 homozygous for 129^{Met} cases, basal ganglia are moderately affected while in heterozygous type 2 cases or type 2 homozygous for 129^{Val}, the basal ganglia are involved severely. Type 3 MV cases are characterized by the presence of kuru plaques already seen on routine H & E preparation.

The translation of the Collinge scheme into the Gambetti is not straightforward, probably due to technical differences in the methodology for Western blots. Furthermore, chelation of metal ions performed prion to PK digestion interconverts both type 1 and 2 MM PrP fragments into so called 2^- PrP[23]. Having said this, the Collinge's type 1 MM, type 2 MM, type 3 VV, type 2 MV and type 3 MV are similar to the Gambetti's type MM1, MM2-cortical, VV2, MV1 and MV2, respectively. Thus, it seems that the Collinge sub-classification and the Gambetti sub-classification are, basically, interconvertible. This notion has been supported by recent work which indicates that alterations in electrophoretic mobility can be markedly influenced by pH variations in the brain tissue homogenate. When pH is controlled, it appears that two major subgroups of PrP^{res} can be identified in terms of electrophoretic mobility of the unglycosylated band, corresponding to the types 1 and 2 of the Gambetti *et al.* classification.

2.5. Classical Neuropathology

Creutzfeldt in 1920[24] described one case of a novel neurodegenerative conditions and Jakob described sequentially 5 cases[25,26]. Four Jakob's cases, still on files at the University of Hamburg, were reexamined by Masters and Gajdusek[27] who confirmed that 2 Jakob's cases fulfill modern criteria of CJD while remaining 2 cases represent other not well defined neurological conditions. Of special interest is one of Jakob's cases with amyotrophy which initiated a long-lasting confusion of "amyotrophic type of CJD" that appear to be merely amyotrophic lateral sclerosis with dementia and which is not transmissible[28]. Neuropathological

Figure 2.1. (a) Typical spongiform change. H & E; (b) status spongiosus; (c) Spongiform vacuoles as seen by electron microscopy. Vacuoles contain curled membrane fragments and secondary chambers. Original magnification, ×12 000; (d) Intramyelin vacuole, Original magnification, ×12 000.

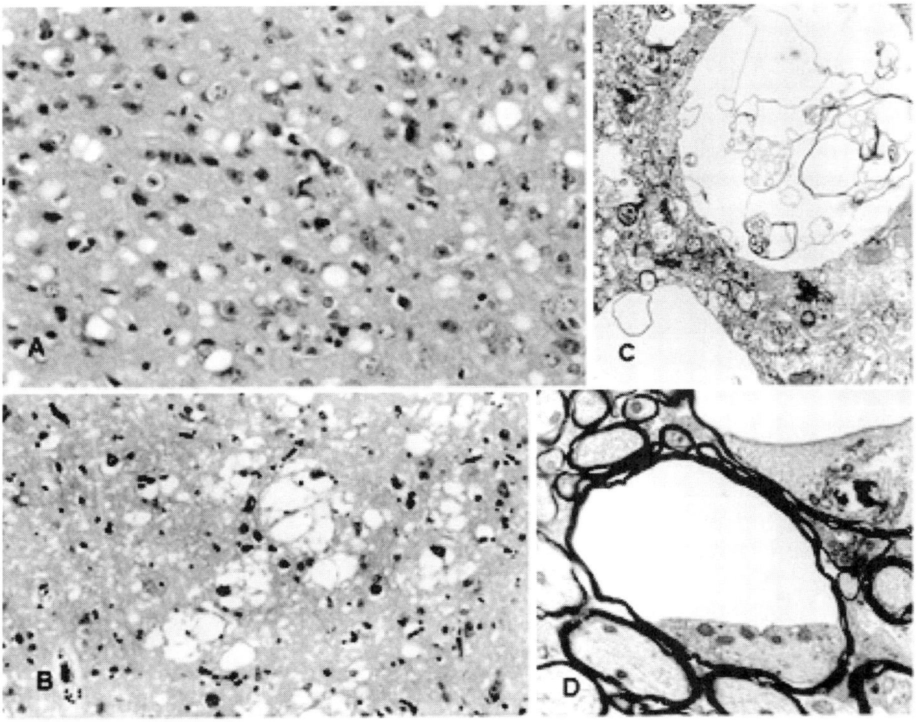

description of Jakob's cases based on studies of thick celloidin-embedded sections stained according to the Nissl technique revealed purely neurodegenerative process encompassing neuronal loss, central chromatolysis and astroglial proliferation with neuronophagia. Parenthetically, spongiform change were not visible by Nissl stain but re-appeared when the coverslips were removed and section re-stained with H & E.

The classical triad of CHD neuropathology consists of vacuolation (spongiform change), neuronal degeneration (neuronal loss) and astrocytosis (Figure 2.1–2.2). The changes are bilaterally symmetrical but may be local and, occasionally, even unilateral[29].

2.6. Structural Changes

2.6.1. Spongiform changes

Most characteristic and even "semi-pathognomic" for CJD is the presence of spongiform change which remain well preserved even

Figure 2.2. (a) Dense astrocytic gliosis in a CJD case as revealed by Cajal gold sublimate method. Courtesy of Prof. Herbert Budka, Vienna, Austria; (b) GFAP-immunopositive astrocytes against a background of severe spongiform change. CJD brain biopsy.

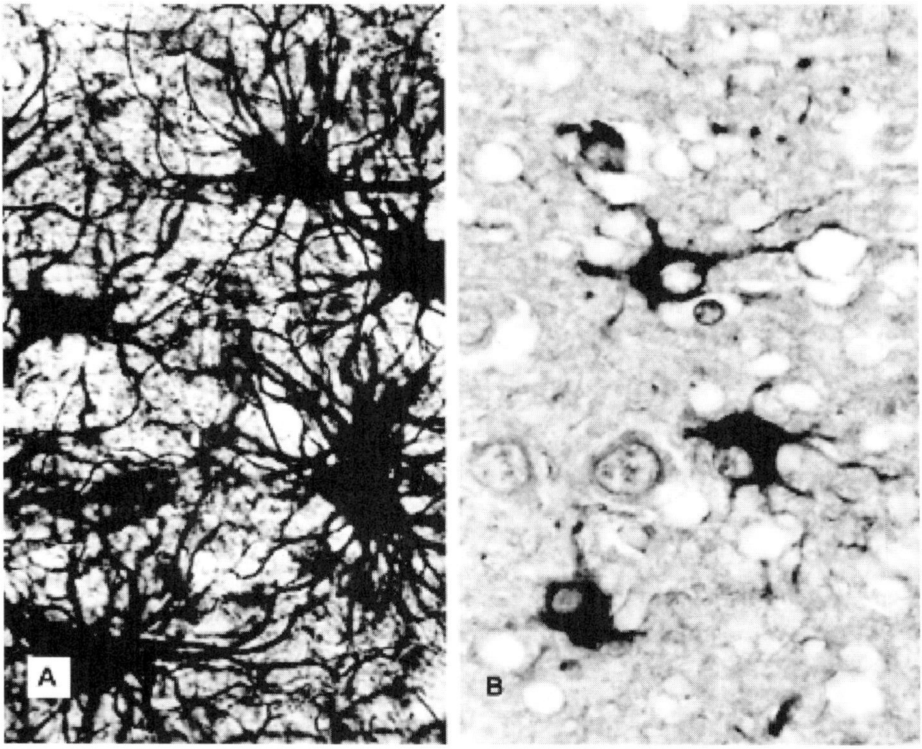

in exhumed cases[30]. Spongiform change consists of small, round or oval vacuoles within neuropil (Figure 2.1); vacuoles are confluent and form typical "morula-like" aggregates. In the cerebral cortex, spongiform change is confined to the deep cortical layers; those vacuoles in the superficial cortical layers are characteristic for fronto-temporal lobar degenerations including Pick disease or are merely artefactual. It must be stressed, that in cases of longer duration spongiform change may be masked by the overall loss of neurons, collapse of the cortical cytoarchitecture and robust proliferation of astrocytes. To this end, Masters and Richardson[31] discriminated "spongiform change" from "spongiform state ("status spongiosus"),' the latter consisting of larger cavities of irregular shape in the neuropil (Figure 2.1b) between dense meshwork of proliferating astrocytes. Status spongiosus is not specific for TSEs and can occur in the end stage of a wide range of neurodegenerative disorders if widespread neuronal degeneration and loss has occurred.

In certain TSE, especially in fatal familial insomnia, vacuolation may be very limited and largely focal; in the latter example it is generally confined to some thalamic nuclei.

Ultrastructurally, vacuoles are always membrane-bound and contain secondary-vacuoles or "chambers" (vacuoles within vacuoles), "curled" membrane fragments and amorphous "fluffy" material of unknown composition (Figure 2.1c). The membranes lining the vacuoles may be simple or multiple[32]. Typical vacuoles originate in neuronal elements—mostly dendrites or, rarely, axons; those described in astrocytes seems to be fixation artifacts. Spongiform vacuoles have been also studies by scanning electron microscopy (SEM)[33] which revealed ulcerations and defective membranes as well as "rough elevated areas" corresponding to "amorphous membranes" as seen by TEM. SEM detected also small blisters that are equivalent to small vesicles by TEM.

The second type of vacuoles are those originated within the myelin sheath (Fig 1d)[34]. These are largely non-specific finding but they may be robust in the panencephalopathic type of CJD in which white matter is predominantly affected and its degeneration does not result from Wallerian degeneration. These intramyelin vacuoles are several times larger than diameter of average myelin fibre and looked "empty". Within distended myelin sheaths, shrunken axons are observed but many bullous swellings contained no axons. Some axons look normal but others were filled with neurofilaments and scanty electron-dense bodies. Still other axons are attached to the innermost myelin lamellae by a thin "neck" probably a mesaxon.

2.6.2. Astrocytosis

Variably severe astrocytosis is observed among almost all neurodegenerative conditions and CJD is no exception[35]. Hypertrophic astrocytes, detected by means of metal impregnation techniques (Holzer, Kanzler or Cajal's—Figure 2.2a) or more recently by immunostaining against glial fibrillary acidic protein (GFAP) (Figure 2.2b), are seen in all vacuolated areas. In cerebral cortex they are particularly prominent in deeper cortical layers, where swollen or gemistocytic forms are frequently observed. When destruction is so severe to lead to the collapse of vacuolated neuropil, proliferating astrocytes may virtually replace all other cellular elements. In such a situation the spongiform change may no longer be recognizable. In the cerebellum the proliferation of the astrocytes known as Bergman glia is frequently observed in a wide range of human TSEs.

Ultrastructurally, hypertophic astrocytes are no much different than those from other conditions. They are characterized by abundant glial filaments within the cytoplasm. Liberski *et al.*[36] described close contacts between astrocytes and oligodendroglial cells; the pathophysiological significance of this phenomenon is uncertain.

There are only two overlapping morphometric studies of astrogliosis in the cerebella (both Bergmann and velate astrocytes) in two cases of the ataxic form of CJD[37,38]. Astrocytes increased from 192.76 ± 117.98 per mm^2 in controls to 278.08 ± 137.73 per mm^2 in CJD. An increase in the cross sectioned nuclear area of Bergmann glia ($32.72 \pm 6.8\ \mu m^2$ vs $42.75 \pm 9.61\ \mu m^2$) and of velate astrocytes ($34.86 \pm 7.29\ \mu m^2$ vs. $39.37 \pm 7.10\ \mu m^2$) was seen when control values were compared with those of CJD values. Of note, the basic three-dimensional geometry of the astrocytic scaffold of the cerebellum was maintained despite severe loss of granule cells. Electron microscopy revealed several subcellular organelles, rare but otherwise typical for reactive astrocytes, single cilia consisting of ciliary shafts, clusters of interchromatin and perichromatin granules, various adhesive plaque junctions and simple and granular nuclear bodies. Of particular interest is the presence of infoldings of plasma membranes in the perivascular regions of astrocytic end-feet. These infoldings were covered by an interrupted or continuous electron-dense undercoat of 30–60 nm in diameter. The latter observation is in agreement with the earlier freeze-etching study of Dubois-Dalcq *et al.*[39] who showed an increased number of astrocyte-specific particles as opposed to their depletion on membranes forming vacuoles.

2.6.3. Amyloid plaques

In GSS, vCJD and in some murine scrapie models of disease, "classical" amyloid plaques are frequent (Figure 2.3), but are often absent in many human TSEs cases[40], in BSE and most ovine scrapie. The presence of amyloid (i.e., a protein in a β-sheeted conformation[12]) can be detected *in situ* by tinctorial stains for amyloids, including birefringence following staining with Congo red, immunohistochemistry for PrP, or, in brain homogenates, in the form of fibrillar PrP aggregates labelled *prion rods*[13]. However, on transmission electron microscopy most of the disease specific PrP^d identified in TSE affected brains by immunocytochemistry is not visibly fibrillar[41].

Typically, "classical" amyloid plaques consist of a congophilic PrP-immunopositive dense core of densely interwoven amyloid fibrils surrounded by different numbers of dystrophic neurites (DN). The amyloid plaque usually exhibits positivity for other tinctorial stain including Periodic acid-Schiff (PAS), Alcian blue and various silver impregnation

Figure 2.3. (a) PrP^d -immunopositive amyloid plaque; (b) An amyloid plaque containing abundant microglial cells immunostained against ferritin. GSS case[42]; (c) an electron micrograph showing a microglial cell (its nucleus is visible in the upper part of the picture) in close contact with amyloid fibrils. Original magnification, ×30 000; (d) A typical stellate kuru plaques. Original magnification, ×12 000; (e) A plaque (upper right) from a GSS case[46]. In the lower left part a huge dystrophic neurite is visible. Original magnification, ×12 000.

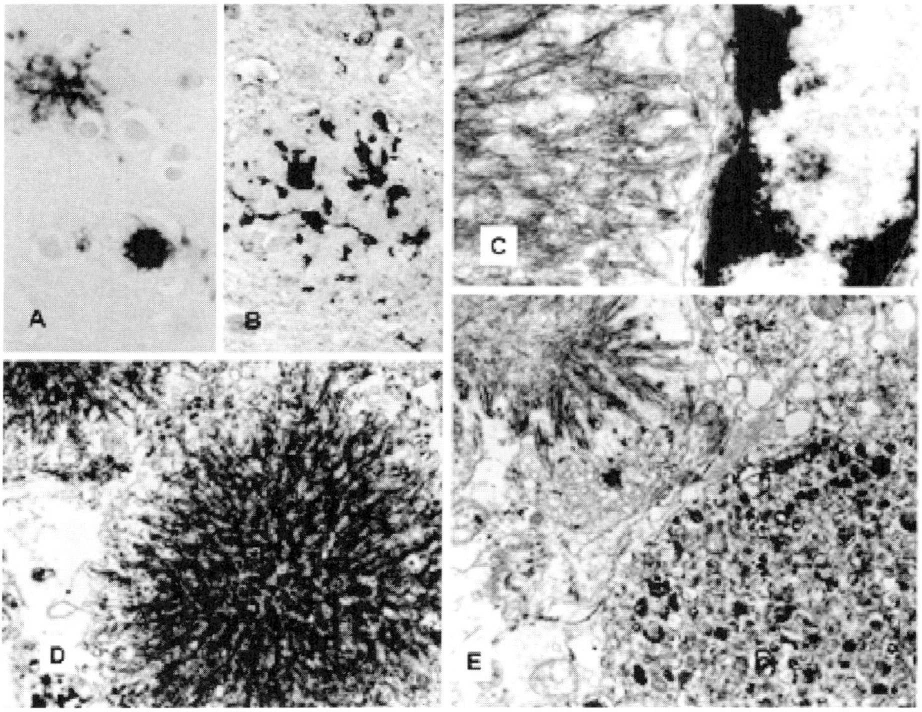

techniques. The PrP-plaque is penetrated by astrocytic processes. Microglial cells were detected in plaques of GSS[42] (Figure 2.3b-c) and in murine scrapie plaques[43]. The exact morphology of the PrP-plaque varies. Thus, the "kuru" plaque consists of stellate core with minimal numbers dystrophic neurites (or non at all) (Figure 2.3e)[44]. Multicentric plaques consisting of numerous such cores of different sizes and shapes *are typical of GSS*[45,46]. In vCJD, the large fibrillary amyloid plaques in the cerebral and cerebellar cortex are surrounded by a corona or "halo" of spongiform change, the so-called "florid" plaque[47].

2.6.4. Neuroaxonal dystrophy

Neuroaxonal dystrophy (NAD) is one form of neuronal degeneration[48] which may occur in apparently normal brains but became "pathological"

Figure 2.4. (a) a cluster of NFP-immunopositive neuritis in a case of CJD; (b) a dystrophic neurite containing numerous neurofilaments and immersed in them lysosomal electron-dense bodies. Original magnification, ×12 000.

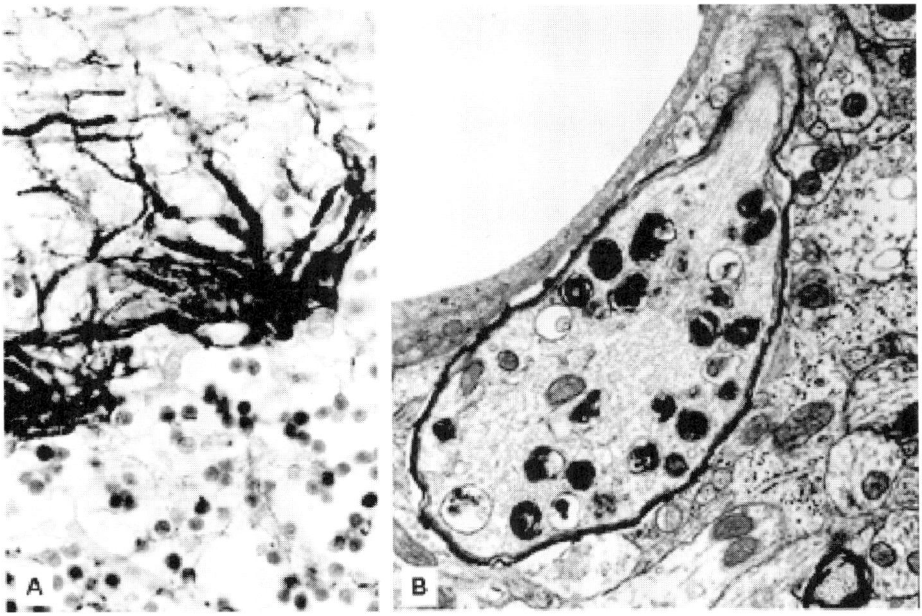

only if found in increased numbers. The ultrastructural correlate of NAD is a dystrophic neurite—a neuronal process (a dendrite or a myelinated axon) filled with abnormal subcellular organelles—like lysosomal electron-dense bodies or masses of neurofilaments (Figure 2.4).

NAD was described in CJD[49], GSS[46] and in numerous experimental scrapie models. Dystrophic neurites in both natural and experimental CJD accumulate phosphorylated neurofilament proteins (Figure 2.4)[50].

2.6.5. Apoptosis and autophagy

As in many neurodegenerative diseases caused by the accumulation of "toxic" proteins such as Alzheimer's disease, neurons in TSEs die via programmed cell death of which only the apoptotic process is relatively well characterized. Three types of programmed cell death (PCD) can be discriminated[51].

1) First type (apoptosis), is characterized ultrastructurally by specific alterations—cell shrinkage, condensation of chromatin and, eventually, formation of so called "apoptotic bodies". The latter are actively phagocytosed by macrophages.

2) A second type is characterized by formation of numerous cytoplasmic autophagic vacuoles that are subsequently fused with lysosomes. The appearance of autophagy is followed by mitochondria dilatation and expansion of Golgi apparatus and endoplastic reticulum. The molecular mechanism is different that that of apoptosis and consists of complex interplay of numerous proteins including the Tor kinase.
3) A third type is similar to the second type, except for the negligible or absent involvement of lysosomes. Ultrastructurally, type 3 cell death is characterized by swelling of intracellular organelles resulting in the formation of large empty spaces within the cytoplasm. Of interest, TSEs are characterized of "spongiform vacuoles" formation within neuronal elements[52]. While the latter have never been linked to the type 3 PCD, the ultrastructural resemblance of "large empty spaces" to "spongiform vacuoles" appear to be worthy of studies.

Using TUNEL methodology, apoptotic neurons have been repeatedly shown in both naturally occurring and experimentally induced TSEs[53–55], and some investigators believe that at least a proportion of "dark neurons" may represent those cells undergoing apoptosis. The latter cells appear shrunken and of homogeneously dark cytoplasm. Their real significance is uncertain and many workers regarded them, however, as merely fixation artefacts. Furthermore, typical autophagic vacuoles have been reported by us[56], and by others[57,58] in CJD- and scrapie-affected rodent brains. Recently, autophagy was found in degenerating synapses in CJD[59].

Autophagy, being an evolutionarily ancient cellular response to intra- and extracellular noxious stimuli, may precede or co-exist with apoptosis, and the process may be induced by apoptotic stimuli. Furthermore, the level of autophagy may define the sensitivity of a given neuronal population to apoptotic stimuli, which may underlie the phenomenon of "selective neuronal vulnerability"[60]. Thus, autophagy and apoptosis are interconnected.

Cellular autophagy is a physiological degradative process employed, like apoptosis, in embryonic growth and development, cellular remodeling and the biogenesis of some subcellular organelles—viz. multilamellar bodies. Nascent immature autophagic vacuoles coalesce with lysosomes to form degraded autophagic vacuoles, and in apoptosis, only excessive or wrongly placed autophagy cause a pathological process. Of note, autophagy is highly enhanced in other brain amyloidoses, Alzheimer disease, Parkinson's disease, Huntington's disease, where the signal for autophagy is huntingtin. Here, we extend these observations using different model of scrapie and CJD[61].

2.6.6. Definition of autophagic vacuoles

Autophagic vacuoles are composed of areas of the cytoplasm sequestrated with single, double or multiple membrane (phagophores) originating from the endoplasmic reticulum. Sequestrated cytoplasm contains ribosomes, small secondary vacuoles, and occasional mitochondria. Some vacuoles presented a homogenously dense appearance.

2.6.7. Formation of autophagic vacuoles in TSE experimental models

Both scrapie models used by us (the 263K and 22C-H) revealed similar frequencies and ultra-structural features of autophagic vacuoles, and will be described together. The various changes were observed simultaneously in different areas of the same sample but the following description is organised according to our interpretation of their chronological evolution. Initially, a part of the neuronal cytoplasm was sequestrated by concentric arrays of double membranes; the enclosed cytoplasm appeared relatively normal except that its density was often increased. In many affected neurons, electron density of the central area drastically increased. The membranes duplicated within the cytoplasm in a labyrinth-like manner and the area sequestrated by these membranes enlarged into a more complex structure consisting of vacuoles, electron-dense areas and areas of normally-looking cytoplasm connected by convoluted membranes (Figure 2.5). Of note, autophagic vacuoles form not only in neuronal perikarya but also in neurites and synapse[59]. We also observed, a large area of the cytoplasm transformed into a collection of autophagic vacuoles of different sizes and it seems that the latter may represent a final stage of autophagy as the number of such cells increased toward the terminal stage of the incubation period. In experimental CJD and GSS, autophagy was also seen albeit with much lower frequency.

2.6.8. Tubulovesicular structures

The *tubulovesicular structures* (TVS) (also known as the *scrapie-associated particles)* are the only structures unique at the level of thin-section electron microscopy for all TSEs so far examined and have been identified in GSS, CJD, many rodent scrapie models, BSE, and in natural scrapie of sheep[49,62–65]. TVS were first described in mice infected intracerebrally scrapie[66]. Of note, in natural scrapie in sheep TVS appeared as membrane-bound accumulations of round particles measuring 35 nm in diameter. The electron-dense core could be demonstrated in some of

Figure 2.5. Typical autophagic vacuole consisting of convoluted membranes sequestrating a part of neuronal cytoplasm. Original magnification, ×12 000.

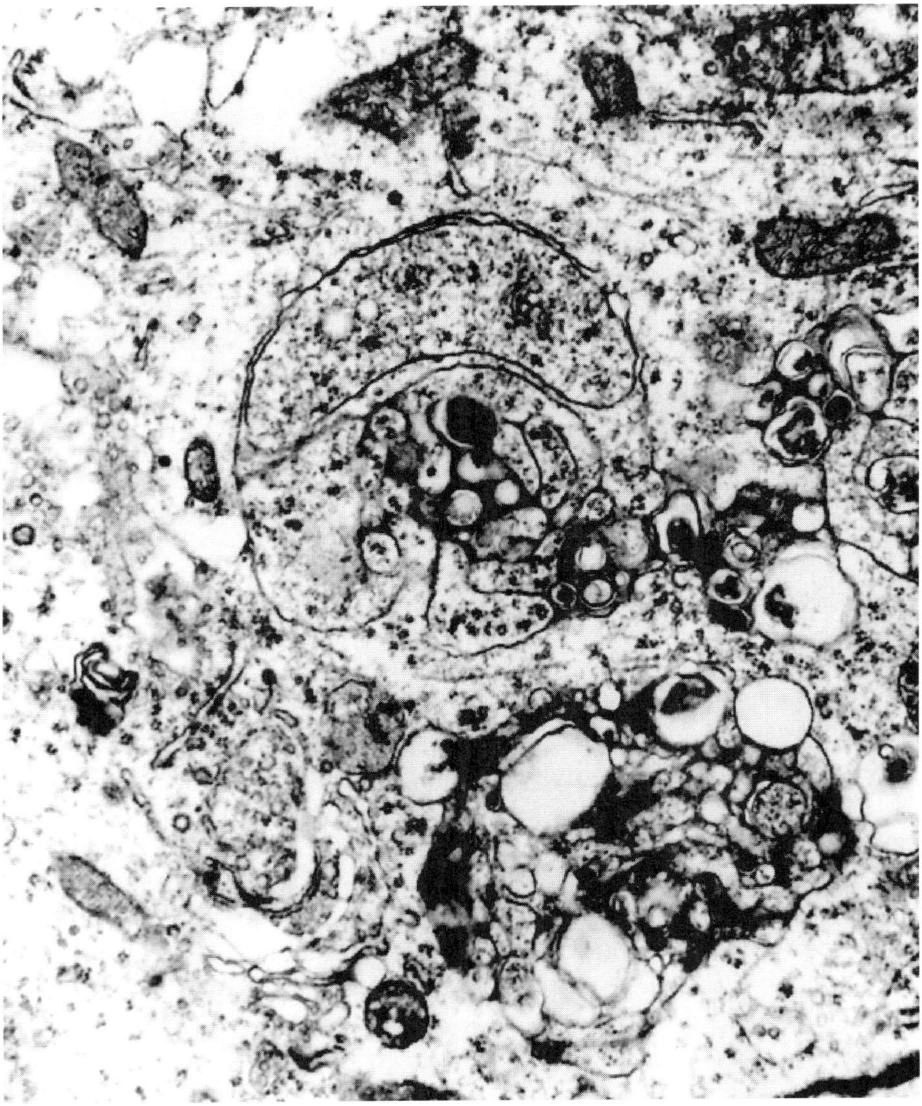

them[67] (Figure 2.2, inset). TVS have been reported in the majority of models of scrapie in rodents studied so far. In humans, TVS were found in CJD[49,68], GSS[63], vCJD and FFI[68].

At low magnification, a process containing TVS is crowded with structures typically of higher electron density than other elements in this field. In cross section, TVS are smaller than synaptic vesicles but larger than microtubules; their profiles are highly variable (Figure 2.6).

Figure 2.6. (a) Low power electron micrograph of distended terminal crowded with TVS. Original magnification, ×12 000; Lower (c) and higher (c) magnification of TVS. Note that TVS are highly pleomorphic. Original magnification, (a), ×12 000; (b), ×50 000.

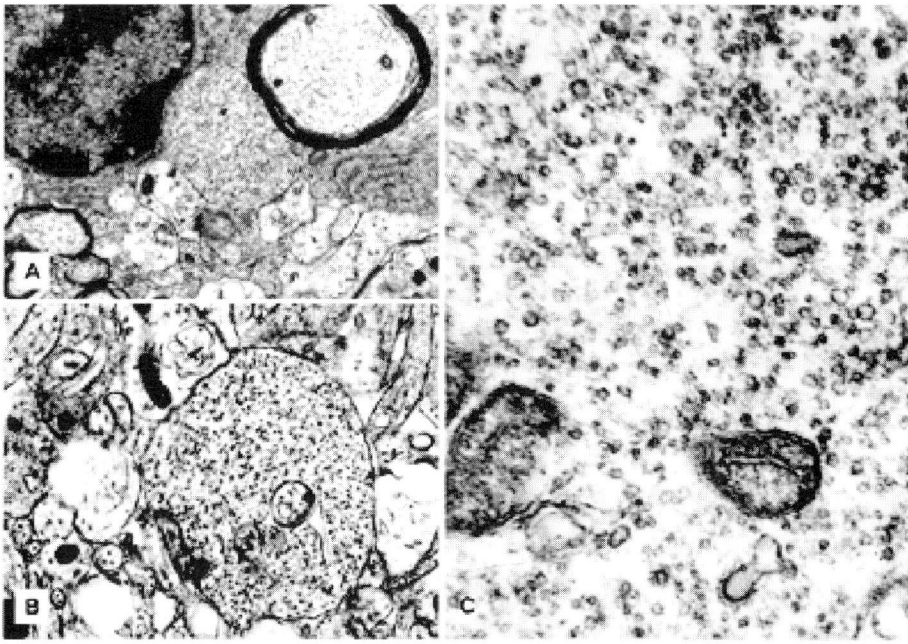

The exact topology of TVS is not entirely clear. In most published electron micrographs TVS appeared as spheres measuring between 20 and 40 nm in diameter. Some investigators suggested that the tubular "arrays" of TVS result from overlapping spherical profiles, but we repeatedly shown short tubular forms of TVS, and it is therefore evident that TVS are pleomorphic structures existing in at least two forms—spheres and short tubules.

Only limited data are available concerning the number and density of TVS through the incubation time of experimental TSEs. In hamsters inoculated with the 263K strain of scrapie, TVS were initially observed as early as 3 weeks after i.c. inoculation, but their number increased only with the onset of disease at 9 to 10 weeks after inoculation[69]. Noteworthy, vacuolation and astrocytosis were detected at 8 weeks post-inoculation and, thus, followed the appearance of TVS.Similar findings were described by[64]. In mice infected with the Fujisaki strain of GSS, TVS were first seen 13 weeks after i.c. inoculation and their numbers increased dramatically at 18 weeks after inoculation, when the first signs of clinical disease were noted[69]. In contrast to 263K, the Fujisaki strain of CJD showed vacuolation and astrocytosis at the same time as the appearance of TVS and the increase in the number of processes containing TVS paralleled the increasing intensity of vacuolation and astrocytosis. TVS were approximately twice as abundant in terminally ill mice following intraocular inoculation as in mice following intracerebral inoculation[69]. By contrast, final intensity of vacuolation and astrocytosis was not dependent on the route of inoculation. In conclusion, TVS appear early in the incubation period, preceding the onset of clinical disease. Furthermore, in scrapie-infected hamsters, TVS preceded the appearance of other neuropathological changes. The approximately 10^3-fold lower infectivity titre of the Fujisaki strain of CJD, compared to the 263K strain, may have accounted for the delayed appearance of TVS in experimental CJD. The apparent correlation between the number of neuronal processes containing TVS and infectivity titre may explain why in cell cultures infected with scrapie, TVS could not be found and why their number in experimental scrapie in sheep or CJD in humans was so low[49,65].

2.6.9. Biochemistry of TVS

The chemical composition of TVS is unknown. Earlier studies reported that ruthenium red enhances contrast of TVS. These staining properties of TVS may be interpreted as an evidence for the presence of glycosyl residues within TVS. To evaluate whether PrP is a part of TVS, we employed immunogold electronmicroscopy[70,71]. In all models TVS-containing processes were readily detected and neither these processes nor TVS themselves were decorated with gold particles. Even if amyloid plaques were observed in close contact with TVS-containing neuronal processes, the plaques were decorated with gold particles while the processes remained unstained[70,71]. At higher magnification, amyloid fibrils were clearly visible with numerous round or short tubular particles attached to them. At such a magnification, these particles were membrane-bound and their diameter was approximately twice that of amyloid fibrils; they were virtually indistinguishable from TVS. TVS located in areas adjacent to plaques in the 87V model and in areas of diffuse PrP immunolabelling in ME7 model were also unlabelled with anti PrP antisera.

2.7. PrP^d Immunohistochemistry

Immunohistochemistry (ICC) became the major diagnostic tool to detect PrP^d[21] along with its more refined equivalent the *histoblot*[72] and *paraffin-embedded blot* (PET) technique[73]. The major caveat of ICC

Figure 2.7. (a) Synaptic pattern of PrP^d accumulation (b) Perivacuolar PrP^d pattern of immunostaining; (c) Perineuronal PrP^d pattern of immunostaining. Note that large neuron in the middle has radiating processes decorated with PrP^d deposits; (20) Stripes of PrP^d immunoreactivity running perpendicularly to the surface of the cerebellum.

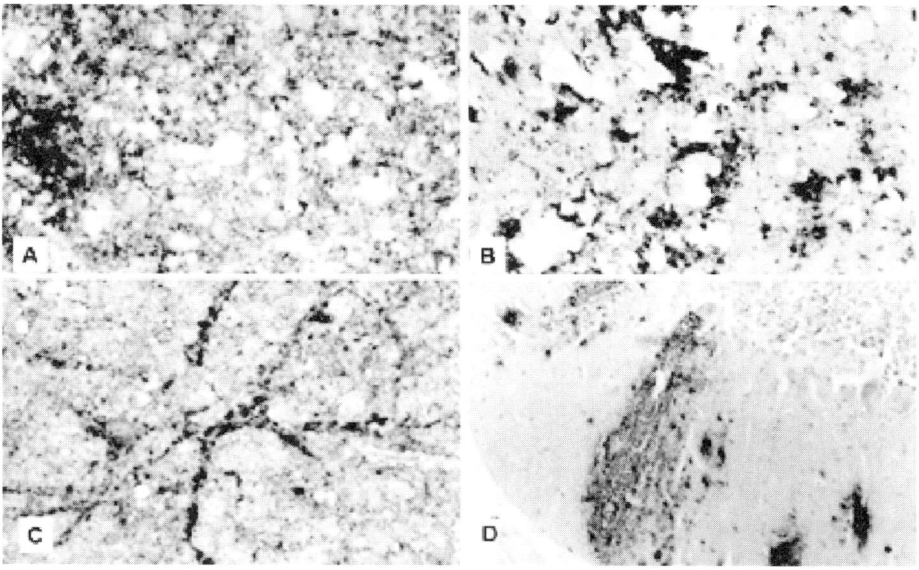

is to get rid of PrP^c as all available antibodies, including widely used and commercially available 3F4 and 6H8 antibodies, cannot discriminate between PrP^c and misfolded PrP^d. To this end, different unmasking techniques are used—hydrating or hydrolytic autoclaving, microwaving, formic acid or guanidinum thiocyanate incubation; a combination of several of preatreatments may give the best results.

Several patterns of PrP^d expression are revealed by ICC[21]. Those include: the most difficult to visualize, synaptic (Figure 2.7a), perivacuolar (Figure 2.7b), perineuronal (Figure 2.7c) and plaque-like. In certain familial forms of CJD, including the E200K mutation, stripes of PrP^d immunopositivity running perpendicularly to the surface of the cerebellar cortex is visible (Figure 2.7d)[74]. If plaques are visualized by a routine neuropathology (H & E, PAS- or Alcian blue stainings) these are labeled "plaques"; if they are detected only by ICC and are not visible by routine techniques, they are called "plaque-like". In chronic wasting disease[6] and in some experimental scrapie models[70,71], subependymal deposits (subependymal plaques) are visible. These correspond, ultrastructurally, to areas of low electron density containing haphazardly-oriented fibrils that, when stained with anti-PrP Abs, are heavily decorated with PrP-conjugated gold particles (Figure 2.8).

Figure 2.8. (a) A loose plaque in the subependymal location. Note amyloid fibrils in electron-lucent spaces; (b) a high magnification showing PrP^d-immunopositive amyloid fibrils floating in electron lucent spaces. Dark deposits in (b) correspond to anti-PrP-conjugated gold-silver enhanced particles[70]. Original magnification, (a), ×4400; (b), ×50 000.

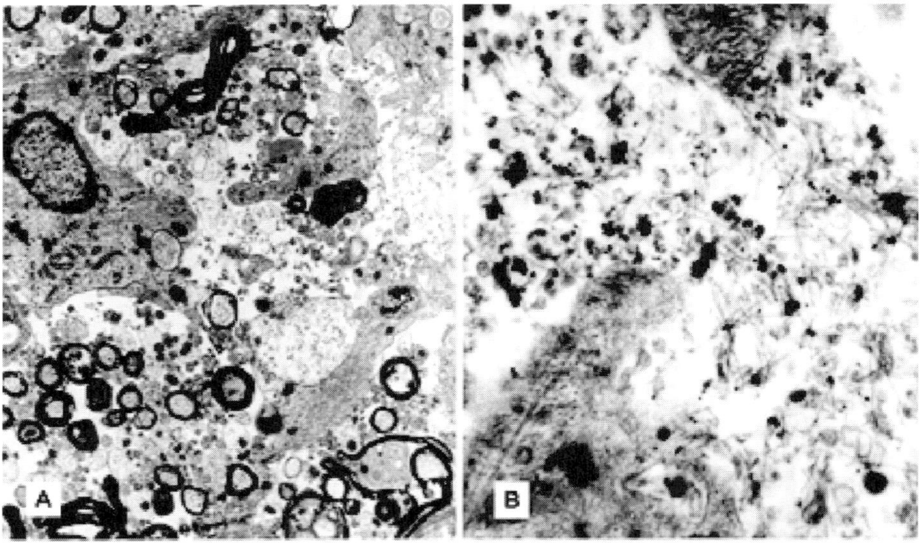

In both models, perivacuolar PrP deposits are also visible. Recently, intraneuronal PrP^d deposits have been described[75–76]. The intraneuronal PrP^d deposits may exist as diffuse type in the neuronal perikaryon; large intracytoplasmic inclusion bodies reminiscent somehow Pick bodies, intracytoplasmic punctuate deposits and somato-synaptic dots. There is inverse relation between PrP^d accumulation in neurons and overall PrP^d-immunostaining; furthermore, cases with robust PrP^d-immunostaining and low intranuclear staining exhibit lower age of onset. Of note, similar intraneuronal PrP^d-immunostaining was observed in celiac, superior mesenteric and stellate ganglia of vCJD cases[77].

PrP^d is also present in the peripheral nervous system as well as in lymphoid tissue, including the spleen[78]. In CJD, PrP^d was detected in satellite cells and neurons of trigeminal ganglia[79], and in an adaxonal location (Schwan cells ?) in nerve fibers. Those deposits were scant and were observed in 1 of 9 CJD cases and in 1 GSS case[80]. The paucity of peripheral PrP^d deposition remains in strong contrast to the widespread expression of PrP^d in experimental scrapie both in sheep and in hamsters[81]. In contrast, using sensitive Western technique, PrP^d was readily detected in spleen in 10 of 28 CJD cases and in the spleen of

8 of 32 CJD cases[82]. Of note, peripheral accumulation of PrP^d seems to be more abundant in uncommon CJD variants—MM2—cortical or MV2 (see section on classification).

2.8. Particularities of disease

2.8.1. Kuru

Kuru[1], the first human neurodegenerative disease classified as a transmissible spongiform encephalopathy (TSE) was first reported to Western medicine in 1957 by Gajdusek and Zigas[83]. The first systematic examination of kuru neuropathology (12 cases) was published by Klatzo *et al.* in 1959[84]. Pathological changes were confined to the brain and spinal cord, with the cerebellum, pontine nuclei, thalamus, and spinal cord bearing most of the burden. Macroscopically, some brains were oedematous and leptomeninges were congested. Microscopically, neuronal changes observed in anterior motor neurons of the spinal cord, in different brain stem nuclei, in the cerebellum, and in the cerebral cortex were non-specific in nature but nonetheless sufficient for Klatzo *et al.* to draw a parallel between kuru and Creutzfeldt-Jakob disease.

Numerous neurons were either shrunken and hyperchromatic or, to the contrary, pale with dispersion of Nissl substance or intracytoplasmic vacuoles not unlike those seen in scrapie, BSE, and chronic wasting disease, but which are rather infrequent in human TSE. In the striatum, some neurons were vacuolated to such a degree that they looked "moth-eaten". Neuronophagia was observed. A few binucleated neurons were visible and torpedo formation was noticed in the Purkinje cell layer, along with empty baskets that marked the presence of degenerated Purkinje cells[85]. In the medulla, neurons of the vestibular nuclei and the lateral cuneatus were frequently affected; the spinal nucleus of the trigeminal nerve and nuclei of VIth, VIIth, and motor nucleus of the VIth cranial nerves were affected less frequently while nuclei of the XIIth cranial nerve, the dorsal nucleus of Xth cranial nerve and nucleus ambiguous were relatively spared. In the cerebral cortex, the deeper layers were affected more than the superficial layers, neurons in the hippocampal formation were normal. In the cerebellum, the paleocerebellar structure (vermis and flocculo-nodular lobe) was most severely affected, and spinal cord pathology was most severe in the corticospinal and spinocerebellar tracts. Astro- and microglial proliferation was widespread; the latter formed rosettes and appeared as rod- or amoeboid types or as macrophages (gitter cells). Myelin degradation was observed in 10 of 12 cases. Interestingly, the significance of vacuolar changes was not appreciated by Klatzo *et al.*[84], but "*small spongy spaces*", were noted in 7 of 13 cases studied by Beck and Daniel[85].

The most striking neuropathologic feature of kuru is the presence of numerous amyloid plaques, described as "*spherical bodies with a rim of radiating filaments*" and found in 6 of 12 cases studied by Klatzo *et al.*[84], and in "about three quarters" of the 13 cases of Beck and Daniel[85]; they became known as "kuru plaques". These measured 20–60 μm in diameter, were round or oval and consisted of a dark-stained core with delicate radiating periphery surrounded by a pale "halo". Kuru plaques were most numerous in the granular cell layer of the cerebellum, basal ganglia, thalamus, and cerebral cortex in that order of frequency. It is noteworthy that in Gerstmann-Sträussler-Scheinker disease (GSS) plaques are located in both the granular cell and molecular layers, whereas in CJD plaques are confined to granular cell layer. Kuru plaques are metachromatic and stain with PAS, Alcian blue, and Congo-red, and a proportion of them are weakly argentophilic when impregnated according to Belschowsky or von Braunmühl techniques. Klatzo *et al.*[84] reported that plaques were most readily visualized by Holmes' silver impregnation method. Of historical interest, another unique disease reported by Seitelberger[86] as "*A peculiar hereditary disease of the central nervous system in a family from lower Austria*" (germ. *Eigenartige familiar-hereditare Krankenheit des Zentralnervensystems in einer niederoosterreichen Sippe*) was mentioned by Neumann *et al.*[87] who was thus the first person to suggest a connection between kuru and GSS and kuru.

Recently, a renewed interest in kuru pathology has been provoked by the appearance of a variant form of CJD (vCJD) resulting from infection by the agent of bovine spongiform encephalopathy (BSE), that is also characterized by numerous plaques, including "florid" plaques—a kuru plaque surrounded by a corona of spongiform vacuoles. Hainfellner *et al.*[44] analyzed by means of modern immunohistochemistry the case of a young male kuru victim (Kupenota) from the South Fore region whose brain tissue had transmitted disease to chimpanzees, and McLean *et al.*[88] examined a series of 11 cases of kuru still in the archives of the University of Melbourne. In contrast to the classical studies described above, both papers stressed the presence of typical spongiform change present in deep layers (III–V) of the cingulate, occipital, enthorrinal and insular cortices, and in the subiculum. Spongiform change was also observed in the putamen and caudate, and some putaminal neurons contained intraneuronal vacuoles. Spongiform change was prominent in the molecular layer of the cerebellum, in peraqueductal gray matter, basal pontis, central tegmental area, and inferior olivary nucleus. The spinal cord showed only minimal spongiform change.

Immunohistochemical studies revealed that misfolded PrP was present not only in kuru plaques (Figure 2.9a), already demonstrated to be PrP^{Sc}-immunoreactive by Piccardo *et al.*[89], PrP^{Sc} but also in synaptic and perineuronal sites, and in the spinal cord the *substantia gelatinosa*was particularly affected, as in iatrogenic CJD cases following peripheral inoculation[90].

2.8.2. Gerstmann-Sträussler-Scheinker (GSS) disease

Gerstmann-Sträussler-Scheinker disease was described in 1936 and then it was re-evaluated by Seitelberger[86] and von Braunmühl[91] and transmitted to chimpanzees by Masters *et al.*[3]. In classical papers of Seitelberger[86] and Boellaard *et al.*[92] the similarities of GSS neuropathology to that of kuru was stressed and preconceived the transmission experiments of Masters *et al.*[3]. Following discovery of a mutation (Pro102Leu) in the *PRNP* gene by Hsiao *et al.*[93] many new mutations and new families have been described[94]. A detailed study of the first GSS family was published by Hainfellner *et al.*[45].

The detailed description of all diverse families is beyond the scope of this chapter and was reviewed elsewhere[94,95] and only general overview will be provided. The hallmark of GSS is multicentric plaque (Figure 2.9b–c). In contrast to CJD, where kuru plaques occur predominantly in the granular and the Purkinje cell layer, multicentric plaques are present in the molecular layer of the cerebellum[45,94,96]. As the name "multicentric" implies, GSS plaques are composed of several cores of different sizes merging together. In contrast to classical kuru plaques, multicentric plaques are neuritic, i.e., they are surrounded by dystrophic neurites immunolabeled against neurofilament proteins. In several families, dystrophic neurites contained also paired helical filaments (PHF)—identical to those found in dystrophic neurites in Alzheimer's disease[96,97]. At the level of electron microscopy, separate cores are readily discernible but overlapping. Multicentric plaques fulfill the criteria for amyloid—like kuru plaques they are congophilic and birefringent under polarized light. In addition, diffuse PrP^{d} deposits may be demonstrated by PrP-immunohistochemistry. PrP^{d} in diffuse (or amorphous) plaques is not yet fibrillised. Spongiform change is variable in GSS and its presence depends on the presence of 21 kDa PrP fragments. In classic description,

Figure 2.9. (a) A row of PrP^d-immunupositive kuru plaques[44]; (b) A typical multicentric plaque stained against PrP^d; (c) an electron micrograph of a cluster of several amyloid plaques from a GSS case[46], note a microglial cell in a vicinity; original magnification, ×4400.

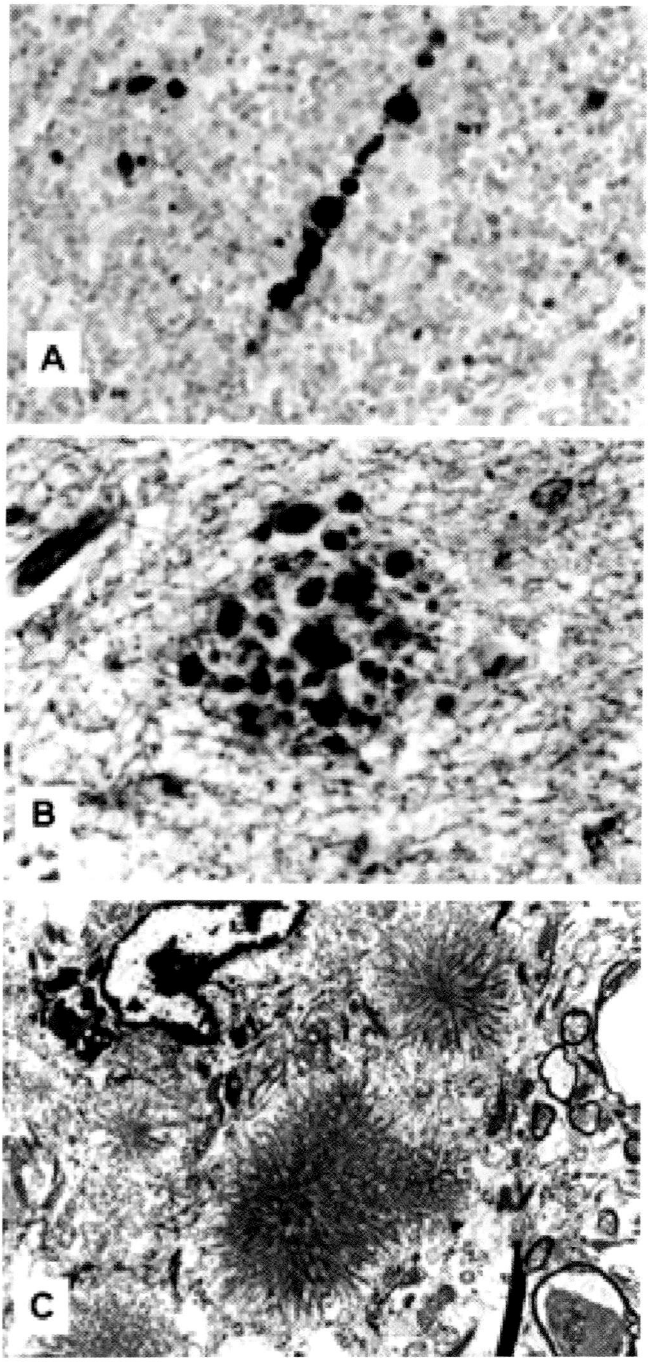

GSS was regarded as "system degenerations"[86] and indeed, several white matter long tracts were found to degenerate in GSS.

2.8.3. Variant Creutzfeldt-Jakob disease

In 1996, Will *et al.*[47] reported a novel from of human prion disease which is now known as vCJD. Clinically, this disorder presents as a progressive neuropsychiatric syndrome with psychiatric and/or sensory symptoms at disease onset followed by ataxia, other movement disorders (particularly myoclonus), visual abnormalities and cognitive impairment, resulting in a terminal akinetic and mute state. Since its original description, 146 cases of vCJD have been identified in the UK, 6 cases in France and 1 in each of Canada, Ireland, Italy and the US. In the UK, the rate of new cases of vCJD has declined significantly over the past 12–18 months. Unlike most other forms of prion disease, this disorder predominantly affects young adults with an average age of onset around 28 years and a duration of illness of around 13 months. All patients who have been genotyped are methionine homozygotes at codon 129 in the prion protein gene.

The neuropathology of vCJD is distinct from other human prion diseases, and is characterized by multiple florid plaques. Florid plaques (Figure 2.10a) are fibrillary structure with a dense core surrounded by a pale region of radiating fibrils, and surrounded by a ring of spongiform change[47]. These plaques can be demonstrated using silver impregnation techniques and the PAS and Alcian blue stains[98]. Florid plaques occur in all layers of the cerebral cortex, but are most conspicuous at the bases of the gyri in the occipital and cerebellar cortex. They are also numerous in the molecular layer of the cerebellum.

Ultrastructural studies of the florid plaques in variant CJD have demonstrated the masses of radiating fibrils at the periphery, with abnormal neurites similar to those seen at the periphery of the Aβ plaques in Alzheimer's disease (Figure 2.10c)[99]. Neurofibrillary tangles and paired helical filaments have not been identified in variant CJD, and immunocytochemistry for tau gives negative results. Immunoelectron microscopy showed PrP accumulation both in the amyloid fibrils and in some of the abnormal cell membranes surrounding the plaques[100].

Spongiform change is widespread but often patchy within the cerebral cortex, mostly in a microvacuolar pattern or in relation to amyloid plaques. In contrast, confluent spongiform change is always present in the caudate nucleus and putamen, and focal spongiform change is present in most of the thalamic nuclei, the hypothalamus and globus pallidus, but the posterior thalamic nuclei (including the pulvinar) are

Figure 2.10. (a) A typical florid plaque from a case of vCJD, composed of a fibrillary amyloid core surrounded by small spongiform vacuoles; (b) PrP immunocytochemistry shows an intense positive reaction in florid plaques in vCJD. Many smaller plaques and amorphous PrP deposits are also demonstrated; (c) Perineuronal and linear periaxonal patterns of the PrP-accumulation in the basal ganglia in a vCJD case; (d) PrP-immunoreactivity in the tonsil of a vCJD case, showing staing of follicular dendritic cells within a germinal centre; (e) An electronmicrograph showing the florid plaque.

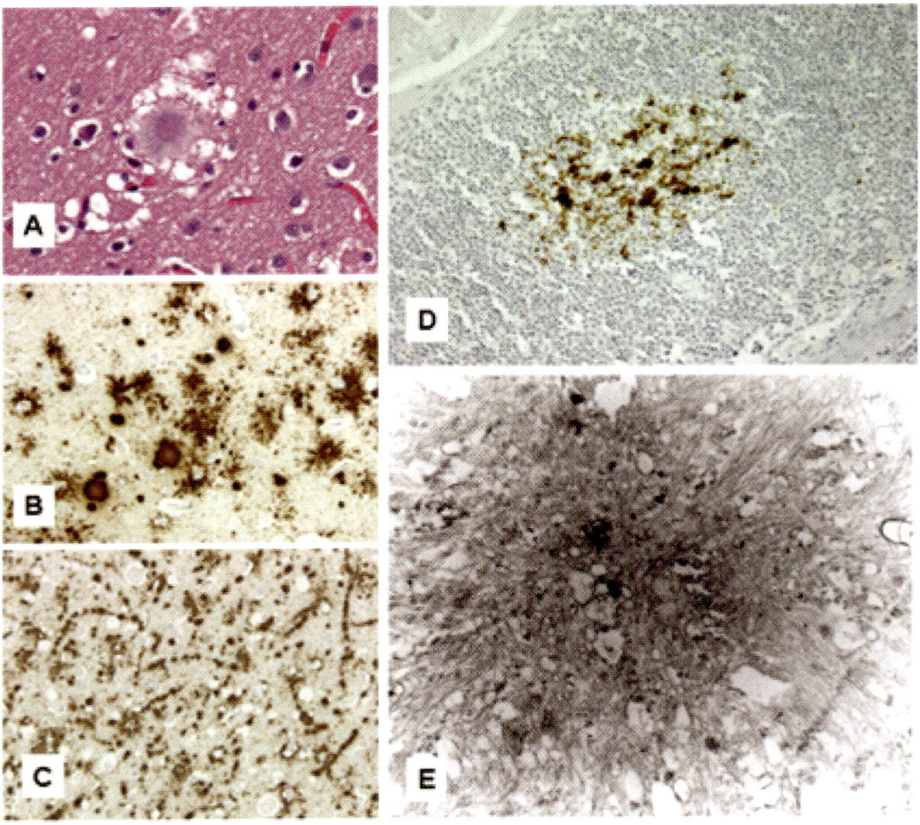

spared. Mild spongiform change is detected in the periaqueductal grey matter in the midbrain, in the pontine nuclei and in cerebellar cortex, often associated with amyloid plaques.

Neuronal loss in the cerebral cortex is most severe in the primary visual cortex. Neuronal loss in the basal ganglia is most evident in cases with severe and confluent spongiform change. In the thalamus, neuronal loss was most severe in the posterior nuclei, particularly in the pulvinar, which also showed marked astrocytosis[101]. The severe astrocytosis in the posterior thalamus was best visualized on immunocytochemistry for glial fibrillary acidic protein[101]. This technique also demonstrated astrocytosis in relation to areas of severe neuronal loss and less frequently around the margins of amyloid plaques in other brain regions. Neuronal loss and astrocytosis were not conspicuous in the pons, medulla and spinal cord, but were variable in the cerebellum, sometimes most severe in the vermis.

2.8.4. Immunocytochemistry

The florid plaques in the cerebral and cerebellar cortex give an intense positive reaction on immunocytochemistry for PrP[98] (Figure 2.10b). Smaller "cluster plaques" are revealed by immunocytochemistry for PrP in all cases. PrP immunocytochemistry also shows a widespread amorphous pericellular deposition of PrP around small neurons in the cerebral and cerebellar cortex. In the basal ganglia there is a predominantly perineuronal and periaxonal pattern of PrP accumulation (Figure 2.10c). A synaptic pattern of immunoreactivity with occasional plaques is detected in the thalamus. In the brainstem and spinal cord, PrP positivity is present at all levels in the gray matter, particularly in the substantia gelatinosa. No PrP accumulation was detected in either the leptomeninges or the dura mater.

2.8.5. Quantitative neuropathology in variant CJD

Quantitative studies on the first cases of variant CJD confirmed the initial morphological descriptions and indicated that the measurable histological accumulation of abnormal PrP deposits in the cerebellum was far greater than in sporadic CJD cases[102]. Furthermore, the selective involvement of the posterior thalamus was demonstrated, with levels of astrocytosis far in excess of sporadic CJD cases[102]. Subsequent quantitative studies have demonstrated that in the cerebral cortex the spongiform change is consistently most pronounced in the occipital cortex, but the relationship between the spongiform change and the presence of PrP amyloid plaques varies in different brain regions[103,104]. Analysis of the spatial patterns of abnormal PrP deposition in variant CJD has found no significant differences between different regions of the cerebral cortex[105]. In addition to quantitative histology, there is the prospect of developing textural analysis techniques to investigate the differences in patterns of abnormal PrP deposition[106].

2.8.6. Non-CNS tissues

PrP accumulation is identified in the retina and optic nerve[107], spinal dorsal root ganglia and in the trigeminal ganglia, but peripheral sensory

Table 2.1. Pathological diagnostic features of variant CJD

1. Multiple florid plaques in H&E sections; numerous small cluster plaques in PrP stained sections.
2. Amorphous pericellular and perivascular PrP accumulation in the cerebral and cerebellar cortex.
3. Severe spongiform change; perineuronal and axonal PrP accumulation in the caudate nucleus and putamen.
4. Marked astrocytosis and neuronal loss in the posterior thalamic nuclei and midbrain.
5. Reticular and perineuronal PrP accumulation in the grey matter of the brainstem and spinal cord.
6. Predominance of di-glycosylated PrP^{RES} in central nervous system and lymphoid tissues.
7. PrP^{RES} accumulation in germinal centres within lymphoid tissues throughout the body.

and motor nerves contain no detectable PrP. Immunocytochemistry for PrP in other organs (adrenal gland, thyroid gland, parathyroid gland, skeletal muscle, bladder, testes, pelvic organs (vagina, cervix, uterus, Fallopian tubes and ovaries), heart, lung, liver, kidney, oesophagus, stomach, pancreas, gall bladder, salivary gland and skin is negative[98,108].

In contrast, PrP accumulation is identified in follicular dendritic cells and macrophages within many germinal centres in the tonsils (Figure 2.10d) and within germinal centres in the appendix, Peyer's patches in the ileum, spleen and lymph nodes from the cervical, mediastinal, para-aortic and mesenteric regions and the thymus[98,108,109].

Neuropathological studies were important in identifying variant CJD as a novel human prion disease[47]. The diagnostic pathological features of variant CJD are summarized in Table 2.1.

Although florid PrP amyloid plaques have been described in cases of iatrogenic CJD following dura mater graft procedures[110], their number and distribution in the brain is more restricted than in variant CJD. In addition, the biochemical features of abnormal PrP in the brain in variant CJD on Western blot examination is relatively uniform[108] (in contrast to sporadic CJD, where multiple PrP isotypes have been identified[111,112]). These findings reinforce the need for detailed characterisation of human prion diseases by clinical, pathological and biochemical studies, which can be reinforced by experimental transmission to demonstrate infectivity and to undertake strain typing studies. This has been reinforced by the identification of a recent case of vCJD in a recipient of a blood transfusion from a donor who has died earlier (after donation) from vCJD[113]. Strain typing studies on the first cases of variant CJD allowed the early recognition of the similarities in the transmissible agent in BSE and variant CJD.

These findings have been supported by other independent workers, reinforcing the link between these disorders[14,114,115,116]. As BSE has now been identified in many other countries across the world, it is possible that more cases of vCJD will be identified outside the UK. The future for vCJD in the UK remains uncertain, since it is not clear if the other codon 129 subgroups will be susceptible to this disease. If so, increased numbers of cases might occur over an unknown time period, indicating a need for continued surveillance for CJD, at least in the near future.

2.9. Acknowledgments

This paper is supported in part by NeurPrion Network of Excellence, MolMed Centre of Excellence, SEEC, the British Council and the State Research Committee.

2.10. References

1. D. C. Gajdusek, C. J. Gibbs, M. P., and Alpers M. P., Experimental transmission of a kuru-like syndrome to chimpanzees. *Nature* **209**, 794–796 (1966).

2. C. J. Gibbs Jr, D. C. Gajdusek, D. M. Asher, M. P. Alpers, E. Beck, P. M. Daniel, and W. B. Matthews, Creutzfeldt-Jakob disease (spongiform encephalopathy): transmission to chimpanzee. *Science* **161**, 388–389 (1968).

3. C. L. Masters, D. C, Gajdusek, and C. J. Gibbs Jr, Creutzfeldt-Jakob disease virus isolations from the Gerstmann-Sträussler syndrome. With an analysis of the various forms of amyloid plaque deposition in the virus induced spongiform encephalopathies. *Brain* **104**, 559–588 (1981).

4. E. Lugaresi, R. Medori, P. Montagna, A. Baruzzi, P. Cortelli, A. Lugaresi, P. Tinuper, M. Zucconi, and P. Gambetti, Fatal familial insomnia and dysautonomia with selective degeneration of thalamic nuclei. *New Engl. J. Med.* **315**, 997–1004 (1986).

5. D. Burger and G. R. Hartsough, A scrapie-like disease of mink, in: *Report of a Scrapie Seminar* (United States Department of Agriculture, paper 27, 1964), pp. 225–227.

6. P. P. Liberski, D. C. Guiroy, E. S. Williams, A. Walis, and H. Budka, Deposition patterns of disease-associated prion protein in captive mule deer brains with chronic wasting disease. *Acta Neuropathol. (Berl.)* **102**, 496–500 (2001).

7. G. A. H. Wells, A. C. Scott, C. T. Johnson, R. F. Gunning, R. D. Hancock, M. Jeffrey, M. Dawson, and R. Bradley, A novel progressive spongiform encephalopathy in cattle. *Vet. Rec.* **121**, 419–420 (1987).

8. C. Weissmann, A "unified theory" of prion propagation. *Nature* **352**, 679–683 (1991).

9. R. Gabizon, G. Telling, Z. Meiner, M. Halimi, I. Kahana, and S. B. Prusiner, Insoluble wild-type and protease-resistant mutant prion protein in brains of patients with inherited prion diseases. *Nature Med.*, **2**, 59–64 (1996).

10. B. Oesch, D. Westaway, M. Walchlii, M. P. McKinley, S. B. H. Kent, R. Aebersold, R. A. Barry, P. Tempst, D. B. Teplow, L. E. Hood, S. B. Prusiner, and C. Weissmann, A cellular gene encodes scrapie PrP 27–30 protein. *Cell* **40**, 735–746 (1985).

11. S. B. Prusiner, Novel proteinaceous infection particles cause scrapie. *Science* **216**, 136–144 (1982).

12. P. P. Liberski, and M. Jaskolski, Prion diseases: a dual view of the prion hypothesis as seen from a distance. *Acta Neurobiol Exp.* **62,**197– 226 (2002).

13. S. B. Prusiner, M. P. McKinley, K. A. Bowman, D. C. Bolton, P. E. Bendheim, D. C. Groth, and G. G. Glenner, Scrapie prions aggregate to form amyloid-like birefringent rods. *Cell* **35**, 349–358 (1983).

14. J. Collinge, K. C. L. Sidle, J. Meads, J. Ironside, and A. F. Hill, Molecular analysis of prion strain variation and the aetiology of "new variant" CJD. *Nature* **383**, 685–690 (1996).

15. A. F. Hill, M. Desbruslais, S. Joiner, K. C. L. Sidle, I. Gowland, J. Collinge, L. J. Doey, and P. Lantos, The same prion strain causes vCJD and BSE. *Nature* **389**, 448–450 (1997).

16. H. A. Kretzschmar, L. E. Stowring, D. Westaway, W. H. Stubblebine, S. B. Prusiner, and S. J. DeArmond, Molecular cloning of human prion protein cDNA. *DNA*, **4**, 315–324 (1986).

17. S. Mead, M. P. Stumpf, J. Whitfield, J. A. Beck, M. Poulter, T. Campbell, J. B. Uphill, D. Goldstein, M. Alpers, F. M. Fisher, and J. Collinge, Balancing selection at the prion protein gene consistent with prehistoric kurulike epidemics. *Science* **300**, 640–643 (2003).

18. P. M. Daniel, Creutzfeldt-Jakob disease, *J. Clin. Pathol.* **25**, 97–101 (1972).

19. B. Brownell and D. R. Oppenheimer, An ataxic form of subacute presenile polioencephalopathy (Creutzfeldt-Jakob disease), *J. Neurol. Neurosurg. Psychiat.* **28**, 350–361 (1965).

20. H. Siedler H and N. Malamud, Creutzfeldt-Jakob disease. Clinicopathological report of 15 cases and review of the literature (with special reference to a related disorder designated as subacute spongiform encephalopathy), *J. Neuropathol. Exp. Neurol.* **22**, 381–402 (1963).

21. H. Budka, M. W. Head, J. W. Ironside, P. Gambetti, P. Parchi, M. Zeidler, and F. Tagliavini, Prion Disorders: CJD: Sporadic Creutzfeldt-Jakob disease, in *Neurodegeneration: The Molecular Pathology of Dementia and Movement Disorders*, edited by D. Dickson (ISN Neuropath Press, Basel, 2003), pp. 287–297.

22. A. F. Hill, S. Joiner, J. D. F. Wadsworth, K. C. L. Sidle, J. E. Bell, H. Budka, J. W. Ironside, and J. Collinge, Molecular classification of sporadic Creutzfeldt-Jakob disease. *Brain* **126**, 1333–1346 (2003).

23. J. D. F. Wadsworth, A. F. Hill, S. Joiner, G. S. Jackson, A. R. Clarke, J. Collinge, Strain-specific prion-protein conformation determined by metal ions. *Nature Cell Biol.* **1**, 55–59 (1999).

24. H. G. Creutzfeldt, Uber eine egenartige herdformige Erkrankung des Zentralnervensystems. *Z. Ges. Neurol. Psychiat.* **57**, 1–18 (1920).

25. A. Jakob, Uber eigenartige Erkrankungen des Zentralnervensystems mit bemerkenswertem anatomischem Befunde (spastische Pseudosclerose-Encephalopathie mit disseminierten Degenerationsherden). *Deutsch Z. Nervenheilk.* **70**, 132–146 (1921).

26. A. Jakob, Spastische Pseudosclerose, in: *Monographien aus dem Gesamtgebiete der Neurologie und Psychiatrie, Die Ekstrapyramidalen Erkrankungen*, Vol. 37, edited by O. Foerster, K. Wilmanns (Julius Springer, Berlin, 1923), pp. 215–245.

27. C. L. Masters and D. C. Gajdusek, The spectrum of Creutzfeldt-Jakob disease and the virus induced subacute spongiform encephalopathies, in: *Recent Advances in Neuropathology*, edited by T. J. Smith and J. B. Cavanagh (Churchill Livingstone, Edinburgh, 1982), pp. 139–163.

28. A. M. Salazar, C. L. Masters, D. C. Gajdusek, C. J. Gibbs Jr., Syndromes of amyotrophic lateral sclerosis and dementia: relation to transmissible Creutzfeldt-Jakob disease. *Ann. Neurol.*, **14**, 17–26 (1983).

29. H. Yamanouchi, H. Budka, and K. Vass, Unilateral Creutzfeldt-Jakob disease. *Neurology* **36**, 1517–1520 (1986).

30. R. Tinter, P. Brown, T. Hedley-Whyte, E. B. Raaport, C. P. Piccardo, and D. C. Gajdusek, Neuropathologic verification of Creutzfeldt-Jakob disease in the exhumed American recipient of human pituitary growth hormone: epidemiologic and pathogenetic implications. *Neurology* **36**, 932–936 (1986).

31. C. L. Masters and E. P. Richardson, Subacute spongiform encephalopathy (Creutzfeldt-Jakob disease). The nature and progression of spongiform change. *Brain* **101**, 333–344 (1978).

32. M. Jeffrey, J. R. Scott, A. Williams, and H. Fraser, Ultrastructural features of spongiform encephalopathy transmitted to mice from three species of bovidae. *Acta Neuropathol. (Berl.)* **84**, 559–569 (1992).

33. S. M. Chou, W. N. Payne, C. J. Gibbs Jr., and D. C. Gajdusek, Transmission and scanning electron microscopy of spongiform change in Creutzfeldt-Jakob disease. *Brain* **103**, 885–904 (1980).

34. P. P. Liberski, R. Yanagihara, C. J. Gibbs Jr, and D. C. Gajdusek, White matter ultrastructural pathology of experimental Creutzfeldt-Jakob disease in mice. *Acta Neuropathol (Berl.)* **79**, 1–9 (1989).

35. P. P. Liberski, R. Kordek, P. Brown, and D. C. Gajdusek, Astrocytes in transmissible spongiform encephalopathies (prion diseases), in: *Astrocytes in Brain Aging and Neurodegeneration,* edited by H. M. Schipper (R. G. Landes Company, Austin, Texas, USA, 1998) pp. 127–164.

36. P. P. Liberski, P. Brown, L. Cervenakova, and D. C. Gajdusek, Interactions between astrocytes and oligodendroglia in human and experimental Creutzfeldt-Jakob disease and scrapie. *Exp. Neurol.* **144**, 227–234 (1997).

37. M. Lafarga, M. T. Berciano, I. Suarez, C. F. Viadero, M. A. Andres, and J. Berciano, Cytology and organization of reactive astroglia in human cerebellar cortex with severe loss of granule cells: a study on the ataxic form of Creutzfeldt-Jakob disease. *Neuroscience* **40**, 337–352 (1991).

38. M. Lafarga, M. T. Berciano, I. Suarez, M. A. Andres, and J. Berciano, Reactive astroglia-neuron relationships in the human cerebellar cortex: a quantitative, morphological and immunohistochemical study in Creutzfeldt-Jakob disease. *Int. J. Dev. Neurosci.* **11**, 199–213 (1993).

39. M. Dubois-Dalcq, M. Rodriguez, T. S. Reese, C. J. Gibbs Jr, and D. C. Gajdusek, Search for a specific marker in the neural membranes of scrapie mice. *Lab. Invest.* **36**, 547–553 (1977).

40. H. Budka, A. Aguzzi, P. Brown, J. M. Brucher, O. Bugiani, F. Gullotta, M. Haltia, J. J. Hauw, J. W, Ironside, K. Jellinger, *et al.*, Neuropathological diagnostic criteria for Creutzfeldt-Jakob disease (CJD) and other human spongiform encephalopathies. *Brain Pathol.*, **4**, 459–466 (1995).

41. M. Jeffrey, C. M. Goodsir, N. Fowler, J. Hope, M. E. Bruce, and P. A. McBride, Ultrastructural immuno-localization of synthetic prion protein peptide antibodies in 87V muine scrapie. *Neurodegeneration* **5**, 101–109 (1996).

42. M. Barcikowska, P. P. Liberski, J. Boellaard, P. Brown, D. C. Gajdusek, and H. Budka, Microglia is a component of the prion protein amyloid plaque in the Gerstmann-Sträussler-Scheinker syndrome. *Acta Neuropathol. (Berl.)*, **85**, 623–627 (1993).

43. M. Jeffrey, C. Goodsir, M. E. Bruce, P. A. McBride, and C. Farquhar, Morhogenesis of amyloid plaque in 87V murine scrapie. *Neuropathol. Appl. Neurobiol.* **20**, 535–542 (1994).

44. J. A. Hainfellner, P. P. Liberski, D. C. Guiroy, L. Cervenakova, P. Brown, D. C. Gajdusek, and H. Budka, Pathology and immunocytochemistry of a kuru brain. *Brain Pathol.* **7**, 547–554 (1997).

45. J. Hainfellner, S. Brantner-Inhaler, L. Cervenakova, P. Brown, T. Kitamoto, J. Tateishi, H. Diringer, P. P. Liberski, H. Regele, M. Feucht, N. Mayr, K. Wessely,

K. Summer, F. Seitelberger, and H. Budka, The original Gerstmann-Sträussler-Scheinker family of Austria: divergent clinicopathological phenotypes but constant PrP genotype. *Brain Pathol.* **5**, 201–213 (1995).

46. P. P. Liberski and H. Budka, Ultrastructural pathology of Gerstmann-Sträussler-Scheinker disease. *Ultrastr. Pathol.* **19**, 23–36 (1995).

47. R. G. Will, J. W. Ironside, M. Zeidler, S. N. Cousens, K. Esstibeiro, A. Alperovitch, S. Poser, M. Pocchiari, A. Hofman, and P. G. Smith PG, A new variant of Creutzfeldt-Jakob disease in the UK. *Lancet* **347**, 921–925 (1996).

48. P. P. Liberski, R. Yanagihara, C. J. Gibbs Jr., and D. C. Gajdusek, Scrapie as a model for neuroaxonal dystrophy: ultrastructural studies. *Exp. Neurol.* **106**, 133–144 (1989).

49. P. P. Liberski, H. Budka, E. Sluga, M. Barcikowska, and H. Kwiecinski, Tubulovesicular structures in Creutzfeldt-Jakob disease. *Acta Neuropathol. (Berl.)* **84**, 238–243 (1992).

50. P. P. Liberski and H. Budka, Neuroaxonal pathology in Creutzfeldt-Jakob disease. *Acta Neuropathol.* **97**, 329–334 (1999).

51. J. R. Fraser, What is the basis of transmissible spongiform encephalopathy induced neurodegeneration and can it be repaired? *Neuropathol. Appl. Neurobiol.* **28**, 1–11 (2002).

52. M. Jeffrey, I. A. Goodbrand, and A. Goodsir, Pathology of the transmissible spongiform encephalopathies with special emphasis on ultrastructure. *Micron* **26**, 277–298 (1995).

53. D. Jesionek-Kupnicka, J. Buczynski, R. Kordek, T. Sobow, I. Kloszewska, W. Papierz, and P. P. Liberski, Programmed cell death (apoptosis) in Alzheimer's disease and Creutzfeldt-Jakob disease. *Folia Neuropathol.* **35**, 233–235 (1997).

54. D. Jesionek-Kupnicka, R. Kordek, J. Buczynski, and P. P. Liberski, Apoptosis in relation to neuronal loss in experimental Creutzfeldt-Jakob disease in mice. *Acta Neurobiol. Exp.* **61**, 13–19 (2001).

55. A. Dorandeu, L. Wingertsmann, F. Chretien, M-B. Delisle, C. Vital, P. Parchi, P. Montagna, E. Lugaresi, J. W. Ironside, H. Budka, P. Gambetti, and F. Gray, Neuronal apoptosis in fatal familial insomnia. *Brain Pathol.* **8**, 531–537 (1998).

56. P. P. Liberski, R. Yanagihara, C. J. Gibbs Jr, and D.C. Gajdusek, Neuronal autophagic vacuoles in experimental scrapie and Creutzfeldt-Jakob disease. *Acta Neuropathol.* **83**, 134–139 (1992).

57. J. W. Boellaard, M. Kao, W. Schlote, and H. Diringer, Neuronal autophagy in experimental scrapie. *Acta Neuropathol.* **82**, 225–228 (1991).

58. J. W. Boellaard, W. Schlote, and J. Tateishi, Neuronal autophagy in experimental Creutzfeldt-Jakob disease. *Acta Neuropathol.* **78**, 410–418 (1989).

59. B. Sikorska, P. P. Liberski, P. Giraud, N. Kopp, and P. Brown, Autophagy is a part of ultrastructural synaptic pathology in Creutzfeldt-Jakob disease: a brain biopsy study. *Int. J. Biochem Cell Biol.*, in press (2004).

60. S. J. DeArmond, S.-L. Yang, A. Lee, R. Bowler, A. Taraboulos, D. Groth, and S. B. Prusiner, Three scrapie prion isolates exhibit different accumulation patterns of the prion protein scrapie isoform. *Proc. Natl. Acad. Sci., USA* **90**, 6449–6453 (1993).

61. P. P. Liberski, B. Sikorska, J. Bratosiewicz-Wasik, D. Carleton Gajdusek, and P. Brown, Neuronal cell death in transmissible spongiform encephalopathies (prion diseases) revisited: from apoptosis to autophagy—a review. *Int. J. Biochem Cell Biol.*, in press (2004).

62. P. P. Liberski, R. Yanagihara, C. J. Gibbs Jr, and D. C. Gajdusek, Tubulovesicular structures in experimental Creutzfeldt-Jakob disease and scrapie. *Intervirology* **29**, 115–120 (1988).

63. P. P. Liberski and H. Budka, Tubulovesicular structures in Gerstmann-Sträussler-Scheinker disease. *Acta Neuropathol. (Berl.)* **88**, 491–492 (1994).

64. M. Jeffrey, J. R. Fraser, Tubulovesicular particles occur early in the incubation period of murine scrapie. *Acta Neuropathol. (Berl.)* **99**, 525–528 (2000).

65. P. H. Gibson and L. A. Doughty, An electron microscopic study of inclusion bodies in synaptic terminals of scrapie-infected animals. *Acta Neuropathol. (Berl.)* **77**, 420–425 (1989).

66. J. F. David-Ferreira, K. L. David-Ferreira, C. J. Gibbs Jr., and J. A. Morris, Scrapie in mice: ultrastructural observations in the cerebral cortex. *Proc. Soc. Exp. Biol. Med.* **127**, 313–320 (1968).

67. A. Bignami and H. B. Parry, Aggregations of 35-nanometer particle associated with neuronal cytopathic changes in natural scrapie. *Science* **171**, 389–390 (1971).

68. P. P. Liberski, N. Streichenberger, P. Giraud, M. Soutrenon, D. Meyronnet, B. Sikorska, and N. Kopp, Ultrastructural pathology of prion diseases revisited: brain biopsy studies. *Neuropathol. Appl. Neurobiol.* in press (2005).

69. P. P. Liberski, R. Yanagihara, C. J. Gibbs Jr, and D. C. Gajdusek, Appearance of tubulovesicular structures in Creutzfeldt-Jakob disease and scrapie precedes the onset of clinical disease. *Acta Neuropathol. (Berl.)* **79**, 1–9 (1989).

70. P. P. Liberski, M. Jeffrey, and C. Goodsir, Tubulovesicular structures are not labeled using antibodies to prion protein (PrP) with the immunogold electron microscopy techniques. *Acta Neuropathol. (Berl.)* **93**, 260–264 (1997).

71. P. P. Liberski, M. Jeffrey, and C. Goodsir, Electron microscopy in prion research: tubulovesicular structures are not composed of prion protein (PrP) but may be intimately associated with PrP amyloid fibrils, in: *Prions and brain diseases in animals and humans*, edited by D. R. O. Morrison (Plenum Press, New York and London, 1998), pp. 77–86.

72. A. Tarabulos, K. Jendroska, D. Serban, S.-L. Yang, S. J. DeArmond, and S. B. Prusiner, Regional mapping of prion protein in brain. *Proc. Natl. Acad. Sci., USA* **89**, 7620–7624 (1992).

73. W. J. Schulz-Schaeffer, S. Tschoke, N. Kranefuss, W. Drose, D. Hause-Reitner, A. Giese, M. H. Groschup, and H. A. Kretzschmar, The paraffin-embedded tissue blot detects PrP^{Sc} early in the incubation time in prion diseases. *Am. J. Pathol.* **156**, 51–56 (2000).

74. C. Jarius, G. G. Kovacs, G. Belay, J. A. Hainfellner, E. Mitrova, and H. Budka, Distinctive cerebellar immunoreactivity for the prion protein in familial (E200K) Creutzfedt-Jakob disease. *Acta Neuropathol.* **105**, 449–454 (2003).

75. G. Kovacs, T. Voigtlander, J. A. Hainfellner, and H. Budka, Distribution of intraneuronal immunoreactivity for the prion protein in human prion diseases. *Acta Neuropathol* **104**, 320–326 (2002).

76. G. G. Kovacs, E. Lindeck-Pozza, L. Chimelli, A. Q. C. Araujo, A. A. Gabbai, T. Strobel, M. Glatzel, A. Aguzzi, and H. Budka, Creutzfeldt-Jakob disease and inclusion body myisitis: abundant disease-associated prion protein in muscle. *Ann. Neurol.* **55**, 121–125 (2004).

77. S. Haik, B. A. Faucheux, V. Sazdovitch, N. Privat, J.-L. Kemeny, A. Perret-Liaudet, and J.-J. Hauw, The sympathetic nervous system is involved in variant Creutzfeldt-Jakob disease. *Nature Med.* **9**, 1121–1123 (2003).

78. H. Budka, Histopathology and immunohistochemistry of human transmissible spongiform encephalopathies (TSEs). *Arch. Virol.* **16**, 135–142 (2000).

79. D. C. Guiroy, S. K. Shankar, C. J. Gibbs Jr., J. A. Messenheimer, S. Das, and D. C. Gajdusek, Neuronal degeneration and neurfilament accumulation in the trigeminal ganglia in Creutzfeldt-Jakob disease. *Ann. Neurol.* **25**, 102–106 (1989).

80. J. A. Hainfellner and H. Budka, Disease associated prion protein may deposit in the peripheral nervous system in human transmissible spongiform encephalopathies. *Acta Neuropathol.* **98**, 458–460 (1999).

81. M. H. Groschup, M. Beekes, P. A. McBride, M. Hardt, J. A. Hainfellner, and H. Budka, Deposition of disease-associated prion protein involves the peripheral nervous system in experimental scrapie. *Acta Neuropathol.* **98**, 453–457 (1999).

82. M. Glatzel, E. Abela, M. Maissen, and A. Aguzzi, Extraneural pathologic prion protein in sporadic Creutzfeldt-Jakob disease. *N. Engl. J. Med.* **349**, 1812–1820 (2003).

83. D. C. Gajdusek and V. Zigas, Degenerative disease of the central nervous system in New Guinea. The endemic occurrence of "kuru" in the native population. *New Engl. J. Med.* **257**. 974–978 (1957).

84. I. Klatzo, D. C. Gajusek and V. Zigas, Evaluation of pathological findings in twelve cases of kuru, in: *Encephalitides*, edited by L. Van Boagert, J. Radermecker, J. Hozay, and A. Lowenthal (Elsevier Publ. Comp, Amsterdam, 1959), pp 172–190.

85. E. Beck and P. M. Daniel, Kuru and Creutzfeldt-Jakob disease: neuropathological lesions and their significance, in: *Slow Transmissible Diseases of the Nervous System*, edited by S. B. Prusiner and W. J. Hadlow (Academic Press, New York, 1979), Vol 1, pp. 253–270.

86. F. Seitelberger, Eigenartige familiar-hereditare Krankheit des Zetralnervensystems in einer niederosterreichischen Sippe. *Wien Klin Wochen* **74**, 687–691 (1962).

87. M. A. Neuman, D. C. Gajdusek and V. Zigas, Neuropathologic findings in exotic neurologic disorder among natives of the Highlands of New Guinea. *J. Neuropathol. Exp. Neurol.* **23**. 486–507 (1964).

88. C. A. McLean, J. W. Ironside, M. P. Alpers, P. W. Brown, L. Cervenakova, R. M. D. Anderson, and C. L. Masters, Comparative neuropathology of kuru with the new variant of Creutzfeldt-Jakob disease: evidence for strain of agent predominating over genotype of host. *Brain Pathol.* **8**, 428–437 (1998).

89. P. Piccardo, J. Safar, M. Ceroni, D. C. Gajdusek, C. J. Gibbs Jr, Immunohistochemical localization of prion protein in spongiform encephalopathies and normal brain tissue. *Neurology* **40**, 518–522 (1990).

90. I. A. Goodbrand, J. W. Ironside, D. Nicolson, and J. E. Bell, Prion protein accumulations in the spinal cords of patients with sporadic and growth hormone-associated Creutzfeldt-Jakob disease. *Neurosci. Lett.* **183**, 127–130 (1995).

91. A. von Braunmühl, Uber eine eigenartige hereditaer-familiaere Erkrankung des Zentralnervensystems. *Arch. Psychiatr. Z. Neurol.* **191**, 419–449 (1954).

92. J. W. Boellaard and W. Schlote, Subakute spongiforme Encephalopathie mit multiformer Plaquebildung. Eigenartige familiar-hereditare Kranknheit des Zentralnervensystems [spino-cerebellare Atrophie mit Demenz, Plaques and plaqueahnlichen im Klein- and Grosshirn (Gerstmann, Sträussler, Scheinker). *Acta Neuropathol. (Berl.)* **49**, 205–212 (1980).

93. K. Hsiao, H. F. Baker, T. J. Crow, M. Poulter, F. Owen, J. D. Terwilliger, D. Westaway, J. Ott, and S. B. Prusiner, Linkage of a prion protein missense variant to Gerstmann-Sträussler syndrome. *Nature* **338**, 342–345 (1989).

94. B. Ghetti, O. Bugiani, F. Tagliavini, and P. Piccardo, Prion disorders: Gerstmann-Sträussler-Scheinker disease, in *Neurodegeneration: The Molecular Pathology of Dementia and Movement Disorders*, edited by D. Dickson (ISN Neuropath Press, Basel, 2003), pp. 318–325.

95. S. J. DeArmond, J. W. Ironside, E. Bouzamondo-Berstein, D. Peretz, and J. R. Fraser, Neuropathology of prion diseases, in: *Prion Biology and Diseases*, edited by S. B. Prusiner—2nd ed. (Cold Spring Harbor, New York, 2004), pp. 777–856.

96. P. P. Liberski, I. Kloszewska, I. Boellaard, W. Papierz, A. Omulecka, and H. Budka, Dystrophic neurites of Alzheimer's disease and Gerstmann-Sträussler-Scheinker's disease dissociate from the formation of paired helical filaments. *Alzheimer's Res.* **1**, 89–93 (1995).

97. B. Ghetti, S. R. Dlouhy, G. Giaccone, O. Bugiani, B. Frangione, M. R. Farlow, and F. Tagliavini, Gerstmann-Sträussler-Scheinker diseases and the Indiana kindred. *Brain Pathol.* **5**, 61–95 (1995).

98. J. W. Ironside, L. McCardle, A. Horsburgh, Z. Lim, and M. W. Head, Pathological diagnosis of variant Creutzfeldt-Jakob disease. *APMIS* **11**, 79–87 (2002).

99. P. P. Liberski, J. Ironside, L. McCardle, A. Sherring, Ultrastructural analysis of the florid plaque in variant Creutzfeldt-Jakob disease. *Folia Neuropathol.* **38**, 167–170 (2000).

100. J. G. Fournier, N. Kopp, N. Streichenberger, F. Escaig-Haye, J. Langeveld, and P. Brown, Electron microscopy of brain amyloid plaques from a patient with new variant Creutzfeldt-Jakob disease. *Acta Neuropathol.* **99**, 637–642 (2000).

101. M. Zeidler, R. J. Sellar, D. A. Collie, R. Knight, G. Stewart, M. A. Macleod, J. W. Ironside, S. Cousens, A. C. Colchester, D. M. Hadley, R. G. Will, and A. F. Colchester, The pulvinar sign on magnetic resonance imaging in variant Creutzfeldt-Jakob disease. *Lancet* **355**, 1412–1418 (2000).

102. J. W. Ironside, K. Sutherland, J. E. Bell, L. McCardle, C. Barrie, K. Estebeiro, M. Zeidler, and R. G. Will, A new variant of Creutzfeldt-Jakob disease: neuropathological and clinical features. *Cold Spring Harb. Symp. Quant. Biol.* **61**, 523–530 (1996).

103. R. A. Armstrong, N. J. Cairns, J. W. Ironside, and P. L. Lantos, Quantification of vacuolation ("spongiform change"), surviving neurones and prion protein deposition in eleven cases of variant Creutzfeldt-Jakob disease. *Neuropathol. Appl. Neurobiol.* **28**, 129–135 (2002).

104. R. A. Armstrong, N. J. Cairns, J. W. Ironside, and P. L. Lantos, Quantitative variations in the pathology of 11 cases of variant Creutzfeldt-Jakob disease (vCJD). *Pathophysiology* **8**, 235–241 (2002).

105. R. A. Armstrong, N. J. Cairns, J. W. Ironside, and P. L. Lantos, The spatial patterns of prion protein deposits in cases of variant Creutzfeldt-Jakob disease. *Acta Neuropathol.* **104**, 665–669 (2002).

106. W. H. Nailon and J. W. Ironside, Variant Creutzfeldt-Jakob disease: immunocytochemical studies and image analysis. *Microsc. Res. Tech.* **50**, 2–9 (2000).

107. M. W. Head, V. Northcott, K. Rennison, D. Ritchie, L. McCardle, T. J. Bunn, N. F. McLennan, J. W. Ironside, A. B. Tullo, and R. E. Bonshek, Prion protein accumulation in eyes of patients with sporadic and variant Creutzfeldt-Jakob disease. *Invest. Ophthalmol. Vis. Sci.* **44**, 342–346 (2003).

108. J. W. Ironside, M. W. Head, J. E. Bell, L. McCardle, and R. G. Will, Laboratory diagnosis of variant Creutzfeldt-Jakob disease. *Histopathology* **37**, 1–9 (2000).

109. A. F. Hill, R. J. Butterworth, S. Joiner, G. Jackson, M. N. Rossor, D. J. Thomas, A. Frosh, N. Tolley, J. E. Bell, M. Spencer, A. King, S. Al-Sarraj, J. W. Ironside, P. L. Lantos, and J. Collinge, Investigation of variant Creutzfeldt-Jakob disease and other human prion diseases with tonsil biopsy samples. *Lancet* **353**, 183–189 (1999).

110. S. Shimizu, K. Hoshi, T. Muramoto, M. Homma, J. W. Ironside, S. Kuzuhara, T. Sato, T. Yamamoto, and T. Kitamoto, Creutzfeldt-Jakob disease with florid-type plaques after cadaveric dura mater grafting. *Arch. Neurol.* **56**, 357–363 (1999).

111. P. Parchi, A. Giese, S. Capellari, P. Brown, W. Schulz-Schaeffer, O. Windl, I. Zerr, H. Budka, N. Kopp, P. Piccardo, S. Poser, A. Rojiani, N. Streichemberger, J. Julien, C. Vital, B. Ghetti, P. Gambetti, and H. Kretzschmar, Classification of sporadic Creutzfeldt-Jakob disease based on molecular and phenotypic analysis of 300 subjects. *Ann. Neurol.* **46**, 224–233 (1999).

112. G. Puoti, G. Giaccone, G. Rossi, B. Canciani, O. Bugiani, and F. Tagliavini, Sporadic Creutzfeldt-Jakob disease: co-occurrence of different types of PrP(Sc) in the same brain. *Neurology* **53**, 2173–2176 (1999).

113. C. A. Llewlyn, P. E. Hewitt, R. S. Knight, K. Amar, S. Cousens, J. Mackenzie, R. G. Will. Possible transmission of variant Creutzfeldt-Jakob disease by blood transfusion. *Lancet* **363**, 417–421 (2004).

114. M. E. Bruce, R. G. Will, J. W. Ironside, I. McConnell, D. Drummond, A. Suttie, L. McCardle, A. Chree, J. Hope, C. Birkett, S. Cousens, H. Fraser, and C. J. Bostock, Transmissions to mice indicate that "new variant" CJD is caused by the BSE agent. *Nature* **389**, 498–501 (1997).

115. M. R. Scott, R. Will, J. Ironside, H. O. Nguyen, P. Tremblay, S. J. De Armond, and S. B. Prusiner, Compelling transgenetic evidence for transmission of bovine spongiform encephalopathy prions to humans. *Proc. Natl. Acad. Sci. USA* **96**, 15137–15142 (1999).

116. S. Notari, S. Capellari, A. Giese, I. Westner, A. Baruzzi, B. Ghetti, P. Gambetti, H. A. Kretzschmar, and P. Parchi. Effects of different experimental conditions on the PrPSc core generated by protease digestion: implications for strain typing and molecular classification of CJD. *J. Biol. Chem.* **279**, 16797–16804 (2004).

Chapter 3

CENTRAL PATHOGENESIS OF PRION DISEASES

Ursula Unterberger, Till Voigtländer, and Herbert Budka
Institute of Neurology, Medical University of Vienna, AKH 4J, Währinger Gürtel 18–20, A-1097 Vienna, Austria

3.1. Introduction

While a great deal of research has been done during the last years to elucidate modes and spread of prion infection, and significant advances have been achieved in those fields, the way *how* prions harm the central nervous system still remains enigmatic. The clinical picture in prion diseases is dominated by central neurological symptoms, which are presently believed to be due to an early synaptic dysfunction[1] and, at later stages, neuronal loss. What we do not know is, how are neurons actually affected by the disease? Why are some neurons more readily destroyed than others? What is the role of the cellular prion protein, PrP^C, in the whole process? And how does the pathological prion protein, PrP^{Sc}, contribute?

Historically, there were two major approaches to prion pathogenesis: once, the so-called "gain of function-hypothesis", which put forward a possible neurotoxic effect of an abnormally folded, not further degradable protein that is deposited in considerable amounts in the brains of affected individuals. On the other hand, one could argue, that the continuous conversion of PrP^C to PrP^{Sc} might lead to decreased availability and/or functional impairment of the former, so that its assumed neuroprotective effects are lost. This was central to the "loss of function-hypothesis". Both theories had their advocates. Although none of them has definitely proven right or wrong, the situation is certainly much too complex to be satisfyingly explained by one simple model.

What seems now clear is the incapacity of PrP^{Sc} accumulation alone for causing symptomatic disease[2,3]. PrP deficient mice are generally resistant to scrapie[4,5]. When PrP^{Sc} deposition is induced in such animals by grafting neural tissue overexpressing PrP^{C} into their brains and intracerebrally inoculating them with scrapie prions, the grafts accumulate PrP^{Sc}, which also spills over to the host brain. But while the grafts develop severe, scrapie-like neurodegeneration, the brain tissue devoid of PrP^{C} shows no damage at all[2]. Furthermore, if neuronal PrP^{C} is depleted in mice with ongoing neuroinvasive prion infection, non-neural replication and accumulation of prion infectivity continues, but early cerebral histopathological changes are reversed, and neuronal loss and progression to clinical disease are prevented[3]. Thus, expression of the normal prion protein must play a crucial role in the development of neurodegeneration after prion infection. This knowledge implies another question: must PrP^{C} necessarily be present on all types of brain cells (i.e. neurons, astrocytes, and oligodendrocytes) to confer 1. susceptibility to clinical prion disease, 2. formation of PrP^{Sc}, and 3. transmission of infectivity and disease?

To address this issue, several transgenic mouse models have been generated expressing PrP^{C} selectively in neurons[6], astrocytes[7], and oligodendrocytes[8], respectively. Subsequent inoculation and transmission experiments revealed that mere neuron-specific expression of hamster PrP^{C} suffices to support prion infection and disease development[6], while restriction of murine PrP^{C} to oligodendrocytes does not[8]. The role and impact of astrocyte-specific PrP^{C} expression is discussed controversially. Raeber and colleagues reported that transgenic mice expressing hamster PrP^{C} selectively in astrocytes are susceptible to prion infection[7]. On the ultrastructural level, brains of these mice show TSE-typical neuronal lesions despite lack of neuronal PrP^{C}[9], suggesting that deposition of PrP^{Sc} in intimate proximity to neurons and their processes is sufficient to induce TSE pathology. However, the studies mentioned before, although using a different approach (neuroectodermal grafting[2] or postnatal neuron-specific downregulation of PrP^{C} expression[3]) argue against this hypothesis, since close proximity of PrP^{Sc} to the neuronal cell surface in these models did not induce obvious morphological alterations, neuronal loss, or clinical disease[2,3]. Whether these differences reflect distinct pathogenic mechanisms determined by neuronal and astrocytic PrP^{C} expression[9] or certain strain properties (hamster 263K versus mouse RML), or if they rely on varying expression levels of astrocytic prion protein in the mouse models used (or on other still unknown factors), is presently not clear.

Finally, it is important to note that mice with selective genetic elimination of the prion protein gene (Prnp) open reading frame (ORF) *within* the borders of exon 3 (leaving the splice acceptor site of exon 3 intact) display *per se* only relatively subtle behavioral and biochemical changes[10,11,12,13,14,15], so that the severe damage of neuronal tissue in scrapie mice can hardly be related just to loss of PrP^C function. In contrast, mice with a deletion of the Prnp-ORF including at least the splice acceptor site of exon 3 develop loss of cerebellar Purkinje cells and late-onset ataxia[16,17]. This phenotype has meanwhile been linked to the expression of a PrP^C-like protein, called Doppel, which is encoded by a Prnp-like gene, named Prnd, located immediately downstream of the Prnp locus[18]. Under physiological conditions, neuronal expression of Doppel is silenced post-developmentally. In PrP knockout mice with deletion of the splice acceptor site of exon 3, however, its expression is re-activated under the control of the Prnp promoter due to an intergenic splicing between Prnp and Prnd[18], leading to Doppel-mediated neuronal cell death.

In sum, cell culture and *in vivo* results, taken together, suggest a combination of various mechanisms resulting in neuronal death. Still other aspects of the interaction of cellular and pathological prion protein will have to be taken into consideration for elucidating prion pathogenesis. At present, we can only collect as many pieces of the puzzle as possible, try to put them together, and struggle for the final breakthrough.

3.2. Local distribution of PrP^{Sc} accumulation and neural tissue damage: implications for central prion pathogenesis

If one considers morphological features of the prion-affected brain with regard to pathogenic mechanisms, primarily two questions arise: 1. what determines patterns and distribution of PrP^{Sc} deposition and neural tissue damage, and 2. is there a direct relation between PrP^{Sc} deposition and histopathological signs, like astrogliosis (which is a rather unspecific reaction), spongiform change (which is highly specific for prion diseases), and, finally, neuronal loss?

Presence and patterns of histopathological changes vary greatly between individual cases and disease subtypes[19,20]. In sporadic Creutzfeldt-Jakob disease (sCJD), the morphologic and clinical phenotype was shown to depend on the physicochemical properties of PrP^{Sc} and the genetic background of the patient[21,22]. In fatal familial insomnia (FFI), there may be little or no spongiform change at all[23], while this disease is specifically characterized by prominent thalamic atrophy with profound astrogliosis[19,20]. In any case, local PrP deposition seems to require the presence of intact neuronal elements; in pre-existing brain lesions, like scarred infarctions with prominent gliosis, PrP^{Sc} does not accumulate[19,20].

One factor rendering neurons more sensitive to prion-mediated toxicity might be the level of PrP^{C} expression. $PrP^{0/+}$ mice, which express about half the normal level of PrP^{C}, display a delayed onset of clinical disease[5,24]. On the other hand, even cell types strongly expressing PrP, like cerebellar Purkinje cells, are relatively resistant to cell death after prion-infection[25,26]. Selective vulnerability of parvalbumin-expressing GABAergic neurons was found both in human[27,28] and experimental prion diseases[29]. This vulnerability was detectable already early in the incubation period and thus represents one of the earliest changes ever described after experimental inoculation[29]. Interestingly, FFI differs in this phenomenon from all other human transmissible spongiform encephalopathies (TSEs)[30]. Other vulnerabilities include that of the granular layer of the cerebellum that is frequently depleted in sporadic CJD[20], and the variable involvement of the basal nucleus of Meynert[31,32]. However, it was shown that in transmitted prion diseases, the respective strain of agent plays a central role with regard to incubation time and neuropathology[33,34,35]. Varying strain-dependent lesion profiles in syngenic animals are suggestive of a strain-specific targeting of different neuronal populations[33,26].

In conclusion, PrP^{Sc} deposition and tissue damage is in a way influenced by host factors, like genotype and individual and selective vulnerability of neuronal subsets, as well as by PrP^{Sc} properties or, in transmitted prion diseases, the strain of agent. So far, the molecular mechanisms underlying these effects remain obscure.

Concerning the second question, originally, it was proposed that disease-associated histopathological changes in the brain correlate well with PrP^{Sc} deposition[36,37]. Reactive astrogliosis was found to follow PrP^{Sc} accumulation by one to two weeks in hamster scrapie. Accordingly, it was suggested that etiology and pathogenesis of prion diseases are directly related to PrP^{Sc} [37]. Later investigations showed somewhat different results. Not in all cases do amount and distribution of PrP^{Sc} actually correspond with type and severity of local tissue damage[23,38]. In some TSEs, like FFI[39] and Gerstmann-Sträussler-Scheinker disease (GSS)[40], lesions may develop without PrP^{Sc} accumulation. In a time course study in mice with experimental Creutzfeldt-Jakob disease, spongiform change preceded PrP deposition in various brain regions[38]. A study using brains of the same series found severe loss of a subpopulation of GABAergic neurons in the cerebral cortex after 5-9 weeks after the infection, i. e. clearly before the appearance of cortical PrP^{Sc} [29] (see above). Furthermore, experiments revealing, under certain conditions, a brain containing loads of pathological prion protein but devoid of any accompanying tissue damage[2] must be taken into consideration in this context. A consistent relationship between PrP^{Sc} deposition and brain tissue damage has never been proven and, with a growing amount of data arguing against it, is becoming less and less likely.

3.3. Oxidative stress and antioxidant stress defense

Research on oxidative stress and its contribution to central pathogenesis of prion diseases has developed enormously within the last decade. Accordingly, a complex picture has emerged, which includes the topics neurotoxicity of PrP^{Sc} (or the prion protein fragment PrP106-126, respectively), copper binding by PrP^{C}, potential direct and indirect antioxidant functions of this molecule, and oxidative stress markers in prion disease. These subjects are closely related to each other; for systematic reasons, they will be presented separately in the following section.

3.3.1. Neurotoxicity of PrP^{Sc} and the prion protein fragment PrP106-126

Until the early nineties of the last century, central pathogenesis of prion diseases was discussed primarily on the basis of morphological observations analyzing the extent and spatial pattern of spongiform change, neuronal loss, astrogliosis, and deposition of PrP^{Sc} (see also section 3.2). In 1993, Forloni et al described a peptide corresponding to amino acid residues 106-126 of the human prion protein, which has a high intrinsic tendency to aggregate into fibrils, thereby mimicking one key feature of PrP^{Sc} [41]. This peptide was the only one amongst a variety of peptides tested that was capable of inducing cell death in primary neuronal cultures after chronic exposure. Although the discovery of the "neurotoxic" property of PrP106-126 has provided prion researchers with a highly valuable tool to unravel pathogenic mechanisms, it must be emphasized that this fragment has in fact never been detected in any form of TSE (natural, acquired or experimental). Thus, it should be regarded as an experimental system for prion research, but not as an actual part of the disease.

During the following years, neurotoxic mechanisms were studied extensively using either PrP106-126 or preparations of purified PrP^{Sc}. Research was focused mainly on two aspects: 1. molecular and structural characteristics of PrP106-126, which might be essential for its toxic effect to neurons, and 2. cellular factors and additional cell types apart from neurons involved in and crucial for neurotoxicity.

Regarding the first topic, it was observed that PrP106-126 displays a variety of conformations. Influenced by environmental conditions, these conformations range from α-helical to β-sheet secondary structure[42]. β-sheet conformation occurs preferentially at acidic pH[42,43], but it also depends on an intact hydrophobic core sequence[44]. Finally, secondary structure is modulated by binding of copper and zinc ions to the peptide[45]. It is important to note that copper binding of PrP106-126 should not be mistaken with copper binding of PrP^C (as described later), as it is located at a completely different site. PrP106-126 aggregates immediately after dissolution in acidic buffer, resulting in partial resistance to digestion with proteinase K and pronase[43]. Aggregation into fibrils, in turn, is an essential prerequisite for the neurotoxicity of PrP106-126[44].

It soon turned out that neurotoxicity requires more than mere structural properties of PrP^{Sc} or its related peptide PrP106-126. As mentioned in the introduction, the first crucial factor for neuronal death is the expression of the cellular prion protein, PrP^C. Neurons that do not express PrP^C are resistant to the *in vitro* toxicity of PrP106-126[46], as well as the *in vivo* neurodegeneration mediated by PrP^{Sc}[4,5]. On the other hand, cerebellar cell cultures of mice overexpressing PrP^C are more sensitive to the toxicity of PrP106-126[47]. Another necessity for a toxic effect of the peptide is the co-existence of neurons and microglia. Neurons are resistant to PrP106-126-mediated cell death after depletion of contaminating microglia[48]. Overexpression of PrP^C increases microglial activation in response to PrP106-126 and enhances its toxicity against neurons[47]..

While this interdependence between neurons, microglia, and PrP^C expression with regard to the *in vitro* toxicity of PrP106-126 and PrP^{Sc} has been confirmed in several experiments and is also supported by experimental data from *in vivo* models of prion disease, the pathological relevance of an astrocytic contribution to the neurotoxicity of PrP106-126 or PrP^{Sc} is discussed controversially. Based on results of a co-culture model of PrP^C-deficient neurons and PrP^C-expressing astrocytes, Brown proposed that PrP^C-ablated neurons develop an increased sensitivity to glutamate toxicity in the presence of PrP^C positive astrocytes, thus becoming even more dependent on a sufficient astrocyte-mediated protection against glutamate. Incubation of these co-cultures with PrP106-126 inhibited this protective property of astrocytes, resulting in an increased glutamate-mediated damage to the sensitized PrP^C-depleted neurons[49], a pathway, by which astrocytes could indirectly contribute to the *in vitro* toxicity of PrP106-126 and the neurodegeneration seen in prion diseases. At first glance, this suggestion might be in line with inoculation experiments of a transgenic mouse model expressing PrP^C selectively in astrocytes[7,9]. However, its relevance *in vivo* has recently been questioned by a conditional knockout model with specific postnatal depletion of neuronal PrP^{C} [3]. In this model, inoculation with RML prions resulted in a progressive accumulation of PrP^{Sc}, which was first converted from neuronal PrP^{C}, but later, after the neuronal Prnp-gene had been eliminated, was solely derived from astrocytes. Neither obvious neuronal damage nor any signs of clinical disease could be observed in the infected animals. Thus, the PrP-negative neurons survived despite an intense formation of PrP^{Sc} and the postulated inhibition of glutamate detoxification by astrocytes, indicating that either neurons are *in vivo* not primed for increased glutamate sensitivity in a similar way as observed *in vitro*, or that they can escape glutamate toxicity by at least one different clearance pathway. Therefore, the role of astrocytes in central pathogenesis *in vivo* needs to be further determined.

On the molecular level, several pathways have been described which could mediate the neurotoxicity of PrP106-126 or PrP^{Sc}. These include for instance activation of NMDA receptors[50,51] and activation of the 5-lipoxygenase pathway[52]. However, it will take further *in vivo* experiments to confirm the pathogenic relevance of the suggested mechanisms.

3.3.2. Copper binding of PrP^{C}

The scientific approach described above aimed at analyzing the central pathogenesis of prion diseases on the basis of the neurotoxic properties of PrP^{Sc} and PrP106-126. Another more indirect strategy tried to explore the physiological roles of PrP^{C} first. In this context, much attention was devoted to the unstructured N-terminal region of PrP^{C} containing a unique octapeptide repeat sequence. This octarepeat region belongs to the best-conserved parts of the mammalian prion protein, although a recent report describes a higher degree of variability than previously anticipated[53]. First insights into a function of this peculiar repeat sequence came from studies in cell free systems. Using peptides corresponding to three to four octarepeats of mammalian PrP, Hornshaw and colleagues demonstrated specific and preferential binding of divalent copper ions as compared to other metals[54,55]. Like the discovery of the neurotoxic effect of PrP106-126, this observation opened new perspectives in prion research and initiated a series of experiments to unravel this new functional property of PrP^{C}. Meanwhile, copper binding by either single octapeptide sequences, PrP fragments containing the octapeptide region, or full length PrP has been confirmed in further cell free assays[56], in cell culture systems[57], and *in vivo*[58]. Copper binding at the N-terminal domain of full length PrP^{C} occurs with positive cooperativity[58,59,60] and with a binding affinity compatible to the physiological concentration range of extracellular copper[58,60]. Stoichiometric data have been discussed controversially, ranging from two[61] to six[58] copper atoms bound per prion protein molecule. Conflicting results might be explained by the use of different analytical techniques as well as different PrP constructs by individual researchers. At present, the most favored view is that PrP^C binds up to five copper ions at the octapeptide region[60].

Copper binding has been reported to influence structural and biochemical properties of PrP^C or parts of it. Incubation of a PrP fragment comprising the octapeptide region with copper ions induced an α-helical conformation in the C-terminal part[56]. Conversely, incubation of aged recombinant PrP or cellular PrP in the microsomal fraction of brain extracts with divalent copper ions initiated a conformational shift from α-helical to β-sheet structure, accompanied by the formation of aggregates, detergent insolubility and resistance against proteinase K digestion, all biochemical characteristics of the pathological prion protein, PrP^{Sc} [62]. However, further studies using a conformation-dependent immunoassay revealed that the protease resistant PrP generated by copper treatment exhibits a different structure compared to PrP^{Sc} [63], arguing against any gain of direct infectivity of copper converted PrP. Nevertheless, PrP conversion by a physiological or even excessive load of copper ions might play a supportive role in the development of prion diseases.

Another significant finding was the rapid and reversible stimulation of endocytosis of PrP^C from the cell surface by copper ions[57]. This may have important physiological implications, as PrP^C could serve as a recycling receptor for copper uptake from the extracellular milieu[57], thus protecting the cell against the toxicity of free copper cations.

3.3.3. Direct and indirect antioxidant functions of PrP^C

Already in 1997, first studies were published which indicated that the role of PrP^C in the cellular defense against oxidative stress might go further than the mere binding of copper. Early results showed that elevated levels of PrP^C correlate with enhanced cellular resistance to oxidative stress[64], while lack of PrP^C results in higher sensitivity[65]. Primarily, the improved cellular resistance to oxidative stress was ascribed to an auxiliary function of PrP^C supporting known cellular antioxidant defense mechanisms. *In vitro*, as well as *in vivo*, studies have provided a net of confirmatory results. Increased activities of copper/zinc superoxide dismutase (Cu,Zn SOD) and glutathione peroxidase (GPx) were found in neuronal cells expressing higher levels of PrP^C [64,66]. PrP deficient cultured neurons, on the other hand, have significantly reduced glutathione reductase (GR) activity, combined with enhanced sensitivity to the toxic effects of hydrogen peroxide[67]. Similarly, mice lacking PrP^C show attenuated Cu,Zn SOD activity[58,65,68,15], but, maybe compensatory, high manganese superoxide dismutase (Mn SOD) activity[65,15], paralleled by a higher rate of protein oxidation and lipid peroxidation[68,69]. In contrast, brain lysates from mice markedly *over*expressing PrP^C revealed an increased enzyme activity for Cu,Zn SOD and GPx[70]. However, these mice also show attenuated resistance to oxidative stress, indicating that overexpression of PrP^C may also have detrimental side effects[70].

Despite these apparently unequivocal results, the exact mechanism of a PrP^C dependent rise in cellular antioxidant activity in general and Cu,Zn SOD activity in particular is still discussed controversially. Brown and colleagues described a close correlation between the expression level of PrP^C, brain copper content[58], cellular copper uptake[70], and copper incorporation into Cu,Zn SOD[70]. Accordingly, they suggested that the activity of Cu,Zn SOD is regulated post-translationally by its copper supply[70]. Enhanced cellular copper binding in response to elevated PrP^C expression levels, as well as increased antioxidant enzyme activity have also been observed by other researchers, whereas a change of copper delivery into the cell could not be demonstrated[66]. Therefore, it was proposed that PrP^C activates a so far unknown signal transduction pathway to stimulate cellular oxidative stress defense[66]. The physiological relevance of PrP^C expression for cellular copper metabolism and antioxidant protection has been questioned in general by Waggoner et al who failed to detect any correlation between PrP^C expression level, ionic copper, and enzyme activity of the copper dependent enzymes Cu,Zn SOD and cytochrome c oxidase in the brain[71]. Thus, the link between PrP^C expression, copper uptake and the activity of Cu,Zn SOD and other stress defense enzymes remains currently elusive.

In 1999, again Brown and colleagues widened the spectrum of potential antioxidant activities of the prion protein by demonstrating an intrinsic SOD-like activity of native and recombinant mouse and chicken PrP^C folded in the presence of copper ions[72]. This enzymatic SOD-like activity depended on an intact octapeptide region[72] and on the amount of bound copper[73]. A study comparing the SOD activity of brain lysates from wild type mice before and after PrP^C depletion confirmed a direct contribution of the prion protein to total SOD activity dependent on PrP^C expression level and the predominant PrP^C glycotype in the respective brain region, suggesting a direct but more subtle and differential contribution of the prion protein to protection against oxidative stress[74].

In an attempt to further elucidate the physiological function of PrP^C in response to oxidative stress, we analyzed the neuronal PrP^C expression profile in human neurodegenerative disorders, in which damage by free radicals is thought to play a pivotal pathogenic role. We hypothesized that the proposed protective properties of the prion protein against oxidative stress should be reflected in its cellular upregulation. In TSEs and Alzheimer's[75], Parkinson's, and diffuse Lewy body disease, as well as in progressive supranuclear palsy and multiple system atrophy[76], we could indeed demonstrate a significant increase of intraneuronal PrP immunoreactivity. Conversely, in motor neuron disease, PrP expression was lost in anterior horn neurons of the spinal chord, i.e. in those neurons, which selectively degenerate during the course of the disease[76]. This indirect approach provided additional evidence for a relevant function of PrP^{C} in the cellular defense against oxidative stress.

Other studies revealed that the intrinsic antioxidant activity of the prion protein is closely related to the interdependence between metal binding and PrP conformation. When copper ions were replaced by manganese ions, the prion protein lost, upon aging, its enzymatic activity and its regular conformation, thereby developing a partial resistance against digestion with proteinase K[77]. Furthermore, co-incubation of PrP^{C} with the neurotoxic fragment PrP106-126 inhibited the SOD-like activity of the prion protein[73], suggesting that conversion of PrP^{C} to PrP^{Sc} might compromise antioxidant protection, thus leading to oxidative damage and neurodegeneration[78].

3.3.4. Oxidative stress markers in prion disease

The potential link between oxidative stress and prion diseases is marked by two categories of pathophysiologic events: 1. enhanced production of reactive oxygen species (ROS) and reactive nitrogen species (RNS) followed by direct oxidative cell damage, and 2. impairment of the cellular oxidative stress defense.

Increased generation of ROS has primarily been reported in *in vitro* experiments using PrP106-126. Analysis of primary neuronal cultures, which contained residual amounts of microglia, demonstrated that neurotoxicity of PrP106-126 depended on a combination of neuronal PrP^{C} expression and enhanced production of oxygen radicals by activated microglia[48]. Recently, it was shown that the mere co-incubation of PrP106-126 with copper ions in a cell free system results in the generation of hydrogen peroxide, which is further converted to highly reactive hydroxyl radicals following the Fenton principle[79]. Elevated levels of ROS/RNS have also been described *in vivo*. Brains of scrapie mice contain significantly elevated levels of nitric oxide (NO) and superoxide[80]. The latter was reported to originate at least in part from mitochondria[81]. However, the role of NO in prion diseases needs to be further investigated, since other researchers found a marked decrease of the activity of neuronal nitric oxide synthase (nNOS) in brains of scrapie-infected mice and hamsters, as well as in prion infected neuroblastoma cells[82,83].

Direct oxidative damage to lipids, proteins and DNA has been reported in several models of prion disease. Scrapie-infected cell cultures revealed markedly increased peroxidation of lipids[84]. Similarly, enhanced lipid peroxidation was shown in mitochondria isolated from brains of scrapie hamsters[85]. In mice with end-stage scrapie, our group demonstrated immunohistochemically a widespread neuronal labeling for nitrotyrosine (NT), a common marker for protein oxidation. In this study, neuronal protein damage was present in all three scrapie strains (RML, ME7, 22A) analyzed[86]. In addition, a marked increase in heme oxygenase-1 (HO-1) immunoreactivity was observed in experimental prion disease[86,87]. HO-1 expression is generally used as a sensitive marker for oxidative stress. Differential induction of HO-1 transcription has also been reported in primary neuronal and astroglial cultures exposed to the neurotoxic fragment PrP106-126[88]. We observed oxidative damage to nucleic acids in brains of patients with sporadic and familial Creutzfeldt-Jakob disease. Immunohistochemical analysis revealed a strong, predominantly diffuse but sometimes also conglomerate-like nuclear and cytoplasmic staining for 8-hydroxyguanosine/8-hydroxydeoxyguanosine (8-OHG/8-OHdG) in neurons, indicating an oxidative damage to DNA as well as RNA. Interestingly, staining intensity was correlated with disease duration but not with the deposition pattern of PrP^{Sc} [89].

Abnormal production of ROS/NOS and subsequent oxidative damage in prion diseases seems to be supported by a parallel, direct impairment of the cellular oxidative stress defense. Neuronal cultures treated with PrP106-126 revealed a significant decrease in glutathion reductase (GR) activity[67]. Similarly, upon infection with prions, a hypothalamic neuronal cell line showed a dramatic decrease in the activity of superoxide dismutase, as well as glutathion-dependent antioxidant systems[84]. Biochemical analysis of different experimental models of prion disease revealed further, sometimes conflicting results regarding the alteration of antioxidant defense *in vivo*. While one study, using scrapie-infected mice, reported diminished Cu,Zn SOD activity[80], other studies on hamster and mouse scrapie did not detect any change of this enzyme[81,85]. More consistently, a significant reduction of mitochondrial Mn SOD activity was described[81,85], pointing towards mitochondria as a potential source of ROS during the course of prion disease. GPx and GR, however, were increased in hamsters[85] but normal in mice[81]. Finally, a dramatic decrease in the intrinsic SOD-like activity of PrP^C was reported in scrapie mice[80]. This finding is especially interesting, since it provides a close link between a specific function of PrP^C and its potential inhibition by the conversion to PrP^{Sc}. In this context, one further observation is noteworthy: by analyzing PrP^{C} expression in human prion diseases, we detected increased intraneuronal immunoreactivity for PrP but not PrP^{Sc}[75,90]. If indeed cellular defense against oxidative stress includes upregulation of PrP^{C}, antioxidant defense in prion diseases might result in a vicious circle, where accumulation of PrP^{Sc} promotes generation of ROS, which in turn induce PrP^{C} expression; newly synthesized PrP^{C} again expedites the formation of PrP^{Sc} which increases the oxidative burden, and so on[75].

Little is known about the molecular mechanisms underlying the impaired cellular response to oxidative stress in TSEs. Since ROS and RNS can both attack a broad variety of lipids, proteins and carbohydrates, not to mention DNA and RNA molecules, it is tempting to assume that members of the antioxidant stress defense system are directly damaged by free radicals, either specifically or just by chance. This was described for instance for Cu,Zn SOD[91] and GPx[92]. Concerning TSEs, a similar mechanism for PrP^{C} was recently reported[93]. McMahon and colleagues demonstrated a site-specific cleavage of the N-terminal octapeptide region of PrP upon exposure to ROS. This cleavage occurred in a copper- and pH-dependent manner and might influence the further degradation of PrP^{C}, especially since abnormal cleavage of PrP^{Sc} is as well confined to the octapeptide region. Brown proposed an additional mechanism for the impairment of antioxidant defense specific for prion diseases[94]. Analyzing possible direct protein-protein interactions, he described a site-specific binding of both the neurotoxic fragment PrP106-126 and PrP^{Sc} to amino acid residues 112-119 of PrP^{C}. In culture systems, interaction between PrP106-126 and PrP^{C} inhibited copper binding of the prion protein, thus rendering the cells indirectly more susceptible to copper toxicity. Furthermore, cellular copper uptake and SOD-like enzymatic activity of PrP^{C} were inhibited, two events, which might directly and indirectly compromise antioxidant defense in prion disease[94]. Impaired cellular copper binding and modulation of cellular copper content upon prion infection has recently been confirmed *in vitro*[95].

Aberrant metal binding by the prion protein has been suggested as a central molecular event related to the significant decline in antioxidant protection by PrP^{C}. Independent analysis of brains of scrapie infected mice by two groups revealed significant perturbations of divalent metal ions with a strong decrease in copper content[80,96] and a major increase in manganese content already in early stages of the disease[96]. These changes were paralleled by a significant decrease in copper binding by PrP^{C} and a proportional decline in its SOD-like activity[80,96]. Similar results were obtained in brains of patients with sporadic CJD[97]. Apart from a decrease in copper and a rise in manganese content of brain tissue, there was a severe alteration in metal occupancy of purified PrP with a striking elevation of manganese ions, paralleled by a modest increase in zinc and a significant decline in copper ions. Again, these changes correlated with a dramatic, concomitant loss of intrinsic antioxidant activity of the prion protein. However, it is still too early to judge whether and to which extent imbalances in metal ions contribute causally to the pathogenesis of prion diseases, or whether they reflect only secondary changes in metal occupancy of the prion protein subsequent to alterations in PrP^C conformation, the latter initiated by other disease-related molecular mechanisms.

Oxidative stress, defined as an imbalance between burden of ROS/NOS and cellular antioxidant defense, is often regarded as an "either-or" mechanism, where either the disproportional generation of oxygen species predominates over the decline in antioxidant protection or vice versa. In prion diseases, the situation is different. Both mechanisms are linked together by the pathological isoform of PrP, PrP^{Sc}. Thus, they act cooperatively instead of alternatively. Formation of PrP^{Sc} enhances the production of ROS and, subsequently, the rate of oxidative damage to brain tissue. In parallel, the disease-immanent interaction between PrP^{Sc} and PrP^C inhibits specific functions of the normal prion protein including binding and detoxification of copper, cellular uptake and possibly cellular delivery of copper, as well as intrinsic SOD-like activity, altogether a loss of function which compromises the antioxidant defense system directly and indirectly. In sum, oxidative stress in prion diseases is a combination of two equally important processes, the increase in oxidative damage and the decrease in protection.

In general, oxidative stress is a mechanism of cell damage, although not a direct mechanism of cell death. Nevertheless, it may induce neuronal death by initiating one of two pathways: apoptosis or necrosis. Both pathways have been described in the nervous system in response to oxidative damage[98,99] depending on the individual disease entity. The pathway in prion diseases is discussed in the following section.

3.4. Apoptosis as the mechanism of neuronal death in TSEs

Two essentially distinct modes of cell death have been described: 1. *necrosis*, in most instances initiated by a sudden cell or tissue injury and followed by a marked inflammatory response, and 2. *apoptosis* or *programmed cell death*, initially regarded as a physiologic type of cell death during tissue development lacking any overt inflammatory reaction[100]. Meanwhile, various conditions have been discovered including loss of trophic support, disturbance of calcium and potassium homeostasis, and cytolethal toxic damage, which all can induce post-developmental apoptosis[99]. Taking into account the different cell biological and molecular events in necrosis and apoptosis, with loss of cellular energy and reduction of macromolecular synthesis on the one and unaffected cellular energy and increased macromolecular synthesis on the other side, necrosis can be described as a process of passive atrophy and apoptosis as one of active degeneration[99]. However, although essentially separated, both modes of cell death show some molecular and cell biological overlap[101]. The ladder type of DNA fragmentation, for instance, in principle one biochemical hallmark of apoptosis, is a specific sign for internucleosomal DNA strand breaks, but it is neither an obligatory requirement for nor a feature solely restricted to programmed cell death. Accordingly, DNA fragmentation was absent in some models otherwise typical for apoptotic cell death[102,103,104], but present in certain forms of necrosis[105,106]. Similarly, the terminal deoxynucleotidyltransferase-mediated dUTP nick end-labeling (TUNEL) technique (also called in situ end-labeling, ISEL)[107], originally introduced to visualize DNA fragmentation on a histological level, marks DNA strand breaks independent of their origin and is therefore unable to discriminate unequivocally between apoptosis and necrosis[108,99]. Furthermore, mitochondrial damage, initially regarded as a mere feature of necrotic cell death, has nowadays "lost" its discriminatory specificity and is thought to be additionally involved in some apoptotic pathways.

In sum, single observations like DNA fragmentation or mitochondrial impairment lack any cogency to classify a certain mode of cell death. Only the combination of several distinct cell death features to an integrative picture offers the possibility to define a pathway with sufficient reliability. This fact has to be kept in mind when assessing the modes of neuronal death in prion diseases.

In 1993, Forloni and colleagues pioneered the research on cell death in TSEs[41]. By analyzing the neurotoxic properties of the prion protein fragment PrP106-126 in primary rat hippocampal neurons, the authors detected morphological changes and a ladder type of DNA fragmentation in degenerating neurons, two features primarily suggestive of apoptotic cell death. In line with these results, DNA fragmentation and DNA strand breaks have been demonstrated with biochemical and/or histological methods (TUNEL labeling) *in vitro* in primary as well as in stable neuronal culture systems incubated with either PrP106-126 or PrP^{Sc}[50,109,110], in several mouse models inoculated with different murine scrapie strains and a mouse-adapted human prion strain[111,112,113,114,115], in scrapie-infected sheep[116], and in several forms of human prion disease including FFI[39] and sporadic, familial, iatrogenic, and variant CJD[108,117]. In addition to DNA breaks, upon evaluation by light and/or electron microscopy, some of the cell culture and mouse studies showed morphological changes reminiscent of apoptotic cell death, supporting the hypothesis of apoptosis as the primary pathway in these TSE models[109,110,111,113]. In contrast, a study using human biopsy and autopsy tissue failed to detect any histological hallmarks of apoptosis (i.e. extremely condensed nuclei, peripheral chromatin condensation, and formation of apoptotic bodies) in TUNEL-labeled neurons despite a high degree of TUNEL-positivity in certain brain areas of single cases[108]. Although these results might be explained by additional DNA damage due to the time interval between agony, death, autopsy and tissue fixation, they have questioned the use of in situ end-labeling as an appropriate screening method for cell death in autopsy tissue in general[108]. In order to evaluate the position of apoptosis in prion pathogenesis, time course experiments have been performed to determine the temporal and spatial correlation between DNA fragmentation, as indicated by TUNEL staining, and neuronal loss. Depending on prion strain and investigated brain region (including the retina), TUNEL-labeled neurons were observed in two studies at either the terminal[111], or an advanced, but pre-terminal[111,113] stage of the disease. Prominent TUNEL positivity of retinal[111] and hippocampal CA1[113] neurons preceded a massive cell loss in precisely congruent tissue areas in terminally ill animals. A third study in mice confirmed the temporal increase in TUNEL-positive neurons towards the final stage of the disease[115]. But unlike the former two studies, these authors observed a striking discrepancy between a high extent of neurodegeneration and a relatively small number of in situ end-labeled neuronal nuclei.

In a similar attempt to clarify the significance of apoptosis for neuronal cell death in human prion diseases, several studies analyzed the spatial correlation between the degree of neuronal loss and apoptosis (as revealed by TUNEL staining) in autopsy tissue. The results, however, were incongruent, since two studies from the same group reported a close correlation between the two parameters[39,117], while a third study described only a weak and inconsistent correlation[108]. As outlined above, these differences might again reflect the limited suitability of DNA fragmentation as a marker for apoptosis in human autopsy tissue. A more indirect approach to detect apoptosis-related DNA strand breaks was used by Bürkle and colleagues[110]. By exposing primary mouse neuronal cultures chronically to PrP106-126, the authors observed a specific, strong nuclear labeling for poly(ADP ribose) 30–48 hours after the beginning of the experiment in the same subset of cells that showed a morphological phenotype typical of apoptosis. The delay in nuclear staining indicated that this activation of poly(ADP-ribose) polymerase (PARP) occurred primarily in response to DNA fragmentation and not as a result of early toxic effects, like the formation of ROS, induced by the prion protein fragment[110].

Unlike these earlier experiments, several more recent studies on apoptosis in prion diseases focused on cell biological events upstream of the final DNA breakage. They describe a variety of pathways possibly involved in the initiation and perpetuation of prion–related apoptotic cascades. However, the current picture is complex and far from being complete. *In vitro*, several analyses were performed in different neuronal culture systems treated with either PrP106-126 or PrP^{Sc}. The central and common finding in all these experiments was the activation of members of the caspase protease family, especially the activation of the executioner caspase-3[118,119,120,121]. Nevertheless, the pathways triggering this caspase activation differed considerably. O'Donovan et al proposed depolarization of the mitochondrial membrane as the initiating event in PrP106-126-induced apoptosis in SH-SY5Y cells. The depolarization was followed by a release of cytochrome c from the outer mitochondrial membrane with subsequent activation of caspases and a release of mitochondrial calcium with activation of calpains, a second protease family involved in programmed cell death[119]. Other researchers, using the same experimental setup, described a p38 MAP kinase-dependent activation of caspase-3 as an underlying mechanism in PrP106-126 induced apoptosis[121]. Hetz et al suggested increased endoplasmic reticulum (ER) stress and calcium release from the ER as further pathways triggering caspase-12 dependent caspase-3 activation in scrapie-infected N2a cells[120]. To support their hypothesis, the authors also investigated brain tissue from a murine scrapie model and patients with sporadic and variant CJD. In line with their *in vitro* results, they demonstrated a significant up-regulation in the expression of selected ER-stress-associated chaperones as well as the generation of active fragments of caspase-12[120].

Considering all results outlined above, the currently available information indicates a crucial role of apoptosis in prion-related neuronal cell death and suggests that it indeed represents the primary cell death pathway in TSEs.

3.5. Neuroinflammation in prion disease

3.5.1. The microglial response

The presence of an inflammatory response to prion infection in the brain was doubted for a long time. We know now that there *are* in fact some inflammatory features, but that they differ from those seen in other infective disorders involving the brain. A classical, conspicuous perivascular leukocyte accumulation is normally absent in TSEs. Nevertheless, in contradiction to earlier results[122], a mild to moderate T-lymphocyte recruitment, mainly of the CD8+ phenotype, was shown in scrapie mice[123]. The contribution of this cell type to the neurodegenerative process is questionable, since immunologically deficient mice depleted of T-lymphocytes readily develop scrapie[124]. More consistently, microglia, which represent the mononuclear phagocyte system in the CNS, were shown to be activated in human[125,126] and murine[122,123,114] prion disease. In these studies, microglia were characterized by the upregulation of major histocompatibility complex (MHC) class I[123] and II antigens[122,125], type 3 complement receptor (CR3)[114,122], leukocyte common antigen (LCA)[122], and the macrophage marker HAM56[126], suggesting a florid response, which represents a modified inflammatory reaction[122]. However, the glial response in general was reported to vary between different agent strains and infected species, respectively[127].

The effects microglia elicit *in vitro* indicate a potent role of these cells in prion pathogenesis. The neurotoxicity of the human prion protein fragment PrP106-126 was shown to depend on the presence of microglia which respond to infection by enhanced secretion of oxygen radicals, resulting in oxidative damage to co-cultured neurons[48] (see also section 3.3). This phenomenon is by several researchers believed to be one of the keys (beside diminished anti-oxidative resistance) to central pathogenesis in TSEs[128] and was discussed in detail earlier in this chapter.

In vivo findings are also consistent with the proposal that microglia may have an impact on neural tissue damage. In the brains of scrapie infected mice, the microglial response is largely confined to regions with vacuolation and PrP deposition[122]. As time course studies have elucidated, microglia activation precedes neuronal death and is thus unlikely to be just a sequel of cell destruction in prion disease[113,114]. On the other hand, Perry and colleagues have argued that the immunological phenotype of microglia *in vivo* closely resembles that of macrophages having ingested apoptotic inflammatory cells, i. e. a profile including the absence of pro-inflammatory cytokines (mainly IL6, IL1β, and TNF-α) and concomitant domination of anti-inflammatory pathways[129,130], which may be in accordance with the widespread neuronal apoptosis occurring in TSEs[39,111,113,117]. However, the reports on the cytokine expression pattern in prion disease are incongruent[131,132], maybe due to different animal models and/or detection methods. *In vitro* infection experiments brought even more contradictory results, revealing an upregulation of pro-inflammatory cytokines like IL6[133,134,135], and IL1β[133]. This might be explained by the "artificial" setup, where an acute response to the acute addition of a neurotoxic peptide is monitored, while the natural process in prion disease is characterized by the slow accumulation of PrP^{Sc} with consecutive neurodegeneration[129]. A recent *in vivo* study has demonstrated a relatively late onset of generally low levels of cerebral cytokine gene expression in the ME7/CV murine scrapie model[136]. Together with the fact that tumor necrosis factor α (TNF-α)-, as well as IL6-deficient mice are fully susceptible to prion disease when challenged intracerebrally with the ME7 strain[137], this renders a crucial role of pro-inflammatory cytokines in central prion pathogenesis unlikely. Interestingly, it has been proposed that although in prion disease microglia do primarily not express significant levels of typical pro-inflammatory cytokines, they could nevertheless be in a "primed" state, and, if further stimulated by peripheral infections, secrete inflammatory mediators which promote the neurodegenerative process[129,138,139,140].

Taken together, virtually all results speak in favor of an important role of microglia in the disease process in TSEs, but we do not yet know with certainty, which of the involved pathways can finally be disastrous for the neuronal cell. Even more vague are assumptions about the direct causes of microglial activation. As microglia are closely related to peripheral macrophages, it seems logical that they are responsible for the phagocytosis of cells and/or extracellular material like deposited PrP. *In vitro*, microglia were shown to internalize to some extent fibrillar PrP106-126, which in turn interferes in a way with the phagocytic process[141]. If this cell type primarily responds to PrP^{Sc} with activation and cytokine secretion *in vivo* is still not clear.

3.5.2. Complement activation

The complement system is involved primarily in the elimination of invading foreign cells and in the initiation of inflammation. Complement activation is generally tightly regulated in order to inhibit excessive activation and consecutive complement-mediated injury[142]. It has so far been described to take place in various neurodegenerative disorders including Alzheimer's disease[143,144]and Huntington's disease[144]. For Creutzfeldt-Jakob and Gerstmann-Sträussler-Scheinker disease, it was demonstrated that factors of the early complement cascades, namely C1q, C4, C3, C3b, C3c, and C3d, co-localize with amyloid plaques[145]. In a recent study that was done in our laboratory, active complement compounds, like C1q and C3b, were detected in extracellular PrP deposits in CJD, and the most important effector of the complement system, the membrane attack complex or MAC, was found to be present in neurons in human TSEs[146]. Localization of early and late complement components correlates well with the severity of disease-specific pathology; in the respective study, areas showing neuronal TUNEL reactivity overlapped with those exhibiting MAC deposits in most of the cases that were examined. Brains without correlation exhibited advanced pathology with only a few remaining neurons which displayed MAC immunoreactivity. C3b, a member of the early complement cascade, was seen also in better preserved regions. However, the mere presence of active complement components in the brain does certainly not prove their unequivocal role in central pathogenesis of TSEs. It is not surprising that some complement activation should happen under the given circumstances, as the complement system can be activated via oxidative stress[142,147,148,149], which is supposed to be a central event in prion pathogenesis. In turn, complement components stimulate the cellular production of reactive oxygen species[150,151], so that the complement system may contribute to a vicious cycle of events threatening neuronal cells. On the other hand, mice which are temporarily depleted of C3, as well as mice deficient of C1q, develop full blown scrapie if infected via the i. c. route[152], and mice lacking early complement components display, despite delayed onset of disease, a cerebral histopathological picture equal to that of wild type mice after intraperitoneal inoculation[153]. Thus, complement-mediated cell toxicity and cell lysis may be part of the pathomechanisms causing cell death in prion disease, but the results obtained from knockout mice render a pivotal role of the complement system rather unlikely.

3.6. A possible role for the transmembrane form of PrP

The prion protein exists in multiple topologic variants, including a secretory form which is completely translocated through lipid bilayers[154], and two transmembrane forms[155]. A possible relation between these variants and neurodegenerative disease was first investigated by Hegde et al[40]: certain mutations within the PrP coding region dramatically alter the ratio of the topological forms in favor of one of the transmembrane forms, termed ^{Ctm}PrP. In mice carrying such mutations, ^{Ctm}PrP was found to be associated with severe neurodegeneration with some neuropathological features typical of prion disease, while at the same time PrP^{Sc} was virtually absent in analyzed brain tissue, suggesting elevated ^{Ctm}PrP as the primary cause of pathology. Accordingly, in human Gerstmann-Sträussler-Scheinker disease, the A117V mutation was proposed to lead to a relative preference for the synthesis of the transmembrane forms of PrP. The analyzed GSS brains contained indeed increased levels of ^{Ctm}PrP, whereas no protease-resistant PrP^{Sc} was detectable[40]. Subsequently, a pathogenic mechanism involving ^{Ctm}PrP has been suggested also for other inherited prion diseases, which, in contrast to GSS, feature accumulation of PrP^{Sc}, and for infectiously acquired prion diseases[156]. In an elegant inoculation study, double transgenic mice, carrying both a murine MoPrP and a hamster SHaPrP transgene, were inoculated with mouse RML prions. In this model, accumulation of PrP^{Sc} during the incubation period triggered in parallel the *de novo* formation of hamster ^{Ctm}PrP as possible cause for disease development and neurodegeneration[156]. On the other hand, only mutations within or near the transmembrane domain of the PrP sequence were found to enhance the formation of ^{Ctm}PrP[157,40], while pathogenic mutations in other regions showed no effect with respect to the formation of transmembrane forms of the prion protein[157]. Furthermore, cell culture and *in vivo* studies, where the amount of ^{Ctm}PrP is not altered after prion infection, render a general, obligatory role of this molecule in TSEs unlikely[158]. Thus, the pathogenic relevance of ^{Ctm}PrP for most prion diseases is still unclear. Similarly, the mechanism by which transmembrane prion protein might evoke cell death is presently not known. While some authors denied any significant accumulation of ^{Ctm}PrP in the endoplasmic reticulum (ER)[40,156], others reported it to be retained in the ER and suggested, that it could stimulate the activation of pro-apoptotic, ER stress-response pathways[159,160]. An involvement of ER stress and caspase-12 activation in the brain was recently demonstrated for sporadic and variant CJD[120].

Given all the inconsistencies mentioned above, further research is necessary to delineate whether accumulation of PrP^{Sc} in prion diseases indeed induces the enhanced formation of ^{Ctm}PrP and, if yes, whether ^{Ctm}PrP really represents a potent neurotoxic agent in TSEs.

3.7. Conclusions

Despite intensive research activities devoted to the clarification of the mechanism of central pathogenesis in transmissible spongiform encephalopathies (TSEs), there is currently no well-established model to explain nerve cell death in these diseases. This might be due to the fact that, for a long time, efforts have been focused mainly on finding a single cause instead of looking at a complex scenario. Thus, we know by now some factors which contribute either significantly or partially to neurodegeneration in prion diseases, and others of minor or no importance.

The first essential prerequisite in TSEs is the presence of the cellular prion protein, PrP^{C}, in the central nervous system. Lack of its expression renders brain tissue resistant to prion infection. A second mechanism with at least some importance is the coincidence of increased oxidative stress and reduced antioxidant protection. Reactive oxygen species seem to be generated by different means: directly by pathological prion protein (PrP^{Sc}) molecules or aggregates, and possibly derivates thereof, and indirectly by PrP^{Sc}-activated microglia. Nevertheless, this oxidative stress scenario does not suffice to explain the pathogenesis of prion diseases entirely (although it seems only plausible that such a fundamental threat should not go without consequences for the cell). In fact, the degree of its contribution to neuronal cell death *in vivo* is at present still unclear and might vary between different forms of TSEs. Other mechanisms have been investigated like direct toxic effects of PrP^{Sc} deposits. However, a decisive role for PrP^{Sc} aggregates is unlikely, as they cannot even be detected in all cases of TSE. Other abnormal forms of PrP may take its place under certain circumstances, as has been proposed for a transmembrane form, called ^{Ctm}PrP. This might be true for relatively rare familial prion diseases, but even there a pathogenic potential of ^{Ctm}PrP has not been definitely proven. Further pathways have recently been suggested, including anti- and pro-apoptotic effects of intracellular and cytosolic PrP^{C}, or loss of PrP^{C}-mediated neurotrophic signaling resulting in reduced neuronal viability. Since research on these topics is at its beginning, and some of the data are inconsistent, it is currently impossible to judge fairly on their importance for central pathogenesis.

So, what *does* actually kill the neuron? As neither the loss of function-, nor the gain of function-hypothesis is solely valid any more, a concerted action of various pathogenic mechanisms is most likely to be the clue. At the moment, each answer appears to raise still more questions. But as often in nature, the solution might be just as fascinating as complex.

3.8. References

1. C. Cunningham, R. Deacon, H. Wells, D. Boche, S. Waters, C. Picanco Diniz, H. Scott, J. N. P. Rawlins, and V. H. Perry, Synaptic changes characterize early behavioural signs in the ME7 model of murine prion disease. *Eur. J. Neurosci.* **17**, 2147–2155 (2003).

2. S. Brandner, S. Isenmann, A. Raeber, M. Fischer, A. Sailer, Y. Kobayashi, S. Marino, C. Weissmann, and A. Aguzzi, Normal host prion protein necessary for scrapie-induced neurotoxicity. *Nature* **379**, 339–343 (1996).

3. G. Mallucci, A. Dickinson, J. Linehan, P.-C. Klöhn, S. Brandner, and J. Collinge, Depleting neuronal PrP in prion infection prevents disease and reverses spongiosis. *Science* **302**, 871–874 (2003).

4. H. Büeler, A. Aguzzi, A. Sailer, R.-A. Greiner, P. Autenried, M. Aguet, and C. Weissmann, Mice devoid of PrP are resistant to scrapie. *Cell* **73**, 1339–1347 (1993).

5. J. C. Manson, A. R. Clarke, P. A. McBride, I. McConnell, and J. Hope, PrP gene dosage determines the timing but not the final intensity or distribution of lesions in scrapie pathology. *Neurodegeneration* **3**, 331–340 (1994).

6. R. E. Race, S. A. Priola, R. A. Bessen, D. Ernst, J. Dockter, G. F. Rall, L. Mucke, B. Chesebro, and M. B. Oldstone, Neuron-specific expression of a hamster prion protein minigene in transgenic mice induces susceptibility to hamster scrapie agent. *Neuron* **15**, 1183–1191 (1995).

7. A. J. Raeber, R. E. Race, S. Brandner, S. A. Priola, A. Sailer, R. A. Bessen, L. Mucke, J. Manson, A. Aguzzi, M. B. A. Oldstone, C. Weissmann, and B. Chesebro, Astrocyte-specific expression of hamster prion protein (PrP) renders PrP knockout mice susceptible to hamster scrapie. *EMBO J.* **16**, 6057–6065 (1997).

8. M. Prinz, F. Montrasio, H. Furukawa, M. E. van der Haar, P. Schwarz, T. Rülicke, O. T. Giger, K.-G. Häusler, D. Perez, M. Glatzel, and A. Aguzzi, Intrinsic resistance of oligodendrocytes to prion infection. *J. Neurosci.* **24**, 5974–5981 (2004).

9. M. Jeffrey, C. M. Goodsir, R. E. Race, and B. Chesebro, Scrapie-specific neuronal lesions are independent of neuronal PrP expression. *Ann. Neurol.* **55**, 781–792 (2004).

10. H. Büeler, M. Fischer, Y. Lang, H. Bluethmann, H.-P. Lipp, S. J. DeArmond, S. B. Prusiner, M. Aguet, and C. Weissmann, Normal development and behaviour of mice lacking the neuronal cell-surface PrP protein. *Nature* **356**, 577–582 (1992).

11. J. C. Manson, A. R. Clarke, M. L. Hooper, L. Aitchison, I. McConnell, and J. Hope, 129/Ola mice carrying a null mutation in PrP that abolishes mRNA production are developmentally normal. *Mol. Neurobiol.* **8**, 121–127 (1994).

12. J. Collinge, M. A. Whittington, K. C. L. Sidle, C. J. Smith, M. S. Palmer, A. R. Clarke, and J. G. R. Jefferys, Prion protein is necessary for normal synaptic function. *Nature* **370**, 295–297 (1994).

13. I. Tobler, S. E. Gaus, T. Deboer, P. Achermann, M. Fischer, T. Rülicke, M. Moser, B. Oesch, P. A. McBride, and J. C. Manson, Altered circadian activity rhythms and sleep in mice devoid of prion protein. *Nature* **380**, 639–642 (1996).

14. G. R. Mallucci, S. Ratté, E. A. Asante, J. Linehan, I. Gowland, J. G. R. Jefferys, and J. Collinge, Post-natal knockout of prion protein alters hippocampal CA1 properties, but does not result in neurodegeneration. *EMBO J.* **21**, 202–210 (2002).

15. D. R. Brown, R. S. J. Nicholas, and L. Canevari, Lack of prion protein expression results in a neuronal phenotype sensitive to stress. *J. Neurosci. Res.* **67**, 211–224 (2002).

16. S. Sakaguchi, S. Katamine, N. Nishida, R. Moriuchi, K. Shigematsu, T. Sugimoto, A. Nakatani, Y. Kataoka, T. Houtani, S. Shirabe, H. Okada, S. Hasegawa, T. Miyamoto,

and T. Noda, Loss of cerebellar Purkinje cells in aged mice homozygous for a disrupted PrP gene. *Nature* **380**, 528–531 (1996).

17. D. Rossi, A. Cozzio, E. Flechsig, M. A. Klein, T. Rülicke, A. Aguzzi, and C. Weissmann, Onset of ataxia and Purkinje cell loss in PrP null mice inversely correlated with Dpl level in brain. *EMBO J.* **20**, 694–702 (2001).

18. R. C. Moore, I. Y. Lee, G. L. Silverman, P. M. Harrison, R. Strome, C. Heinrich, A. Karunaratne, S. H. Pasternak, M. A. Chishti, Y. Liang, P. Mastrangelo, K. Wang, A. F. Smit, S. Katamine, G. A. Carlson, F. E. Cohen, S. B. Prusiner, D. W. Melton, P. Tremblay, L. E. Hood, and D. Westaway, Ataxia in prion protein (PrP)-deficient mice is associated with upregulation of the novel PrP-like protein doppel. *J. Mol. Biol.* **292**, 797–817 (1999).

19. H. Budka, Histopathology and immunohistochemistry of human transmissible spongiform encephalopathies (TSEs). *Arch. Virol. [Suppl.]* **16**, 135–142 (2000).

20. H. Budka, Neuropathology of prion diseases. *Br. Med. Bull.* **66**, 121–130 (2003).

21. P. Parchi, A. Giese, S. Capellari, P. Brown, W. Schulz-Schaeffer, O. Windl, I. Zerr, H. Budka, N. Kopp, P. Piccardo, S. Poser, A. Rojiani, N. Streichemberger, J. Julien, C. Vital, B. Ghetti, P. Gambetti, and H. Kretzschmar, Classification of sporadic Creutzfeldt-Jakob disease based on molecular and phenotypic analysis of 300 subjects. *Ann. Neurol.* **46**, 224–233 (1999).

22. A. F. Hill, S. Joiner, J. D. F. Wadsworth, K. C. L. Sidle, J. E. Bell, H. Budka, J. W. Ironside, and J. Collinge, Molecular classification of sporadic Creutzfeldt-Jakob disease. *Brain* **126**, 1333–1346 (2003).

23. G. Almer, J. A. Hainfellner, T. Brücke, K. Jellinger, R. Kleinert, G. Bayer, O. Windl, H. A. Kretzschmar, A. Hill, K. Sidle, J. Collinge, and H. Budka, Fatal familial insomnia: a new Austrian family. *Brain* **122**, 5–16 (1999).

24. H. Büeler, A. Raeber, A. Sailer, M. Fischer, A. Aguzzi, and C. Weissmann, High prion and PrPSc levels but delayed onset of disease in scrapie-inoculated mice heterozygous for a disrupted PrP gene. *Mol. Med.* **1**, 19–30 (1994).

25. H. A. Kretzschmar, S. B. Prusiner, L. E. Stowring, and S. J. DeArmond, Scrapie prion proteins are synthesized in neurons. *Am. J. Pathol.* **122**, 1–5 (1986).

26. A. Giese and H. A. Kretzschmar, Prion-induced neuronal damage—the mechanisms of neuronal destruction in the subacute spongiform encephalopathies. *Curr. Top. Microbiol. Immunol.* **253**, 203–217 (2001).

27. M. Guentchev, J.-A. Hainfellner, G. R. Trabattoni, and H. Budka, Distribution of parvalbumin-immunoreactive neurons in brain correlates with hippocampal and temporal cortical pathology in Creutzfeldt-Jakob disease. *J. Neuropathol. Exp. Neurol.* **56**, 1119–1124 (1997).

28. P. V. Belichenko, J. Miklossy, B. Belser, H. Budka, and M. R. Celio, Early destruction of the extracellular matrix around parvalbumin-immunoreactive interneurons in Creutzfeldt-Jakob disease. *Neurobiol. Dis.* **6**, 269–279 (1999).

29. M. Guentchev, M. H. Groschup, R. Kordek, P. P. Liberski, and H. Budka, Severe, early and selective loss of a subpopulation of GABAergic inhibitory neurons in experimental transmissible spongiform encephalopathies. *Brain Pathol.* **8**, 615–623 (1998).

30. M. Guentchev, J. Wanschitz, T. Voigtländer, H. Flicker, and H. Budka, Selective neuronal vulnerability in human prion diseases. Fatal familial insomnia differs from other types of prion diseases. *Am. J. Pathol.* **155**, 1453–1457 (1999).

31. T. Arendt, V. Bigl, and A. Arendt, Neurone loss in the nucleus basalis of Meynert in Creutzfeldt-Jakob disease. *Acta Neuropathol.* **65**, 85–88 (1984).

32. L. Cartier, R. Verdugo, C. Vergara, and S. Galvez, The nucleus basalis of Meynert in 20 definite cases of Creutzfeldt-Jakob disease. *J. Neurol. Neurosurg. Psychiatry* **52**, 304–309 (1989).

33. M. E. Bruce, TSE strain variation. *Br. Med. Bull.* **66**, 99–108 (2003).

34. M. E. Bruce, I. McConnell, H. Fraser, and A. G. Dickinson, The disease characteristics of different strains of scrapie in Sinc congenic mouse lines: implications for the nature of the agent and host control of pathogenesis. *J. Gen. Virol.* **72**, 595–603 (1991).

35. H. Fraser, Diversity in the neuropathology of scrapie-like diseases in animals. *Br. Med. Bull.* **49**, 792–809 (1993).

36. S. J. DeArmond, W. C. Mobley, D. L. DeMott, R. A. Barry, J. H. Beckstead, and S. B. Prusiner, Changes in the localization of brain prion proteins during scrapie infection. *Neurology* **37**, 1271–1280 (1987).

37. K. Jendroska, F. P. Heinzel, M. Torchia, L. Stowring, H. A. Kretzschmar, A. Kon, A. Stern, S. B. Prusiner, and S. J. DeArmond, Proteinase-resistant prion protein accumulation in Syrian hamster brain correlates with regional pathology and scrapie infectivity. *Neurology* **41**, 1482–1490 (1991).

38. R. Kordek, J. A. Hainfellner, P. P. Liberski, and H. Budka, Deposition of the prion protein (PrP) during the evolution of experimental Creutzfeldt-Jakob disease. *Acta Neuropathol.* **98**, 597–602 (1999).

39. A. Dorandeu, L. Wingertsmann, F. Chrétien, M.-B. Delisle, C. Vital, P. Parchi, P. Montagna, E. Lugaresi, J. W. Ironside, H. Budka, P. Gambetti, and F. Gray, Neuronal apoptosis in fatal familial insomnia. *Brain Pathol.* **8**, 531–537 (1998).

40. R. S. Hegde, J. A. Mastrianni, M. R. Scott, K. A. DeFea, P. Tremblay, M. Torchia, S. J. DeArmond, S. B. Prusiner, and V. R. Lingappa, A transmembrane form of the prion protein in neurodegenerative disease. *Science* **279**, 827–834 (1998).

41. G. Forloni, N. Angeretti, R. Chiesa, E. Monzani, M. Salmona, O. Bugiani, and F. Tagliavini, Neurotoxicity of a prion protein fragment. *Nature* **362**, 543–546 (1993).

42. L. De Gioia, C. Selvaggini, E. Ghibaudi, L. Diomede, O. Bugiani, G. Forloni, F. Tagliavini, and M. Salmona, Conformational polymorphism of the amyloidogenic and neurotoxic peptide homologous to residues 106-126 of the prion protein. *J. Biol. Chem.* **269**, 7859–7862 (1994).

43. C. Selvaggini, L. De Gioia, L. Cantu, E. Ghibaudi, L. Diomede, F. Passerini, G. Forloni, O. Bugiani, F. Tagliavini, and M. Salmona, Molecular characteristics of a protease-resistant, amyloidogenic and neurotoxic peptide homologous to residues 106–126 of the prion protein. *Biochem. Biophys. Res. Commun.* **194**, 1380–1386 (1993).

44. M. F. Jobling, L. R. Stewart, A. R. White, C. McLean, A. Friedhuber, F. Maher, K. Beyreuther, C. L. Masters, C. J. Barrow, S. J. Collins, and R. Cappai, The hydrophobic core sequence modulates the neurotoxic and secondary structure properties of the prion peptide 106-126. *J. Neurochem.* **73**, 1557–1565 (1999).

45. M. F. Jobling, X. Huang, L. R. Stewart, K. J. Barnham, C. Curtain, I. Volitakis, M. Perugini, A. R. White, R. A. Cherny, C. L. Masters, C. J. Barrow, S. J. Collins, A. I. Bush, and R. Cappai, Copper and zinc binding modulates the aggregation and neurotoxic properties of the prion peptide PrP106-126. *Biochemistry* **40**, 8073–8084 (2001).

46. D. R. Brown, J. Herms, and H. A. Kretzschmar, Mouse cortical cells lacking cellular PrP survive in culture with a neurotoxic PrP fragment. *Neuroreport* **5**, 2057–2060 (1994).

47. D. R. Brown, Prion protein-overexpressing cells show altered response to a neurotoxic prion protein peptide. *J. Neurosci. Res.* **54**, 331–340 (1998).

48. D. R. Brown, B. Schmidt, and H. A. Kretzschmar, Role of microglia and host prion protein in neurotoxicity of a prion protein fragment. *Nature* **380**, 345–347 (1996).

49. D. R. Brown, Prion protein peptide neurotoxicity can be mediated by astrocytes. *J. Neurochem.* **73**, 1105–1113 (1999).

50. W. E. Muller, H. Ushijima, H. C. Schroder, J. M. Forrest, W. F. Schatton, P. G. Rytik, and M. Heffner-Lauc, Cytoprotective effect of NMDA receptor antagonists on prion protein (PrionSc)-induced toxicity in rat cortical cell cultures. *Eur. J. Pharmacol.* **246**, 261–267 (1993).

51. D. R. Brown, J. W. Herms, B. Schmidt, and H. A. Kretzschmar, PrP and beta-amyloid fragments activate different neurotoxic mechanisms in cultured mouse cells. *Eur. J. Neurosci.* **9**, 1162–1169 (1997).

52. L. R. Stewart, A. R. White, M. F. Jobling, B. E. Needham, F. Maher, J. Thyer, K. Beyreuther, C. L. Masters, S. J. Collins, and R. Cappai, Involvement of the

5-lipoxygenase pathway in the neurotoxicity of the prion peptide PrP106-126. *J. Neurosci. Res.* **65**, 565–572 (2001).

53. T. van Rheede, M. M. W. Smolenaars, O. Madsen, and W. W. de Jong, Molecular evolution of the mammalian prion protein. *Mol. Biol. Evol.* **20**, 111–121 (2003).

54. M. P. Hornshaw, J. R. McDermott, and J. M. Candy, Copper binding to the N-terminal tandem repeat regions of mammalian and avian prion protein. *Biochem. Biophys. Res. Commun.* **207**, 621–629 (1995).

55. M. P. Hornshaw, J. R. McDermott, J. M. Candy, and J. H. Lakey, Copper binding to the N-terminal tandem repeat region of mammalian and avian prion protein: structural studies using synthetic peptides. *Biochem. Biophys. Res. Commun.* **214**, 993–999 (1995).

56. T. Miura, A. Hori-i, and H. Takeuchi, Metal-dependent alpha-helix formation promoted by the glycine-rich octapeptide region of prion protein. *FEBS Lett.* **396**, 248–252 (1996).

57. P. C. Pauly and D. A. Harris, Copper stimulates endocytosis of the prion protein. *J. Biol. Chem.* **273**, 33107–33110 (1998).

58. D. R. Brown, K. Qin, J. W. Herms, A. Madlung, J. Manson, R. Strome, P. E. Fraser, T. Kruck, A. von Bohlen, W. Schulz-Schaeffer, A. Giese, D. Westaway, and H. Kretzschmar, The cellular prion protein binds copper in vivo. *Nature* **390**, 684–687 (1997).

59. J. H. Viles, F. E. Cohen, S. B. Prusiner, D. B. Goodin, P. E. Wright, and H. J. Dyson, Copper binding to the prion protein: structural implications of four identical cooperative binding sites. *Proc. Natl. Acad. Sci. USA* **96**, 2042–2047 (1999).

60. M. L. Kramer, H. D. Kratzin, B. Schmidt, A. Römer, O. Windl, S. Liemann, S. Hornemann, and H. Kretzschmar, Prion protein binds copper within the physiological concentration range. *J. Biol. Chem.* **276**, 16711–16719 (2001).

61. J. Stöckel, J. Safar, A. C. Wallace, F. E. Cohen, and S. B. Prusiner, Prion protein selectively binds copper(II) ions. *Biochemistry* **37**, 7185–7193 (1998).

62. K. Qin, D.-S. Yang, Y. Yang, M. A. Chishti, L.-J. Meng, H. A. Kretzschmar, C. M. Yip, P. E. Fraser, and D. Westaway, Copper(II)-induced conformational changes and protease resistance in recombinant and cellular PrP. Effect of protein age and deamidation. *J. Biol. Chem.* **275**, 19121–19131 (2000).

63. E. Quaglio, R. Chiesa, and D. A. Harris, Copper converts the cellular prion protein into a protease-resistant species that is distinct from the scrapie isoform. *J. Biol. Chem.* **276**, 11432–11438 (2001).

64. D. R. Brown, B. Schmidt, and H. A. Kretzschmar, Effects of oxidative stress on prion protein expression in PC12 cells. *Int. J. Dev. Neurosci.* **15**, 961–972 (1997).

65. D. R. Brown, W. J. Schulz-Schaeffer, B. Schmidt, and H. A. Kretzschmar, Prion protein-deficient cells show altered response to oxidative stress due to decreased SOD-1 activity. *Exp. Neurol.* **146**, 104–112 (1997).

66. W. Rachidi, D. Vilette, P. Guiraud, M. Arlotto, J. Riondel, H. Laude, S. Lehmann, and A. Favier, Expression of prion protein increases cellular copper binding and antioxidant enzyme activities but not copper delivery. *J. Biol. Chem.* **278**, 9064–9072 (2003).

67. A. R. White, S. J. Collins, F. Maher, M. F. Jobling, L. R. Stewart, J. M. Thyer, K. Beyreuther, C. L. Masters, and R. Cappai, Prion protein-deficient neurons reveal lower glutathione reductase activity and increased susceptibility to hydrogen peroxide toxicity. *Am. J. Pathol.* **155**, 1723–1730 (1999).

68. F. Klamt, F. Dal-Pizzol, M. L. Conte da Frota Jr., R. Walz, M. E. Andrades, E. G. da Silva, R. R. Brentani, I. Izquierdo, and J. C. Fonseca Moreira, Imbalance of antioxidant defense in mice lacking cellular prion protein. *Free Radic. Biol. Med.* **30**, 1137–1144 (2001).

69. B.-S. Wong, T. Liu, R. Li, T. Pan, R. B. Petersen, M. A. Smith, P. Gambetti, G. Perry, J. C. Manson, D. R. Brown, and M.-S. Sy, Increased levels of oxidative stress markers detected in the brains of mice devoid of prion protein. *J. Neurochem.* **76**, 565–572 (2001).

70. D. R. Brown and A. Besinger, Prion protein expression and superoxide dismutase activity. *Biochem. J.* **334**, 423–429 (1998).

71. D. J. Waggoner, B. Drisaldi, T. B. Bartnikas, R. L. B. Casareno, J. R. Prohaska, J. D. Gitlin, and D. A. Harris, Brain copper content and cuproenzyme activity do not vary with prion protein expression level. *J. Biol. Chem.* **275**, 7455–7458 (2000).

72. D. R. Brown, B.-S. Wong, F. Hafiz, C. Clive, S. J. Haswell, and I. M. Jones, Normal prion protein has an activity like that of superoxide dismutase. *Biochem. J.* **344**, 1–5 (1999).

73. D. R. Brown, C. Clive, and S. J. Haswell, Antioxidant activity related to copper binding of native prion protein. *J. Neurochem.* **76**, 69–76 (2001).

74. B.-S. Wong, T. Pan, T. Liu, R. Li, P. Gambetti, and M.-S. Sy, Differential contribution of superoxide dismutase activity by prion protein in vivo. *Biochem. Biophys. Res. Commun.* **273**, 136–139 (2000).

75. T. Voigtländer, S. Klöppel, P. Birner, C. Jarius, H. Flicker, S. Verghese-Nikolakaki, T. Sklaviadis, M. Guentchev, and H. Budka, Marked increase of neuronal prion protein immunoreactivity in Alzheimer's disease and human prion diseases. *Acta Neuropathol.* **101**, 417–423 (2001).

76. G. G. Kovacs, P. Zerbi, T. Voigtländer, M. Strohschneider, G. Trabattoni, J. A. Hainfellner, and H. Budka, The prion protein in human neurodegenerative disorders. *Neurosci. Lett.* **329**, 269–272 (2002).

77. D. R. Brown, F. Hafiz, L. L. Glasssmith, B.-S. Wong, I. M. Jones, C. Clive, and S. J. Haswell, Consequences of manganese replacement of copper for prion protein function and proteinase resistance. *EMBO J.* **19**, 1180–1186 (2000).

78. D. R. Brown and J. Sassoon, Copper-dependent functions for the prion protein. *Mol. Biotechnol.* **22**, 165–178 (2002).

79. S. Turnbull, B. J. Tabner, D. R. Brown, and D. Allsop, Copper-dependent generation of hydrogen peroxide from the toxic prion protein fragment PrP106-126. *Neurosci. Lett.* **336**, 159–162 (2003).

80. B.-S. Wong, D. R. Brown, T. Pan, M. Whiteman, T. Liu, X. Bu, R. Li, P. Gambetti, J. Olesik, R. Rubenstein, and M.-S. Sy, Oxidative impairment in scrapie-infected mice is associated with brain metals perturbations and altered antioxidant activities. *J. Neurochem.* **79**, 689–698 (2001).

81. D. W. Lee, H. O. Sohn, H. B. Lim, Y. G. Lee, Y. S. Kim, R. I. Carp, and H. M. Wisniewski, Alteration of free radical metabolism in the brain of mice infected with scrapie agent. *Free Radic. Res.* **30**, 499–507 (1999).

82. H. Ovadia, H. Rosenmann, E. Shezen, M. Halimi, I. Ofran, and R. Gabizon, Effect of scrapie infection on the activity of neuronal nitric-oxide synthase in brain and neuroblastoma cells. *J. Biol. Chem.* **271**, 16856–16861 (1996).

83. G. I. Keshet, H. Ovadia, A. Taraboulos, and R. Gabizon, Scrapie-infected mice and PrP knockout mice share abnormal localization and activity of neuronal nitric oxide synthase. *J. Neurochem.* **72**, 1224–1231 (1999).

84. O. Milhavet, H. E. M. McMahon, W. Rachidi, N. Nishida, S. Katamine, A. Mangé, M. Arlotto, D. Casanova, J. Riondel, A. Favier, and S. Lehmann, Prion infection impairs the cellular response to oxidative stress. *Proc. Natl. Acad. Sci. USA* **97**, 13937–13942 (2000).

85. S.-I. Choi, W.-K. Ju, E.-K. Choi, J. Kim, H.-Z. Lea, R. I. Carp, H. M. Wisniewski, and Y.-S. Kim, Mitochondrial dysfunction induced by oxidative stress in the brains of hamsters infected with the 263 K scrapie agent. *Acta Neuropathol.* **96**, 279–286 (1998).

86. M. Guentchev, T. Voigtländer, C. Haberler, M. H. Groschup, and H. Budka, Evidence for oxidative stress in experimental prion disease. *Neurobiol. Dis.* **7**, 270–273 (2000).

87. Y.-G. Choi, J.-I. Kim, H.-P. Lee, J.-K. Jin, E.-K. Choi, R. I. Carp, and Y.-S. Kim, Induction of heme oxygenase-1 in the brains of scrapie-infected mice. *Neurosci. Lett.* **289**, 173–176 (2000).

88. M. Rizzardini, R. Chiesa, N. Angeretti, E. Lucca, M. Salmona, G. Forloni, and L. Cantoni, Prion protein fragment 106-126 differentially induces heme oxygenase-1 mRNA in cultured neurons and astroglial cells. *J. Neurochem.* **68**, 715–720 (1997).

89. M. Guentchev, S. L. Siedlak, C. Jarius, F. Tagliavini, R. J. Castellani, G. Perry, M. A. Smith, and H. Budka, Oxidative damage to nucleic acids in human prion disease. *Neurobiol. Dis.* **9**, 275–281 (2002).

90. G. G. Kovacs, T. Voigtländer, J. A. Hainfellner, and H. Budka, Distribution of intraneuronal immunoreactivity for the prion protein in human prion diseases. *Acta Neuropathol.* **104**, 320–326 (2002).

91. T. Ookawara, N. Kawamura, Y. Kitagawa, and N. Taniguchi, Site-specific and random fragmentation of Cu,Zn-superoxide dismutase by glycation reaction. Implication of reactive oxygen species. *J. Biol.Chem.* **267**, 18505–18510 (1992).

92. M. Asahi, J. Fujii, K. Suzuki, H. G. Seo, T. Kuzuya, M. Hori, M. Tada, S. Fujii, and N. Taniguchi, Inactivation of glutathione peroxidase by nitric oxide. Implication for cytotoxicity. *J. Biol. Chem.* **270**, 21035–21039 (1995).

93. H. E. M. McMahon, A. Mangé, N. Nishida, C. Créminon, D. Casanova, and S. Lehmann, Cleavage of the amino terminus of the prion protein by reactive oxygen species. *J. Biol. Chem.* **276**, 2286–2291 (2001).

94. D. R. Brown, PrPSc-like prion protein peptide inhibits the function of cellular prion protein. *Biochem. J.* **352**, 511–518 (2000).

95. W. Rachidi, A. Mangé, A. Senator, P. Guiraud, J. Riondel, M. Benboubetra, A. Favier, and S. Lehmann, Prion infection impairs copper binding of cultured cells. *J. Biol.Chem.* **278**, 14595–14598 (2003).

96. A. M. Thackray, R. Knight, S. J. Haswell, R. Bujdoso, and D. R. Brown, Metal imbalance and compromised antioxidant function are early changes in prion disease. *Biochem. J.* **362**, 253–258 (2002).

97. B.-S. Wong, S. G. Chen, M. Colucci, Z. Xie, T. Pan, T. Liu, R. Li, P. Gambetti, M.-S. Sy, and D. R. Brown, Aberrant metal binding by prion protein in human prion disease. *J. Neurochem.* **78**, 1400–1408 (2001).

98. N. A. Simonian and J. T. Coyle, Oxidative stress in neurodegenerative diseases. *Annu. Rev. Pharmacol. Toxicol.* **36**, 83–106 (1996).

99. P. S. Sastry and K. S. Rao, Apoptosis and the nervous system. *J. Neurochem.* **74**, 1–20 (2000).

100. J. F. Kerr, A. H. Wyllie, and A. R. Currie, Apoptosis: a basic biological phenomenon with wide-ranging implications in tissue kinetics. *Br. J. Cancer* **26**, 239–257 (1972).

101. M. Raffray and G. M. Cohen, Apoptosis and necrosis in toxicology: a continuum or distinct modes of cell death? *Pharmacol. Ther.* **75**, 153–177 (1997).

102. G. M. Cohen, X. M. Sun, R. T. Snowden, D. Dinsdale, and D. N. Skilleter, Key morphological features of apoptosis may occur in the absence of internucleosomal DNA fragmentation. *Biochem. J.* **286**, 331–334 (1992).

103. E. Falcieri, A. M. Martelli, R. Bareggi, A. Cataldi, and L. Cocco, The protein kinase inhibitor staurosporine induces morphological changes typical of apoptosis in MOLT-4 cells without concomitant DNA fragmentation. *Biochem. Biophys. Res. Commun.* **193**, 19–25 (1993).

104. Z. F. Zakeri, D. Quaglino, T. Latham, and R. A. Lockshin, Delayed internucleosomal DNA fragmentation in programmed cell death. *FASEB J.* **7**, 470–478 (1993).

105. M. K. Collins, J. Marvel, P. Malde, and A. Lopez-Rivas, Interleukin 3 protects murine bone marrow cells from apoptosis induced by DNA damaging agents. *J. Exp. Med.* **176**, 1043–1051 (1992).

106. K. Fukuda, M. Kojiro, and J. F. Chiu, Demonstration of extensive chromatin cleavage in transplanted Morris hepatoma 7777 tissue: apoptosis or necrosis? *Am. J. Pathol.* **142**, 935–946 (1993).

107. Y. Gavrieli, Y. Sherman, and S. A. Ben-Sasson, Identification of programmed cell death in situ via specific labeling of nuclear DNA fragmentation. *J. Cell Biol.* **119**, 493–501 (1992).

108. I. Ferrer, Nuclear DNA fragmentation in Creutzfeldt-Jakob disease: does a mere positive in situ nuclear end-labeling indicate apoptosis? *Acta Neuropathol.* **97**, 5–12 (1999).

109. H. M. Schätzl, L. Laszlo, D. M. Holtzman, J. Tatzelt, S. J. DeArmond, R. I. Weiner, W. C. Mobley, and S. B. Prusiner, A hypothalamic neuronal cell line persistently infected with scrapie prions exhibits apoptosis. *J. Virol.* **71**, 8821–8831 (1997).

110. A. Bürkle, H. A. Kretzschmar, and D. R. Brown, Poly(ADP-ribose) immunostaining to detect apoptosis induced by a neurotoxic fragment of prion protein. *Histochem. J.* **31**, 711–716 (1999).

111. A. Giese, M. H. Groschup, B. Hess, and H. A. Kretzschmar, Neuronal cell death in scrapie-infected mice is due to apoptosis. *Brain Pathol.* **5**, 213–221 (1995).

112. P. J. Lucassen, A. Williams, W. C. Chung, and H. Fraser, Detection of apoptosis in murine scrapie. *Neurosci. Lett.* **198**, 185–188 (1995).

113. A. Williams, P. J. Lucassen, D. Ritchie, and M. Bruce, PrP deposition, microglial activation, and neuronal apoptosis in murine scrapie. *Exp. Neurol.* **144**, 433–438 (1997).

114. A. Giese, D. R. Brown, M. H. Groschup, C. Feldmann, I. Haist, and H. A. Kretzschmar, Role of microglia in neuronal cell death in prion disease. *Brain Pathol.* **8**, 449–457 (1998).

115. D. Jesionek-Kupnicka, J. Buczynski, R. Kordek, and P. P. Liberski, Neuronal loss and apoptosis in experimental Creutzfeldt-Jakob disease in mice. *Folia Neuropathol.* **37**, 283–286 (1999).

116. D. W. Fairbairn, K. G. Carnahan, R. N. Thwaits, R. V. Grigsby, G. R. Holyoak, and K. L. O'Neill, Detection of apoptosis induced DNA cleavage in scrapie-infected sheep brain. *FEMS Microbiol. Lett.* **115**, 341–346 (1994).

117. F. Gray, F. Chretien, H. Adle-Biassette, A. Dorandeu, T. Ereau, M. B. Delisle, N. Kopp, J. W. Ironside, and C. Vital, Neuronal apoptosis in Creutzfeldt-Jakob disease. *J. Neuropathol. Exp. Neurol.* **58**, 321–328 (1999).

118. A. R. White, R. Guirguis, M. W. Brazier, M. F. Jobling, A. F. Hill, K. Beyreuther, C. J. Barrow, C. L. Masters, S. J. Collins, and R. Cappai, Sublethal concentrations of prion peptide PrP106-126 or the amyloid beta peptide of Alzheimer's disease activates expression of proapoptotic markers in primary cortical neurons. *Neurobiol. Dis.* **8**, 299–316 (2001).

119. C. N. O'Donovan, D. Tobin, and T. G. Cotter, Prion protein fragment PrP-(106-126) induces apoptosis via mitochondrial disruption in human neuronal SH-SY5Y cells. *J. Biol. Chem.* **276**, 43516–43523 (2001).

120. C. Hetz, M. Russelakis-Carneiro, K. Maundrell, J. Castilla, and C. Soto, Caspase-12 and endoplasmic reticulum stress mediate neurotoxicity of pathological prion protein. *EMBO J.* **22**, 5435–5445 (2003).

121. A. Corsaro, S. Thellung, V. Villa, D. R. Principe, D. Paludi, S. Arena, E. Millo, D. Schettini, G. Damonte, A. Aceto, G. Schettini, and T. Florio, Prion protein fragment 106-126 induces a p38 MAP kinase-dependent apoptosis in SH-SY5Y neuroblastoma cells independently from the amyloid fibril formation. *Ann. N Y Acad. Sci.* **1010**, 610–622 (2003).

122. A. E. Williams, L. J. Lawson, V. H. Perry, and H. Fraser, Characterization of the microglial response in murine scrapie. *Neuropathol. Appl. Neurobiol.* **20**, 47–55 (1994).

123. S. Betmouni, V. H. Perry, and J. L. Gordon, Evidence for an early inflammatory response in the central nervous system of mice with scrapie. *Neuroscience* **74**, 1–5 (1996).

124. D. E. McFarlin, M. C. Raff, E. Simpson, and S. H. Nehlsen, Scrapie in immunologically deficient mice. *Nature* **233**, 336 (1971).

125. A. Sasaki, J. Hirato, and Y. Nakazato, Immunohistochemical study of microglia in the Creutzfeldt-Jakob diseased brain. *Acta Neuropathol.* **86**, 337–344 (1993).

126. H. Mühleisen, J. Gehrmann, and R. Meyermann, Reactive microglia in Creutzfeldt-Jakob disease. *Neuropathol. Appl. Neurobiol.* **21**, 505–517 (1995).

127. C. A. Baker, Z. Y. Lu, I. Zaitsev, and L. Manuelidis, Microglial activation varies in different models of Creutzfeldt-Jakob disease. *J. Virol.* **73**, 5089–5097 (1999).

128. D. R. Brown, Mayhem of the multiple mechanisms: modelling neurodegeneration in prion disease. *J. Neurochem.* **82**, 209–215 (2002).

129. V. H. Perry, C. Cunningham, and D. Boche, Atypical inflammation in the central nervous system in prion disease. *Curr. Opin. Neurol.* **15**, 349–354 (2002).

130. V. A. Fadok, D. L. Bratton, A. Konowal, P. W. Freed, J. Y. Westcott, and P. M. Henson, Macrophages that have ingested apoptotic cells in vitro inhibit proinflammatory cytokine production through autocrine/paracrine mechanisms involving TGF-β, PGE2, and PAF. *J. Clin. Invest.* **101**, 890–898 (1998).

131. A. Williams, A.-M. Van Dam, D. Ritchie, P. Eikelenboom, and H. Fraser, Immunocytochemical appearance of cytokines, prostaglandin E2 and lipocortin-1 in the CNS during the incubation period of murine scrapie correlates with progressive PrP accumulations. *Brain Res.* **754**, 171–180 (1997).

132. C. Cunningham, D. Boche, and V. H. Perry, Transforming growth factor β1, the dominant cytokine in murine prion disease: influence on inflammatory cytokine synthesis and alteration of vascular extracellular matrix. *Neuropathol. Appl. Neurobiol.* **28**, 107–119 (2002).

133. J.-M. Peyrin, C. I. Lasmézas, S. Haïk, F. Tagliavini, M. Salmona, A. Williams, D. Richie, J.-P. Deslys, and D. Dormont, Microglial cells respond to amyloidogenic PrP peptide by the production of inflammatory cytokines. *Neuroreport* **10**, 723–729 (1999).

134. C. Bate, S. Reid, and A. Williams, Killing of prion-damaged neurones by microglia. *Neuroreport* **12**, 2589–2594 (2001).

135. C. Bate, R. S. Boshuizen, J. P. M. Langeveld, and A. Williams, Temporal and spatial relationship between the death of PrP-damaged neurones and microglial activation. *Neuroreport* **13**, 1695–1700 (2002).

136. A. R. Brown, J. Webb, S. Rebus, R. Walker, A. Williams, and J. K. Fazakerley, Inducible cytokine gene expression in the brain in the ME7/CV mouse model of scrapie is highly restricted, is at a strikingly low level relative to the degree of gliosis and occurs only late in the disease. *J. Gen. Virol.* **84**, 2605–2611 (2003).

137. N. A. Mabbott, A. Williams, C. F. Farquhar, M. Pasparakis, G. Kollias, and M. E. Bruce, Tumor necrosis factor alpha-deficient, but not interleukin-6-deficient, mice resist peripheral infection with scrapie. *J. Virol.* **74**, 3338–3344 (2000).

138. M. I. Combrinck, V. H. Perry, and C. Cunningham, Peripheral infection evokes exaggerated sickness behaviour in pre-clinical murine prion disease. *Neuroscience* **112**, 7–11 (2002).

139. V. H. Perry, T. A. Newman, and C. Cunningham, The impact of systemic infection on the progression of neurodegenerative disease. *Nat. Rev. Neurosci.* **4**, 103–112 (2003).

140. V. H. Perry, The influence of systemic inflammation on inflammation in the brain: implications for chronic neurodegenerative disease. *Brain Behav. Immun.* **18**, 407–413 (2004).

141. J. Ciesielski-Treska, N. J. Grant, G. Ulrich, M. Corrotte, Y. Bailly, A.-M. Haeberle, S. Chasserot-Golaz, and M.-F. Bader, Fibrillar prion peptide (106-126) and scrapie prion protein hamper phagocytosis in microglia. *Glia* **46**, 101–115 (2004).

142. C. D. Collard, R. Lekowski, J. E. Jordan, A. Agah, and G. L. Stahl, Complement activation following oxidative stress. *Mol. Immunol.* **36**, 941–948 (1999).

143. P. Eikelenboom, C. Bate, W. A. Van Gool, J. J. M. Hoozemans, J. M. Rozemuller, R. Veerhuis, and A. Williams, Neuroinflammation in Alzheimer's disease and prion disease. *Glia* **40**, 232–239 (2002).

144. P. Gasque, Y. D. Dean, E. P. McGreal, J. VanBeek, and B. P. Morgan, Complement components of the innate immune system in health and disease in the CNS. *Immunopharmacology* **49**, 171–186 (2000).

145. T. Ishii, S. Haga, S. Yagishita, and J. Tateishi, The presence of complements in amyloid plaques of Creutzfeldt-Jakob disease and Gerstmann-Straussler-Scheinker disease. *Appl. Pathol.* **2**, 370–379 (1984).

146. G. G. Kovacs, P. Gasque, T. Ströbel, E. Lindeck-Pozza, M. Strohschneider, J. W. Ironside, H. Budka, and M. Guentchev, Complement activation in human prion disease. *Neurobiol. Dis.* **15**, 21–28 (2004).

147. W. Vogt, B. Damerau, I. von Zabern, R. Nolte, and D. Brunahl, Non-enzymic activation of the fifth component of human complement, by oxygen radicals. Some properties of the activation product, C5b-like C5. *Mol. Immunol.* **26**, 1133–1142 (1989).

148. C. D. Collard, A. Väkevä, M. A. Morrissey, A. Agah, S. A. Rollins, W. R. Reenstra, J. A. Buras, S. Meri, and G. L. Stahl, Complement activation after oxidative stress. Role of the lectin complement pathway. *Am. J. Pathol.* **156**, 1549–1556 (2000).

149. M. L. Hart, M. C. Walsh, and G. L. Stahl, Initiation of complement activation following oxidative stress. In vitro and in vivo observations. *Mol. Immunol.* **41**, 165–171 (2004).

150. S. Adler, P. J. Baker, R. J. Johnson, R. F. Ochi, P. Pritzl, and W. G. Couser, Complement membrane attack complex stimulates production of reactive oxygen metabolites by cultured rat mesangial cells. *J. Clin. Invest.* **77**, 762–767 (1986).

151. J. Elsner, M. Oppermann, W. Czech, G. Dobos, E. Schöpf, J. Norgauer, and A. Kapp, C3a activates reactive oxygen radical species production and intracellular calcium transients in human eosinophils. *Eur. J. Immunol.* **24**, 518–522 (1994).

152. N. A. Mabbott, M. E. Bruce, M. Botto, M. J. Walport, and M. B. Pepys, Temporary depletion of complement component C3 or genetic deficiency of C1q significantly delays onset of scrapie. *Nat. Med.* **7**, 485–487 (2001).

153. M. A. Klein, P. S. Kaeser, P. Schwarz, H. Weyd, I. Xenarios, R. M. Zinkernagel, M. C. Carroll, J. S. Verbeek, M. Botto, M. J. Walport, H. Molina, U. Kalinke, H.

Acha-Orbea, and A. Aguzzi, Complement facilitates early prion pathogenesis. *Nat. Med.* **7**, 488–492 (2001).

154. B. Hay, S. B. Prusiner, and V. R. Lingappa, Evidence for a secretory form of the cellular prion protein. *Biochemistry* **26**, 8110–8115 (1987).

155. B. Hay, R. A. Barry, I. Lieberburg, S. B. Prusiner, and V. R. Lingappa, Biogenesis and transmembrane orientation of the cellular isoform of the scrapie prion protein. *Mol. Cell. Biol.* **7**, 914–920 (1987).

156. R. S. Hegde, P. Tremblay, D. Groth, S. J. DeArmond, S. B. Prusiner, and V. R. Lingappa, Transmissible and genetic prion diseases share a common pathway of neurodegeneration. *Nature* **402**, 822–826 (1999).

157. R. S. Stewart and D. A. Harris, Most pathogenic mutations do not alter the membrane topology of the prion protein. *J. Biol. Chem.* **276**, 2212–2220 (2001).

158. R. S. Stewart and D. A. Harris, Mutational analysis of topological determinants in prion protein (PrP) and measurement of transmembrane and cytosolic PrP during prion infection. *J. Biol. Chem.* **278**, 45960–45968 (2003).

159. R. S. Stewart, B. Drisaldi, and D. A. Harris, A transmembrane form of the prion protein contains an uncleaved signal peptide and is retained in the endoplasmic reticulum. *Mol. Biol. Cell* **12**, 881–889 (2001).

160. D. A. Harris, Trafficking, turnover and membrane topology of PrP. *Br. Med. Bull.* **66**, 71–85 (2003).

Chapter 4

HEREDITARY PRION PROTEIN AMYLOIDOSES

Bernardino Ghetti[1], Pedro Piccardo[1,2], Orso Bugiani[3], Gianluigi Forloni[4], Michela Morbin[3], Mario Salmona[4], and Fabrizio Tagliavini[3]

[1]*Department of Pathology and Laboratory Medicine, Indiana University School of Medicine, Indianapolis, IN 46202, USA*
[2]*Center for Biologics Evaluation and Research, Food and Drug Administration, Rockville, MD 20852, USA*
[3] *Instituto Nazionale Neurologico "Carlo Besta", Division of Neuropathology, 20133 Milan, Italy*
[4]*Istituto di Ricerche Farmacologiche "Mario Negri", Department of Neuroscience and Biochemistry and Molecular Pharmacology, 20157 Milan, Italy*

4.1. Introduction

The term prion protein (PrP) amyloidoses is used to describe a group of diseases in which large amounts of PrP degradation products accumulate as fibrillary deposits leading to amyloid formation[1,2]. Accumulation of abnormal PrP isoforms occurs without significant amyloid deposition in the brain parenchyma during the course of most subacute spongiform encephalopathies, which include sporadic, familial, and iatrogenic Creutzfeldt-Jakob disease (CJD) as well as sporadic and familial fatal insomnia (FFI)[3–6]. Typically, PrP-amyloid is seen in Gerstmann-Sträussler-Scheinker disease (GSS), PrP cerebral amyloid angiopathy (PrP-CAA), variant CJD (vCJD), and kuru[1–3].

The hereditary PrP amyloidoses include GSS and PrP-CAA[1–8]. GSS is autosomal dominant and is characterized clinically by motor abnormalities and intellectual deterioration, and pathologically by PrP-amyloid deposits[9–25]. To date, at least 54 families affected by GSS have been studied (Table 4.1). GSS has been found in Australia, Austria, Canada, Denmark, France, Germany, Hungary, Ireland, Israel, Italy, Japan, Mexico, Poland, United Kingdom, and United States. It is difficult to determine the exact incidence of GSS for two main reasons: (i) the disease has been reported only in a few countries and (ii) the disease may be underreported due to its clinical similarity to olivopontocerebellar atrophy, spinocerebellar ataxia, Parkinson disease, amyotrophic lateral sclerosis, Huntington disease or Alzheimer disease (AD)[1–8]. PrP-CAA is very rare and has been characterized clinically, pathologically, biochemically, and genetically in only one patient[7].

Table 4.1. Genetic mutations in the *PRNP* gene, haplotypes, nucleotide changes and amino acid changes associated with GSS and PrP-CAA. Note the number of families associated with each of the mutations.

Point Mutations in *PRNP* Associated with GSS & PrP-CAA			
Haplotype	Nucleotide change	Amino acid change	Number of families
P102L-129M	CCG → CTG	P → L	27
P102L-129M -219K			1
P102L-129V			3
P105L-129V	CCA → CTA	P → L	5
A117V-129V	GCA → GTG	A → V	8
G131V-129M	GGA → GTA	G → V	1
Y145Stop-129M	TAT → TAG	Y → Stop	1
H187R-129V	CAC → CGC	H → R	1
F198S-129V	TTC → TCC	F → S	2
D202N-129V	GAC → AAC	D → N	2
Q212P-129M	CAG → CCG	Q → P	1
Q217R-129V	CAG → CGG	Q → R	1
M232T	ATG → ACG	M → T	1

4.2. Parenchymal and vascular PrP amyloidosis

Among the hereditary PrP amyloidoses, a parenchymal amyloidosis is seen in association with GSS, whereas, a vascular PrP amyloidosis is found in PrP-CAA. Currently, ten missense mutations in the *Prion*

Figure 4.1. Cerebral hemisphere of a patient with GSS reveals PrP-immunoreactivity in the neocortex, caudate nucleus, putamen, globus pallidus and amygdala. PrP immunohistochemistry using antibody 3F4.

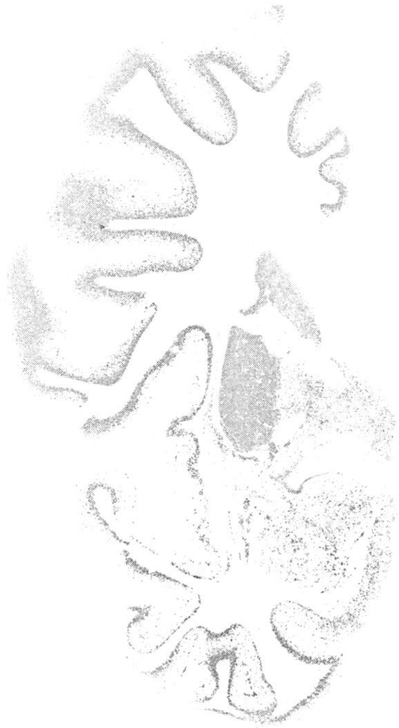

Protein gene (*PRNP*) are known to be associated with GSS (Table 4.1). In addition, a GSS-like phenotype has been seen associated with insertional mutations characterized by eight or nine additional 24-base pair repeats[3–4]. Of the known polymorphisms in the *PRNP* gene, those at codons 129 and 219 have been shown to influence the clinical and pathologic phenotypes of GSS[3–4,9].

The neuropathologic diagnosis of GSS is based on the presence of PrP-amyloid and pre-amyloid (diffuse) deposits, the distribution and extent of which differ widely between families (Figure 4.1)[1–4].

Amyloid is accompanied by glial proliferation as well as a loss of neuronal processes and perikarya. Spongiform degeneration is not consistently found. In some families, numerous neurofibrillary lesions (neurofibrillary tangles, neuropil threads), indistinguishable from those found in AD, are present in the neocortex and subcortical nuclei (Figure 4.2)[18,22–23,26].

PrP-CAA has been described in association with a nonsense mutation at codon 145 of *PRNP* (Table 4.1)[2,7–8]. The clinical phenotype is characterized by memory disturbance and disorientation evolving into a severe dementia[2,7–8]. Deposition of PrP amyloid in cerebral vessels walls and abundant neurofibrillary lesions are observed (Figure 4.3)[2,7–8]. In

Figure 4.2. Neocortex of a patient with GSS associated with a *PRNP* F198S mutation. Fluorescent unicentric and multicentric plaques as well as numerous neurofibrillary tangles are present in several cortical layers. Thioflavin S method.

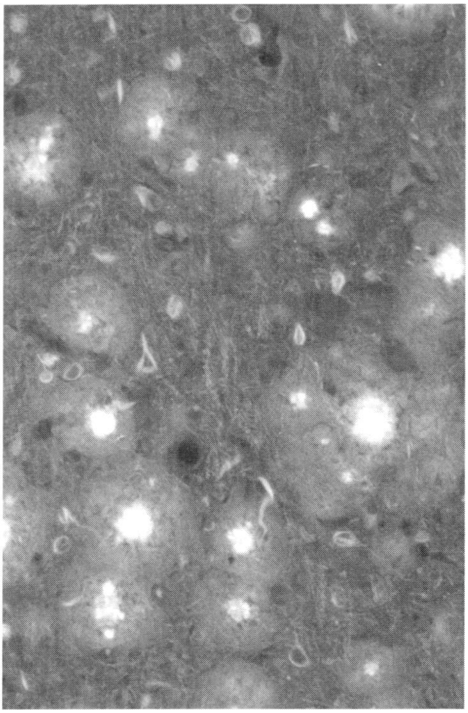

Figure 4.3. Cerebellar (A) and cerebral (B) cortices of a patient with PrP-CAA associated with the *PRNP* Y145Stop mutation. (A) Vessels of the cerebellar cortex have fluorescent amyloid deposits. (B) Vessels of the cerebral cortex have PrP deposits. (A) thioflavin S method and (B) PrP-immunohistochemistry.

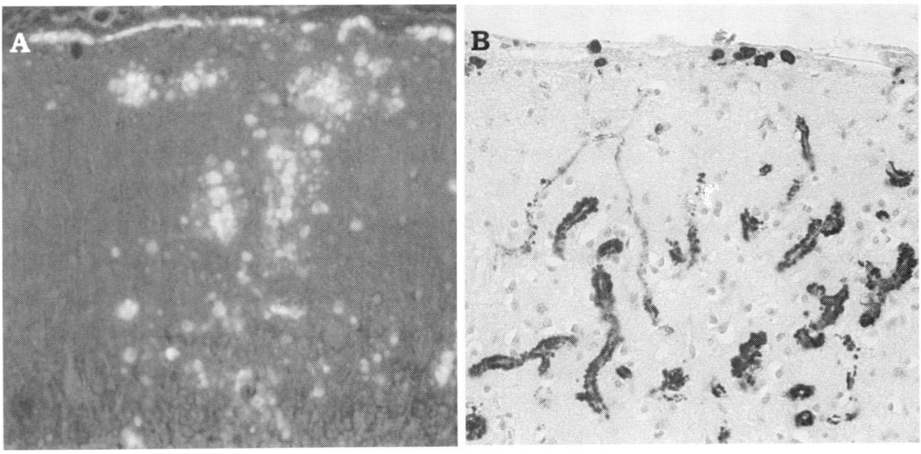

PrP-CAA, fibrils are seen adjacent to and within the vessel wall. PrP-amyloid deposits are immunolabeled with antibodies to PrP spanning the region 90-147 (i.e. mid-region) and not with antibodies to amyloid β (Aβ). The amyloid deposits immunoreact with an antiserum to the C-terminus, suggesting that normal PrP or C-terminal fragments co-exist with vascular amyloid[7].

4.3. Gerstmann-Sträussler-Scheinker disease

4.3.1. *PRNP* P102L-129M[1–6,17,27–28]

Clinical Presentation. The *PRNP* P102L-129M is the most common haplotype among those associated with GSS. The clinical phenotype is characterized by a progressive cerebellar syndrome with ataxia, dysarthria, and incoordination of saccadic eye movements as well as pyramidal and pseudobulbar signs. Behavioral and cognitive dysfunctions are seen and in most instances, they evolve into dementia or akinetic mutism. Clinical symptoms start in the fourth to sixth decades of life with a disease duration of a few months to six years. In some cases, amyotrophy and an electromyographic pattern of denervation may be seen early in the course of the disease. A computed tomography (CT) may show brain atrophy and a single photon emission computed tomography (SPECT) might reveal hypoperfusion of the frontal cortex and cerebellum. Myoclonus and pseudoperiodic sharp wave discharges in the electroencephalogram (EEG), characteristics of CJD, are observed in some individuals, whose disease may have a very rapid clinical course ranging from 5 to 9 months and a picture indistinguishable from that of CJD.

Neuropathology. The *PRNP* P102L-129M haplotype is associated with two neuropathologic phenotypes. The first is that of typical GSS; the second is a combination of GSS and CJD features, namely amyloid plaques and spongiform degeneration. The typical GSS features include unicentric and/or multicentric PrP plaques that are most numerous in the molecular layer of the cerebellum, but they are also found in the cerebral gray matter. Astrocytic proliferation is present in areas with the most severe PrP deposition. Neuronal loss appears to be more severe in the cases with spongiform degeneration than in those cases with PrP plaques only.

4.3.2. *PRNP* P102L-129M-219K[9]

Clinical presentation. The clinical presentation is characterized by either dementia or cerebellar signs. Magnetic resonance imaging (MRI) may show severe cerebral atrophy.

Neuropathology. Mild PrP deposition is present in the cerebral and cerebellar cortices as well as the basal ganglia; however, amyloid and spongiform changes are not observed.

4.3.3. *PRNP* P102L-129V[29]

Clinical presentation. The clinical course and duration are different from that associated with the P102L-129M haplotype. These differences are evidenced by the presence of seizures and long-tract signs and the absence of dementia. In addition, this mutation is associated with a clinical course that may last up to 12 years.

Neuropathology. There is a moderate to severe loss of fibers in the corticospinal, spinocerebellar, and gracilis tracts. In addition, diffuse PrP deposits are present in the substantia gelatinosa. PrP-amyloid plaques are frequently seen in the cerebellar cortex and to a lesser extent in the neocortex. No spongiform degeneration is seen.

4.3.4. *PRNP* P105L-129V[30–31]

Clinical presentation. Spastic gait, hyperreflexia, and the Babinski sign are prominent in the initial stages. Extrapyramidal signs such as fine finger tremor and rigidity of limbs may be observed. Paraparesis progresses to tetraparesis and is accompanied by emotional incontinence and dementia. Myoclonus, pseudoperiodic sharp wave discharges in EEG or severe cerebellar signs have not been reported. T2-weighted MRI shows hypointensity of the striatum. An electromyogram (EMG) shows a pattern of denervation, while evoked potentials reveal conduction delays in the posterior funiculi and corticospinal tract. The age at the onset of the clinical signs is in the fourth and fifth decades of life; the duration of the disease ranges from six to twelve years.

Neuropathology. PrP-amyloid plaques and diffuse deposits are frequently seen in the neocortex, especially the motor area, striatum, and thalamus, but rarely seen in the cerebellum. Neurofibrillary tangles are seen in some cases and may occur in varying amounts. In addition, axonal loss occurs in the pyramidal tracts. No spongiform changes are seen.

4.3.5. *PRNP* A117V-129V[22,27,32–34]

Clinical presentation. The *PRNP* A117V-129V haplotype is associated with a variety of clinical phenotypes ranging from typical GSS disease to classic AD. The age at onset of clinical signs is in the second to seventh decade of life; the duration ranges from one to eleven years. In some individuals, marked extrapyramidal signs with Parkinsonian features occurs early in the course of the disease followed by other neurological symptoms. Additional signs that may be seen include pyramidal signs, amyotrophy, myoclonus, emotional lability and pseudobulbar signs. Behavioral and personality disturbances, such as mood swings, aggressive behavior, and paranoia, frequently present long before neurological signs and symptoms. The phenotypic variability observed among affected individuals even occurs within the same family. EEGs are either normal or non-specifically abnormal, but no pseudoperiodic sharp wave discharges are seen. Results from CT scans varied from normal to moderate cerebral atrophy.

Neuropathology. Cerebral atrophy is seen in some cases. Variable amounts of PrP-amyloid plaques and diffuse deposits are widespread throughout the cerebral cortex, hippocampus, basal ganglia and thalamus; however, in the cerebellum, they may be absent or present in variable amounts. Pyramidal tract degeneration may be present. Spongiform degeneration may also be seen, but if so, it is focally present in the cerebrum or cerebellum. Neuronal loss, when present, may be severe in the substantia nigra. Neurofibrillary tangles have been seen in individuals that had a long disease duration.

4.3.6. *PRNP* G131V-129M[35]

Clinical presentation. Changes in personality, decrease in cognitive performance, apraxia, tremor, and increased tendon reflexes are the presenting signs. The onset of the disease is early in the fifth decade of life and the disease has a duration of nine years. MRIs may show cerebral and cerebellar atrophy. EEGs do not show pseudoperiodic sharp wave discharges. In the late stages, dementia becomes progressively more severe and ataxia develops.

Neuropathology. PrP-amyloid plaques and diffuse deposits are seen in the cortex and subcortical nuclei as well as in the cerebellum. Neurofibrillary tangles are seen in the Ammon's horn and in the entorhinal cortex. No spongiform degeneration is seen.

4.3.7. *PRNP* H187R-129V[36]

Clinical presentation. The clinical phenotype is characterized by early progressive cognitive impairment, cerebellar ataxia and dysarthria followed by myoclonus, seizures and occasionally pyramidal and extrapyramidal signs. Neuroimaging shows a severe, widespread atrophy of the cerebrum and cerebellum. The age at onset is in the fourth to six decade of life with duration of seven to 18 years.

Neuropathology. PrP deposition is seen in the neocortex, hippocampus and cerebellum with the latter two also having PrP-amyloid plaques.

The cortical deposits have a round or elongated, "curly" appearance. Neurofibrillary tangles are also seen in the hippocampus. Atrophy and astrogliosis of the subcortical white matter are present. No spongiform degeneration is seen.

4.3.8. *PRNP* F198S-129V[1–4,8,18,20–26,37–38]

Clinical presentation. A gradual loss of short-term memory, clumsiness in walking evolving into ataxia, bradykinesia, rigidity, mild tremor, dysarthria, and cognitive impairment evolving into dementia are the main clinical characteristics. In the early stages of clinical presentation, T2-weighted MRI shows cerebellar atrophy and a reduced signal in the substantia nigra and red nucleus. Signs of cognitive impairment and eye-movement abnormalities may be detected before the onset of clinical symptoms. Psychotic depression has been observed in several patients. The symptoms may progress slowly over five years or rapidly over as little as one year. The age at onset of clinical signs ranges from late in the fourth decade to early in the eighth decade of life. Patients homozygous for valine at codon 129 have clinical signs more than ten years earlier, on average, than heterozygous patients. The duration of the disease ranges from two to twelve years.

Neuropathology. The neuropathologic phenotype is characterized by a severe PrP deposition, in the form of PrP-amyloid plaques and diffuse deposits, and by the presence of numerous neurofibrillary tangles (Figures. 4.1–2, 4.4–5). This PrP deposition is the most severe seen associated with GSS.

Unicentric and multicentric PrP-amyloid plaques and diffuse deposits are distributed in varying degrees throughout most gray structures of the brain (Figures. 4.2, 4.4–5). The core of the amyloid plaque is immunoreactive to antibodies raised against the midregion of PrP, but are

Figure 4.4. Cerebellum of a patient with GSS associated with the *PRNP* F198S mutation. Numerous plaques are seen in the molecular and granule cell layers. (A) thioflavin S method and (B) PrP-immunohistochemistry.

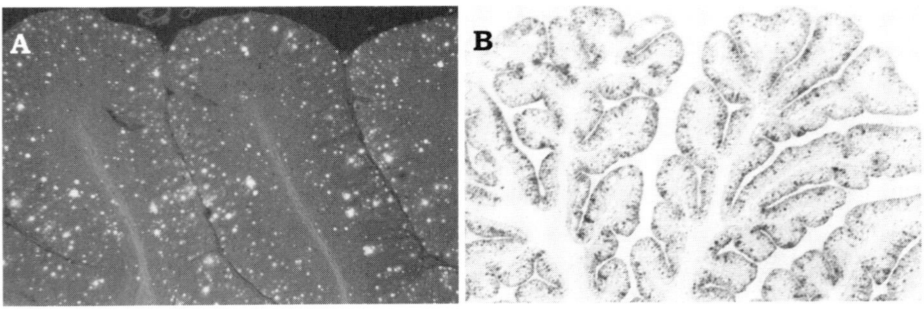

Figure 4.5. Cerebellum of a patient with GSS associated with the *PRNP* F198S mutation. (A) Unicentric and multicentric amyloid deposits and (B) bundles of fibrils radiating out from a central core of an amyloid plaque. (A) PrP-immunohistochemistry and (B) electron microscopy.

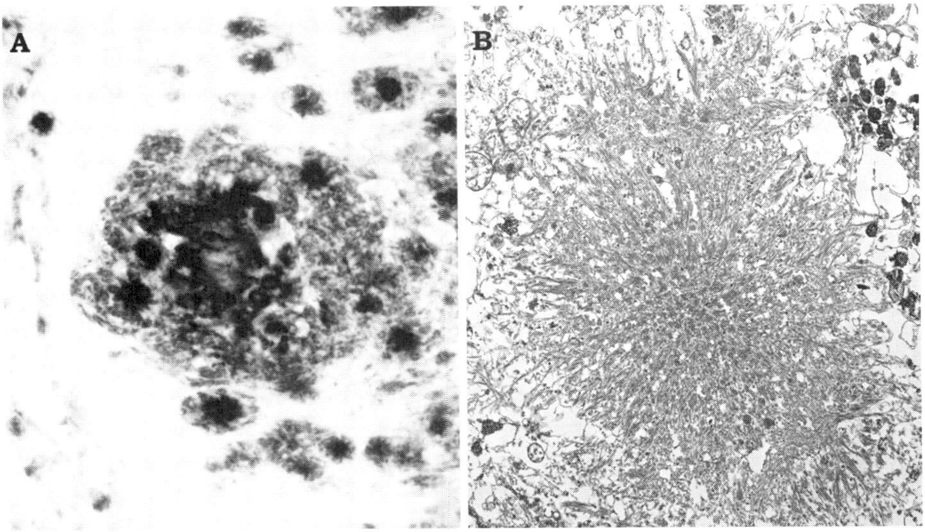

unreactive or weakly reactive to antibodies raised against the amino and carboxy terminal regions of PrP. In contrast, there is immunopositivity to antibodies raised against the amino and carboxy terminal regions of PrP in the area adjacent to the amyloid core. Nonfibrillar PrP deposits appear as diffusely immunolabeled areas in the neuropil. By electron microscopy, the amyloid is composed of bundles of fibrils radiating from a central core; each fibril measures 8–10 nm in diameter.

Amyloid deposition is severe in the frontal, insular, temporal and parietal cortices with the highest concentration of deposits in layers I, IV, V, and VI. In the hippocampus, plaques occur predominantly within the stratum lacunosum-moleculare of the CA1 sector and subiculum. In the cerebellum, amyloid plaques are most numerous (Figure 4.4). Amyloid deposits are surrounded by astrocytes, astrocytic processes and microglial cells. In the neocortex, many amyloid cores are associated with abnormal neurites causing them to appear morphologically similar to neuritic plaques of AD. Neurofibrillary tangles and neuropil threads, which are immunoreactive to antibodies raised against the tau protein, are found in cortical and subcortical grey nuclei as well as in the midbrain and pons. They are most numerous in areas of neocortex that have severe PrP-amyloid deposition. Iron deposition in the globus pallidus, striatum, red nucleus and substantia nigra is seen. Spongiform changes are rarely observed. Some of these neuropathologic characteristics have also been observed in clinically non-symptomatic individuals with this haplotype.

4.3.9. *PRNP* D202N-129V[8,27]

Clinical presentation. Cognitive impairment leading to dementia and cerebellar signs are the main clinical features. The age at onset is early in the eighth decade and the duration is six years.

Neuropathology. Abundant PrP-amyloid deposits are present in the cerebrum and cerebellum and neurofibrillary tangles are seen in the cerebral cortex. No spongiform degeneration is observed.

4.3.10. *PRNP* Q212P-129M[8,27]

Clinical presentation. Gradual development of incoordination and slurring of speech are the presenting signs, followed by dysarthria, and ataxia. The age at onset is late in the sixth decade and the duration is eight years. Dementia has not been reported.

Neuropathology. PrP-amyloid deposition is mild throughout the central nervous system including the cerebellum, which is significantly less affected than that of individuals with any other GSS-associated haplotype. There is degeneration of myelinated fibers in the anterior and lateral corticospinal tracts in the spinal cord. No spongiform degeneration is seen.

4.3.11. *PRNP* Q217R-129V[1–4,8,23]

Clinical presentation. The phenotype is characterized by gradual memory loss, progressive gait disturbances, Parkinsonism and dementia. The neurological signs may be preceded by episodes of mania or depression that respond to antidepressant medications, lithium and neuroleptics. The age at onset of clinical signs varies from the fifth to seventh decade. The duration of the disease is two to six years.

Neuropathology. The neuropathologic phenotype, similar to that associated with the F198S-129V haplotype, is characterized by a severe PrP deposition, in the form of PrP-amyloid plaques and diffuse deposits, and by the presence of numerous neurofibrillary tangles. PrP-amyloid deposits are numerous in the cerebrum and cerebellum. Neurofibrillary tangles are abundant in the cerebral cortex, amygdala, substantia innominata, and thalamus. Lewy bodies may be found in the substantia nigra. No spongiform degeneration is seen.

4.3.12. *PRNP* M232T-129M/V[39–40]

Clinical presentation. Cerebellar signs and spastic paraparesis are the initial symptoms followed by dementia. The age at onset is in the fifth decade and the duration is six years.

Neuropathology. PrP-amyloid plaques and diffuse deposits are seen in the neocortex, subcortical nuclei and cerebellum. It is unclear whether spongiform changes are present.

4.4. Prion protein cerebral amyloid angiopathy

4.4.1. *PRNP* Y145STOP-129M[2,7,8,41]

Clinical presentation. The clinical phenotype is characterized by memory disturbance, disorientation and a progressive dementia. The EEG did not show pseudoperiodic sharp waves. The age at onset is in the fourth decade and the duration is 21 years.

Neuropathology. Diffuse atrophy of the cerebrum, enlargement of the lateral ventricles, neuronal loss and gliosis are severe. PrP-amyloid deposits are present in the walls of small and medium-sized parenchymal and leptomeningeal blood vessels and in the perivascular neuropil. PrP-amyloid fibrils are seen adjacent to and within the vessel wall. Neurofibrillary tangles, neuropil threads and dystrophic neurites are numerous in the cerebral gray matter. These neurofibrillary lesions are immunoreactive to phosphorylation dependent and phosphorylation independent anti-tau antibodies. No spongiform changes are seen.

4.5. Inherited prion disease with variable phenotypes

4.5.1. Ins 192bp-129V and Ins 216bp-129M[3]

Clinical presentation. The phenotypes associated with the insertional mutations are highly variable; however, individuals with eight or nine additional repeats have a GSS-like syndrome characterized by the presence of mental deterioration, cerebellar and extrapyramidal signs. In addition, these individuals often lack the pseudoperiodic slow wave complexes on EEG examination. It appears that the age at onset and duration of the disease are related to the number of inserted repeats. These specific mutations are associated with an age at onset in the third to six decade of life and disease duration of five months to 13 years.

Neuropathology. The majority of individuals with these haplotypes have a phenotype characterized by PrP-amyloid plaques in the molecular layer of the cerebellum and frequently in the cerebral gray matter. In addition, various degrees of spongiosis, gliosis and neuronal loss may be present in the neocortex.

4.6. PrP Characterization in GSS

Studies have shown that the pattern of the pathologic PrP isoforms (PrP^{sc}) in GSS variants is different from that seen in other prion diseases[27,42–45] (Figure 4.6). Patients affected by CJD or FFI present full-length and N-truncated proteinase-K (PK) resistant PrP fragments, while patients affected by GSS present full-length as well as N- and C-terminal truncated, non-glycosylated, and PK-resistant PrP peptides[27,42–44,46]. In GSS, these fragments can be detected in non-enzymatically digested brain homogenates, although they are more prominent after PK digestion. This suggests that they are generated *in vivo* by a GSS-specific proteolytic pathway. In addition, the quantity of fragments present does not correlate with the amyloid burden[27,42–44]. It is of interest that the molecular mass of N- and C-truncated PrP fragments may vary according to the specific *PRNP* mutation present[27,42–44].

The lowest molecular weight N- and C-truncated fragments of PrP in PK-treated brain extracts are 7 kDa (A117V-129V), 8 kDa (P102L-129M, G131V-129M, F198S-129V, D202N-129V, Q217R-129V), or 10 kDa (Q212P-129M)[27,42–44] (Figure 4.6). The 8-kDa peptides associated with P102L-129M and F198S-129V have the major N terminus starting at residues 78–82 and 74 respectively[43–44]. The 7-kDa fragment

Figure 4.6. Western blot patterns of PrP associated with Gerstmann-Sträussler-Scheinker disease variants.

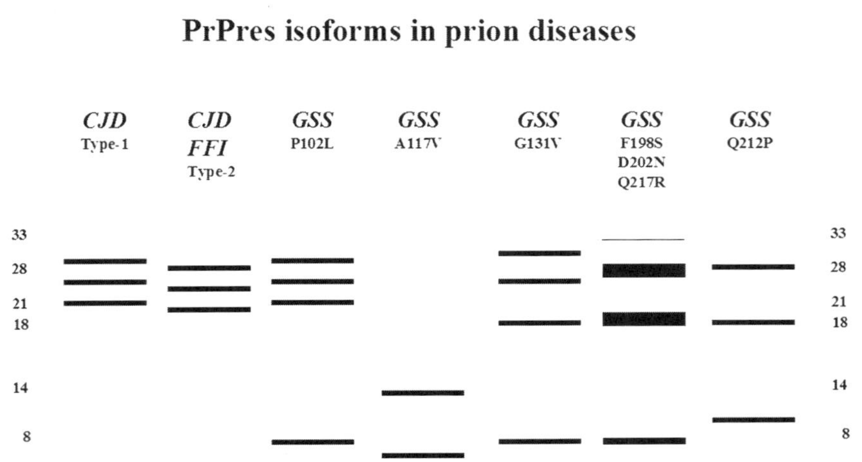

associated with A117V-129V has the major N-terminus cleavage site at residue 90[44]. Studies carried out on GSS A117V have shown presence of PrP^{sc} in some patients but not in others[44–45]. It is important to note that when PrP^{sc} is present in GSS A117V, the amount is significantly smaller than that found in other GSS variants. It has been proposed that a transmembrane PrP isoform plays a central role in the pathogenesis of GSS A117V[45].

The results related to the characteristics of the N and C-truncated fragments coupled with the concept that the pattern of digestion may depend on the tertiary structure of PrP argue that the type of conformational isomers present varies according to the *PRNP* mutation present. Another interesting finding is the presence of 21-30 kDa fragments in individuals with P102L-129M and spongiform degeneration[27,43]. This pattern was originally described in individuals with CJD, which is characterized by the presence of spongiform degeneration, neuronal loss and gliosis[3–6].

Thus, in all GSS variants, PK digestion leads to an increase in the 7–10 kDa fragments, a disappearance of the high molecular weight isoforms of ca. 27–35 kDa and a change of the stoichiometry of the various PrP peptides[27,42,44]. These findings indicate that patients with GSS accumulate PrP isoforms that can be cleaved to smaller, insoluble, PK-resistant fragments. The 7–10-kDa PrP isoforms seen in patients with GSS are detected by antibodies directed to the mid-region of PrP and include residues 109-112. It has been shown that in the human brain, the normal cellular form of PrP (PrP^{c}) is endogenously cleaved at residues 111(H) or 112(M), suggesting that patients with GSS have an alternative metabolic pathway, leading to the accumulation of N- and C-truncated PrP fragments[27,42,44].

Finally, it should be emphasized that in GSS F198S, the 8 kDa peptide contains part of the octarepeat region, but the 7 kDa peptide present in GSS A117V does not[44]. The significance of the accumulation of peptides with different N-termini in GSS is unclear at this time. However, it has been suggested that Cu^{2+} binds to a structure defined by two of the octarepeats containing the sequence PHGGGWGQ and it has been hypothesized that this binding could induce conformational changes in PrP[47]. In addition, short peptides corresponding to the octapeptide repeat motif of PrP have been reported to bind Cu^{2+} ions[48–49]. Moreover, it has been shown that PrP fragments can be transformed from a predominantly α-helical monomeric form to an oligomeric β-sheet-rich secondary structure[50–51]. Therefore, PrP^{sc} fragments with a distinct structure could have various neurotoxic properties or a tendency to form aggregates, providing a possible mechanism underlying the difference in phenotypic presentation among GSS variants.

4.7. Amyloid characterization in GSS and PrP-CAA

4.7.1. GSS

Studies carried out in A117V, F198S, and Q217R have shown that PrP-amyloid filaments are composed of 7 kDa PrP peptides[52–53]. These peptides extend from residue ca 85 to 153 in A117V-129V, 81 to 150 in F198S-129V, and 81 to 146 in Q217R-129V[52–53]. Individuals with the F198S-129V haplotype were also found to have an 11 kDa PrP fibrillogenic peptide spanning residues 58–150[54]. Additional studies have shown that the N- and C-truncated peptides, present in patients with *PRNP* mutations A117V, F198S and Q217R, originate only from the mutant allele, suggesting that these mutations are a dominant factor for amyloidogenesis[52–53].

The allelic origin of PrP accumulating in the brain of patients with other inherited prion diseases, including CJD and FFI, has been investigated. These studies showed that only mutant PrP is detected in FFI and CJD D178N, whereas both wild-type and mutant PrP are present in CJD V210I[55–57]. Therefore, the recruitment of wild-type PrP isoforms by mutant PrP may depend on mutation-specific PrP configurations that allow the interaction between the mutant and normal PrP species.

In conclusion, data obtained from the study of purified amyloid fractions indicate that in GSS A117V, GSS F198S and GSS Q217R the minimal PrP segment essential for amyloidogenesis is a N- and C-truncated fragment spanning residues ca. 88 to 146[53–54]. The major part of this peptide corresponds to the end of the flexible N-terminal domain of PrP^c, which has been thought to undergo conformational changes in the transition of PrP^c into disease-specific species[58–59]. It is of interest that in PrP-CAA, the C-terminus of the genetically truncated PrP is similar to the C-terminus of the amyloid peptides found in GSS[7]. The plasticity of the minimal amyloid fragment and its tendency to adopt β-sheet secondary structure has been supported by studies done using synthetic peptides[60].

4.7.2. PrP-CAA

By immunoblot analysis of proteins extracted from purified amyloid fractions, it appears that the smallest amyloid subunit migrates as a band of ca. 7.5 kDa[7]. This band is strongly immunoreactive with antibodies to epitopes located within residues 90–147 of PrP and is nonimmunoreactive with antisera to N- and C-termini[7]. In addition, amyloid fractions contain higher molecular weight PrP peptides migrating as poorly resolved bands of 12–16 and 22–30 kDa[7]. These bands are immunoreactive to the same antibodies that recognize the 7.5 kDa band, suggesting that they represent primarily polymers of amyloid protein[7].

4.8. *In vitro* studies with synthetic amyloid peptides

To investigate the physicochemical properties and kinetics of assembly of the PrP-amyloid protein that constitutes the plaques in GSS (GSS-amyloid protein), a peptide homologous to residues 82-146 of human PrP (PrP 82-146_{wt}: GQPHGGGWGQGGGTHSQWN-KPSKPKTNMKHMAGAAAAGAVVGGLGGYMLGSAMSRPIIHFGSDYE) and scrambled sequences thereof were synthesized. In particular, besides an entirely scrambled sequence (PrP 82-146_{scr}: EADQFALG-GSKHGNGMQQVAGHGGSMGAKAWGANGHPSGTGIPTAKMVPYKI-YGGGWAG MGRPSS), partially scrambled peptides with selective changes in the 106-126 or the 127-146 region (PrP 82-$146_{106-126scr}$: GQPHGGGWGQGGGTHSQWNKPSKPNGAKALMGGHGATKVVGA-AAGYMLGSAMSRPIIHFGSDYE; PrP 82-$146_{127-146scr}$: GQPHGGG-WGQGGGTHSQWNKPSKPKTNMKHMAGAAAAGAVVGGLGSMYPA-SHGLMEDFYGIGSIR) were generated. The regions to be modified were selected based on previous studies showing that a peptide spanning residues 106-126 (PrP 106-126) adopted a stable β-sheet secondary structure and formed amyloid fibrils similar to those observed in GSS[60–64]. Furthermore, a peptide corresponding to residues 127-147 (PrP 127-147) also formed filamentous structures resembling scrapie-associated fibrils; however, this peptide possessed a lower amyloidogenic potential than PrP 106-126. Shorter fragments spanning residues 127-135 or 135-147 were non-fibrillogenic[65]. It is worth noting that PrP 106-126 was highly toxic to primary neuronal cultures; this is at variance with PrP 127-147[65].

Using these peptides, it was possible to investigate the role of various regions of PrP in the physicochemical and fibrillogenic properties of the GSS-amyloid protein. These properties were assessed using a variety of methods including electron microscopy, X-ray diffraction and Fourier transform infrared spectroscopy (FTIR)[64].

4.8.1. Solid state features of PrP 82-146_{wt}: from protofilaments to amyloid fibrils

The aggregation of PrP 82-146_{wt} was analyzed at various time points. As part of this analysis, electron microscopy showed two distinct types of fibrils. The first type of fibrils (protofilaments) were long, straight, unbranched, and 5.5-nm in diameter and had a high propensity to adhere

Figure 4.7. Electron micrographs of aggregates generated by PrP 82-146$_{wt}$ as revealed by negative staining of peptide suspensions. A, B and C correspond to 1, 24 and 72 h incubation, respectively. The sample contained essentially 5.5 nm protofilaments after 1 h, a mixed population of fibrillary structures after 24 h and only 9.8 nm fibrils after 72 h.

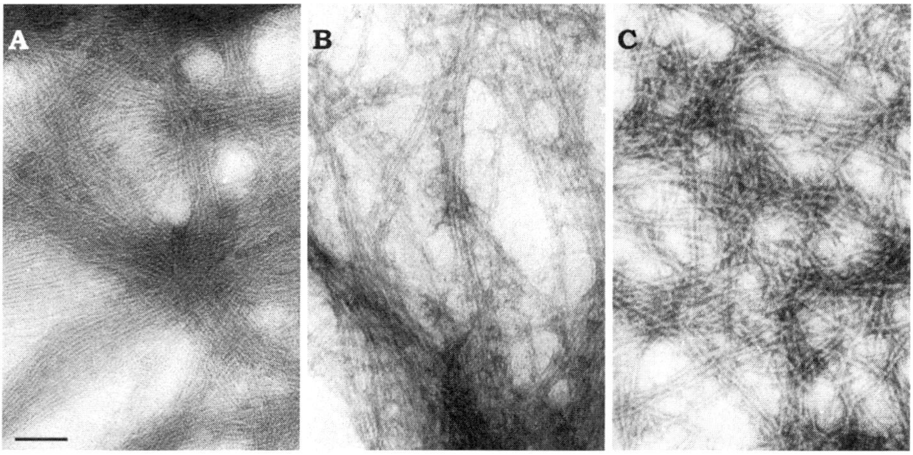

to each other often organizing into bundles of various sizes. The second type were 9.8-nm in diameter forming meshworks or star-like structures. In the early phases of aggregation, fibrils of the first type were the most numerous; however, their quantity decreased over time. On the contrary, fibrils of the second type were scarce in the early phases and increased over time. After 72 h, only fibrils of the second type were detectable (Figure 4.7, Table 4.2).

These findings suggest that PrP 82-146$_{wt}$ aggregation is a stepwise process in which protofilaments transform into mature amyloid fibrils. Macromolecular assemblies of PrP 82-146$_{wt}$ were analyzed by X-ray diffraction under three different conditions: lyophilized (L), vapour hydrated (VH), and solubilized/dried (S/D) (Table 4.2). The samples, after the S/D treatment, gave an oriented diffraction pattern with a sharp and strong reflection at Bragg spacing 0.477 nm on the meridian, while after L and VH treatments the samples showed a circular reflection at ~0.47 nm[66–67]. The spacing at ~0.47 nm corresponds to the hydrogen–bonding distance between parallel or antiparallel β chains. The sharpness of this reflection indicated that PrP 82-146$_{wt}$ form a periodic array of β chains that run normal to the long axis of the elongated assembly. An oriented sample of PrP 82-146$_{wt}$, after solubilization and drying, showed a low–angle reflection at 5.9 nm spacing corresponding to 4.8 nm for a solid cylinder and 3.5 nm for a tubular cylinder corresponding to a fibril diameter of ~7–10 nm. This observation is in agreement with the 9.8 nm-width fibrils observed during the electron microscopic analysis.

Table 4.2. Structural features of PrP 82-146$_{wt}$

	Time of incubation (h)		Fibril diameter (nm)	
Electron Microscopy	1		5.5 ± 1.8 9.5 ±1.1	
	24		5.6 ± 1.8 9.6 ± 1.2	
	72		9.7 ± 1.1	
	168		9.8 ± 1.1	
X-ray diffraction	Condition[a]	L	VH	S/D
	Spacing[b]	0.484 C	0.477 C	0.477 M
	Forward scatter[c]	+	+	+5.9
	Exposure (h)	43	40	70
FTIR	α-Helix	β-sheet	Turn	Random coil
	22 ± 5%	54 ± 7%	24 ± 4%	≤2%

[a]L, lyophilized; *VH*, vapor hydrated; *S/D*, solubilized and dried.
[b]Bragg spacing in nm; *C*; circular reflection, *M*; meridional reflection.
[c]Forward scatter refers to the central scattering observed near the beam stop, likely arising from the structure factor of the macromolecular assembly.

FTIR spectra of PrP 82-146$_{wt}$ assemblies were characterized by a narrow peak at 1630 cm^{-1} (Amide I region), a peak at 1535 cm^{-1} (Amide II region) and a band at 3400 cm^{-1} (Figure 4.8: peaks 1, 2, and 3, respectively), a profile indicating the presence of an extended β-sheet structure[68]. The absence of a peak at 1675–1695 cm^{-1} suggested a parallel β-sheet organization of fibrils[68]. The Amide I infrared band was quantitatively resolved to identify the different secondary structures. The analysis showed that PrP 82-146$_{wt}$ aggregates are primarily composed of β-sheet (54%) and turn (24%) which is consistent with their amyloid-like properties (Figure 4.8 and Table 4.2).

4.8.2. The integrity of the C-terminal region of PrP 82-146$_{wt}$ is central to secondary structure and aggregation properties

To investigate the determinants of the physicochemical properties of the GSS-amyloid protein, comparative studies were carried out with PrP 82-146$_{wt}$ and partially scrambled peptides PrP 82-146$_{106-126scr}$ and PrP 82-146$_{127-146scr}$. The ability to form macro aggregates was first

Figure 4.8. FTIR spectra of PrP 82-146$_{wt}$ aggregates. Band 1 (Amide I region) is representative of symmetric carbonyl stretch, Band 2 (Amide II region) of N-H bond deformation, and Band 3 of N-H symmetric stretch.

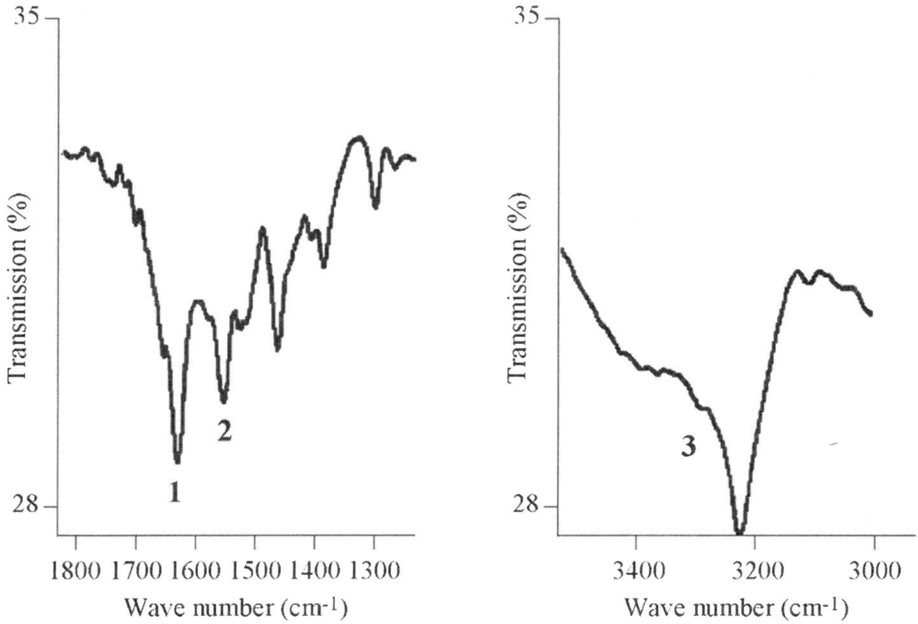

analyzed by sedimentation experiments. Peptide solutions were incubated at 37°C for various times ranging from 0 to 96 h, and then were centrifuged. The supernatants were analyzed by HPLC to determine the percentage of peptide still in solution after centrifugation. The study showed that the kinetics of aggregation of PrP 82-146$_{wt}$ was linear during the first 3 days yielding 25%, 40% and 60% of sedimentable peptide after 24, 48 and 72 h, respectively. Thereafter, the aggregation rate increased and more than 90% of peptide was found in the pellet after 96 h. The modification of the amino acid sequence 106-126 or 127-146 resulted in remarkable changes in the aggregation kinetics, although with opposite effects (i.e. PrP 82-146$_{106-126scr}$ had an increased aggregation rate and PrP 82-146$_{127-146scr}$ had a decreased aggregation rate). Almost 60% of PrP 82-146$_{106-126scr}$ was sedimentable after 24 h and more than 85% after 96 h. By contrast, the aggregation ability of PrP 82-146$_{127-146scr}$ was considerably reduced, paralleling that of the totally scrambled peptide (Figure 4.9).

To assess the degree of protease resistance, pre-aggregated peptides were subjected to proteinase K (PK) digestion at 37°C for 30 min; the samples were then centrifuged and pellets were dissolved in formic acid and analyzed by HPLC. The extent of proteolysis was calculated

Figure 4.9. Time-course of the aggregation of PrP 82-146$_{wt}$ and analogues. Aliquots of 0.5 mM peptides in 20 mM Tris-HCl were incubated at 37°C for 24, 48, 72 and 96 h and then centrifuged at 13,000 × g for 10 min at 4°C. Supernatants were analyzed by reverse-phase HPLC and peptide concentrations at different times were expressed as percentages of the corresponding values determined at zero time. The values corresponding to each time point are the mean ± SD of 5 experiments.

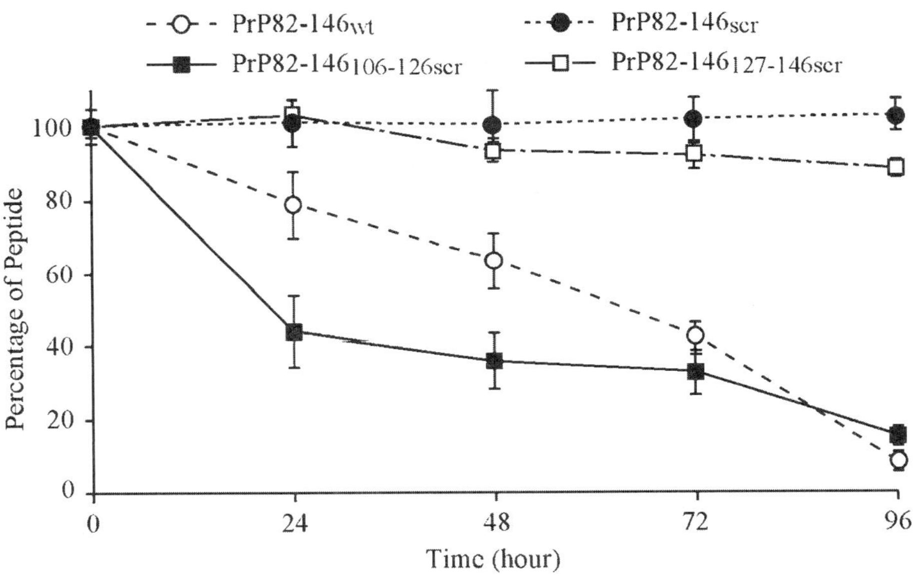

as percentage of peptide present in the pellet compared to undigested controls. PrP 82-146$_{wt}$ was partially resistant to PK (47 ± 3% as compared to control values) at variance with PrP 82-146$_{scr}$ that was almost completely degraded. The modification of the amino acid sequence 106-126 or 127-146 resulted in significant changes in protease resistance, although with opposite effects as observed for the aggregation properties. In fact, PrP 82-146$_{127-146scr}$ was largely sensitive to PK digestion, as only 15 ± 2% of peptide was detected in the protease resistant fraction. Conversely, the peptide PrP 82-146$_{106-126scr}$ showed an unexpectedly high PK resistance (78 ± 8%).

The ultrastructure and staining properties of aggregates generated by PrP 82-146$_{wt}$ and analogues thereof were analyzed by electron microscopy and polarized light microscopy after Congo red staining at various incubation times ranging from 1 to 168 h. The electron microscopic study was carried out on both negatively-stained peptide suspensions and positively-stained, resin-embedded pellets. While PrP 82-146$_{wt}$ readily formed amyloid fibrils as reported above (Figure 4.10 A), the modification of the 106-126 or 127-146 regions had profound qualitative and quantitative effects on fibrillogenesis. In early phases,

Figure 4.10. Electron micrographs of the aggregates generated by PrP 82-146$_{wt}$ and analogues, as revealed by positive staining of ultrathin sections of the pellets after 7-day incubation. A, PrP 82-146$_{wt}$; B, PrP 82-146$_{106-126scr}$; C, PrP 82-146$_{127-146scr}$.

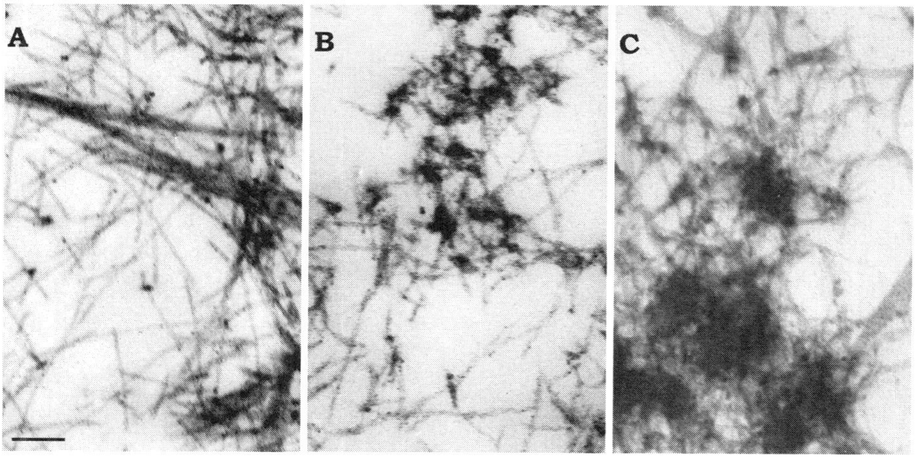

PrP 82-146$_{106-126scr}$ generated primarily amorphous aggregates that were associated with a few, relatively short, irregular fibrils with average diameter of 7.3 nm. With time, the density of fibrillar assemblies progressively increased, and more regular, ~7.7 nm-diameter fibrils were detected. These fibrils were often paired or organized into bundles or loose meshworks that were birefringent after Congo red staining and were associated with a substantial amount of amorphous aggregates up to the end-point of the experiment (Figure 4.10 B). PrP 82-146$_{127-146scr}$ did not generate amyloid fibrils even after long incubation times. Peptide preparations contained primarily electron dense amorphous material intermingled with a relatively small number of short filamentous structures that were usually assembled into twisted bundles without the ultrastructural features and staining properties of amyloid (Figure 4.10 C).

In conclusion, the physicochemical characteristics of the synthetic peptide PrP 82-146$_{wt}$ (Table 4.3) account for the massive deposition of PrP amyloid that occurs in GSS. The analysis of the aggregation process revealed a complex fibril assembly of PrP 82-146$_{wt}$, suggesting a dynamic process that includes the generation of protofilaments in early stages and later the formation of fibrils. Both 106-126 and 127-146 sequences are important determinants of the physicochemical characteristics of PrP 82-146$_{wt}$; however in the intact peptide, the C-terminal segment 127-146 plays a crucial role for the secondary structure and aggregation properties.

Table 4.3. Features of PrP 82-146$_{wt}$ and analogues

	PrP 82-146$_{wt}$	PrP 82-146$_{106-126scr}$	PrP 82-146$_{127-146scr}$
β-sheet content (%)	54	52	Not determined
X-ray diffraction patterns	β	β	Not determined
PK-resistant fraction (%)	47	78	15
Peptide assemblies (168h)	+++	++	±
Amyloid fibrils Amorphous aggregates	−	++	+++

4.9. Acknowledgements

Supported in part by US PHS grant P30 AG 10133 and EU grants QLRT 2001-00283, QLK2 2002-81523, NeuroPrion FOOD-CT-2004-506579 "HeteroPrion project". The authors thank Bradley S. Glazier for his editorial assistance. No official endorsement of this article by the FDA is intended or should be inferred.

4.10. References

1. B. Ghetti, P. Piccardo, B. Frangione, O. Bugiani, G. Giaccone, K. Young, F. Prelli, M. R. Farlow, S. R. Dlouhy, and F. Tagliavini, Prion protein amyloidosis. *Brain Pathol.* **6**, 127–145 (1996).

2. B. Ghetti, P. Piccardo, B. Frangione, O. Bugiani, G. Giaccone, K. Young, F. Prelli, M. R. Farlow, S. R. Dlouhy, and F. Tagliavini, Prion protein hereditary amyloidosis: Parenchymal and vascular. *Seminars Virol.* **7**, 189–200 (1996).

3. Q. Kong, W. K. Surewicz, R. B. Petersen, W. Zou, S. G. Chen, P. Gambetti, P. Parchi, S. Capellari, L. G. Goldfarb, P. Montagna, E. Lugaresi, P. Piccardo, and B. Ghetti, Inherited Prion Diseases, in *Prion Biology and Diseases,* edited by S. B. Prusiner 2004), pp. 673–776.

4. B. Ghetti and P. Gambetti, Human Prion Diseases, in *Advances in Cell Aging and Gerontology,* (JAI Press Inc., 1999), pp. 135–187.

5. J. Collinge, Prion diseases of humans and animals: Their causes and molecular basis. *Ann. Rev. Neurosci.* **24**, 519–550 (2001).

6. S. B. Prusiner, An introduction to prion biology and diseases, in *Prion Biology and Disease,* edited by S. B. Prusiner (Cold Spring Harbor Laboratory Press, New York, 2004), pp. 1–88.

7. B. Ghetti, P. Piccardo, M. G. Spillantini, Y. Ichimiya, M. Porro, F. Perini, T. Kitamoto, J. Tateishi, C. Seiler, B. Frangione, O. Bugiani , G. Giaccone, F. Prelli, M. Goedert, S. R. Dlouhy, and F. Tagliavini, Vascular variant of prion protein cerebral amyloidosis with τ-positive neurofibrillary tangles: The phenotype of the stop codon 145 mutation in *PRNP*. *Proc. Natl. Acad. Sci. USA* **93**, 744–748 (1996).

8. B. Ghetti, O. Bugiani, F. Tagliavini, and P. Piccardo, Gerstmann-Sträussler-Scheinker Disease, in *Neurodegeneration: The Molecular Pathology of Dementia and Movement Disorders,* edited by D. W. Dickson (ISN Neuropath Press, Basel, 2003), pp. 318–325.

9. H. Furukawa, T. Kitamoto, Y. Tanaka, and J. Tateishi, New variant prion protein in a Japanese family with Gerstmann-Sträussler syndrome. *Mol. Brain Res.* **30**, 385–388 (1995).

10. L. Dimitz, Bericht des Vereines für Psychiatrie and Neurologie in Wien. (Vereinsjahr 1912/1913). Sitzung vom 11. Juni 1912. *Jahrb Psychiat Neurol* **34**, 384– (1913).

11. J. Gerstmann, Über ein noch nicht beschriebenes Reflexphanomen bei einer Erkrankung des zerebellaren Systems. *Wien Med Wochensch* **78**, 906–908 (1928).

12. J. Gerstmann, E. Sträussler, and I. Scheinker, Über eine eigenartige hereditär-familiäre Erkrankung des Zentralnervensystems. Zugleich ein Beitrag zur Frage des vorzeitigen lokalen Alterns. *Z Gesante Neurol Psychiat* **154**, 736–762 (1936).

13. A. von Braunmühl, Über eine eigenartige hereditär-familiäre Erkrankung des Zentralnervensystems. *Archiv Psychiatr Nervenkr* **191**, 419–449 (1954).

14. F. Seitelberger, Eigenartige familiär-hereditäre Krankheit des Zentralennervensystems in einer niederösterreichischen Sippe. *Wiener Klinische Wochenschrift* **74**, 687–691 (1962).

15. F. Seitelberger, Sträussler's disease. *Acta Neuropathol.* **7**, 341–343 (1981).

16. H. A. Kretzschmar, G. Honold, F. Seitelberger, M. Feucht, P. Wessely, P. Mehraein, and H. Budka, Prion protein mutation in family first reported by Gerstmann, Sträussler, and Scheinker. *Lancet* **337**, 1160 (1991).

17. J. A. Hainfellner, S. Brantner-Inthaler, L. Cervena kova, P. Brown, T. Kitamoto, J. Tateishi, H. Diringer, P. P. Liberski , H. Regele, M. Feucht, N. Mayr, P. Wessely, K. Summer, F. Seitelberger, and H. Budka, The original Gerstmann-Sträussler-Scheinker family of Austria: Divergent clinicopathological phenotypes but constant PrP genotype. *Brain Pathol.* **5**, 201–211 (1995).

18. B. Ghetti, F. Tagliavini, C. L. Masters, K. Beyreuther, G. Giaccone, L. Verga, M. R. Farlow, P. M. Conneally, S. R. Dlouhy, B. Azzarelli, and O. Bugiani, Gerstmann-Sträussler-Scheinker disease. II. Neurofibrillary tangles and plaques with PrP-amyloid coexist in an affected family. *Neurology* **39**, 1453–1461 (1989).

19. C. L. Masters, D. C. Gajdusek, and C. J. Gibbs, Jr., Creutzfeldt-Jakob disease virus isolations from the Gerstmann-Sträussler syndrome with an analysis of the various forms of amyloid plaque deposition in the virus-induced spongiform encephalopathies. *Brain* **104**, 559–588 (1981).

20. B. Ghetti, M. Farlow, P. M. Conneally, B. Azzarelli, C. Masters, G. Giaccone, F. Tagliavini, and O. Bugiani, Amyloid plaques and neurofibrillary tangles of Gerstmann-Straussler disease. *Neurology* **38 (Suppl 1)**, 266–(1988).

21. M. R. Farlow, R. D. Yee, S. R. Dlouhy, P. M. Conneally, B. Azzarelli, and B. Ghetti, Gerstmann-Sträussler-Scheinker disease. I. Extending the clinical spectrum. *Neurology* **39**, 1446–1452 (1989).

22. B. Ghetti, S. R. Dlouhy, G. Giaccone, O. Bugiani, B. Frangione, M. R. Farlow, and F. Tagliavini, Gerstmann-Sträussler-Scheinker disease and the Indiana kindred. *Brain Pathol.* **5**, 61–75 (1995).

23. B. Ghetti, F. Tagliavini, G. Giaccone, O. Bugiani, B. Frangione, M. R. Farlow, and S. R. Dlouhy, Familial Gerstmann-Sträussler-Scheinker disease with neurofibrillary tangles. *Mol. Neurobiol.* **8**, 41–48 (1994).

24. O. Bugiani, G. Giaccone, L. Verga, B. Pollo, B. Frangione, M. Farlow, F. Tagliavini, and B. Ghetti, βPP participates in PrP-amyloid plaques of Gerstmann-Sträussler-Scheinker disease, Indiana kindred. *J Neuropathol Experimen Neurol* **52**, 64–70 (1993).

25. G. Giaccone, L. Verga, O. Bugiani, B. Frangione, D. Serban, S. B. Prusiner, M. R. Farlow, B. Ghetti, and F. Tagliavini, Prion protein preamyloid and amyloid deposits in the brains of patients with Gerstmann-Sträussler-Scheinker disease, Indiana kindred.*Proc. Natl. Acad. Sci. USA* **89**, 9349–9353 (1992).

26. F. Tagliavini, G. Giaccone, F. Prelli, L. Verga, M. Porro, J. Trojanowski, M. Farlow, B. Frangione, B. Ghetti, and O. Bugiani, A68 is a component of paired helical filaments of Gerstmann-Sträussler-Scheinker disease, Indiana kindred. *Brain Res.* **616**, 325–328 (1993).

27. P. Piccardo, S. R. Dlouhy, P. M. J. Lievens, K. Young, T. D. Bird, D. Nochlin, D. W. Dickson, H. V. Vinters, T. R. Zimmerman, I. R. A. Mackenzie, S. J. Kish, L. C. Ang, C. De Carli, M. Pocchiari, P. Brown, C. J. Gibbs Jr, D. C. Gajdusek, O. Bugiani, J. Ironside, F. Tagliavini, and B. Ghetti, Phenotypic variability of Gerstmann-Sträussler-Scheinker disease is associated with prion protein heterogeneity. *J. Neuropathol. Experiment. Neurol.* **57**, 979–988 (1998).

28. P. Piccardo, B. Ghetti, D. W. Dickson, H. V. Vinters, G. Giaccone, O. Bugiani, F. Tagliavini, K. Young, S. R. Dlouhy, C. Seiler, C. K. Jones, A. Lazzarini, L. I. Golbe, T. R. Zimmerman, S. L. Perlman, D. C. McLachlan, P. H. George-Hyslop, and A. Lennox, Gerstmann-Sträussler-Scheinker disease (PRNP P102L): Amyloid deposits are best recognized by antibodies directed to epitopes in PrP region 90-165. *J. Neuropathol. Experiment. Neurol.* **54**, 790–801 (1995).

29. K. Young, H. B. Clark, P. Piccardo, S. R. Dlouhy, and B. Ghetti, Gerstmann-Sträussler-Scheinker disease with the *PRNP* P102L mutation and valine at codon 129. *Mol. Brain Res.* **44**, 147–150 (1997).

30. T. Kitamoto, N. Amano, Y. Terao, Y. Nakazato, T. Isshiki, T. Mizutani, and J. Tateishi, A new inherited prion disease (PrP-P105L mutation) showing spastic paraparesis. *Ann. Neurol.* **34**, 808–813 (1993).

31. M. Yamada, Y. Itoh, A. Inaba, Y. Wada, M. Takashima, S. Satoh, T. Kamata , R. Okeda, T. Kayano, N. Suematsu, T. Kitamoto, E. Otomo, M. Matsushita, and H. Mizusawa, An inherited prion disease with a PrP P105L mutation—Clinicopathologic and PrP heterogeneity. *Neurology* **53**, 181–188 (1999).

32. G. R. Mallucci, T. A. Campbell, A. Dickinson, J. Beck, M. Holt, G. Plant, K. W. dePauw, R. N. Hakin, C. E. Clarke, S. Howell, G. A. B. Davies-Jones, M. Lawden, C. M. Smith, P. Ince, J. W. Ironside, L. R. Bridges, A. Dean, I. Weeks, and J. Collinge, Inherited prion disease with an alanine to valine mutation at codon 117 in the prion protein gene. *Brain* **122**, 1823–1837 (1999).

33. D. Nochlin, S. M. Sumi, T. D. Bird, A. D. Snow, C. M. Leventhal, K. Beyreuther, and C. L. Masters, Familial dementia with PrP-positive amyloid plaques: A variant of Gerstmann-Sträussler syndrome. *Neurology* **39**, 910–918 (1989).

34. C. Tranchant, K. Doh-ura, J. M. Warter, G. Steinmetz, Y. Chevalier, A. Hanauer, T. Kitamoto, and J. Tateishi, Gerstmann-Sträussler-Scheinker disease in an Alsatian family—Clinical and genetic studies. *J. Neurol. Neurosurg. Psychiat.* **55**, 185–187 (1992).

35. P. K. Panegyres, K. Toufexis, B. A. Kakulas, L. Cernevakova, P. Brown, B. Ghetti, P. Piccardo, and S. R. Dlouhy, A new *PRNP* mutation (G131V) associated with Gerstmann-Sträussler-Scheinker disease. *Arch. Neurol.* **58**, 1899–1902 (2001).

36. C. M. Butefisch, P. Gambetti, L. Cervenakova, K. Y. Park, M. Hallett, and L. G. Goldfarb, Inherited prion encephalopathy associated with the novel *PRNP* H187R mutation—A clinical study. *Neurology* **55**, 517–522 (2000).

37. S. R. Dlouhy, K. Hsiao, M. R. Farlow, T. Foroud, P. M. Conneally, P. Johnson, S. B. Prusiner, M. E. Hodes, and B. Ghetti, Linkage of the Indiana kindred of Gerstmann-Sträussler-Scheinker disease to the prion protein gene. *Nat Genet.* **1**, 64–67 (1992).

38. K. Hsiao, S. R. Dlouhy, M. R. Farlow, C. Cass, M. Da Costa, P. M. Conneally, M. E. Hodes, B. Ghetti, and S. B. Prusiner, Mutant prion protein in Gerstmann-Sträussler-Scheinker disease with neurofibrillary tangles. *Nat Genet.* **1**, 68–71 (1992).

39. P. P. Liberski, M. Barcikowska, L. Cervenakova, J. Bratosiewicz, M. Marczewska, P. Brown, and D. C. Gajdusek, A case of sporadic Creutzfeldt-Jakob disease with a Gerstmann- Sträussler-Scheinker phenotype but no alterations in the *PRNP* gene. *Acta Neuropathol.* **96**, 425–430 (1998).

40. P. P. Liberski, J. Bratosiewicz, M. Barcikowska, L. Cervenakova, M. Marczewska, P. Brown, and D. C. Gajdusek, A case of sporadic Creutzfeldt-Jakob disease with a Gerstmann- Sträussler-Scheinker phenotype but no alterations in the *PRNP* gene. *Acta Neuropathol.* **100**, 233–234 (2000).

41. T. Kitamoto, R. Iizuka, and J. Tateishi, An amber mutation of prion protein in Gerstmann-Sträussler syndrome with mutant PrP plaques. *Biochem. Biophys. Res.Comm.* **192**, 525–531 (1993).

42. P. Piccardo, C. Seiler, S. R. Dlouhy, K. Young, M. R. Farlow, F. Prelli, B. Frangione, O. Bugiani, F. Tagliavini, and B. Ghetti, Proteinase-k-resistant prion protein isoforms in Gerstmann-Sträussler-Scheinker disease (Indiana kindred). *J Neuropathol Experiment Neurol* **55**, 1157–1163 (1996).

43. P. Parchi, S. G. Chen, P. Brown, W. Q. Zou, S. Capellari, H. Budka, J. Hainfellner, P. F. Reyes, G. T. Golden, J. J. Hauw, D. C. Gajdusek, and P. Gambetti, Different patterns of truncated prion protein fragments correlate with distinct phenotypes in P102L Gerstmann-Sträussler-Scheinker disease. *Proc. Natl. Acad. Sci. USA* **95**, 8322–8327 (1998).

44. P. Piccardo, J. J. Liepnieks, A. William, S. R. Dlouhy, M. R. Farlow, K. Young, D. Nochlin, T. D. Bird, R. R. Nixon, M. J. Ball, C. DeCarli, O. Bugiani, F. Tagliavini, M. D. Benson, and B. Ghetti , Prion proteins with different conformations accumulate in Gerstmann-Sträussler-Scheinker disease caused by A117V and F198S mutations. *Am. J. Pathol.* **158**, 2201–2207 (2001).

45. R. S. Hegde, J. A. Mastrianni, M. R. Scott, K. A. Defea, P. Tremblay, M. Torchia, S. J. DeArmond, S. B. Prusiner, and V. R. Lingappa, A transmembrane form of the prion protein in neurodegenerative disease. *Science* **279**, 827–834 (1998).

46. S. G. Chen, D. B. Teplow, P. Parchi, J. K. Teller, P. Gambetti, and L. Autilio-Gambetti, Truncated forms of the human prion protein in normal brain and in prion diseases. *J. Biol. Chem.* **270**, 19173–19180 (1995).

47. J. Stöckel, J. Safar, A. C. Wallace, F. E. Cohen, and S. B. Prusiner, Prion protein selectively binds copper(II) ions. *Biochemistry* **37**, 7185–7193 (1998).

48. M. P. Hornshaw, J. R. Mcdermott, J. M. Candy, and J. H. Lakey, Copper binding to the n-terminal tandem repeat region of mammalian and avian prion protein: structural studies using synthetic peptides. *Biochem. Biophys. Res.Comm.* **214**, 993–999 (1995).

49. T. Miura, A. Hori-i, and H. Takeuchi, Metal-dependent alpha-helix formation promoted by the glycine-rich octapeptide region of prion protein. *FEBS Let.s* **396**, 248–252 (1996).

50. H. Zhang, K. Kaneko, J. T. Nguyen, T. L. Livshits, M. A. Baldwin, F. E. Cohen, T. L. James, and S. B. Prusiner, Conformational transitions in peptides containing two putative alpha-helices of the prion protein. *J. Mol. Biol.* **250**, 514–526 (1995).

51. H. Zhang, J. Stöckel, I. Mehlhorn, D. Groth, M. A. Baldwin, S. B. Prusiner, T. L. James, and F. E. Cohen, Physical studies of conformational plasticity in a recombinant prion protein. *Biochemistry* **36**, 3543–3553 (1997).

52. F. Tagliavini, F. Prelli, M. Porro, G. Rossi, G. Giaccone, M. R. Farlow, S. R. Dlouhy, B. Ghetti, O. Bugiani, and B. Frangione, Amyloid fibrils in Gerstmann-Sträussler-Scheinker disease (Indiana and Swedish kindreds) express only PrP peptides encoded by the mutant allele. *Cell* **79**, 695–703 (1994).

53. Tagliavini, P. M. J. Lievens, C. Tranchant, J. M. Warter, M. Mohr, G. Giaccone, F. Perini, G. Rossi, M. Salmona, P. Piccardo, B. Ghetti, R. C. Beavis, O. Bugiani, B. Frangione, and F. Prelli, A 7-kDa prion protein (PrP) fragment, an integral component of the PrP region required for infectivity, is the major amyloid protein in Gerstmann-Sträussler-Scheinker disease A117V. *J. Biol. Chem.* **276**, 6009–6015 (2001).

54. F. Tagliavini, F. Prelli, J. Ghiso, O. Bugiani, D. Serban, S. B. Prusiner, M. R. Farlow, B. Ghetti, and B. Frangione, Amyloid protein of Gerstmann-Sträussler-Scheinker disease (Indiana kindred) is an 11 kd fragment of prion protein with an N-terminal glycine at codon 58. *EMBO J.l* **10**, 513–519 (1991).

55. R. Gabizon, G. Telling, Z. Meiner, H. Halimi, I. Kahana, and S. B. Prusiner, Insoluble wild-type and protease-resistant mutant prion protein in brains of patients with inherited prion disease. *Nat Med.* **2**, 59–64 (1996).

56. S. G. Chen, P. Parchi, P. Brown, S. Capellari, W. Zou, E. J. Cochran, C. L. Vnencak-Jones, J. Julien, C. Vital, J. Mikol, E. Lugaresi, L. Autilio-Gambetti, and P. Gambetti, Allelic origin of the abnormal prion protein isoform in familial prion diseases. *Nat Med.* **3**, 1009–1015 (1997).

57. M. C. Silvestrini, F. Cardone, B. Maras, P. Pucci, D. Barra, M. Brunori, and M. Pocchiari, Identification of the prion protein allotypes which accumulate in the brain of sporadic and familial Creutzfeldt-Jakob disease patients. *Nat Med.* **3**, 521–525 (1997).

58. T. L. James, H. Liu, N. B. Ulyanov, S. Farr-Jones, H. Zhang, D. G. Donne, K. Kaneko, D. Groth, I. Mehlhorn, S. B. Prusiner, and F. E. Cohen, Solution structure of a 142-residue recombinant prion protein corresponding to the infectious fragment of the scrapie isoform. *Proc. Natl. Acad. Sci. USA* **94**, 10086–10091 (1997).

59. R. Riek, S. Hornemann, G. Wider, R. Golckshuber, and K. Wuthrich, NMR characterization of the full-length recombinant murine prion protein, mPrP(23-231). *FEBS Lett.* **413**, 282–288 (1997).

60. F. Tagliavini, F. Prelli, L. Verga, G. Giacconi, R. Sarma, P. Gorevic, B. Ghetti, F. Passerini, E. Ghibaudi, G. Forloni, M. Salmona, O. Bugiani, and B. Frangione, Synthetic peptides homologous to prion protein residues 106-147 form amyloid-like fibrils *in vitro*. *Proc. Natl. Acad. Sci. USA* **90**, 9678–9682 (1993).

61. C. Selvaggini, L. De Gioria , L. Cantu, E. Ghibaudi, L. Diomede, F. Passerini, G. Forloni, O. Bugiani, F. Tagliavini, and M. Salmona, Molecular characteristics of a

protease-resistant, amyloidogenic and neurotoxic peptide homologous to residues 106-126 of the prion protein. *Biochem. Biophys. Res.Comm.* **194**, 1380–1386 (1993).

62. L. De Gioria, C. Selvaggini , E. Ghibaudi, L. Diomede, O. Bugiani, G. Forloni, F. Tagliavini, and M. Salmona, Conformational polymorphism of the amyloidogenic and neurotoxic peptide homologous to residues 106-126 of the prion protein. *J. Biol. Chem.* **269**, 7859–7862 (1994).

63. M. Salmona, P. Malesani, L. De Gioria, S. Gorla, M. Bruschi, A. Molinari, F. Della Vedova, B. Pedrotti, M. S. Marrari, T. Awan, O. Bugiani, G. Forloni, and F. Tagliavini, Molecular determinants of the physicochemical properties of a critical prion protein region comprising residues 106-126. *Biochemistry* **342**, 207–214 (1999).

64. M. Salmona, M. Morbin, T. Massignan, L. Colombo, G. Mazzoleni, R. Capobianco, L. Diomede, F. Thaler, L. Mollica, G. Musco, J. J. Kourie , O. Bugiani, D. Sharma, H. Inouye, D. A. Kirschner, G. Forloni, and F. Tagliavini, Structural properties of Gerstmann-Sträussler-Scheinker disease amyloid protein. *J. Biol. Chem.* **278**, 48146–48153 (2003).

65. G. Forloni, N. Angeretti, R. Chiesa, E. Monzani, M. Salmona, O. Bugiani, and F. Tagliavini, Neurotoxicity of a prion protein fragment. *Nature* **362**, 543–546 (1993).

66. R. Oldengourgh and W. C. Phillips, Small permanent magnet for field up to 2.6T. *Review Sci. Inst.* **57**, 2362–2365 (1986).

67. L. Makowski, Preparation of magnetically oriented specimens for diffraction experiments, in Brookhaven Symposium, Synchrotron Radiation, in *Biology*, edited by R. M. Sweet and A. D. Woodhead (Plenum Press, New York, 1988), pp. 341–347.

68. M. Balbimie, R. Grothe, and D. S. Eisenberg, An amyloid-forming peptide from the yeast prion Sup35 reveals a dehydrated β-sheet structure for amyloid. *Proc. Natl. Acad. Sci. USA* **98**, 2375–2380 (2001).

Chapter 5

MOUSE BEHAVIOURAL STUDIES AND WHAT THEY CAN TEACH US ABOUT PRION DISEASES

Colm Cunningham

CNS Inflammation Group, School of Biological Sciences, University of Southampton, SO16 7PX, UK

5.1. Introduction

Prion diseases comprise an extraordinary family of diseases with a novel infectious agent. Unsurprisingly, the nature of this agent has stolen much of the limelight in the study of this chronic neurodegenerative disease. This has meant rapid advances in our understanding of the formation of aberrant PrP from the normal cellular form of the protein and has prompted much investigation into how fragments of this protein may affect the viability or functioning of cultured neurons. That deletion of the PrP gene prevents the transmission of prion disease to mice, and mutations in the PrP gene cause human forms of prion disease have provided strong evidence for the prion hypothesis. However, rather less is understood of neuronal death in the disease. Which neurons die first? How do they die? Which neurons must die before the animal succumbs to terminal disease and death? Obviously these are key problems in understanding the progression of prion disease and are not trivial ones to solve. If one studies the whole animal rather than the isolated cell this imposes restrictions on the molecular detail one can hope to reveal about a disease process. However, dysfunction at the behavioural level can reveal neuroanatomical circuitry that is no longer functioning and can provide us with ways into understanding the disease that studies of neurons in culture cannot. While mice are often considered to have a limited behavioural repertoire, recent work has shown that behavioural tests can reveal signs of prion disease long before the clinical phase of disease. This work has brought our understanding of behavioural changes up to date with what we already knew about some aspects of the neuropathology of disease, but has also placed us in a position to identify areas of dysfunction so far unsuspected. In this chapter, behavioural and comparative neuropathological studies in murine models of prion disease will be discussed. The focus will be on early behavioural changes rather than late neurological signs since, by the terminal stages of disease, the pathology tends to be rather widespread and it becomes difficult to relate specific behavioural impairments to discrete pathological features. This chapter will discuss:

- Similarities between murine and human prion diseases.
- The behavioural profile of mouse-adapted prion disease strains
- Neuroanatomical substrates of behavioural dysfunction.
- How behaviour can be used in therapeutic and transgenic studies of prion disease.

5.2. Clinical symptoms in human disease

Mouse models of human disease allow us to study the *in vivo* disease in a way that is obviously not possible in patients. They allow us to harvest tissue at specific time points during disease in order to make assessments of alterations in protein expression as disease progresses and to observe the early stages of neuronal damage or death and other features of neuropathology. What is sometimes overlooked is that they also offer the opportunity to watch the animal itself as it develops disease. The human prion diseases Gerstmann-Straussler-Scheinker (GSS) syndrome, fatal familial insomnia (FFI) and Creutzfeldt-Jakob disease (CJD) in both its sporadic and variant forms are fatal neurodegenerative diseases and as such those who succumb to them present with psychiatric and/or neurological symptoms. These include alterations in behaviour, cognition and movement as well as autonomic nervous system dysfunction. It has become increasingly clear that rodents infected with transmissible forms of prion disease also show such behavioural impairments.

One emerging feature in prion disease is that the psychiatrist is now often the first medical practitioner to come into contact with patients with prion diseases. This is due to the gradual appearance in the population of young adults suffering from new variant CJD, the disease entity apparently caused by the ingestion of BSE-infected meat[1]. These patients are affected at a much younger age than is seen with other prion diseases and show a pattern of behavioural abnormalities that are less often seen in familial or sporadic forms of prion disease[2]. These behavioural abnormalities include a number of features of depression and precede the appearance of dementia and typical clinical signs.

Since the number of cases of variant CJD is still relatively small, the picture of early psychiatric and neurological features is still emerging, and currently forms a somewhat heterogeneous combination of different "first" symptoms. Since there may be multiple subtle impairments at presentation, the impairments observed might depend on how and by whom the patient is first examined, and it may also be difficult to retrospectively pinpoint the first manifestation. However, the first larger studies of these patients have shown differences in the clinical patterns shown by vCJD and sporadic CJD. Patients with suspected sporadic CJD show varied presentation with anything from ataxia to cognitive impairment to hallucinations, but most often neurological deficits such as visual disturbances and cerebellar dysfunction, and behavioural changes such as apathy are the first symptoms, while the rapidly progressive dementia is a hallmark[3]. The balance is more clearly weighted towards psychiatric symptoms in the new variant form of disease. In the study of Spencer et al.,[2] on the first 100 patients diagnosed with vCJD, psychiatric symptoms preceded neurological ones in 63 cases while the opposite was true in only 15 cases. The most common psychiatric symptoms included dysphoria, withdrawal, anxiety, irritability, insomnia and loss of interest. This combination of affective (emotional) symptoms led, in many cases, to an initial diagnosis of depression. However, the frequent development of cognitive impairment including memory loss, impaired concentration and disorientation signalled a possible neurological disorder underlying these affective changes and at least one neurological symptom was present in 57% of cases within 2 months of clinical onset. The neurological signs include gait disturbance and impaired coordination, myoclonus, incontinence, upper motor neuron signs, sensory changes such as paraesthesia ("pins and needles") and numbness, as well as visual symptoms, but in general these appear late in disease progression. Indeed the late stages of many human prion diseases tend to show a convergence in symptoms. The terminal stages of most prion diseases are marked by cerebellar ataxia, pyramidal and extra-pyramidal motor signs, myoclonus, profound dementia, severe autonomic dysfunction[3] and ultimately, akinetic mutism[4]. Most of these diseases are described in the literature as being of short duration compared to other dementias. That is to say that the symptomatic stage of the disease lasts from a few months to, at most, a few years. However the rapidity of deterioration once clinical signs become apparent probably masks a slow degenerative disease that has been progressing without obvious signs before overt neurological or psychiatric symptoms become apparent.

This rapid onset after "clinically silent" incubation periods is also a frequently described feature of murine models of prion disease. Infected mice have classically been described as asymptomatic until late stages of disease, when they begin to show motor in-coordination and changes in gait, followed by urinary incontinence, priapism, decreased locomotion, ruffled fur and hunched posture. These are neurological signs of motor and autonomic dysfunction. To ask whether there are also psychiatric and subtle cognitive symptoms in prion-diseased mice is obviously slightly more complicated than in humans. However, we can use behavioural measures to determine what is occurring before the onset of neurological symptoms of disease. It has been a major aim of studies in our laboratory and those of our collaborators, to develop behavioural assays that inform on prion disease progression and to link these to observed patterns of pathology. Initially the model chosen in which to monitor behavioural impairments was the ME7 mouse-adapted scrapie strain. This strain is one of the best characterised of those currently studied in mice. At terminal stages of disease one can observe widespread vacuolation of the tissue, PrP^{Sc} deposition, and microglial and astrocytic activation[5]. This is most obvious in the hippocampus and thalamus. Less is known about the patterns of neuronal death. It has been known for some time that mice infected with this strain show marked neuronal death in the CA1 of the hippocampus, which is preceded by dendritic and synaptic damage/loss in the stratum radiatum[6] but the focus of pathological studies in this strain has remained mostly in the hippocampus.

5.3. Murine behaviour phase 1: species-typical, affective, reward-seeking behaviours

We have shown early behavioural changes in ME7-inoculated animals[7–9] and have demonstrated that the first pathological changes observed in this strain correspond temporally with the earliest behavioural changes in prion disease[10]. These animals show a progressive decline in glucose-induced polydipsia (increased drinking when presented with a palatable glucose solution instead of the normal drinking water), in burrowing of food pellets and in nest building, all of which generally become statistically significant at approximately 12 weeks (approximately 50% of disease duration)[7–10]. The appearance of this group of behavioural impairments is highly reproducible and these tests are extremely useful as non-invasive indicators of disease progression, but what do these tests actually assay? Can they tell us anything about the nature of the ongoing disease process?

The normal performance of these tests is species-typical. That is to say that mice will spontaneously become polydipsic (i.e. drink more than normal) when presented with a palatable glucose solution instead of their usual drinking water. They naturally will burrow food pellets out of a tube in the home cage if presented with this option and will spontaneously build a nest in the home cage if presented with the materials with which to do this. These are "instinctive" behaviours for mice but in the early stages of ME7-induced prion disease they no longer behave in this instinctive way. What drives these behaviours? The most simplistic explanation of these behaviours is that there is a benefit or a reward element in each case. Mice, both male and female, will construct nests in the home cage. In our experiments mice are presented with a pressed cotton "nestlet", which they normally shred into small pieces and use these to build a deep saucer-shaped nest. Poorer nests consist of a flat cushion of material and in the worst cases the nestlet is incompletely torn or completely untouched. The spontaneous nest-building in mice is thought to be primarily for thermoregulatory purposes since standard housing temperatures for mice (21°C) are lower than that which is ideal for thermal comfort (i.e. when placed in a temperature gradient C57BL/6 mice choose to remain at 27°C[11]). Mice challenged with bacterial endotoxin (lipopolysaccharide, LPS) decrease locomotor activity as part of the sickness behaviour response to infection. Sickness behaviour is a coordinated set of neural responses that represents a reorganisation of priorities in order to adapt to the onset of infection. This response includes decreases in locomotor and social activity, grooming, food consumption, anorexic and fever responses. The response is driven by the central nervous system expression of pro-inflammatory cytokines[12]. Mice also cease nest-building activities as part of this sickness behavioural response, but if challenged with the same dose of LPS in a cold environment (6°C) they now resume nest-building activities despite continued-lower locomotor activity[13]. Thus, there is a point at which the benefit of nest building is outweighed by the cost or motivation required to perform this activity, but when deemed not strictly necessary it is sacrificed. Since prion-diseased mice are not exhibiting typical sickness behaviour (in early stages of disease they appear normal and have no obvious change in activity or body-temperature) the explanation of decreased nesting is not immediately obvious. This will be discussed further below.

Another dysfunctional behaviour we have observed in prion-diseased mice is decreased consumption of a glucose solution. Presentation of a 10% glucose solution induces polydipsia, or higher than normal liquid consumption, in mice. From approximately 10–12 weeks into ME7-induced prion disease this polydipsic effect declines. We have not measured water consumption per se in these animals and thus cannot comment on whether they drink more or less than normal animals. There is some evidence that drinking water consumption shows similar effects to sucrose solution[14,15]. Although the evidence presented by Outram et al. and by Suckling et al. for decreased water consumption is less convincing, similar observations have also been made more recently by Dell-Omo et al[16]. In our hands, even upon decline in polydipsia these mice are consuming glucose solution at or above the normal daily water consumption for C57BL/6 mice and so it appears that the test does not inform on reduced ingestive behaviour, but rather reduced reward-seeking behaviour. The decreased consumption of palatable glucose solutions has been a widely used measure of anhedonia[17] (the failure to derive pleasure from normally pleasurable activities), but is often complicated by the anorexic effects that are associated with typical sickness behaviour-inducing treatments. However, using intracranial self-stimulation paradigms, it has been shown that mice will perform bar-presses to induce pleasurable hypothalamic stimulation[18]. As the intensity of the cranial stimulation is decreased both untreated and LPS-treated animals decrease their responses to it at the same rate, demonstrating a gradual loss of interest in the stimulus as it becomes less intense/pleasurable. However, on increasing the stimulus from zero, normal animals clearly respond to a less intense signal than that required to induce bar presses in LPS-treated animals. This experiment demonstrates that animals respond differentially to the rewarding stimulus depending on the cost:benefit ratio and those animals experiencing anhedonia are less likely to engage in reward-seeking behaviour.

It has been noted by some researchers that strains of prion disease, including 79A and ME7, produce vacuolar pathology in the hypothalamus[19] and ME7 in particular causes glucose intolerance and obesity[20]. To my knowledge no clear demonstration of neuronal loss in the hypothalamus has been presented. Though often associated with cell loss, it remains unclear exactly what tissue vacuolation signifies, and we have thus preferred to use microglial activation as a means of assessing regions of the brain undergoing dysfunction or degeneration. Microglia are exquisitely sensitive to neuronal death, damage or even altered homeostasis[21]. We have used this method to identify the stratum radiatum of the hippocampus, the medial and lateral septum and the entorhinal cortex as the first areas affected when behavioural impairments first become apparent in the ME7 model of prion disease[10]. Using this method, we fail to find marked pathology in the hypothalamic nuclei, even at terminal stages of disease. The hypothalamus is known to be a crucial centre in the control of ingestive behaviours since electrolytic lesions of the lateral hypothalamic area tend to produce decreased consumption (adipsia/aphagia) while ventromedial lesions cause hyperphagia and obesity[22]. However, in ME7 induced prion disease we observe no marked pathology in this region, no clear adipsia but clear weight gain. This group of findings is not consistent with any particular hypothalamic pathology. Therefore, it appears that while ingestive behaviour *per se* maybe somewhat affected in this model, reward-oriented consumption is decreased rather more. This is an important distinction since it necessarily relates decreases in glucose consumption to motivational and emotional states in the animal, which may have parallels in the human disease. The majority of variant CJD cases and many sporadic cases present with dysphoria, withdrawal and loss of interest (i.e. depression) as the first symptoms[2].

To date, the test that has showed most sensitivity to early prion-associated dysfunction is burrowing. It remains unclear exactly what burrowing assays. In this test, a tube of material such as food pellets is placed in the cage of a single mouse, and almost all mice spontaneously dig some or all of the material out of the tube. This test, devised by R. Deacon[8], was based on the food-hoarding paradigm in rats. However, mice will perform this activity even with non-edible materials. What is clear is that mice appear to find this activity rewarding; a typical C57BL/6 mouse will burrow many times its own weight in food pellets in 2 hours and will repeat this performance on many consecutive days. Though they may sometimes sleep in the emptied tube, mice will empty a full tube even if there is already an empty tube close by in the same cage. Prion-diseased mice become impaired on this burrowing task early in the disease, without showing any evidence of motor dysfunction. As discussed above, the withdrawal from activities that the individual usually finds pleasurable is a clear sign of an affective (or emotional) disturbance. From excitotoxic lesion studies we know that this burrowing task, as well as nesting, requires an intact hippocampus[23]. In addition we know that both of these tasks and glucose consumption can be decreased by administration of LPS, although glucose consumption and, in particular, burrowing are more sensitive to these challenges than nesting (Deacon et al., manuscript submitted). Thus, nest construction may be retained during bacterial infection because of its thermoregulatory importance while more "hedonistic" behaviours such as burrowing and glucose consumption are lost in the sickness-induced reorganisation of priorities. Motivation has been defined as a central state that organises perception and action; whereby an altered motivational state may produce a reduced attention to external events[24], and this seems appropriate to describe changes such as that seen in burrowing, nesting and glucose consumption after challenges with LPS and in the ME7 model of prion disease.

However, changes in motivation caused by prion disease are likely to be different to those caused by LPS. While I have made many comparisons between behavioural alterations caused by LPS and by prion disease there are fundamental differences between these. In the case of LPS these animals are exhibiting sickness behaviour and show a general depression of many activities, including locomotion in the open field, with a likely depression of most activities that involve motor activity. It is clear that prion-diseased animals do not become hypoactive in the open field (indeed they show open field hyperactivity) and thus the behavioural spectra overlap but differ fundamentally. As discussed above, sickness behaviour is driven by CNS expression of pro-inflammatory cytokines[12] and we have shown that the expression of cytokines in the ME7 model of prion disease is absent or lower than that produced by peripheral injections with LPS, and is also qualitatively different in that ME7 also induces synthesis of the anti-inflammatory cytokine TGFβ1[25,26]. The prion-disease behavioural profile is not a sickness profile and is much more likely to reflect neuronal dysfunction and synaptic loss, unlike the response to LPS, which is a temporary and homeostatic mechanism to adapt to the onset of bacterial infection. However, some of the sites of dysfunction, particularly the hippocampus[27], are likely to be common to both mechanisms of behavioural change.

Hippocampal circuitry appears to be important in linking external stimuli with the innate motivations and motor programs of species-typical behaviours, in the same way that objects and actions are processed in a co-ordinated fashion during exploration and other spatial learning[28]. Thus, although comparisons between impaired nesting in mice and altered behaviours in demented patients are not obvious, it is of interest that apathy or loss of interest is a consistent early observation in new variant CJD[2]. A decrease in performance of a "domestic" task like nesting may have parallels in the impairments in "activities of daily living" observed in Alzheimer's disease[29]. Though this type of assessment has only occasionally been applied in rapidly progressing dementias such as prion diseases[30] it is undeniable that people with these conditions become less functional in the home and workplace before becoming obviously demented. Hippocampal-lesioned mice remain healthy and show no obvious disadvantages in the home cage, where they are presented with food, water and shelter daily, but are more susceptible to "natural" pressures such as predation[31] in altered environments. With a lowered exploratory drive, poor knowledge of spatial layout and the locations of resources such as food, water and shelter, these mice easily succumb to natural pressures. Hippocampal damage is prominent in Alzheimer's disease patients but there is some debate as to whether this is true in CJD. It is generally believed that CA1 hippocampal neurons are not lost during CJD[32], but there are reports of pathology in this area[33,34] and synaptic loss without neuronal death has been shown to be sufficient for impairment on hippocampal tasks in mice[10]. Without the support of carers, people suffering from dementia would also eventually face life-threatening difficulties. While the environments we inhabit are fundamentally different, both mice and humans must provide nourishment, shelter and security for themselves and the ability to do this is impaired in dementia, whether it is prion disease in mice or in humans.

5.4. Phase 2: altered activity and cognition

The next phase of mouse-adapted scrapie disease is marked by changes in open field activity and in cognition. This overlaps the decreases in burrowing to some extent but does not become marked until the 14–18 week period. During a 3-minute open field test, ME7-infected mice show marked hyperactivity in an open field arena measuring 50 cm (length) × 30 cm (width) × 20 cm (depth). This activity reaches as high as four fold increases compared to the activity of normal brain homogenate-injected animals and is apparent both in distance travelled and in number of rears (where the animal raises itself onto its hind paws, with or without the assistance of the arena wall). Our laboratory is not the first to describe changes in motor activity in mouse adapted prion diseases, but there has been much variability in the findings of various research groups and it may be informative to examine these differences. In 1980 McFarland and Hotchin described increased open field activity at 100 and 150 days post-inoculation with the Chandler strain of mouse-adapted scrapie using an open field arena similar to that used in our studies[35]. However, even as early as 1976 Suckling et al. demonstrated that mice, also inoculated with the Chandler strain show a progressive decrease in home cage activity when monitored continuously over the course of the disease, with an excitable phase just preceding the clinical phase[15]. A more recent study by Dell'Omo et al concurs with these data[16]. Their data show clearly decreased home-cage activity of ME7 injected animals at weeks 11 and 15 post-inoculation in the striatum, with less clear decreases at other points in the disease. Interestingly, like in the studies of Suckling et al., the home cage activity begins to increase in ME7 in the later stages such that overall nocturnal activity is almost equal to that of controls by terminal stages of disease. In the same study the 139A strain remains more active in the home cage than controls, while 301C, a mouse-adapted BSE strain, showed lower home cage activity throughout. It should be noted that the control animals in these studies were naïve and thus not inoculated with NBH as is customary with prion disease studies and it is unclear whether one can assume that animals injected with normal brain homogenate are truly "normal". It is also of note that to facilitate home cage monitoring, these animals were singly housed for the duration of the disease progression and this is known to cause multiple alterations in murine behaviour, including effects on locomotor activity[11].

As well as home cage activity, these authors[16] also examined open field activity and reported increased open field walking and wall rearing in ME7- and 139A-injected animals at 19 weeks post-inoculation. It is informative that in different 2-minute intervals during a 20-minute open field test there were clear differences between successive periods. In general ME7-injected animals were considerably more active in the first 2 minutes in the open field, with a gradual lessening of their hyperactivity over the duration of the open field measurements. When combined with the home cage findings of these authors and our own open field measurements this suggests that activities may be altered in multiple ways that reflect more than simply locomotor hyper- or hypoactivity. Thus it is possible, at least in the ME7 model, that these animals are hypoactive in the home cage, which may inform on apathy, loss of interest in exploration of the home cage, or indeed hypersomnia, but may be more active in the open field as a result of an altered response to novel environments. Indeed disturbances in sleep patterns are a defining characteristic of another prion disease, fatal familial insomnia, albeit, in this case insomnia is observed. It is also of interest that deletion of PrP in mice causes changes in circadian activity[36].

The increased activity in the open field may represent a cognitive impairment, which causes these mice to continually explore an open field arena that normal animals gradually habituate to and explore less each week. Rearing, in particular, is generally considered to be an exploratory behaviour and such increased exploration of a somewhat familiar environment may inform on some impairment in cognition. There may also be some anxiety response to the novel environment. Little has been done to investigate anxiety in prion diseased mice but a more detailed analysis of multiple activities in the open field can distinguish between exploration and locomotion and thus inform on changes in emotional/anxiety states in mice[37]. Typical fear/anxiety responses include increased defecation, freezing and significant thigmotaxis in a novel environment (rodents tend to stay in close proximity with the walls of the arena in which they are placed) while upon habituation to this environment mice show more exploratory behaviour. Since both simple locomotor activity around the edges of the arena and more directed exploratory activity are totalled in the "distance travelled" or "squares crossed" readouts of most open field studies (including our own), we currently do not have a clear indication of anxiety in this model. One study that did examine central versus peripheral locomotion in the open field reported no differences between prion-infected and normal animals[35]. Using defecation as an index of emotionality, we also found no evidence from the open field of any major increase in emotionality (unpublished observations). However anxiety remains one relatively frequent early symptom in variant CJD and the indications from our early murine behavioural studies are that there are early affective problems. There are a number of relatively specific tests for anxiety (elevated plus-shaped maze, light dark box, hyponeophagia) and in combination with automated open field measurements with multiple readouts, such as those currently being used in our laboratory, these should allow examination of anxiety in some detail.

With no clear indication as to whether anxiety has a role to play, it is tempting to speculate on whether cognitive changes may underpin the observed open field hyperactivity. It is clear at this stage of the disease that these animals are showing signs of cognitive impairment. The initial studies of learning in scrapie were performed using the passive avoidance test. Passive avoidance involves a mild footshock as stimulus and avoidance of the location of the stimulus is the expected response in normal animals. Testing can be assessed in various ways. In a one trial test, animals are exposed to a footshock when entering a dark compartment of the apparatus and then 24 hours later are placed in the apparatus again and the time taken to revisit this compartment (latency) is determined. In multi-trial passive avoidance the animals perform an acquisition session in which they are footshocked at intervals until they move from the dark area and remain in the safe side for a set time. On reaching this set time (criterion) they have "acquired" passive avoidance (acquisition is assessed by the number of shocks required to induce exit from the shock compartment). This measure is useful in its own right but these animals are then assessed again at intervals throughout treatments or disease in order to test whether the learned avoidance of the dark compartment is retained (retention). One trial passive avoidance testing performed by Hunter et al. suggested that mice with preclinical ME7 and 22C did not become cognitively impaired after intraperitoneal inoculation while 139A and 79A were only mildly impaired and this impairment was overcome at higher shock levels[38]. Our laboratories have published impairment of prion-diseased animals on multi-trial passive avoidance[7]. The impairments were noted on retention of avoidance from 16–20 weeks post-inoculation onwards. These tests have been repeated in subsequent studies in our laboratories with similar, though not identical, results[9] and retention has also been shown to be impaired in hamsters inoculated with the 263K strain of scrapie[39]. The results with passive avoidance have been relatively variable, but the stress component of repeated testing on this test make it less than ideal as a repetitive testing strategy. Spatial learning tests exploit the animals' natural exploratory tendencies and have also been used to demonstrate impairments in prion disease. Lysons et al. employed a food-rewarded T-shaped maze test to show that 139A scrapie-infected mice were not impaired on acquisition of learning, as judged by the mice requiring the same number of days training to reach a pre-set "pass-rate" (the criterion)[40]. However these animals were impaired on reversal learning (i.e. the reward is switched to the other arm of the maze and they must re-learn and once again reach a criterion). The authors described these learning abnormalities as a rapid extinction of the original discrimination followed by a normal relearning of the reversal. This effect was pronounced from approximately 68 days before clinical symptoms. We now routinely use spontaneous alternation as a means of assessment of working spatial memory. Normal mice, when placed in the start arm of a T-maze, will generally visit one of the goal arms of the maze, and after being confined in this arm for 30 seconds and then replaced at the starting point will then choose the other goal arm (i.e. they spontaneously alternate their choices). This is the criterion for spontaneous alternation. Normal mice show spontaneous alternation at approximately 80% while ME7-infected mice decline to chance levels (50%). This decrease was first reported at 18 weeks post-inoculation[9] but we now find this impairment as early as 14 weeks and find it continued thereafter (Figure 5.1).

Together these findings suggest clear cognitive impairment from the middle stages of the disease. The human diseases are characterized by rapidly progressive dementia and the studies discussed above demonstrate impairments at least on working and reference memory across a number of mouse and hamster models, with different routes of inoculation. Thus, it appears that in the early to middle stages of mouse-adapted scrapie we can observe depression of mood and reward-seeking behaviours, open field hyperactivity and impaired cognitive performance. In effect, therefore, this mouse model recreates many of the reported psychiatric and cognitive changes of the early stages of variant CJD. A representative sample of behavioural monitoring data carried out in our laboratories is shown in Figure 5.1.

5.5. Neuroanatomical substrates of altered behaviours

These studies of early behavioural changes reveal a pattern of behavioural changes that is not unlike those most commonly observed in

Figure 5.1. Sample data for burrowing, glucose consumption, open field activity and spontaneous alternation in normal brain homogenate and ME7-infected C57BL/6 mice after intra-hippocampal inoculation.

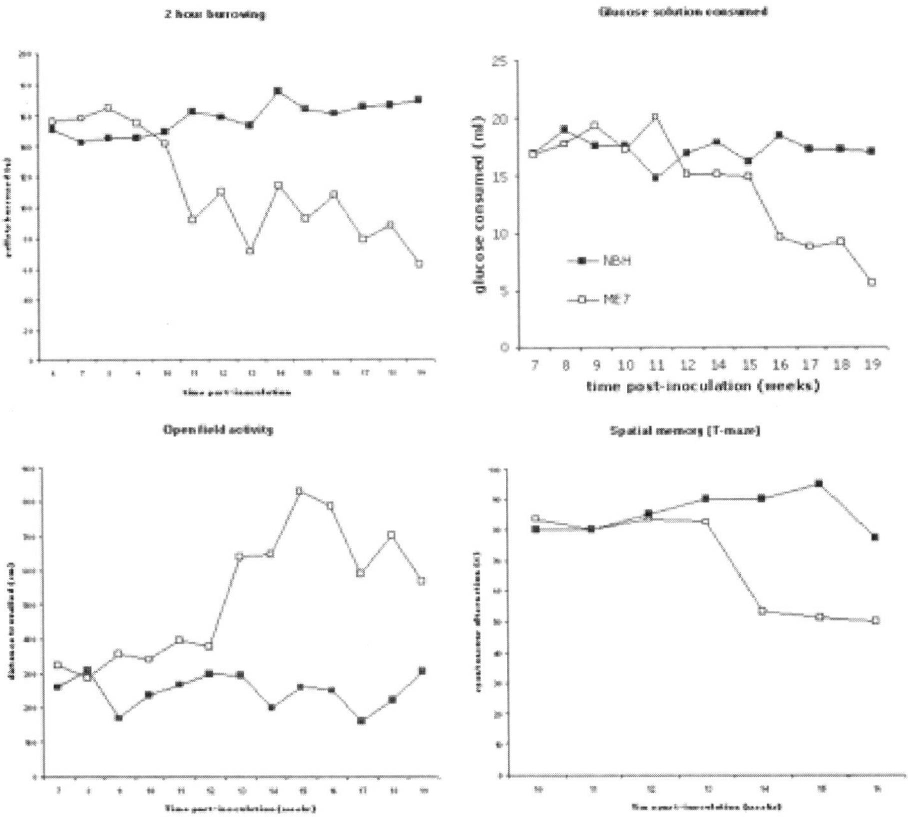

variant CJD: affective changes such as apathy, dysphoria, withdrawal, followed by cognitive impairment and neurological signs. In order to maximise the information obtained from these behavioural studies it is useful to consider the neuroanatomy of these tasks and the neuropathology that accompanies the observed impairment in the disease. Though much has been written about neuropathological signs in prion disease, the read-outs used have been various and their meanings often not clear. The most commonly used readouts of prion disease pathology are deposition of PrP^{Sc}, vacuolation of the tissue, and astrocytosis. While the identification of some of these features in a given structure may provide an indication of aberrant activity in this area, neither vacuolation nor PrP^{Sc} show a strong correlation with neuronal or synaptic loss in humans[41] or in animal models[10]. In order to understand loss of function in brain circuitry we have taken the approach of assessing synaptic and neuronal loss. As discussed above, since the late stage pathology is widespread it is difficult to relate specific impairments to neuronal dysfunction in specific circuitry. Thus, we reasoned that if the earliest signs of neuronal dysfunction could be established and these early signs could be related to neuropathological features in particular brain regions this would in turn facilitate studies of the earliest cellular and molecular components of the pathogenesis. Much had already been published on hippocampal pathology in the ME7 model of prion disease, including quantification of CA1 neuronal loss and descriptions of both pre-synaptic and dendritic damage[6,42–44], but little information on neuronal damage outside this area was available. Using the stereotactic intra-hippocampal injection of ME7 we carried out detailed analyses of the pathology in these mice at 13 weeks post-inoculation. In the first instance we waited to observe statistically significant changes in the early behaviours described above and then, using the same animals, used microglial staining to identify areas of neuronal dysfunction that might underlie the behavioural changes. Microglia are the resident macrophages of the brain and are exquisitely sensitive to neuronal death, damage or even altered homeostasis[21]. In the early stages of ME7, initiated by intra-hippocampal inoculation, we found that the hippocampus, the medial and lateral septum and the entorhinal cortex all showed clear microglial activation. The most marked activation was visible in the stratum radiatum of the hippocampus and thus this formed the focus for closer investigation of cell death and synaptic damage. At the time of the first behavioural impairments there was no evidence of neuronal death in the CA1 of the hippocampus but significant loss of staining for the pre-synaptic marker synaptophysin. Interpreting this as synaptic loss, it seems likely that this loss of connectivity in the stratum radiatum underlies the impairments seen. Thus it appears that the synapse is the primary target of prion-associated pathology[10].

It is a frequent criticism of this work that hippocampal damage might be the expected result of hippocampal injections. Intuitively, one might expect toxicity where one injects a toxin, but the "prion" is not a conventional toxic agent and the first sites of toxicity may not necessarily be where it is first placed. In early studies Fraser, Outram and others have shown that the patterns of vacuolation observed are not affected by the initial brain site of inoculation[19,45]. The pathology that we observe in late stage ME7 injected brain is entirely consistent with what has previously been reported in these animals after intraperitoneal injection[5]. This finding demonstrates that neurons in this area are more susceptible to ME7-induced cell death than those in other areas and thus, the behavioural and early pathological impairments can not be explained by the site of injection. Indeed it is informative to note that the granule cells of the dentate gyrus, which also lie at the injection site, do not die until much later in the disease, if at all, and that there is no significant CA1 neuronal death after intra-hippocampal inoculation with 22L or 22A strains (discussed further below). Thus, proximity to the inoculum clearly does not imply susceptibility to PrP-associated death.

5.6. Applying this approach to other strains

Armed with the knowledge that CA1 synaptic loss is associated with the behavioural changes observed in ME7, we studied the pattern of behavioural changes in a number of other strains. There are multiple strains of mouse-adapted prion disease, which have been characterised by their survival times and vacuolation profiles[5]. However, some authors have suggested that many of these strains are artefacts of passage through mice of different PrP genotypes, and that the real number of distinct strains is far fewer than reported[46]. Though some previous behavioural studies also used strains other than ME7, we planned to study specific strains, which are described as having distinct patterns of pathology, and which have been little studied in behavioural terms. We used the combined behavioural and neuropathological approach to investigate whether strains with different pathologies produce different patterns of behavioural dysfunction. This approach may serve to define susceptible neuronal populations and highlight pathways of degeneration that are common to, or distinct from, other strains. One might have expected that 22L, which shows conspicuous cerebellar pathology[5], would fail to induce those behavioural impairments that we know to be related to hippocampal dysfunction[23]. We repeated our previous behavioural test battery in the strains ME7, 79A, 22L and 22A. Remarkably, all strains produced the same sequence of behavioural changes[47]. There were differences in the precise timing of changes, but no significant difference in the sequence. That is to say 22A, which has a very long incubation period in C57BL/6 mice, did not show changes in burrowing until much later than the other strains, but showed this change before any other impairments. In addition, 22L tended to be slightly more advanced than the other strains in its early increase in open field activity, its late stage decrease in this activity, and its onset of neurological deficits such as muscle weakness and motor co-ordination, but this strain has a slightly shorter incubation period than the others examined. However, the impairments appeared in the same sequence in all strains (phase 1 impairments, followed by phase 2 impairments, followed by neurological and finally terminal clinical signs). Slight differences in timing could be explained by the time-course of incubation. A summary of the changes in each strain is shown in Table 5.1.

Table 5.1.

	Nest	Gluc	Burr	Rear	Sqr	Bar	Scrn	R.rod	S.rod	Sqr▼	Rear▼
ME7	1	1	1	2	2	3	4	4	4	5	5
79A	1	1	1	2	2	4	3	4	4	5	4
22L	1	1	2	1	2	3	3	4	4	4	3

Note: The time at which animals become impaired on each task has been assigned a number in the sequence of the overall behavioural pattern such that the first behaviour, or the first group of behaviours to become impaired is assigned 1. These are assigned on the basis of trends observed rather than on statistically significant decreases/increases. Burr, burrowing; Sqr, squares crossed; Gluc, glucose; Bar, horizontal bar; Scrn, inverted screen; R rod, rotating rod; S rod, static rod; Sqr▼, decreased squares crossed; Rear▼, decreased rears. Reproduced from Cunningham et al.[47]

This finding that all of these strains showed essentially the same behavioural pattern prompted a close re-examination of the pathology of these strains to determine how different they really are. This analysis showed clear differences and some important similarities between the strains. 79A showed marked white matter microglial activation, as was predicted from previous studies of increased vacuolation in white matter[5], while 22L and 22A showed the most severe microglial activation in the cerebellum, consistent with previous reports of severe vacuolation in this region[5]. However, it was clear that all strains showed marked microglial activation in both hippocampus and thalamus. This was most severe in ME7 and least severe in 22L (ME7>79A>22L). Consistent with these patterns of microglial activation, neurodegeneration in the hippocampal CA1 layer was statistically significant in the ME7 and 79A strains, whereas 22L and 22A did not show statistically significant neuronal loss in this region. Conversely only 22L and 22A showed significant cerebellar Purkinje cell loss. These distinct patterns of neuronal death are consistent with vacuolar profiles in these regions in the studies of Bruce et al.[5] and function as an important validation of the intra-hippocampal route of inoculation i.e. the findings uphold the idea that specific strains target specific areas rather than preferentially killing neurons proximal to the injection site. In the studies of Bruce[5] the focus was on the differences between strains with respect to vacuolation profiles and incubation times, but the study also revealed that there are areas of clear overlap. Since those studies focus only on late stage pathological changes, it remains possible that some or all of the early events are similar between strains and that divergence between strains appears only later in disease. We have demonstrated a common early behavioural pattern among the four strains ME7, 79A, 22L and 22A, and believe that a common pathway of neuronal dysfunction may underlie it. Neuropathological assessment of these strains at the point of first behavioural deficits should prove very informative.

5.7. A common pathological profile?

The end stage pathology of all four strains shows clear distinctions but includes significant microglial activation that encompasses the limbic system, brainstem, cortex and cerebellum to greater or lesser degrees. Although all strains showed hippocampal pathology there were clear differences between ME7, 79A, 22L, and 22A at the level of the synaptic loss. There was severe loss of hippocampal CA1 synapses in ME7-injected animals while those in 79A and 22L were better preserved, despite also showing some loss. Thus, although there is significantly less synaptic loss in the CA1 of 22L and 79A, it is sufficient for loss of function with respect to the behavioural tasks examined in this study. Given that 80% of the input to the CA1 stratum radiatum originates from the CA3 it seems reasonable to assume that this projection has largely degenerated in ME7 but is partially intact in 22L and 79A. However we do not observe obvious loss of CA3 pyramidal neurons and we have observed that the mossy fibre projection from the dentate gyrus to the CA3 is intact even at terminal stages of disease. Since there is a clear difference between the strains on CA1 synapse loss but no clear difference on early behavioural tasks we propose that the behavioural impairments in our studies do not result from loss of CA3 projections but from some other projection to the CA1. The activity of the pyramidal cells of CA1 is also regulated by the inhibitory GABAergic hippocampal interneurons, the cholinergic input from the medial septum, and the serotonergic input from the raphe nucleus, among others and the current approach suggests that these projections to the hippocampus should be examined as candidates to explain the convergence of hippocampal behavioural symptoms despite widely divergent overall hippocampal pathology in these strains.

What was most striking and less variable between strains was neuronal loss in the thalamus. There was severe neuronal death observed in the thalamus of all four strains. This loss was particularly consistent in the posterior thalamic nuclei and the ventroposterior nuclei (>50%). Examples of this are shown in Figure 5.2. The neurons of this region may constitute a group that is more susceptible to prion-associated neuronal death and certainly represents the area of most obvious overlap between the 4 strains studied so far by our selves. This area has also been reported to be affected in the 87V, 139A and 22C strains, on the basis of vacuolation and astrocytosis[5,48] but neuronal death in this region has been the subject of very little discussion. This is particularly surprising

Figure 5.2. NeuN staining demonstrating clear neuronal loss in the thalamus of ME7, 79A and 22L inoculated animals. Approximate regions of the thalamus are depicted on the ME7 panel as follows: dLGN, dorsal lateral geniculate nucleus; LP, lateroposterior nuclei (mediorostral and laterorostral); VP, ventroposterior nuclei (ventroposterolateral and ventroposteromedial); Po, posterior thalamic group.

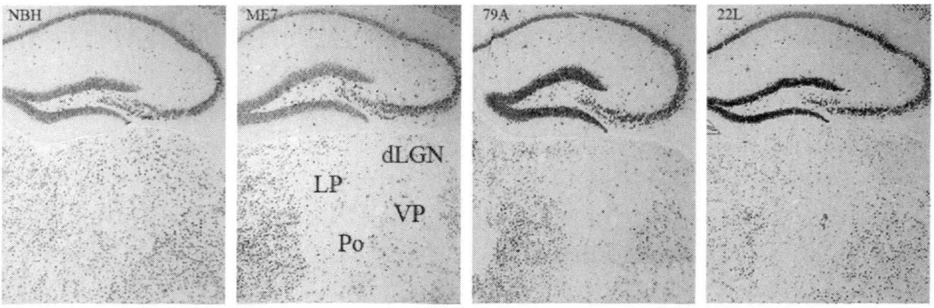

given that severe thalamic pathology is widely reported in human forms of prion disease (see below).

The thalamus is a highly complex region of the brain that is not well understood. A detailed discussion of its many functions and the consequences of dysfunction is beyond the scope of this chapter, but some aspects of its function that could be involved in behavioural changes and clinical symptoms will be considered. The thalamus is a major link between the external world and the cortex. It receives afferents from spinal pathways to the ventroposterior nucleus (touch, pain, movement, temperature), and cerebellothalamic tracts from the deep cerebellar dentate nucleus to the ventral anterior and ventrolateral thalamus. Thus, dysfunction in thalamic areas could underlie positive and negative sensory symptoms such as paraesthesia, pain or numbness that are often seen in vCJD and other prion diseases[49]. Since severe deficits, such as tremor and ataxia, are caused by lesions to the dentate nucleus, the loss of its projection areas in the thalamus may also have a role in the motor coordination problems ultimately experienced by most patients with prion diseases. The cerebellar dentate nucleus shows severe microglial activation in all four strains we have examined and this may explain the similar dysfunction on typical cerebellar tasks despite marked differences in pathology in the purkinje and granule cell layer[47].

In addition to sensory and motor functions, there is considerable reciprocity in connections between the thalamus and cortex and consequently the thalamus also has an integrative role in higher mental function. Thalamic infarcts and space-occupying lesions in humans are known to cause multiple cognitive and emotional impairments as well as the more traditional sensory and motor disturbances (see[50] for review). Despite the potentially wide-ranging effects of thalamic degeneration, this has been little studied in prion diseases. Recent interest in thalamic involvement in vCJD has been driven by the discovery of the "Pulvinar sign", which has become one the key diagnostic aids in vCJD[41]: most vCJD patients display a characteristic distribution of symmetrical T2 hyperintensity of the pulvinar (posterior) nucleus on MRI imaging in the axial plane. This hyperintensity correlates best with gliosis in humans[51] and in animal models[52], but widespread neuronal loss is observed in these patients at post-mortem. These nuclei are highly developed in the human brain but are absent or very much reduced in rodents. Sporadic CJD also shows MRI abnormalities in the posterior thalamus[53,54], but concomitant striatal MRI abnormalities distinguish this signature from that of vCJD. In FFI, the neuropathological change common to all cases is severe degeneration of the anteroventral and mediodorsal thalamic nuclei[55]. These nuclei may show up to 90% loss[56]. Clinically, the key aspects of FFI are inability to generate EEG sleep patterns and sympathetic autonomic nervous system hyperactivity leading to autonomic dysfunction (which is also a key feature of late stage disease in mouse-adapted scrapie). The damaged nuclei constitute the limbic part of the thalamus and interconnect limbic and paralimbic cortex with the hypothalamus, which appears to become dysfunctional upon release from cortical control.

It is important to note that the medial and anterior thalamic (limbic thalamus), but not the pulvinar show pathology in FFI and in thalamic variant sporadic CJD, whereas the posterior and ventral thalamic nuclei are lost in vCJD, with sparing of the medial and anterior nuclei. Therefore one cannot generalise about the thalamus as a point of convergence in prion disease progression. However, it is clear that loss of thalamic neurons may produce significant loss of connectivity and can have effects on sensory, motor, cognitive, emotional and autonomic functions. This short treatment of the thalamus provides more questions than answers, but thalamic pathology has been documented in man and observed but largely neglected in murine models and it may be very informative to examine this aspect of pathology in more detail. The implication of the thalamus in some of the clinical phenotypes discussed above demonstrates that there are many different neuroanatomical routes to the same behavioural or clinical deficit, and emphasizes that at late stages of prion disease it is difficult to relate neuropathology to symptomology. In the approach described here, by identifying the earliest behavioural alterations in mouse models we hope to uncover the specific pathologies that underlie them.

5.8. The use of behaviour to assess disease progression/therapeutic intervention

Finally, it is useful to consider the use of behaviour as a read-out of disease progression. Monitoring the development of simple, sensitive, reproducible, and objective measures of disease progression is a powerful means of non-invasively assessing therapeutic intervention and/or gene deletion studies. A large battery of genes have been mutated or deleted in attempts to identify key genes in prion disease pathogenesis. While these studies have revealed some key events in peripheral pathogenesis after intraperitoneal injection, very few gene manipulations have produced anything but the mildest of effects on disease incubation time or time to death after intracerebral inoculation. Genes identified as affecting disease incubation time include complement proteins (2 weeks longer)[57] and CD-40 (40 days shorter)[58]. While these studies are judged to implicate the genes in question in pathogenesis or protection, judgement on the basis of increasing survival time, may do no more than prolong later stages of the disease, which in a human context is obviously undesirable. In short, death is not a good readout of disease progression. We have some evidence for this assertion on the basis of studies in our laboratory with macrophage chemoattractant protein 1. This protein is involved in macrophage/microglial activation and chemotaxis. The deletion of this gene had no impact on any of the early behavioural changes but delayed the appearance of the late stage clinical signs by 3-4 weeks (Felton et al., manuscript submitted). This suggests that the gene has no involvement in the progression of disease *per se*, but has an inhibitory effect on some aspect of late stage disease. It is of note that when mice begin to show urinary incontinence in the late stages of disease they usually develop inflammation of the genital area and this peripheral inflammation alone is sufficient to produce the general appearance of sickness. Interference with this process through anti-inflammatory therapy or deletion of key inflammatory genes may lead to confusing signals about the disease status of the animal. The monitoring of specific disease-associated behaviours is a more satisfactory way of assessing this.

We have also used anti-inflammatory drugs in an attempt to assess the role of inflammation in prion disease progression. We have previously used the steroidal anti-inflammatory drug dapsone to show that, despite earlier reports that it increases survival time[59], this drug has no effect whatsoever on the development of behavioural impairments[9]. We have now administered the anti-inflammatory drug nimesulide (a cyclooxygenase 2 inhibitor) in the food pellets (approximately 7.5 mg/day) and monitored disease progression using behavioural testing. As is shown

Figure 5.3. Development of early prion-associated behavioural impairments in animals treated with the cyclooxygenase-2 inhibitor nimesulide from 12 weeks into disease progression.

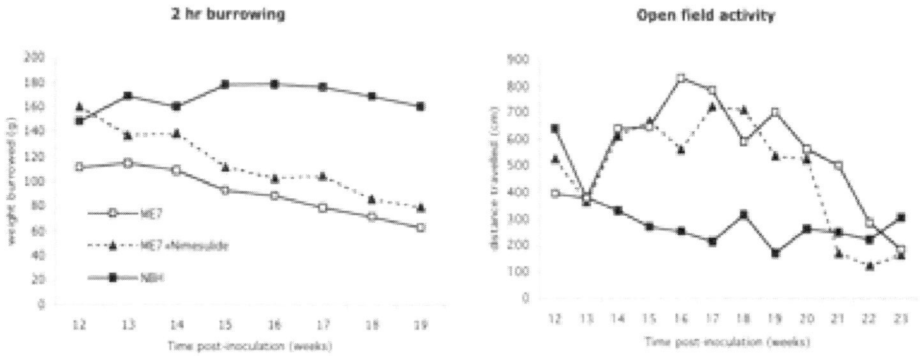

in Figure 5.3, this drug also had no effect on the development of the previously described prion-associated behavioural impairments. We have observed similar effects with the anti-inflammatory drug indomethacin and these data provide further evidence that inflammation does not have a key role in the progression of murine-adapted scrapie.

This method should become a crucial tool in assessing some of the emerging potential treatments for prion diseases. Potential therapeutic agents such as tetrapyrroles, sulphated glycans, branched polyamines and anti-PrP monoclonal antibodies[60–63], have mostly been selected on the basis of their slowing or halting the conversion of PrP^c to PrP^{Sc} *in vitro* systems (see[64] for review), and some of these treatments have also been found to be beneficial in animal models of disease[61]. However, whether they have any major impact on the progression of prion disease is best assessed in an animal model with clear behavioural readouts. Such an approach would also be an addition to the conditional PrP^c knockout study of Mallucci et al.[65]. This study shows that spongiosis is reversed and neuronal death is prevented upon PrP^c knockout halfway through disease progression and one would like to know whether these animals also show a halt or indeed a reversal of behavioural symptoms.

5.9. Conclusion

The behavioural studies discussed here illustrate that alterations in activity, reward-seeking behaviours and cognition in murine prion diseases parallel many of the early affective and cognitive changes in human prion diseases such as vCJD. The generality of this sequence of behavioural changes in our strain studies highlights the utility of these tests as a tool in the assessment of therapeutic strategies in prion diseases. In addition the common behavioural impairments observed between multiple prion strains with distinct neuropathologies suggest that a common early neurodegenerative pathway may underlie these changes. Such a common pathway has the potential to lead researchers to the hypothetical "clinical target areas" proposed by Kimberlin[66].

5.10. Acknowledgements

I would like to thank Dr Rob Deacon and Professors Nick Rawlins and V. Hugh Perry for their advice and input into this chapter and the work contained in it. Special thanks is due to Rob Deacon for his part in my education in mouse behaviour. Thanks also to Tracey Newman and Ian Galea for critical reading of the manuscript. The financial support of the Wellcome Trust is gratefully acknowledged.

5.11. References

1. M. E. Bruce, R. G. Will, J. W. Ironside, I. McConnell, , D. Drummond, A. Suttie, L. McCardle, A. Chree, J. Hope, C. Birkett, S. Cousens, H. Fraser and C. J. Bostock, Transmissions to mice indicate that 'new variant' CJD is caused by the BSE agent. *Nature* **389**, 498–501. (1997).

2. M. D. Spencer, R. S. Knight, and R. G. Will, First hundred cases of variant Creutzfeldt-Jakob disease: retrospective case note review of early psychiatric and neurological features. *Bmj* **324**, 1479–1482. (2002).

3. I. Zerr and S. Poser, Clinical diagnosis and differential diagnosis of CJD and vCJD. With special emphasis on laboratory tests. *Apmis* **110**, 88–98. (2002).

4. A. Otto, I. Zerr, M. Lantsch, K. Weidehaas, C. Riedemann and S. Poser, Akinetic mutism as a classification criterion for the diagnosis of Creutzfeldt-Jakob disease. *J Neurol Neurosurg Psychiatry* **64**, 524–528. (1998).

5. M. E. Bruce, I. McConnell, H. Fraser and A. G. Dickinson, The disease characteristics of different strains of scrapie in Sinc congenic mouse lines: implications for the nature of the agent and host control of pathogenesis. *J Gen Virol* **72**, 595–603. (1991).

6. M. Jeffrey, W. G. Halliday, J. Bell, A. R. Johnston, N. K. MacLeod, C. Ingham, A. R. Sayers, D. A. Brown and J. R. Fraser, Synapse loss associated with abnormal PrP precedes neuronal degeneration in the scrapie-infected murine hippocampus. *Neuropathol Appl Neurobiol* **26**, 41–54. (2000).

7. S. Betmouni, R. M. J. Deacon, J. N. P. Rawlins and V. H. Perry, Behavioural consequences of prion disease targeted to the hippocampus in a mouse model of scrapie. *Psychobiology* **27(1)**, 63–71 (1999).

8. R. M. Deacon, J. M. Raley, V. H. Perry and J. N.Rawlins, Burrowing into prion disease. *Neuroreport* **12**, 2053–2057. (2001).

9. K. Guenther, R. M. Deacon, V. H. Perry and J. N. Rawlins, Early behavioural changes in scrapie-affected mice and the influence of dapsone. *Eur J Neurosci* **14**, 401–409. (2001).

10. C. Cunningham, R. Deacon, H. Wells, D. Boche, S. Waters, C. P. Diniz, H. Scott, J. N. Rawlins and V. H. Perry, Synaptic changes characterize early behavioural changes in the ME7 model of murine prion disease. *Eur. J. Neurosci.* **17**, 2147–2155 (2003).

11. C. J. Gordon, P. Becker and J. S.Ali, Behavioral thermoregulatory responses of single- and group-housed mice. *Physiol Behav* **65**, 255–262. (1998).

12. J. P. Konsman, P. Parnet, and R. Dantzer, Cytokine-induced sickness behaviour: mechanisms and implications. *Trends Neurosci* **25**, 154–159. (2002).

13. A. Aubert, G. Goodall, R. Dantzer and G. Gheusi, Differential effects of lipopolysaccharide on pup retrieving and nest building in lactating mice. *Brain Behav Immun* **11**, 107–118. (1997).

14. G. W. Outram, Changes in drinking and feeding habits of mice with experimental scrapie. *J Comp Pathol* **82**, 415–427. (1972).

15. A. J. Suckling, S. Bateman, C. B. Waldron, H. E. Webb and R. H. Kimberlin, Motor activity changes in scrapie-affected mice. *Br J Exp Pathol* **57**, 742–746 (1976).

16. G. Dell'Omo, E. Vannoni, A. L. Vyssotski, M. A. Di Bari, R. Nonno, U. Agrimi and H. P. Lipp, Early behavioural changes in mice infected with BSE and scrapie: automated home cage monitoring reveals prion strain differences. *Eur J Neurosci* **16**, 735–742. (2002).

17. R. Yirmiya, Endotoxin produces a depressive-like episode in rats. *Brain Res* **711**, 163–174. (1996).

18. T. Borowski, L. Kokkinidis, Z. Merali, and H. Anisman, Lipopolysaccharide, central in vivo biogenic amine variations, and anhedonia. *Neuroreport* **9**, 3797–3802. (1998).

19. G. W. Outram, H. Fraser and D. T. Wilson, Scrapie in mice. Some effects on the brain lesion profile of ME7 agent due to genotype of donor, route of injection and genotype of recipient. *J Comp Pathol* **83**, 19–28. (1973).

20. R. I. Carp, Y. S. Kim, and S. M. Callahan, Scrapie-induced alterations in glucose tolerance in mice. *J Gen Virol* **70**, 827–835. (1989).

21. G. W. Kreutzberg, Microglia: a sensor for pathological events in the CNS. *Trends Neurosci* **19**, 312–318. (1996).

22. A. W. Hetherington, and S. W. Ranson,. The spontaneous activity and food intake of rats with hypothalamic lesions. *Am J Physiol* **136**, 609–617 (1942).

23. R. M. Deacon, A. Croucher and J. N. Rawlins, Hippocampal cytotoxic lesion effects on species-typical behaviours in mice. *Behav Brain Res* **132**, 203–213. (2002).

24. R. C. Bolles and M. S. Fanselow, A perceptual-defensive recuperative model of fear and pain. *Behav Brain Sci* **3**, 291–323 (1980).

25. D. T. Walsh, S. Betmouni and V. H. Perry, Absence of detectable IL-1beta production in murine prion disease: a model of chronic neurodegeneration. *J Neuropathol Exp Neurol* **60**, 173–182. (2001).

26. C. Cunningham, D. Boche and V. H. Perry, Transforming growth factor beta1, the dominant cytokine in murine prion disease: influence on inflammatory cytokine synthesis and alteration of vascular extracellular matrix. *Neuropathol Appl Neurobiol* **28**, 107–119. (2002).

27. T. Cartmell, G. N. Luheshi and N. J. Rothwell, Brain sites of action of endogenous interleukin-1 in the febrile response to localized inflammation in the rat. *J Physiol* **518**, 585–594. (1999).

28. R. M. J. Deacon, and J. N. P Rawlins, Hippocampal lesions, species-typical behaviours and anxiety in mice. *Behav Brain Res* **156**, 241–249 (2005).

29. B. Reisberg, S. Finkel, J. Overall, N. Schmidt-Gollas, S. Kanowski, H. Lehfeld, F. Hulla, S. G. Sclan, H. U. Wilms, K. Heininger, I. Hindmarch, M. Stemmler, L. Poon, A. Kluger, C. Cooler, M. Bergener, L. Hugonot-Diener, P. H. Robert, S. Antipolis and H. Erzigkeit, The Alzheimer's disease activities of daily living international scale (ADL-IS). *Int Psychogeriatr* **13**, 163–181. (2001).

30. R. Nitrini, L. S. Teixeira da Silva, S. Rosemberg, P. Caramelli, P. E. Carrilho, P. Iughetti, M. R. Passos-Bueno, M. Zatz, S. Albrecht and A. LeBlanc, Prion disease resembling frontotemporal dementia and parkinsonism linked to chromosome 17. *Arq Neuropsiquiatr* **59**, 161–164. (2001).

31. S. E. Glickman and B. J. Morrison, Some behavioural and neural correlates of predation susceptibility in mice. *Commun Behav Biol A* **4**, 261–267 (1969).

32. C. Masullo and G. Macchi, Resistance of the hippocampus in Creutzfeldt-Jakob disease. *Clin Neuropathol* **16**, 37–44. (1997).

33. H. Mizusawa, A. Hirano and J. F. Llena, Involvement of hippocampus in Creutzfeldt-Jakob disease. *J Neurol Sci* **82**, 13–26. (1987).

34. M. A. Poon, S. Stuckey and E. Storey, MRI evidence of cerebellar and hippocampal involvement in Creutzfeldt-Jakob disease. *Neuroradiology* **43**, 746–749. (2001).

35. D. McFarland and J. Hotchin, Early behavioral abnormalities in mice due to scrapie virus encephalopathy. *Biol Psychiatry* **15**, 37–44 (1980).

36. I. Tobler, T. Deboer and M. Fischer, Sleep and sleep regulation in normal and prion protein-deficient mice. *J Neurosci* **17**, 1869–1879. (1997).

37. E. Choleris, A. W. Thomas, M. Kavaliers and F. S. Prato, A detailed ethological analysis of the mouse open field test: effects of diazepam, chlordiazepoxide and an extremely low frequency pulsed magnetic field. *Neurosci Biobehav Rev* **25**, 235–260. (2001).

38. A. J. Hunter, M. P. Caulfield and R. H. Kimberlin, Learning ability of mice infected with different strains of scrapie. *Physiol Behav* **36**, 1089–1092. (1986).

39. S. R. Bareggi, D. Braida, M. Gervasoni, G. Carcassola, C. Pollera, C. Verzoni, M. Sala and C. Vergerio, Neurochemical and behavioural modifications induced by scrapie infection in golden hamsters. *Brain Res* **984**, 237–241. (2003).

40. A. M. Lysons and S. J. Woollard, Spatial reversal learning in preclinical scrapie-inoculated mice. *Neuroreport* **7**, 1087–1091 (1996).

41. M. Zeidler, R. J. Sellar, D. A. Collie, R. Knight, G. Stewart, M. A. Macleod, J. W. Ironside, S. Cousens, A. C. Colchester, D. M. Hadley, R. G. Will and A. F. Colchester, The pulvinar sign on magnetic resonance imaging in variant Creutzfeldt-Jakob disease. *Lancet* **355**, 1412–1418. (2000).

42. D. Brown, P. Belichenko, J. Sales, M. Jeffrey and J. R. Fraser, Early loss of dendritic spines in murine scrapie revealed by confocal analysis. *Neuroreport* **12**, 179–183. (2001).

43. P. V. Belichenko, D. Brown, M. Jeffrey and J. R. Fraser, Dendritic and synaptic alterations of hippocampal pyramidal neurones in scrapie-infected mice. *Neuropathol Appl Neurobiol* **26**, 143–149. (2000).

44. M. Jeffrey, C. M. Goodsir, M. E. Bruce, P. A. McBride and J. R. Fraser, In vivo toxicity of prion protein in murine scrapie: ultrastructural and immunogold studies. *Neuropathol Appl Neurobiol* **23**, 93–101. (1997).

45. H. Fraser and A. G. Dickinson, Distribution of experimentally induced scrapie lesions in the brain. *Nature* **216**, 1310–1311 (1967).

46. R. M. Ridley and H. F. Baker, To what extent is strain variation evidence for an independent genome in the agent of the transmissible spongiform encephalopathies? *Neurodegeneration* **5**, 219–231. (1996).

47. C. Cunningham, R. Deacon, K. Chan, D. Boche, J. N. P. Rawlins and V. H. Perry, Neuropathologically distinct strains give rise to similar temporal profiles of behavioural impairments. *Neurobiol. Dis.* **18**, 258–269 (2005).

48. J. I. Kim, W. K. Ju, J. H. Choi, E. Choi, R. I. Carp, H. M. Wisniewski and Y. S. Kim, Expression of cytokine genes and increased nuclear factor-kappa B activity in the brains of scrapie-infected mice. *Brain Res Mol Brain Res* **73**, 17–27. (1999).

49. M. A. Macleod, G. E. Stewart, M. Zeidler, R. Will and R. Knight, Sensory features of variant Creutzfeldt-Jakob disease. *J Neurol* **249**, 706–711. (2002).

50. J. D. Schmahmann, Vascular syndromes of the thalamus. *Stroke* **34**, 2264–2278. Epub 2003 Aug 2221. (2003).

51. H. Urbach, J. Klisch, H. K. Wolf, D. Brechtelsbauer, S. Gass and L. Solymosi, MRI in sporadic Creutzfeldt-Jakob disease: correlation with clinical and neuropathological data. *Neuroradiology* **40**, 65–70. (1998).

52. B. Canciani, M. G. Bruzzone and T. Awan, Neuropathological correlates of magnetic resonance abnormalities in experimental prion disease. *J Neuropathol Exp Neurol* **58**, 560 (1999).

53. H. J. Tschampa, J. W. Herms, W. J. Schulz-Schaeffer, B. Maruschak, O. Windl, U. Jastrow, I. Zerr, B. J. Steinhoff, S. Poser and H. A. Kretzschmar, Clinical findings in sporadic Creutzfeldt-Jakob disease correlate with thalamic pathology. *Brain* **125**, 2558–2566. (2002).

54. H. J. Tschampa, P. Murtz, S. Flacke, S. Paus, H. H. Schild and H. Urbach, Thalamic involvement in sporadic Creutzfeldt-Jakob disease: a diffusion-weighted MR imaging study. *AJNR Am J Neuroradiol* **24**, 908–915. (2003).

55. E. Lugaresi, I. Tobler, P. Gambetti and P. Montagna, The pathophysiology of fatal familial insomnia. *Brain Pathol* **8**, 521–526. (1998).

56. G. Macchi, G. Rossi, A. L. Abbamondi, G. Giaccone, D. Mancia, F. Tagliavini and O. Bugiani, Diffuse thalamic degeneration in fatal familial insomnia. A morphometric study. *Brain Res* **771**, 154–158. (1997).

57. M. A. Klein, P. S. Kaeser, P. Schwarz, H. Weyd, I. Xenarios, R. M. Zinkernagel, M. C. Carroll, J. S. Verbeek, M. Botto, M. J. Walport, H. Molina, U. Kalinke, H. Acha-Orbea and A. Aguzzi, Complement facilitates early prion pathogenesis. *Nat Med* **7**, 488–492. (2001).

58. M. Burwinkel, A. Schwarz, C. Riemer, J. Schultz, F. Van Landeghem and M. Baier, Rapid disease development in scrapie-infected mice deficient for CD40 ligand. *EMBO Rep* **8**, 8 (2004).

59. L. Manuelidis, W. Fritch and I. Zaitsev, Dapsone to delay symptoms in Creutzfeldt-Jakob disease. *Lancet* **352**, 456. (1998).

60. S. Supattapone, H. O. Nguyen, F. E. Cohen, S. B. Prusiner and M. R. Scott, Elimination of prions by branched polyamines and implications for therapeutics. *Proc Natl Acad Sci U S A* **96**, 14529–14534. (1999).

61. A. R. White, P. Enever, M. Tayebi, R. Mushens, J. Linehan, S. Brandner, S. Anstee, J. Collinge and S. Hawke, Monoclonal antibodies inhibit prion replication and delay the development of prion disease. *Nature* **422**, 80–83. (2003).

62. V. Perrier, J. Solassol, C. Crozet, Y. Frobert, C. Mourton-Gilles, J. Grassi and S. Lehmann, Anti-PrP antibodies block PrPSc replication in prion-infected cell cultures by accelerating PrPC degradation. *J Neurochem* **89**, 454–463. (2004).

63. S. A. Priola, A. Raines and W. S. Caughey, Porphyrin and phthalocyanine antiscrapie compounds. *Science* **287**, 1503–1506. (2000).

64. M. W. Head, C. F. Farquhar, N. A. Mabbott and J. R. Fraser, The transmissible spongiform encephalopathies: pathogenic mechanisms and strategies for therapeutic intervention. *Expert Opin Ther Targets* **5**, 569–585. (2001).

65. G. Mallucci, A. Dickinson, J. Linehan, P. C. Klohn, S. Brandner and J. Collinge, Depleting neuronal PrP in prion infection prevents disease and reverses spongiosis. *Science* **302**, 871–874. (2003).

66. R. H. Kimberlin, S. Cole and C. A. Walker, Pathogenesis of scrapie is faster when infection is intraspinal instead of intracerebral. *Microb Pathog* **2**, 405–415. (1987).

Chapter 6

ELECTROPHYSIOLOGICAL APPROACHES TO THE STUDY OF PRION DISEASES

Nikki K. MacLeod, Alex R. Johnston and John C. Curtis

Biomedical Sciences, Hugh Robson Building, Edinburgh EH8 9XD, UK

6.1. Introduction

Recording the electrical activity of the nervous system in both health and disease has a long and distinguished pedigree going back to Luigi Galvani in 1791[1], whose bold and contentious experiments on *animal electricity* led to the modern science of electrophysiology. The discovery that electrical activity could be recorded from the animal brain was made a century later in 1875 by Richard Caton, whose *feeble currents of the brain* were incredibly, discovered at least half a century before the advent of electrical amplification[2]. This electrical activity is what we now know as the *electroencephalogram* (EEG), a term coined by Hans Berger while working on human heads in Jena in the 1920s. Berger was also the first to suggest that the EEG could be used for clinical diagnosis and was the first to record the high amplitude synchronous electrical activity associated with epileptic seizures in the human brain 3. Even at this early state of the art, distinctions could be seen between focal damage to the brain surface and more widespread activity inferred to be projected from subcortical, diencephalic structures. Today, analysis of the EEG is routinely used in the diagnosis of many neurological conditions, often in conjunction with a variety of brain imaging methodologies. In a few cases of focal damage, where the lesion is invisible to current imaging methodologies, detailed EEG analysis is still the only way of securing an initial diagnosis. An excellent account of early electrophysiological

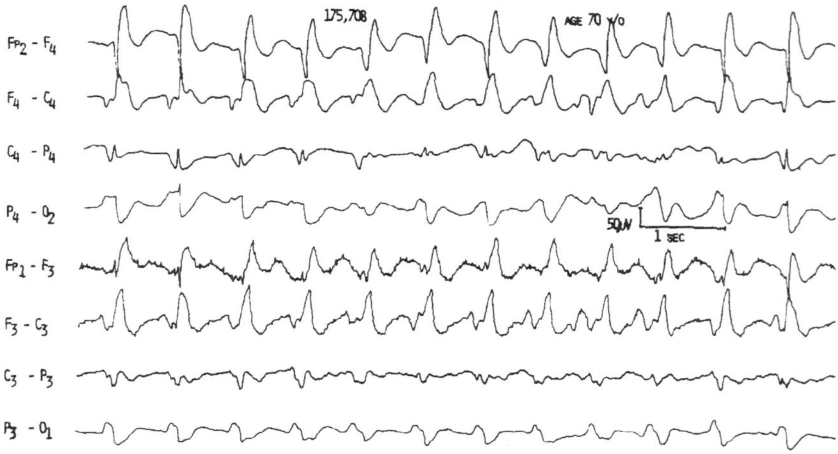

Figure 6.1. Multi-electrode EEG recorded from a patient with sporadic CJD showing periodic synchronous wave discharges across a wide area of cortex.

history can be found in Mary Brazier's delightful book, *The Electrical Activity of the Nervous System*[4].

One area where the EEG plays an essential part of the diagnosis is in cases of sporadic Creutzfeldt-Jakob disease. Classically, the normal EEG patterns gradually disappear and are replaced with widespread bilateral synchronous spike-wave discharges with a periodicity of 1 to 2 second as shown in Figure 6.1. This activity does not constitute a definitive diagnosis, nor does the absence of these periodic discharges necessarily rule out a diagnosis of sCJD, so the use of the EEG must be used in conjunction with other methods of assessment, including an MRI scan and currently, the abnormal appearance of 14-3-3 proteins in the CSF. Interestingly, the appearance of synchronized discharges in the EEG is not a diagnostic feature of new variant CJD and just to complicate matters, the following conditions sometimes also present with similar EEG abnormalities: Alzheimer's disease, multiple cerebral abscesses, metabolic encephalopathy, certain toxic encephalopathies (e.g. lithium), anoxic encephalopathy, progressive multifocal leucoencephalopathy, Lewy body disease[5].

Moving away from the EEG, many other, often much more sophisticated electrophysiological methods have been used in the study of neurodegeneration. In the field of basal ganglia disease for example, electrophysiology has been crucial in laying a foundation for understanding the basic anatomy and physiology of a complex set of inter-related brain areas with complex and often unexpected interactions. The most important of these discoveries was that the circuits linking the various anatomical players in this complex map worked by long axon neurons which inhibited their targets rather than facilitating them as was previously supposed. Control of neuronal circuitry in and around the basal ganglia is achieved largely through a process known as disinhibition, discovered by painstaking electrophysiological studies using single unit recording and electrical stimulation along with the infusion of pharmacological agents into some nuclei while recording from others[6,7]. Further electrophysiological studies from all of the basal ganglia and related nuclei, together with studies using 2-deoxyglucose allowed the development of a model of basal ganglia function that went a long way to clarifying our understanding not only of the normal functioning of these structures but also how their connectivity and activity was altered in hypo- and hyperkinetic diseases such as Parkinsonism and Huntington's chorea[8,9]. This model stood relatively unchallenged for about a decade and is now undergoing subtle revision in order to explain some of the more difficult anomalies that have arisen since its proposal, such as the paradoxical increased firing of subthalamic neurons in MPTP models of Parkinson's disease and the unexpected ameliorating effect of subthalamic stimulation on the condition[10,11]. So, although the original model may have been a shade too simplistic it has however proved to be immensely useful in devising extremely useful surgical interventions for Parkinson's disease, even if we don't yet understand all the fine print[12].

6.2. Methodologies

There are many ways of doing electrophysiology and most of them have been used in one way or another to study prion diseases. As mentioned in the introductory section, sCJD is usually associated with paroxysmal discharges in the EEG. The EEG is an excellent way of obtaining a lot of information very quickly on how entire populations of neurones are behaving, since the technique uses large electrodes, normally placed on the surface of the skull, which record the extracellular fields and population spikes generated by ionic currents flowing around and between individual neurones. The technique can be enhanced by making recordings from the surface of the cortex itself or better still by recording from different depths within the cortex when it becomes possible to plot the lines of current flow and make accurate predictions about current sources and sinks within the tissue, providing detailed knowledge about how the entire population of recorded neurones is behaving. The use of extracellular field potential recording following afferent or efferent electrical stimulation is a natural extension of EEG recording and has been used extensively to study neurodegenerative processes, particularly in the rodent hippocampus.

Figure 6.2. A. Schematic diagram of basal ganglia circuitry. All black neurones are GABAergic inhibitory neurones and the grey subthalamic neurones are glutamatergic excitatory connections. B. Typical extracellular pathway mapping electrophysiological recordings, in this case used to identify the GABAergic inhibitory nature of the nigrothalamic pathway in an anaesthetised rat. bc-brachium conjunctivum, sn - substantia nigra, bmc—bicuculline methyl chloride.

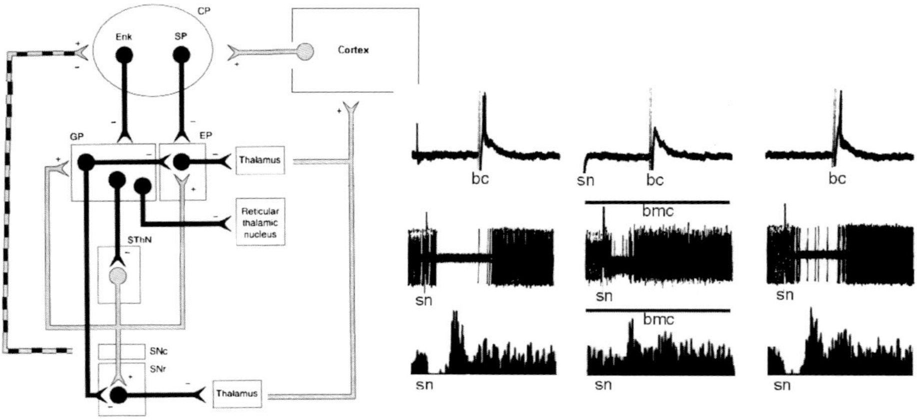

This kind of recording is very useful in giving information about the behaviour of many neurones but divulges little information about the individual neurones involved. To obtain more detailed information about individual neurones, electrophysiologists resort to recording from individual neurones, either extracellularly or intracellularly. Extracellular recording can provide information about synaptic field potentials and how fast a given neuron is firing, either spontaneously, or in response to a stimulus and many sophisticated computational methods are available for analysing this kind of information. This type of recording was used extensively for determining the inhibitory and disinhibitory nature of many neuronal projections within the basal ganglia circuitry discussed above and some typical examples are shown in Figure 6.2B.

More recently electrophysiologists have turned to using intracellular recordings from *in vitro* preparations, either of slices of whole brain tissue or from cultures of isolated cells, using either sharp or patch microelectrodes. This allows the experimenter to sample the membrane properties of individual neurones with the ability regulate membrane potential and the ionic composition of the cell under study and its immediate environment. By careful manipulation of extracellular and intracellular environments in combination with fluorescent dyes it is now possible to obtain a bewildering variety of data from individual neurones, populations of neurones and even individual ion channels. Recent developments have facilitated the combined use of electrophysiological recording and cellular imaging of individual ion concentrations and fluxes and even the use of PCR to monitor transcriptional levels following the sucking out of cellular contents via the patch electrode. An intriguing very recent development is the use of electrophysiology in combination with line-scanning confocal imaging of intracellular calcium levels to measure stochastic process at individual synaptic boutons terminaux. In view of the controversies surrounding the role of prion proteins in events at synapses this particular technique in combination with intracellular ion manipulations may well be the type of technique that finally allows us to answer some of the current enigmas in this field.

6.3. Studies on in scrapie-infected tissue and cells

Very few electrophysiological studies have been carried out on scrapie infected tissue and those that do exist, suggest little uniformity in the pathophysiology of different scrapie experimental models. The first study was conducted using the well-known short incubation hamster model by the group of Professor Jefferys at the University of Birmingham[14]. They were able to show cortical paroxysmal discharges very similar to those routinely observed in cases of human sCJD. This lead the authors to suppose that they were on the right track of electrophysiological indicators that might be relevant to this family of diseases. Intracellular recordings from individual neurones in both the neocortex and the hippocampus of the infected hamsters revealed a series of changes that might underlie the paroxysmal EEG. The most noticeable of these changes were action potential broadening, which was reversed by an inhibitor of voltage-gated calcium channels, and the attenuation of potassium conductances associated with spike repolarisation and after hyperpolarisation. These observations were important and set a marker for later studies on scrapie and PrP-null animal models, many of which have revealed impairment of cellular calcium regulation and calcium-associated potassium channels.

A series of studies in our own lab has used a different scrapie model, namely ME7 scrapie infected mice in which there was previously well documented evidence of hippocampal sclerosis including almost complete loss of CA1 pyramidal cells[15]. Recordings made exclusively from hippocampal pyramidal cells, documented a series of physiological morphological changes that correlated well with the known progression of the disease[16–19]. The most obvious and consistent change to the physiology of CA1 pyramidal cells in the scrapie infected animals was a large increase in membrane input resistance. This near doubling of membrane resistance was a very robust finding and was observed in several

Figure 6.3. Combined electrophysiology and confocal microscopy of synaptic spines in the mouse hippocampus before and after induction of LTP. Upper image series; confocal image and line scans before and after LTP induction; middle traces; sharp electrode recordings of membrane voltage; lower traces are the changes in intracellular calcium levels to pairs of pre-synaptic pulses[13].

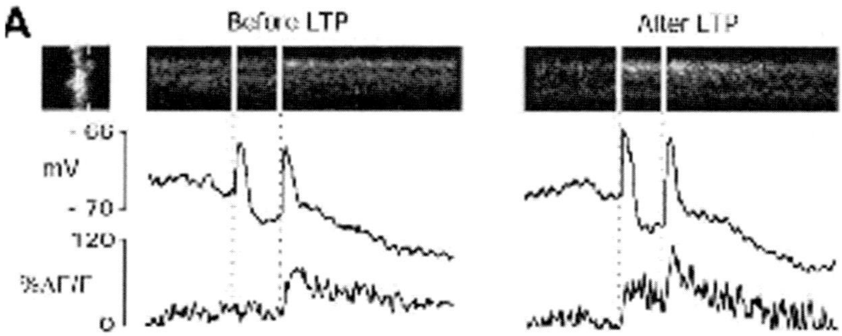

different series of experiments spread over a five year period. Morphological examination of some of the recorded neurones revealed that as the disease progressed there was a gradual reduction in the number of dendritic spines, leading to an almost 50% reduction in cell surface area which correlated well with the doubling of membrane resistance and in the absence of any alteration in membrane time constant was clearly the anatomical substrate underlying the observed physiological change. These changes can easily be seen in the physiological recordings presented in Figure 6.4B and the micrographs shown in Figure 6.5C and D.

The dendritic spines of most central system neurones are the major site of afferent input so the loss of the majority of these spines represents a major insult to the ability of the cell to function as an integrative unit. The physiological consequence of massive spine loss would be expected to be a large attenuation of synaptic potentials, both spontaneous and evoked by stimulation of afferent pathways. Such a loss was seen in all of our studies with this model and was noticed first as a gradual reduction in the amplitude of the field potential generated by stimulation of the Schaffer collateral pathway. As shown in Figure 6.3D, the field potential disappeared completely towards the expected end of the incubation period indicating an almost total loss of this particular input onto the cells.

In a later study we wondered if the well known enhancement of synaptic efficacy known as Long Term Potentiation (LTP) might be affected by the disease. We evoked LTP in these experiments by using the well tried

Figure 6.4. Composite diagram illustrating the changes in hippocampal pyramidal cells in scrapie infected mice. A) Camera Lucida drawings of hippocampal pyramidal cells in control (black) and scrapie infected (blue) animals. Intracellular recordings from these two cells shown in B, are of complex electrophysiological protocols involving a series of inward and outward current pulses applied through the recording microelectrode and the simultaneous application of a stimulus pulse to an electrode placed in the Schaffer collateral afferent pathway. Identical stimulation protocols were applied to both cells. Note the 2-fold increase in membrane resistance and the attenuation of both EPSPs and IPSPs evoked by the afferent stimulation in cells from the scrapie-infected animals. The inset below B, shows 2 traces extracted from the complex recordings shown above to illustrate more clearly the attenuation of the EPSP and IPSPs evoked by the stimulus. C) shows intracellular recordings of spontaneous activity in hippocampal pyramidal cells from control and scrapie infected mice. Note the increased bursting activity in cells from the scrapie infected animals. D) shows recordings of extracellular field potentials from the CA1 Stratum Radiatum following Schaffer collateral stimulation in the hippocampus of control mice and at two stages during the incubation period of the disease. All of these data were obtained from recordings made from hippocampal brain slices maintained *in vitro*.

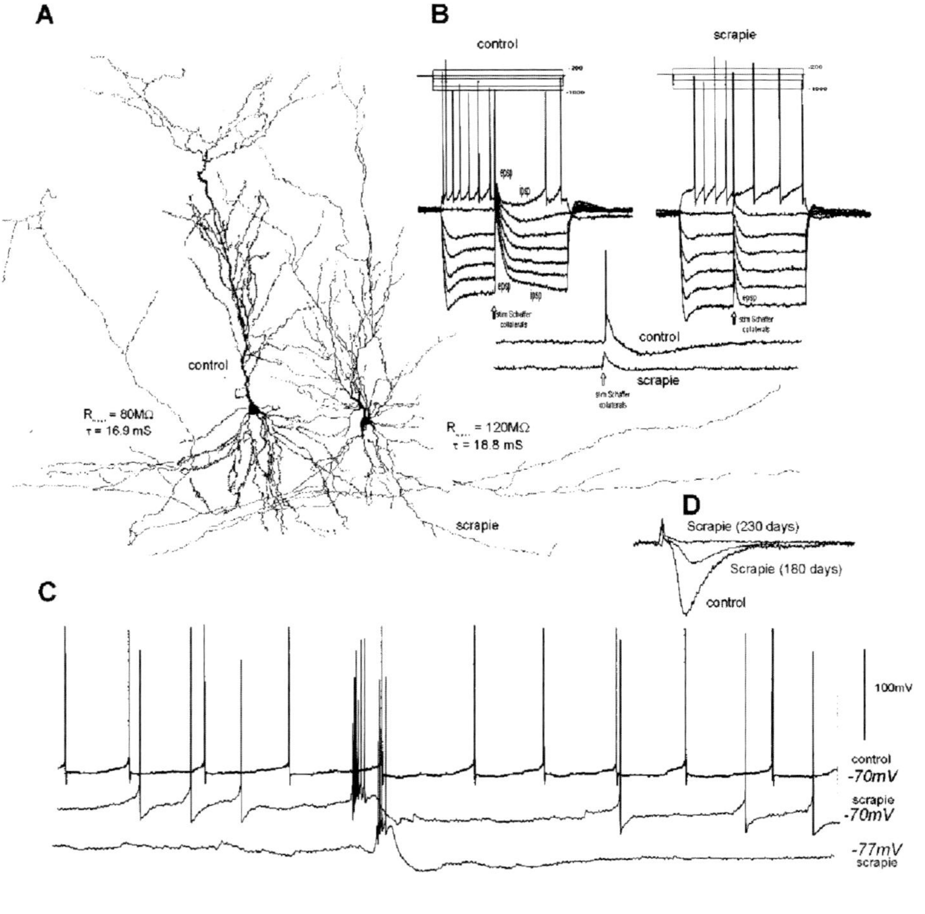

Figure 6.5. Morphology of the ME7 model. A) and B) show the distribution of disease-specific PrP in the mouse hippocampus at 70dpi (28%IP) and in a terminally infected mouse. C) and D) show light micrographs of biocytin-filled hippocampal pyramidal cell dendrites in the neuropil of the Stratum Radiatum of the hippocampus in control (C) and scrapie infected (D) mice. Note the loss of dendritic spines in the scrapie-infected cell (D). E) and F) are electron micrographs of hippocampal neuropil at 155dpi (62%IP) and 180dpi (72%IP) showing degenerate axon terminals and disordered neuropil.

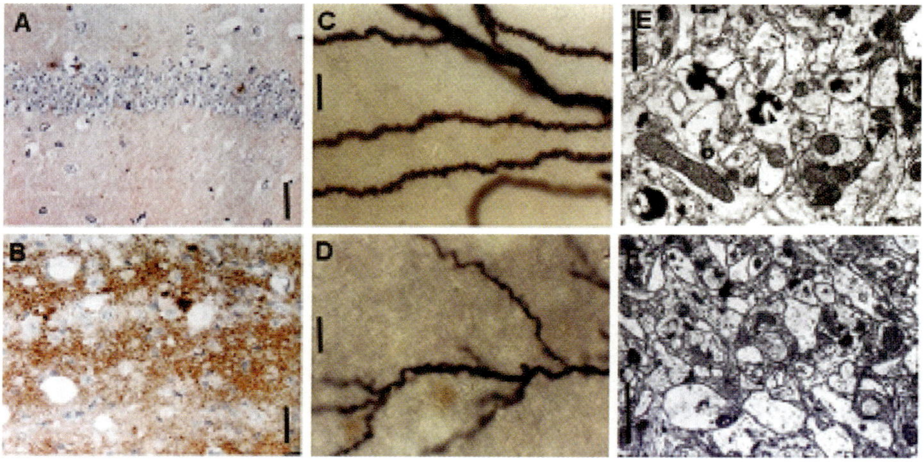

method of applying a one second conditioning burst of pulses at 100Hz and continually monitoring the size of the subsequent synaptic potentials evoked by a standard test pulse. As shown in Figure 6.6, no changes in synaptic plasticity were detectable at around 80 days into the incubation period but LTP had virtually disappeared 40 days later. The same stimulus protocol only evoked Short Term Potentiation (STP). The loss of LTP was first detectable at around 100 days into the incubation period, which correlates beautifully with our parallel morphological study that demonstrated clear evidence of synapse loss at around the same time. The synapse loss shows a gradual decline over the incubation period and correlates well with the accumulation of disease-specific PrP and appearance of degenerating axon terminals[16] (Figure 6.5). It is worth noting that this was a complete loss of LTP rather than a change in threshold since repeated series of conditioning trains of pulses at increasing intensity or duration failed to evoke LTP. This is well illustrated in Figure 6.6B, where only STP is evoked by two separate conditioning trains. The relationship of this pathophysiological observation to the presence or absence of PrP will be discussed in the next section, although it might be helpful here to refer to Figure 6.10A and B which beautifully demonstrate the location of PrP in synapses in the hippocampal neuropil and not in the region of pyramidal cell bodies.

Figure 6.6. Loss of LTP in the CA1 region of ME7 scrapie infected CVF_1 mice. A and B show the combined data from several experiments on control and scrapie-infected hippocampal slices at approximately one third (80-85dpi) and just over half way (125-130dpi) through the incubation period. C Shows the combined results of 4 experiments on scrapie-infected slices in which a second tetanus was applied at twice the stimulus strength used for the test pulses and the first tetanus. The second, larger tetanus still failed to induce LTP. D shows the gradual loss of LTP over the incubation period observed in two sets of experiments expressed as a percentage of the total number of animals in the individual groups. In all graphs, control data are plotted as filled circles or bars and experimental data are plotted as open circles or bars.

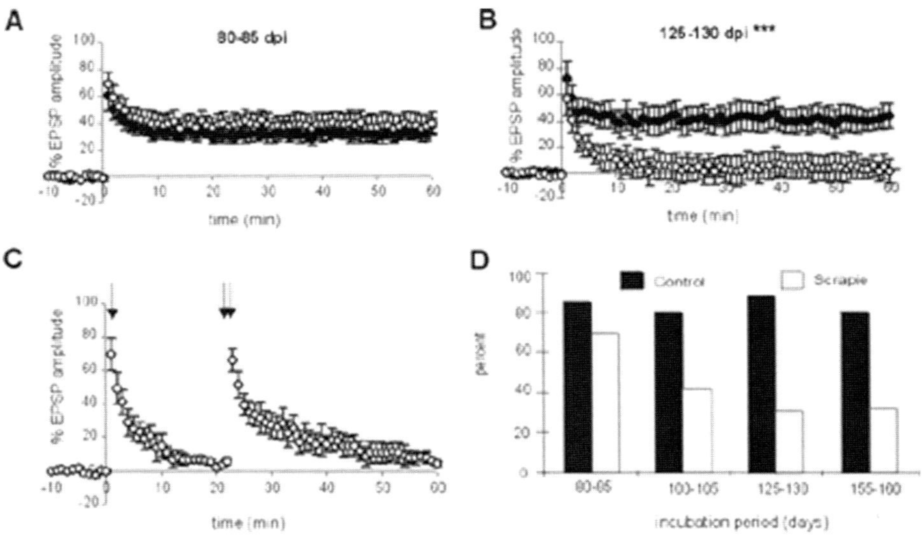

The attenuation of LTP and the loss of the field potential is well supported by data obtained during our intracellular recordings. The recordings shown in Figure 6.3 used a complex protocol in which stimulus pulses were applied to the Schaffer collateral pathway while simultaneously passing a series of hyperpolarising and depolarising pulses through the recording microelectrode (Figure 6.3B). Stimulation of this pathway evokes a complex response involving feed-back and feed-forward inhibition as well as direct synaptic excitation of the cell. As well as the obvious increase in membrane resistance, it is clear from these recordings that both the inhibitory and excitatory synaptic potentials evoked by Schaffer collateral stimulation are massively attenuated in the cell recorded from the scrapie infected animal. Because of the complexity of the local circuitry this could have several explanations, the most likely of which is that direct excitatory synapses are lost because the dendritic spines are lost through deafferentation. The inhibitory synaptic potentials are lost both through direct deafferentation of the primary

Figure 6.7. The effect of scrapie-infection on electrophysiological properties related to potassium currents in CA1 pyramidal cells. A shows superimposed action potentials from control (grey) and scrapie-infected (black) cells. The spike decay time is much shorter in the scrapie-infected cell and the fast and medium AHPs are much deeper fAHP, mAHP). B shows superimposed recordings from control (grey) and scrapie-infected (black) slices and demonstrates the loss of the slow AHP that normally prevents repetitive firing of calcium spikes in this preparation treated with TTX and TEA. Note that as well as permitting repetitive firing, scrapie infection has also removed the afterhyperpolarisation following the depolarising current pulse. C shows the effects of adding 1μM TTX and 10mM TEA on the resting membrane potential in control and scrapie-infected mice. Asterisk denotes a significant difference between data sets (Student's t test, $P < 0.05$). SVF1 mice injected with normal brain homogenate of 301V scrapie.

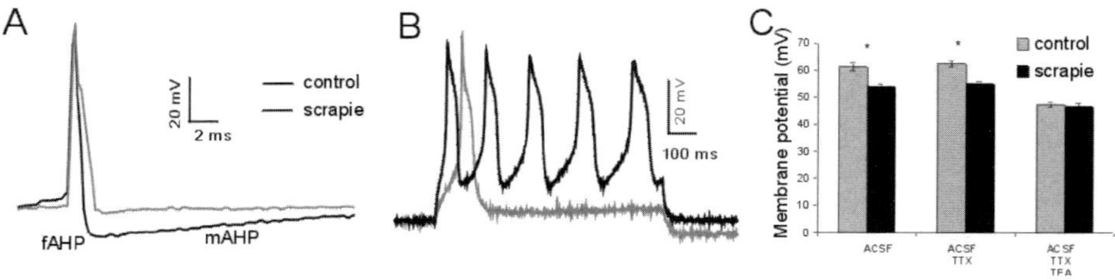

recorded neuron but also because of excitatory deafferentation of the feed-back and feed-forward local inhibitory interneurones.

One other feature noticeable in Figure 6.3 is the alteration to the shape of the action potential wave forms in the cells from scrapie infected animals. The small negative afterhyperpolarising deflections immediately following the action potentials appear to be larger. This change of waveform was confirmed by a careful analysis of many action potentials from the two experimental groups, where it was shown that the fast and medium components of the AHP in the scrapie infected animals were larger than in control animals. This is illustrated in Figure 6.7, which also illustrates the complete loss of a much slower late AHP which normally prevents repetitive firing in scrapie infected animals. In this particular example, the recording was made from a preparation treated with TTX to inhibit sodium currents, so that the wide action potential shown are generated by calcium ions entering the cell via voltage gated calcium channels. The histogram in Figure 6.7C shows one further change noted in scrapie infected animals, namely that the membrane potential recorded in CA1 pyramidal cells was slightly depolarised, possibly indicating attenuation of a leak potassium current such a I_m.

In the only other comprehensive study carried out on hippocampal cells in scrapie infected animals, the alteration of afterhyperpolarising potentials was the only observed difference between cells from scrapie infected and control animals[20]. This may at first seem a curious result when considered in the light of the comprehensive series of changes discussed above but when considered with the knowledge that virtually all different combinations of scrapie strain and animal genotype produce different patterns of neuropathology and disease incubation period. In fact, these are the two parameters that define scrapie *strain*. Several distinct *strains* of sheep scrapie are known and these have been of immense use in the study of this group of diseases. The combination of specific strains of scrapie with particular mouse *Sinc* genotypes produces a variety of clearly distinguishable pathological profiles. In some scrapie models the hippocampus remains unaffected while in others, hippocampal pathology can vary from vacuolation alone to a total loss of pyramidal cells accompanied by extensive tissue shrinkage and gliosis. The 22A strain induces none of the sclerosis and vacuolation seen with the ME7 agent. The bottom line of this story is that all scrapie strain/animal combinations are different and in planed investigation of cellular pathophysiology in scrapie, it is important to choose a model in which it is known that the cells you wish to study are damaged or lost in the disease since it would be possible to waste years of valuable research time investigating something where actually no any significant changes should be expected

However, having said that, our own laboratory carried out a study on the lateral geniculate nucleus of mice in which one eye had been inoculated with scrapie in a well known and much investigated animal model where retinal electrophysiological changes had previously been documented. To our great surprise we found no electrophysiological changes other than the fact that as the disease progressed, LGN relay neurones became harder to locate with our microelectrodes, even for extracellular recording (Figure 6.8)[21]. This, in spite of the fact the LGN undergoes profound pathological change and neurone loss in this model[22,23]. It appears from this data that the cells simply died without showing any detectable pathophysiological changes. However this is extremely unlikely, if impossible and probably reflects the lack of suitable techniques for picking up very subtle pathophysiological changes. Also it is worth noting that neither in the LGN model nor the hippocampal model used in our laboratory was there any sign of paroxysmal activity observed, mirroring the presence of paroxysmal activity in vCJD and it's absence in sCJD. This again emphasises the importance of not assuming that all spongiform encephalopathies (SEs) are the same.

To summarise this section; in an experimental model of scrapie in which electrophysiological recordings were made from neurones with a previously known and clearly demonstrable pathology, profound physiological changes preceded cell death. These changes included loss of measurable afferent inputs, reduced membrane potential, increased

Figure 6.8. Examples of current-voltage experiments from the dLGN of control and scrapie infected mice. A and B show responses of thalamic relay cells to depolarising and hyperpolarising current pulses delivered from holding potentials of –55 and –70 mV respectively. C, D, and E show histograms of properties of the optic tract evoked maximal EPSP in dLGN relay cells at various points throughout the incubation period. No changes were detectable in any parameters of these cells recorded from control and scrapie infected slices. Experimental animals were C57BL mice injected intraocularly with ME7 scrapie or normal brain homogenate. The time course for this particular model was approximately 240–250 days and all recordings shown here are from terminally infected animals and age-matched controls.

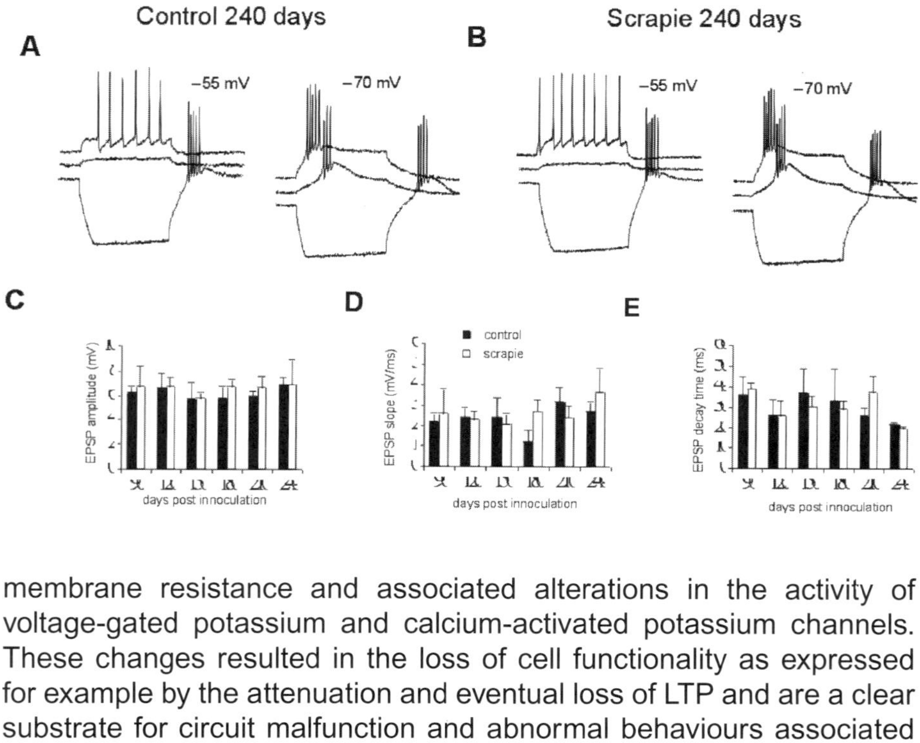

membrane resistance and associated alterations in the activity of voltage-gated potassium and calcium-activated potassium channels. These changes resulted in the loss of cell functionality as expressed for example by the attenuation and eventual loss of LTP and are a clear substrate for circuit malfunction and abnormal behaviours associated with SEs. The particular model chosen and described has clearly set a gold standard for studies of this kind and is absolutely ideal for studies of scrapie neuropathology and pathophysiology.

6.4. Studies on PrP-depleted tissues and cells

6.4.1. Synaptic Properties

In an attempt to discover the elusive function of PrP, several different physiological approaches have been used to study the effect of life without PrP. This has invariably involved the use of genetically engineered mice in which the PrP gene has either been compromised or ablated using a number of different methodologies[24–27]. These mice do not normally become infected when challenged with intracerebral inoculations of scrapie infected brain suggesting that PrP is essential for SE disease to occur[24,28].

In a first, well quoted study, Collinge et al, studied LTP in the hippocampus of PrP-null mice and reported a significant reduction of LTP in the CA3-CA1 pathway[29]. They also demonstrated a loss of functional $GABA_A$-mediated synaptic transmission, which they concluded was responsible for the loss of LTP according to the metaplasticity argument originally proposed by Coan et al.[30]. This suggested that the prior history of activation of the relevant sub-set of NMDA receptors was important in determining whether a particular stimulus pattern would generate LTP. The result suggested that PrP was necessary for normal synaptic transmission to occur at the hippocampal synapses under investigation. This group went on to successfully *rescue* this phenotype by the insertion of multiple copies of human transgene encoding PrP into their experimental mice[31].

Figure 6.9. LTP is attenuated in aged PrP-null mice. Part A shows data from *in vitro* LTP experiments performed on slices from neonatal mice). No differences were seen in hippocampal LTP between juvenile PrP-null and WT mice. Part B shows data from *in vitro* LTP experiments performed on slices from aged mice. In this series of experiments there was a clear attenuation of PTP/LTP in the PrP-null mice compared with WT mice. Part C shows the data from paired pulse experiments carried out on these older animals. There was no difference between the two groups. Part D shows Figure 9B re-drawn on an expanded scale to show in more detail the relative amplitudes of the initial PTP in null and WT animals. PTP is clearly attenuated in the PrP-null mice (ANOVA, $P < 0.001$). Graphs show mean values ± standard error of the mean. Arrows indicate the point of application of conditioning stimuli).

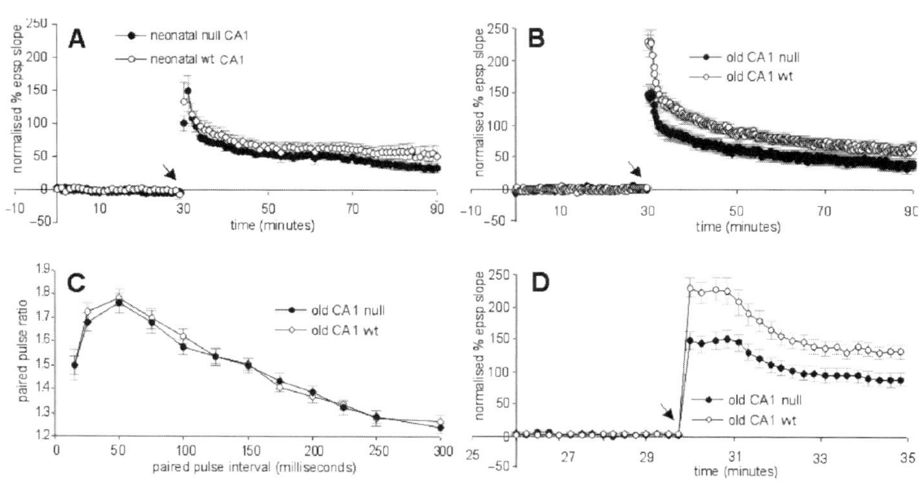

Several laboratories have tried to repeat these experiments with mixed success. Attempts using the same C57BL/129 PrP knockout mouse used in the Collinge experiments were unsuccessful, revealing no changes in LTP or in $GABA_A$ receptor-mediated currents,[32]. In a detailed patch clamp analysis of synaptic transmission in cerebellar Purkinje cells, Herms et al were unable to detect any differences in either spontaneous $GABA_A$ inhibitory post synaptic currents or in climbing fibre evoked excitatory post-synaptic currents between PrP-null and control animals using the same PrP-null construction as the Collinge group[28,33]. Experiments in our own laboratory using an in-bred Ola 129 PrP knockout mouse were able to reproduce the loss of CA1 LTP very convincingly, both in $PrP^{-/-}$ and $PrP^{+/-}$ mice[24,34]. These data show a marked similarity with our LTP experiments in scrapie-infected mice and suggested that there might be a relationship between the ability of PrP^C to function normally and the presence of rapidly increasing levels of PrP^{Sc}. The situation became even more complicated when further experiments in our laboratory on the same line of 129 Ola knockout mice have failed to repeat this observation in either young adult or juvenile mice either *in vitro* or *in vivo*. However, a further recent series of experiments from our own lab has helped to shed some light on the problem. In these experiments on mice aged over 12 months, a clear attenuation of LTP was demonstrated, including a large reduction in the post-tetanic potentiation phase of the experiments. Both these sets of data were recently published together and are illustrated in Figures 6.9[35]. No changes were noticed in the short-term form of plasticity known as Paired Pulse Facilitation (PPF) at these synapses in young adult or aged PrP-null mice which could be taken as confirming that these observations in older animals reflect a compromise of post-synaptic events. The compromise of PTP suggests otherwise. Like PPF, PTP is a phenomenon associated with the elevation of pre-synaptic intracellular calcium levels but unlike PPF, the slower PTP is dependent upon the slow release of calcium stored in mitochondria present in the pre-synaptic boutons[36]. This is potentially a particularly important observation in light of the recent report that a high proportion of mitochondria in young adult PrP-null mice have abnormal morphology, being swollen with fewer cristae[37].

Carleton et al have recently investigated synaptic transmission in the mouse hippocampus of PrP-null mice using plots of input-output curves of the Schaffer collateral evoked field potential demonstrated a facilitation of transmission in the CA3-CA1 pathway, which appeared to be dependent on gene dosage[38]. This observation differs from our own experiments on the input-output properties of the CA1 field EPSP in which we failed to detect a change, possibly as a result of the well-known difficulty associated with making accurate measurements of fibre volley amplitudes from field potential recordings[34]. However, a more recent series of experiments in our laboratory has been able to confirm the Carleton result, indicating indeed that there is facilitation of excitability in the CA3-CA1 pathway (Patterson, Curtis and MacLeod in preparation).

6.4.2. Intrinsic membrane properties of central neurones

Several studies have reported on the intrinsic properties of CA1 pyramidal cells in PrP-null mice[29,39,40]. Most of the biophysical properties of the neurones in PrP-null animals have been shown to be indistinguishable from cells in control animals. However, Colling et al were able to demonstrate a disruption in calcium-activated potassium currents that resulted in the abolition of the fast CTX-sensitive AHP, generated by I_C and the late, slow AHP, generated by I_{AHP}[40]. These are the same two potassium currents that we have shown to be disrupted in CA1 cells by scrapie infection[18,19]. Another striking similarity of these results with our scrapie work is the demonstration of a significant alteration in spike-frequency accommodation in PrP-null mice, such that there is a significant reduction in the number of action potentials evoked during the first 100 ms of a depolarising pulse. An almost identical pattern was seen in the scrapie infected slices and adds to the idea that the enhanced endocytotic cycling or accumulation of abnormal PrP may interfere with the normal functioning of this molecule.

Results obtained with conventional gene knockout technology should always be treated with caution because the absence of a particular sequence of DNA may inadvertently cause the up- or down-regulation of other genes which may have a bearing on the function of the protein under investigation or on the normal development of the animals. To try and overcome this problem a technique has recently been developed using a Cre-loxP system in which the knockout can be specifically targeted to genes in the adult mouse[41]. Using this system to acutely knockout PrP, the Collinge/Jeffreys group have conducted a further study of CA1 pyramidal cell intrinsic properties in which their results are much in line with their previous study, revealing no changes in basic membrane properties other than an attenuation of the medium and slow AHPs following a train of stimulus pulses delivered down the recording microelectrode[42].

We have recently carried out a series of experiments in our laboratory on the intrinsic membrane properties of CA1 pyramidal cells of 129 Ola PrP-null mice and have been able to confirm that there appear to be no differences between PrP-null and control mice. These experiments, using whole cell patch recording showed no detectable difference in membrane potential, membrane resistance, membrane time constant, action potential threshold. However, in contrast to both the Colling and Mallucci studies were unable to confirm the loss of medium and slow AHPs, or of spike frequency adaptation (Curtis and MacLeod, *in preparation*). The difference in our results and those of Collinge et al. might possibly arise from the different PrP-null mouse used or the different calcium buffering properties of whole cell patch electrodes which might mask changes in calcium influx through voltage gated calcium channels.

6.4.3. PrP, calcium and copper

From the foregoing discussion it would seem that a consensus is developing suggesting the absence of PrP has an effect on neuronal calcium, some aspects of which are also altered in cells infected with scrapie. In particular, calcium activated potassium currents have been shown to be attenuated in the majority of published studies and the age-related loss of LTP and PTP described above might well be related to reduced mitochondrial calcium buffering[35]. Direct evidence that calcium homeostasis was altered in PrP-deficient mice was obtained by Herms et al, in which they measured residual calcium levels, depolarisation induced calcium entry and performed direct patch clamp analysis of voltage gated calcium channels in cerebellar Purkinje cells. They noted reductions in basal calcium and in potassium depolarisation induced calcium entry but no changes in voltage gated calcium currents[43] although it had been shown earlier that recombinant PrP elevated free calcium in rat brain synaptosomes via calcium gated calcium channels[44]. In a more recent study on voltage gated calcium channels in cerebellar granule cells prepared from wild type and PrP-null mice, Herms' group have convincingly demonstrated a nifedipine-reversible attenuation of voltage gated calcium channels by recombinant PrP from mouse, bovine and human which depends on the presence of the N-terminal octapeptide repeat region of the protein[45]. Crucially, not only must the octapeptide region be present but the expected four Cu(II) ions must be bound to the protein for the full inhibitory effect to be seen. Clearly this effect does not involve a direct interaction with membrane bound PrP^Cas it is unaltered in PrP-null mice with no membrane bound PrP in contrast to the observed toxicity of the PrP fragment, $PrP_{106-126}$[46]. This interesting work followed on from an earlier study that had suggested that copper might be an important factor in whatever functional role PrP played at central synapses[47–49]. This in spite of an earlier study from this group suggesting that PrP played no role in modulating neuronal excitability or synaptic transmission in cerebellar Purkinje cells[33].

Figure 6.10. Light and electron micrographs to show the distribution of PrP^C in the neuropil of the hamster hippocampus. A is a light micrograph showing the distribution of reaction PrPC revealed by ABC/DAB using 3F4 anti-PrP at a 1:5000 dilution. B is an electron micrograph taken from the stratum oriens in tissue pre-incubated with 3F4 and immunoreactivity revealed with DAB and counterstained with uranyl acetate. A DAB-positive synaptic bouton is seen terminating on a clear, unstained dendritic spine. There is an accumulation of fuzzy dark reaction product associated with the post-synaptic density. Figure generously donated by Nicole Salès, Raymond Hassïg and Ken Moya. Also see Salès, Rodolpho et al, 1998).

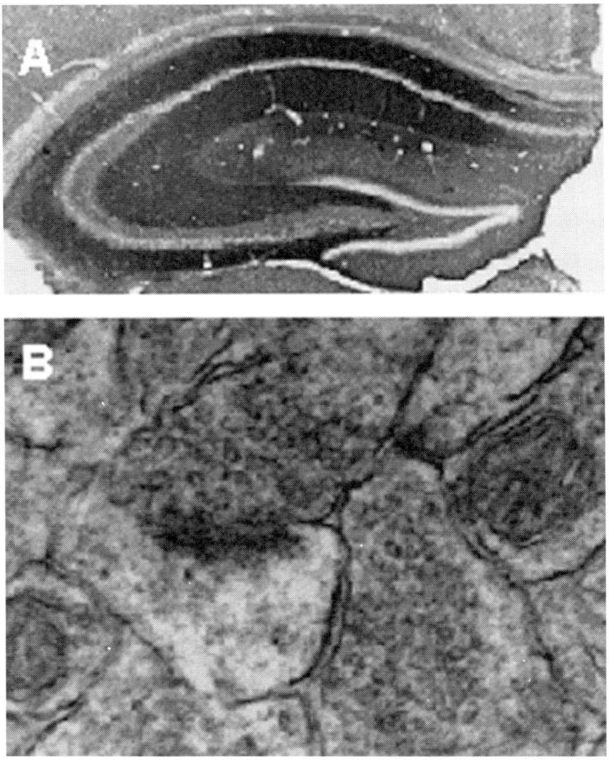

Is there evidence for a functional relationship between PrP, calcium and copper? Several important pieces of circumstantial evidence certainly suggest that there may be and lead to the conclusion of PrP being a multifunctional neuroprotective protein. In the central nervous system PrP is localised in areas of neuropil in close association with synaptic contacts between neurones[50–52]. In the rodent hippocampus, for example, PrP immunoreactivity is not located in the cell bodies of the principle neurones of the pyramidal cell layers and dentate gyrus but in the Stratum Radiatum and Stratum Oriens, where it seems to be localised to synaptic boutons (Figure 6.10). This rather specific localisation alone suggests that PrP may indeed play in role in synaptic function and suggestions have been made concerning possible roles in cell adhesion, endocystosis and copper buffering/transport and free radical dismutation and detoxification, all of which are important in ensuring that synapses work correctly and efficiently. Of course, it should not be forgotten that PrP exists in many other tissue types, where it may have nothing to do with synaptic function but that is outside the scope of this review[53,54].

During intense synaptic activity the concentration of copper within the synaptic cleft increases from around 15mM up to around 100mM as it is released during synaptic activity[55,56]. Copper is a transition metal which in its unbound, Cu(II) state is weakly oxidising, so the binding of Cu(II) to PrP, immediately removing it from active neuropil, maybe one important role, namely as a copper buffer[57]. Once bound to PrP, the copper appears to assume a role as a catalytic site for the dismutation of oxygen radicals with the catalytic activity increasing with increasing numbers of bound Cu(II) ions[58,59]. It has been estimated that PrP contributes around 10–15% of brain superoxide dismutase (SOD) activity[60], although locally at synapses its contribution maybe much higher. Synapses are metabolic powerhouses stuffed with mitochondria which are one of the major sources of free radical generation in all organisms and have been shown to be abnormal in PrP-null mice[37]. Another source of free radicals at synapses might be PrP itself, since a product of the dismutation reaction is hydrogen peroxide along with the reduction of Cu(II) to Cu(I), which can in turn react with H_2O_2 in the Fenton reaction to generate further free radicals. On balance however it appears that PrP^C is neuroprotective since when copper-loaded it does protect cells against free radical toxicity[58] and cells devoid of PrP are more susceptible to oxidative stress[60]. This idea is further supported by the reduced glutathione reductase activity in PrP-null mice; interestingly the toxicity of copper is greatly enhanced in cell cultures depleted of glutathione[58,62]. PrP has a fairly rapid turn over and there is considerable evidence that binding of copper to PrP stimulates endocytosis at nerve terminals[63]. This could be considered as a function in its own right, or as part of a multitude of intimately related roles for a well conserved molecule that sequesters copper, first for itself, then for any other molecule that currently wants it.

What then are the possible links, if any, between voltage-gated calcium channels, copper, free radicals and PrP? Is PrP the link between calcium and copper buffering and protection against oxidative stress? Intriguingly, copper is known to reduce the influx of calcium ions through voltage gated calcium channels[64] and so PrP may actually protect these channels against the inhibitory action of copper and the absence of PrP could therefore lead to attenuation of calcium channel activity and alterations in cellular calcium homeostasis. The same logic could account for the changes in calcium and calcium-activated potassium channels in SE infections due to an increased turn over of PrP^{C}, PrP^{Sc} conversion and consequent loss of function. Also, the same logic could equally apply to the loss of protection against oxidative stress in PrP null animals, especially older ones in which ROS protection is already diminished by the natural aging process. In SEs the loss of this normal function as suggested above would leave synapses particularly sensitive to free radical damage, reinforced by alterations in calcium homeostasis.

6.5. Studies on other neurodegenerative conditions

There isn't space in such a short article to do justice to the extensive use of electrophysiological methods in the study of other neurodegenerative conditions. These include motoneurone disease, Alzheimer's disease, Huntington's disease and Parkinson's disease, where, as discussed above, it has been particularly helpful in understanding the basic circuitry of the basal ganglia both in normal healthy individuals and in disease. However, it seems appropriate here to consider briefly some of the recent work on models of Alzheimer's disease where extensive use has been made of transgenic animals in which the expression of one or more of the proteins whose activity is known to be altered in the human disease has been increased or decreased. These include mice over-expressing APP, Aβ, Presenilin, and *tau*. Because loss of memory is one of the defining characteristics of the human disease, the use of LTP as a measure of the likely success or failure of many of these models has been particularly prevalent. An example of such a series of experiments from our own laboratory is shown in Figure 6.11, which shows a significant facilitation of LTP in aged transgenic rats engineered to over-express the protein presenilin-I[65]. The result shown here is reminiscent of our recent observations on PrP-null mice in that it is age-related, being expressed in aged animals and not younger ones, as is typical of Alzheimer's disease itself. Very similar data has been obtained independently from several laboratories using PS1-over-expressing mouse models[66–69]. All of this data and data obtained using other methodologies suggest that the enhanced LTP phenotype results from an elevation in the storage and release of calcium from intracellular calcium stores such as the endoplasmic reticulum[67,70,71].

Currently, the most successful animal models of Alzheimer's disease are transgenic mouse which over expresses a mutant form of human amyloid precursor protein (APP)[72,73]. These mice develop Alzheimer-type pathology and show age-related alterations in synaptic

Figure 6.11. LTP recorded in slices from control (filled circles) and transgenic (open circles) rats at (A) 6 months and (B) 18 months old. The Schaffer collaterals (A and B) or perforant path (C and D) were conditioned at 30 and 90 minutes (arrows). (A) LTP recorded from CA1 in slices from 6 month old transgenic rats was not significantly different from that recorded in slices from control rats. (B) LTP recorded from CA1 in slices from 18 month old transgenic rats was significantly greater than that recorded in slices from wt rats (P = 0.003, ANOVA). (C) LTP recorded from DG in slices from 6 month old transgenic rats was not significantly different from that recorded in slices from control rats. (D) LTP recorded from DG in slices from 18 month old transgenic rats was not significantly greater than that recorded in slices from control rats.

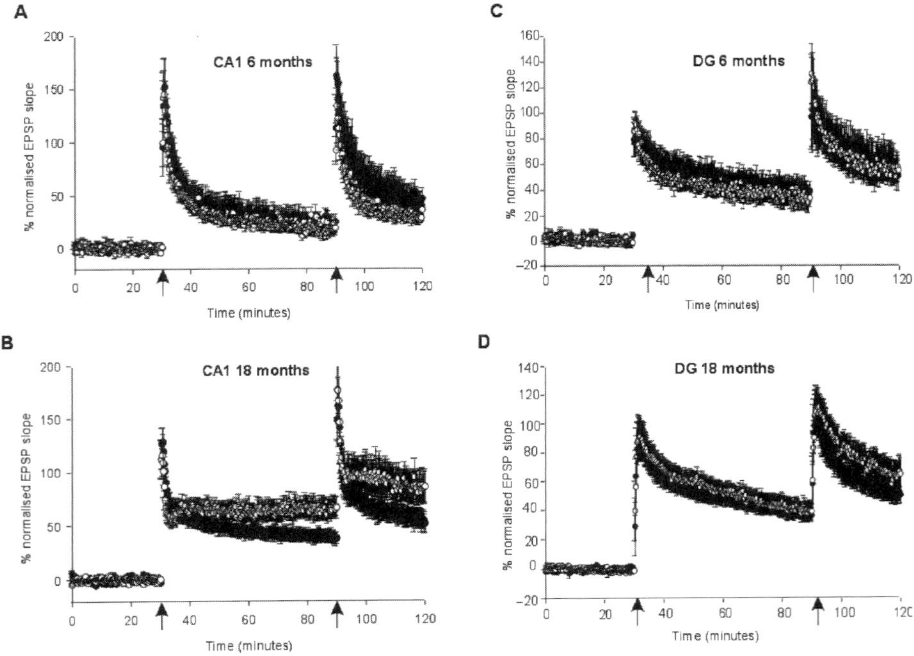

transmission and synaptic plasticity, expressed as a loss or attenuation of LTP in both CA1region and dentate gyrus of the hippocampus[74–76]. Changes reported in hippocampal synaptic plasticity in these models this far have not been entirely consistent. For example, the loss of LTP observed in one set of experiments on younger animals being counter intuitively regained as the animals aged [76]. However, in general, as with the SE-related models, electrophysiology has shown consistently that, alterations in calcium homeostasis appear to lie at the heart of the pathophysiology of these conditions, particularly alterations to calcium regulation in the endoplasmic reticulum and the disruption of associated signal transduction cascades[77].

6.6. Conclusions

Compared with genetic analysis and molecular biology, electrophysiology has played a relatively minor role in attempting to unravel the pathology associated with various neurodegenerative conditions. However in combination with these two methodologies, electrophysiology has provided and will continue to provide useful and pertinent information towards the unravelling of neurodegenerative pathology, helping to out together the pieces if a complex puzzle. Where there are clearly targeted neurones, as in Parkinson's disease, the extensive use of electrophysiology has highlighted the crucial importance of disinhibition as a concept in brain physiology and allowed an almost complete picture to be built up of how the physiology of basal ganglia circuitry is systematically altered by the disease. Curiously, very little direct study on the pathophysiology of degenerating dopaminergic neurones has been carried out, so ironically we know less about this than we do about what happens to hippocampal pyramidal cells when they are dying from scrapie.

Much of the work discussed in this short review has concluded that alterations in calcium transport and homeostasis are central to the pathophysiology of at least scrapie and Alzheimer's disease. However, much is still to be done and in particular, detailed studies employing combinations of free radical spectroscopy, electrophysiology, cellular imaging and single cell PCR applied to the roles of mitochondria and other calcium buffering organelles may well find common ground between the studies of Parkinson's disease, Alzheimer's disease and spongiform encephalopathies as well as other conditions such as motoneurone disease (ALS) and Huntington's disease. Experiments starting to use such a wide ranging multidisciplinary approach are currently getting underway in our own laboratory. Only the future will tell how accurate this prophecy will be.

6.7. Acknowedgments

A special thanks to Cathy Black, Ruth Pybus and Andrew Paterson for carrying out many of the experiments and providing much of the inspiration for the work discussed here.

6.8. References

1. L. Galvani, De viribus electricitaits in motu musculari. *Proc. Bologna Acad.* **24**, 80 (1791).

2. R. Caton, The electric currents of the brain. *Brit. Med. J.* **2**, 278 (1875).

3. H. Berger, Uber das elektrenkephalogramm des menschen. *Arch f. Psychiat.* **87**, 527–570 (1929).

4. M. A. B. Brazier, The Electrical Activity of the Nervous System. Pitman, London (1960).

5. I. Zerr and S. Poser, Clinical diagnosis and differential diagnosis of CJD and vCJD. *APMIS* **110**, 88–98 (2002).

6. G. Chevalier and J. M Deniau,. Disinhibition as a basic process in the expression of striatal functions. *Trends Neurosci* **13**, 277–280 (1990).

7. I. C. Kilpatrick, M. S. Starr, T. A. James and N. K MacLeod, Evidence for the involvement of nigrothalamic GABA neurons in circling behaviour in the rat. *Adv. Biochem. Psychopharmac.* **30**, 205–224 (1981).

8. T. Wichmann and M. R. DeLong, Functional and pathophysiological models of the basal ganglia. *Cur. Opin. Neurobiol.* **6**, 751–758 (1996).

9. R. L. Albin, A. B.Young and J. B. Penney, The Functional-Anatomy of Basal Ganglia Disorders. *Trends. Neurosci.* **12**, 366–375 (1989).

10. M. F. Chesselet and J. M. Delfs, Basal ganglia and movement disorders: An update. *Trends. Neurosci.* **19**, 417–422 (1996).

11. J. Feger, Updating the functional model of the basal ganglia. *Trends. Neurosci.* **20**, 152–153 (1997).

12. M. S. Baron, J. L. Vitek, R. A. Bakay, J. Green, Y. Kaneoke, T. Hashimoto, R. S. Turner, J. L. Woodard, S. A. Cole, W. M. McDonald and M. R. DeLong, Treatment of advanced Parkinson's disease by posterior GPi pallidotomy: 1-Year results of a pilot study. *Ann. Neurol.* **40**, 355–366 (1996).

13. N. J. Emptage, C. A. Reid, A. Fine and T. V.Bliss, Optical quantal analysis reveals a presynaptic component of LTP at hippocampal Schaffer-associational synapses. *Neuron* **38**, 797–804 (2003).

14. J. G. R. Jefferys, R. M. Empson, M. A. Whittington and S. B. Prusiner, Scrapie infection of transgenic mice leads to network and intrinsic dysfuntion of cortical and hippocampal neurones. *Neurobiol. Dis.* **1**, 3–15 (1994).

15. J. R. Scott and H. Fraser, Degenerative hippocampal pathology in mice infected with scrapie. *Acta Neuropathol. (Berl)* **65**, 62–68 (1984).

16. M. Jeffrey, W. G. Halliday, J. Bell, A. R. Johnston, N. K. MacLeod, C. Ingham, A. R. Sayers, D. A. Brown, J. R. Fraser. Synapse loss associated with abnormal PrP preceedes neuronal degeneration in the scrapie infected murine hippocampus. *Neuropathol. Appl. Neurobiol.* **26**, 41–54 (2000).

17. A. R. Johnston, J.R. Fraser, M. Jeffrey and N. K. MacLeod, Synaptic plasticity in the CA1 area of the hippocampus of scrapie-infected mice. *Neurobiol. Dis.* **5**, 188–195 (1998).

18. A. R. Johnston, C. Black, J. Fraser and N. K. MacLeod, Scrapie infection alters the membrane and synaptic properties of mouse hippocampal CA1 pyramidal neurones. *J. Physiol.* **500**, 1–15 (1997).

19. A. R. Johnston, J. R. Fraser, M. Jeffrey and N. K. MacLeod, Alterations in potassium currents may trigger cell death in murine scrapie. *Exp. Neurol.* **151**, 326–333 (1998).

20. P. A. Barrow, C. D. Holmgren, A. J. Tapper and J. G. Jefferys, Intrinsic physiological and morphological properties of principal cells of the hippocampus and neocortex in hamsters infected with scrapie. *Neurobiol Dis* **6**, 406–423 (1999).

21. C. J. Black, A. R. Johnston, J. R. Fraser, and N. Macleod, Electrophysiological properties of dorsal lateral geniculate neurons in brain slices from ME7 scrapie-infected mice. *Exp Neurol* **149**, 253–261 (1998).

22. J. R. Scott and H. Fraser, Transport and targeting of scrapie infectivity and pathology in the optic nerve projections following intraocular infection. *Prog. Clin. Biol. Res.* **317**, 645–652 (1989).

23. R. Curtis, H. Fraser, J. D. Foster and J.R. Scott, The correlation of electroretinographic and histopathological findings in the eyes of mice infected with the 79A strain of scrapie. *Neuropathol. Appl. Neurobiol.* **15**, 75–89 (1989).

24. J. C. Manson, A. R. Clarke, M. L. Hooper, L. Aitchison, I. McConnell and J. Hope 129/Ola mice carrying a null mutation in PrP that abolishes mRNA production are developmentally normal. *Mo.l Neurobiol.* **8**, 121–127 (1994).

25. H. Büeler, A. Aguzzi, A. Sailer, R. A. Greiner, P. Autenried, M. Aguet and C. Weissmann, Mice devoid of PrP are resistant to scrapie. *Cell* **73**, 1339–1347 (1993).

26. G. R. Mallucci, S. Ratte, E. A. Asante, J. Linehan, I. Gowland, J. G. Jefferys and J. Collinge, Post-natal knockout of prion protein alters hippocampal CA1 properties, but does not result in neurodegeneration. *EMBO J.* **21**,202–210 (2002).

27. S. Sakaguchi, S. Katamine, N. Nishida, R. Moriuchi, K. Shigematsu, T. Sugimoto, A. Nakatani, Y. Kataoka, T. Houtani, S. Shirabe, H. Okada, S. Hasegawa, T. Miyamoto and T. Noda, Loss of cerebellar Purkinje cells in aged mice homozygous for a disrupted PrP gene. *Nature* **380**, 528–531 (1996).

28. H. Büeler, M. Fischer, Y. Lang, H. Bluethmann, H.-P. Lipp, S.J. DeArmond, S. B. Prusiner, M. Aguet and C. Weissmann, Normal development and behaviour of mice lacking the neuronal cell-surface PrP protein. *Nature*, **356**, 577–582, (1992).

29. J. Collinge, M. A. Whittington, K. C. Sidle, C. J. Smith, M. S. Palmer, A. R. Clarke, and J. G. Jefferys, Prion protein is necessary for normal synaptic function. *Nature* **370**, 295–7 (1994).

30. E. J. Coan, A. J. Irving, and G. L.Collingridge, Low-frequency activation of the nmda receptor system can prevent the induction of ltp. *Neurosci. Lett.* **105**, 205–210 (1989).

31. M. A. Whittington, K. C. Sidle, I. Gowland, J. Meads, A. F. Hill, M. S. Palmer, J. G. Jefferys, and J. Collinge, Rescue of neurophysiological phenotype seen in PrP null mice by transgene encoding human prion protein. *Nat Genet* **9**, 197–201 (1995).

32. P. M. Lledo, P. Tremblay, S. J. DeArmond, S. B. Prusiner, and R. A. Nicoll, Mice deficient for prion protein exhibit normal neuronal excitability and synaptic transmission in the hippocampus. *Proc Natl Acad Sci U S A* **93**, 2403–7 (1996).

33. J. W. Herms, H. A. Kretzchmar, S. Titz, and B. U. Keller, Patch-clamp analysis of synaptic transmission to cerebellar purkinje cells of prion protein knockout mice. *Eur J Neurosci* **7**, 2508–12 (1995).

34. J. C. Manson, A. R. Clarke, P. A. McBride, I . McConnell and J. Hope PrP dosage and long term potentiation. *Neurodegeneration* **4**, 113–114 (1995).

35. J. Curtis, M. Errington, T. Bliss, K. Voss, and N. Macleod, Age-dependent loss of PTP and LTP in the hippocampus of PrP-null mice. *Neurobiol. Dis.* **13**, 55–62 (2003).

36. Tang, Y.-G. and Zucker, R. S. Mitochondrial involvement in post-tetanic potentiation of synaptic transmission. *Neuron* **18**, 483–491 (1997).

37. G. Miele, M. Jeffrey, D. Turnbull, J. Manson and M.Clinton, Ablation of cellular prion protein expression affects mitochondrial numbers and morphology. *Biochem. Biophys. Res. Commun.* **291**, 372–377 (2002).

38. A. Carleton, P. Tremblay, J. D.Vincent and P. M. Lledo, Dose-dependent, prion protein (PrP)-mediated facilitation of excitatory synaptic transmission in the mouse hippocampus. *Pflugers Arch.* **442**, 223–229 (2001).

39. P.-M. Lledo, G. O. Hjelmstad, S. Mukherji, T. R. Soderling, R. C. Malenka and R. A. Nicoll. Calcium/calmodulin-dependent kinase II and long-term potentiation enhance synaptic transmission by the same mechanism. *Proc. Natl. Acad. Sc. USA* **92**, 11175–11179 (1995).

40. S. B. Colling, J. Collinge, and J. G. R. Jefferys, Hippocampal slices from prion protein null mice: disrupted Ca^{2+}- activated K^{+} currents. *Neurosci. Lett.* **209**, 49–52 (1996).

41. B. Sauer and N. Henderson, Cre-stimulated recombination at loxP-containing DNA sequences placed into the mammalian genome. *Nucl. Acid Res.* **17**, 147–161 (1989).

42. G. Mallucci, A. Dickinson, J. Linehan, P. C. Klohn, S. Brandner and J. Collinge, Depleting neuronal PrP in prion infection prevents disease and reverses spongiosis. *Science.* **302**, 871–874 (2003).

43. J. W. Herms, S. Korte, S. Gall, I. Schneider, S. Dunker and H. A. Kretzschmar, Altered intracellular calcium homeostasis in cerebellar granule cells of prion protein-deficient mice. *J. Neurochem.* **75**, 1487–1492 (2000).

44. S. A. Whatley, J. F. Powell, G. Politopoulou, I. C. Campbell, M. J. Brammer and N. S. Percy, Regulation of intracellular free calcium levels by the cellular prion protein. *Neuroreport* **6**, 2333–2337 (1995).

45. S. Korte, N. Vassallo, M. L. Kramer, H. A. Kretzschmar, and J. Herms, Modulation of L-type voltage-gated calcium channels by recombinant prion protein. *J. Neurochem.* **87**, 1037–1042 (2003).

46. D. R. Brown, J. Herms and H. A. Kretzschmar, Mouse cortical cells lacking cellular PrP survive in culture with a neurotoxic PrP fragment. *Neuroreport* **5**, 2057–2060 (1994).

47. D. R Brown., K. Qin, J. W. Herms, A. Madlung, J. Manson, R. Strome, P. E. Fraser, T. Kruck , A. von Bohlen, W. Schulz-Schaeffer, A. Giese, D. Westaway and H. Kretzschmar, The cellular prion protein binds copper *in vivo*. *Nature* **390**, 684–687 (1997).

48. D. R. Brown, Prion and prejudice: normal protein and the synapse. *Trends Neurosci.* **24**, 85–90 (2001).

49. D. R. Brown, Copper and prion disease. *Brain Res Bull* **55**, 165–173 (2001).

50. N. Sales, K. Rodolfo, R. Hassig, B. Faucheux, L. Di Giamberardino and K. L. Moya, Cellular prion protein localization in rodent and primate brain. *Eur. J. Neurosci.* **10**, 2464–2471 (1998).

51. J. Herms, T. Tings, S. Gall, A. Madlung, A. Giese, H. Siebert, P. Schurmann, O. Windl, N. Brose and H. Kretzschmar, Evidence of presynaptic location and function of the prion protein. *J Neurosci* **19**, 8866–8875 (1999).

52. Fournier, J. G., Escaig, H. and Grigoriev, V. Ultrastructural localization of prion proteins: physiological and pathological implications. *Microsc Res Tech* **50**, 76–88 (2000).

53. M. Horiuchi, N. Yamazaki, T. Ikeda, N.Ishiguro and M. Shinagawa. A cellular form of prion protein (PrP(C)) exists in many non-neuronal tissues of sheep. *J. Gen. Virol.* **76**, 2583–2587 (1995).

54. J.-G. Fournier, F. Escaig-Haye, T. Billette de Villemeur, O. Robain, C. I. Lasmezas, J. P. Deslys, D. Dormont and P. Brown. Distribution and submicroscopic immunogold localization of cellular prion protein (PrPc) in extracerebral tissues. *Cell Tissue Res.* **292**, 77–84 (1998).

55. D. E. Hartter and A. Barnea, Evidence for release of copper in the brain: depolarization-induced release of newly taken-up 67copper. *Synapse* **2**, 412–415 (1988).

56. J. Kardos, I. Kovacs, F. Hajos, M. Kalman, and M. Simonyi, Nerve endings from rat brain tissue release copper upon depolarization. A possible role in regulating neuronal excitability. *Neurosci. Lett.* **103**, 139–144 (1989).

57. B. Halliwell and J. M. Gutteridge, Oxygen toxicity, oxygen radicals, transition metals and disease. *Biochem J* **219**, 1–14 (1984).

58. D. R. Brown, Clive, C. and S. J. Haswell, Antioxidant activity related to copper binding of native prion protein. *J Neurochem* **76**, 69–76 (2001).

59. A. R. White, S. J. Collins, F. Maher, M. F. Jobling, L. R. Stewart, J. M. Thyer, K. Beyreuther, C. L. Masters and R. Cappai, Prion protein-deficient neurons reveal lower glutathione reductase activity and increased susceptibility to hydrogen peroxide toxicity. *Am. J. Pathol.* **155**, 1723–1730 (1999).

60. B. S. Wong, T. Pan, T. Liu, R. Li, P. Gambetti and M. S. Sy, Differential contribution of superoxide dismutase activity by prion protein in vivo. *Biochem. Biophys. Res. Commun.* **273**, 136–139 (2000).

61. D. R. Brown, W. J. Schulz-Schaeffer, B. Schmidt and H. A. Kretzschmar, Prion protein-deficient cells show altered response to oxidative stress due to decreased SOD-1 activity. *Exp. Neurol.* **146**, 104–112 (1997).

62. A. R. White, A. I. Bush, K. Beyreuther, C. L. Masters and R. Cappai, Exacerbation of copper toxicity in primary neuronal cultures depleted of cellular glutathione. *J. Neurochem.* **72**, 2092–2098 (1999).

63. P. C. Pauly and D. A. Harris, Copper stimulates endocytosis of the prion protein. *J. Biol. Chem.* **273**, 33107–33110 (1998).

64. S. C. Nam and P. E. Hockberger, Divalent ions released from stainless steel hypodermic needles reduce neuronal calcium currents. *Pflugers Arch.* **420**, 106–108 (1992).

65. R. Pybus, E. Barnard, P. Estibeiro, J. Mullins, and N. Macleod, Enhanced long-term potentiation in the hippocampus of rats expressing mutant presenillin-1 is age related. *Neurob. Dis.* **12**, 212–224 (2003).

66. A. Parent, D. J. Linden, S. S. Sisodia and D. R. Borchelt, Synaptic transmission and hippocampal long-term potentiation in transgenic mice expressing FAD-linked presenilin 1. *Neurobiol. Dis.* **6**, 56–62 (1999).

67. S. H. Zaman, A. Parent, A. Laskey, M. K. Lee, D. R. Borchelt, S. S. Sisodia and R. Malinow. Enhanced synaptic potentiation in transgenic mice expressing presenilin 1 familial Alzheimer's disease mutation is normalized with a benzodiazepine. *Neurobiol. Dis.* **7**, 54–63 (2000).

68. R. S. Erzurumlu, S. Jhaveri, and G. E. Schneider, Distribution of morphologically different retinal axon terminals in the hamster dorsal lateral geniculate nucleus. *Brain Res.* **461**, 175–181 (1988).

69. I. Schneider, D. Reverse, I. Dewachter, L. Ris, N. Caluwaerts, C. Kuiperi, M. Gilis, H. Geerts, H. Kretzschmar, E. Godaux, D. Moechars, F. Van Leuven and J. Herms, Mutant presenilins disturb neuronal calcium homeostasis in the brain of transgenic mice, decreasing the threshold for excitotoxicity and facilitating long-term potentiation. *J. Biol. Chem.* **276**, 11539–11544 (2001).

70. J. G. Begley, W. Duan, S. Chan, K. Duff, and M. P. Mattson,. Altered calcium homeostasis and mitochondrial dysfunction in cortical synaptic compartments of presenilin-1 mutant mice. *J. Neurochem.* **72**, 1030–1039 (1999).

71. M. A. Leissring, T. R. Yamasaki, W. Wasco, J. D. Buxbaum, I. Parker, F. M. LaFerla. Calsenilin reverses presenilin-mediated enhancement of calcium signaling. *Proc. Natl. Acad. Sci. U S A* **97**, 8590–8593 (2000).

72. D. Games, D. Adams, R. Alessandrini, R. Barbour, P. Borthelette, C. Blackwell, T. Carr, J. Clemens, T. Donaldson, F. Gillespie, T. Guido, S. Hagopian, K. Johnson-Wood, K. Khan, M. Lee, P. Leibowitz, Ivan Lieberburg, S. Little, E. Masliah, L. McConlogue, M. Montoya-Zavala, L. Mucke, L. Paganini, E. Penniman, M. Power, D. Schenk, P. Seubert, B. Snyder, F. Soriano, H. Tan, J. Vitale, S. Wadsworth, B. Wolozin and J. Zhao, Alzheimer-type neuropathology in transgenic mice overexpressing V717F beta-amyloid precursor protein. *Nature* **373**, 523–527 (1995).

73. K. Hsiao, P. Chapman, S. Nilsen, C. Eckman, Y. Harigaya, S.Younkin, F. Yang and G. Cole, Correlative memory deficits, Abeta elevation, and amyloid plaques in transgenic mice. *Science* **274**, 99–102 (1996).

74. J. Larson, G. Lynch, D. Games and P. Seubert, Alterations in synaptic transmission and long-term potentiation in hippocampal slices from young and aged PDAPP mice. *Brain Res* **840**, 23–35 (1999).

75. P. F. Chapman, G. L. White, M. W. Jones, D. Cooper-Blacketer, V. J. Marshall, M. Irizarry, L. Younkin, M. A. Good, T. V. Bliss, B. T. Hyman, S. G. Younkin K. K. Hsiao, Impaired synaptic plasticity and learning in aged amyloid precursor protein transgenic mice. *Nat. Neurosci.* **2**, 271–276 (1999).

76. J. Giacchino, J. R. Criado, D. Games and S. Henriksen, In vivo synaptic transmission in young and aged amyloid precursor protein transgenic mice. *Brain. Res.* **876**, 185–190 (2000).

77. M. P. Mattson, F. M. LaFerla, S. L. Chan, M. A. Leissring, P. N. Shepel and J. D. Geiger, Calcium signaling in the ER: its role in neuronal plasticity and neurodegenerative disorders. *Trends Neurosci* **23**, 222–229 (2000).

Chapter 7

PRION PROTEIN, PRION PROTEIN-LIKE PROTEIN, AND NEURODEGENERATION

Suehiro Sakaguchi

Department of Molecular Microbiology and Immunology, Nagasaki University Graduate School of Biomedical Sciences, Nagasaki 852-8523 and PRESTO, Japan Science and Technology Agency, Saitama, Japan

7.1. Introduction

Gene knockout technology using homologous recombination in embryonic stem cells in mice has made it possible to investigate the physiological functions of a huge number of genes. However, this targeting strategy may alter the integrated complexity of the genome and affect the expression of other genes located near the disrupted gene, thereby profoundly modifying the specific phenotypes resulting from the inactivation of the gene of interest. This neighboring gene effect was strikingly evident in the targeted disruption of the gene for a myogenic basic-helix-loop-helix protein, MRF4, in mice[1]. Three lines of MRF4-null mice, independently generated using different targeting strategies, exhibited tremendously discrepant phenotypes, ranging from complete viability to neonatal lethality, due to the different effects of strategies on the expression of the downstream Myf5 gene[1]. A similar neighborhood effect mediated by an unusual mechanism of intergenic splicing was recently demonstrated in the targeting of prion protein (PrP) gene, *Prnp*[2,3].

The normal cellular isoform of PrP, designated PrP^C, is a membrane glycoprotein anchored by a glycosyl-phosphatidyl-inositol (GPI) moiety and highly expressed in the central nervous system (CNS), particularly in neurons, and to a lesser extent in other non-neuronal tissues including the spleen, kidney, lung, and heart[4,5]. The structural conversion of PrP^C

into the pathogenic isoform, PrP^{Sc}, is known to be an essential process in the pathogenesis of prion diseases, including prion propagation and disease development[6]. Indeed, mice devoid of PrP^{C} (*Prnp*$^{0/0}$) were resistant to the diseases, showing neither the accumulation of PrP^{Sc} nor the propagation of prions in the brain[7–10]. In contrast, marked phenotypic discrepancies were observed among different lines of *Prnp*$^{0/0}$ mice, some lines of the mice developing and growing normally but the others exhibiting progressive ataxia and Purkinje cell degeneration spontaneously at old age[11–13]. It was recently determined that the downstream gene, *Prnd*, encoding a PrP-like protein was ectopically overexpressed in the brain of the latter lines of *Prnp*$^{0/0}$ mice[2,3].

7.2. Neurological phenotypes of mice devoid of PrP

7.2.1. Targeting strategies used in PrP-null mice

Prnp is located on human chromosome 20 and on mouse chromosome 2[5]. Human and mouse *Prnp* consist of two and three exons, respectively[5]. The PrP coding sequence is present only in the last single exon in all species[5].

The lines of *Prnp*$^{0/0}$ mice reported so far have been generated using different targeting strategies as shown in Figure 7.1. The first established

Figure 7.1. Configurations of the targeted *Prnp* alleles in different lines of *Prnp*$^{0/0}$ mice. UTR, untranslational region; TK, thymidine kinase; neo, neomycin phosphotransferase; MT, metallothionein; PGK, phosphoglycerate kinase; HRPT, hypoxanthine phosphoribosyltransferase.

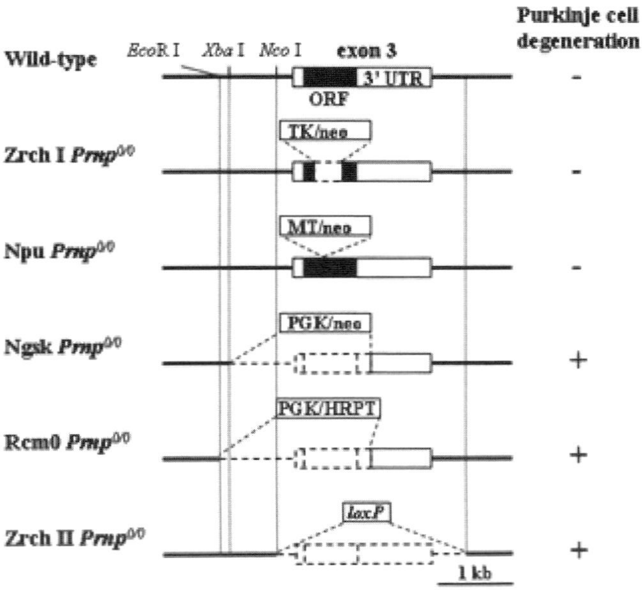

line, Zrch I *Prnp*$^{0/0}$, was generated by replacement of PrP codons 4-187 among a total of 254 codons with the neomycin phosphotransferase (neo) gene under the control of the herpes simplex virus thymidine kinase promoter[11]. This targeting strategy happened to produce a fused mRNA consisting of the neo and the residual *Prnp* sequences in the brains[11]. The second line, Npu *Prnp*$^{0/0}$, contained the disrupted *Prnp* alleles, in which the neo gene under the control of the mouse metallothionein promoter was simply inserted into a unique *Knp* I site in the PrP-coding sequence[12]. In these mice, the residual *Prnp* sequences were not expressed[12]. In a third line of PrP-null mice, Ngsk *Prnp*$^{0/0}$, a 2.1-kb genomic DNA segment including 0.9-kb of intron 2, 10-bp of 5' untranslated region (UTR) of exon 3, the entire PrP open reading frame (ORF), and 0.45-kb of 3' UTR was replaced by the neo gene under the control of the mouse phosphoglycerate kinase (PGK) promoter[14]. A fourth line, Rcm0 *Prnp*$^{0/0}$, was generated by a similar targeting strategy utilized in Ngsk *Prnp*$^{0/0}$ mice[2]. The hypoxanthine phosphoribosyltransferase gene under control of the PGK promoter was used in Rcm0 *Prnp*$^{0/0}$ mice as a selectable marker[2]. In a fifth line of PrP-null mice, Zrch II *Prnp*$^{0/0}$, 0.27-kb of intron 2, the entire exon 3, and 0.6-kb of the 3' flanking DNA segment were targeted by a specific 34-bp *loxP* sequence[15].

7.2.2. Normal embryonic development of PrP-null mice

During embryogenesis, PrP^{C} is ubiquitously expressed both in the nervous system and in various non-neuronal tissues[16]. Büeler et al. reported that Zrch I *Prnp*$^{0/0}$ mice were born in accordance with Mendel's law from a breeding pair of Zrch I heterozygous (*Prnp*$^{0/+}$) mice, showing no obvious defects in those tissues[11]. Normal development was also subsequently demonstrated in the other four lines of *Prnp*$^{0/0}$ mice[2,12,14,15], indicating that PrP^{C} is unnecessary for mammalian embryogenesis.

7.2.3. Learning/memory and synaptic plasticity in PrP-null mice

PrP^{C} is predominantly expressed in pyramidal neurons of the hippocampus[17], which plays an important role in learning and memory. To investigate the role of PrP^{C} in these functions, Büeler et al. subjected Zrch I *Prnp*$^{0/0}$ mice to different behavioral tasks, such as a swimming navigation test and a Y-maze discrimination test[11]. The mutant mice performed these tasks as well as control mice[11]. The swimming navigation test is thought to be effective in evaluating spatial learning ability[18,19], and the Y-maze test in assessing discrimination ability[20]. Nishida et al. also described the normal behavior of Ngsk *Prnp*$^{0/0}$ mice in a swimming navigation test[21]. However, interestingly, they reported that these mutant mice performed very poorly on other tests, including a water-finding test and a conditioned passive-avoidance test[21]. The water-finding test was developed to evaluate latent learning[22], and the passive avoidance task investigates short- and long-term memory[23]. It is possible, therefore, that PrP^C might be involved in certain types of learning and memory.

In electrophysiological studies, Collinge et al. showed that γ-aminobutyric acid type A ($GABA_A$) receptor-mediated fast inhibition was weakened and long-term potentiation (LTP) was impaired in the hippocampal CA1 neurons of Zrch I *Prnp*$^{0/0}$ mice, compared with those of control wild-type mice[24]. They also confirmed that these abnormal phenotypes in Zrch I *Prnp*$^{0/0}$ mice were successfully rescued by re-introduction of multi-copies of a transgene encoding human PrP^C[25]. Loss of LTP was also reported in another line of PrP-null mice, Npu *Prnp*$^{0/0}$ [26]. These results indicate that PrP^C is involved in synaptic transmission and plasticity. The impaired learning and memory observed in Ngsk *Prnp*$^{0/0}$ mice might be attributable to these electrophysiological abnormalities. However, in addition to normal synaptic transmission in Purkinje cells of Zrch I *Prnp*$^{0/0}$ mice[27], the failure to detect impaired LTP in the CA1 region of these mice was reported[28].

7.2.4. Altered circadian rhythm in PrP-null mice

Tobler et al. reported altered sleep patterns and rhythms of circadian activity in both Zrch I *Prnp*$^{0/0}$ and Npu *Prnp*$^{0/0}$ mice[29]. Sleep fragmentation was much more prominent in these PrP-null mice than in wild-type mice. The period of circadian activity was 23.3 h in wild-type mice under a condition of constant darkness. In contrast, it was 23.9 h in the mutant mice. The suprachiasmatic nuclei, a center for circadian rhythm regulation, were morphologically normal in these *Prnp*$^{0/0}$ mice. The authors also showed that these dysregulated circadian activities could be rescued by re-introducing transgenes encoding PrP^C. The results indicate that PrP^C plays an important role in the regulation of circadian rhythm.

7.2.5. Purkinje cell degeneration in PrP-null mice

Zrch I *Prnp*$^{0/0}$ and Npu *Prnp*$^{0/0}$, the first two established lines of PrP-null mice, were reported to grow without any neurological abnormalities[11,12]. However, we found that Ngsk *Prnp*$^{0/0}$ mice began

Figure 7.2. Purkinje cell degeneration in Ngsk *Prnp*$^{0/0}$ mice. Purkinje cells were immunohistochemically stained using anti-calbindin antibodies.

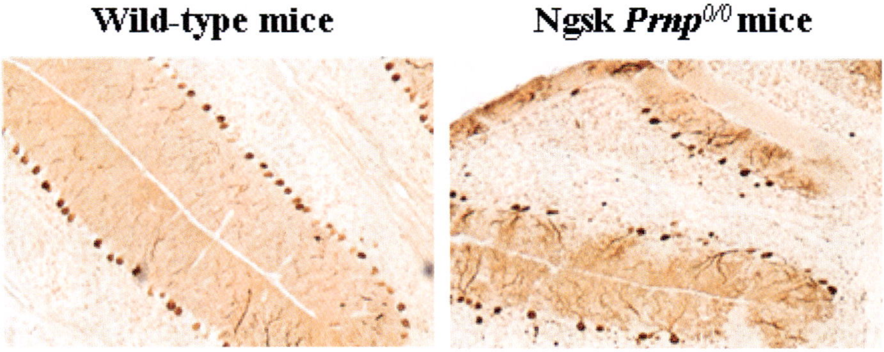

to exhibit progressive ataxia from around 70 weeks old, showing poor performance in a rotorod task, which is a motor coordination test[13]. Macroscopic inspection of the brains of these ataxic mice revealed marked cerebellar atrophy, compared with those of wild-type mice. In these cerebella, Purkinje cells were dramatically decreased in number due to extensive degeneration (Figure 7.2). The molecular layer was much thinner in these mice than in wild-type mice, probably as a result of the loss of the dendritic trees of Purkinje cells (Figure 7.2). Cerebellar granule cells and inferior olive nuclei, two major sources of afferents to Purkinje cells, remained intact. No significant microscopic abnormalities could be detected in the cerebrum of these mice, particularly the cerebral cortex, hippocampus, putamen, thalamus, and hypothalamus. We also subsequently confirmed that the ataxia and Purkinje cell degeneration in Ngsk *Prnp*$^{0/0}$ mice could be successfully rescued by re-introduction of multiple copies of a cosmid transgene encoding mouse PrP^C [30]. Moreover, the same cerebellar phenotypes were reported in other independent lines of *Prnp*$^{0/0}$ mice, Rcm0 *Prnp*$^{0/0}$ and Zrch II *Prnp*$^{0/0}$ [2,15]. These results clearly indicate that PrP^C is necessary for the long-term survival of Purkinje cells.

The cerebellar atrophy in Ngsk *Prnp*$^{0/0}$ mice was not evident until about 40 weeks after birth, and no Purkinje cell loss was detectable in younger Ngsk *Prnp*$^{0/0}$ mice[21], indicating the normal development of Purkinje cells in these mice. In Ngsk *Prnp*$^{0/0}$ mice aged 43 weeks old, on the other hand, torpedo-like varicosities composed of a swollen disorganized myelin sheath were observed along the axons of Purkinje cells, in spite of the fact that the cell bodies and dendritic trees seemed normal[30]. It was previously shown that PrP^C was expressed along the axons and at the axonal terminals of Purkinje cells, but not in the cell bodies and dendrites[17,31]. It is therefore possible that the loss of PrP^C could cause defects in the axonal organization of Purkinje cells, consequently leading to the degeneration of the cells.

7.2.6. Gliosis in PrP-null mice

We also reported another discrepant phenotype between Ngsk *Prnp*$^{0/0}$ and Zrch I *Prnp*$^{0/0}$ mice[32]. In the former, glial cells, including astrocytes and microglia, were markedly activated both in the cerebrum and in the cerebellum in an age-dependent manner. In contrast, the latter showed no such activation of glial cells. However, as in the case of Purkinje cell degeneration, we could confirm that re-introduction of a cosmid transgene encoding mouse PrP^C rescued Ngsk *Prnp*$^{0/0}$ mice from this glial activation[32], indicating that the glial activation in Ngsk *Prnp*$^{0/0}$ mice is attributable to the loss of PrP^C.

Since both astrocytes and microglia were previously shown to express PrP^C abundantly[33,34], it is conceivable that PrP^C expressed on glial cells is involved in the regulation of the glial activation. Interestingly, glial cells were activated much earlier than the point of which Purkinje cell degeneration became detectable in Ngsk *Prnp*$^{0/0}$ mice[32], suggesting the primary role of the activated glial cells in the neurodegeneration. However, it is also conceivable that, aside from Purkinje cell degeneration, undetectable neuronal damages induced by the loss of PrP^C activate glial cells indirectly. Moreover, it is possible that neurons devoid of PrP^C produce a signal(s) to activate surrounding glial cells.

7.2.7. Demyelination in PrP-null mice

We found spongiosis in the spinal cord of Ngsk *Prnp*$^{0/0}$ mice aged 31 weeks[30]. The white matter was widely affected, while the gray matter remained completely intact. Each vacuole in the spongiosis was surrounded by an enlarged myelin sheath. In some cases, splits within a myelin sheath formed vacuoles. Large myelinated fibers were mainly affected. Similar pathologies were detected in the peripheral sciatic nerves of aged Ngsk *Prnp*$^{0/0}$ mice. In addition to many similar vacuoles, large myelinated fibers were prominently reduced in number and remaining axons were thinly myelinated. Onion bulbs, a pathological hallmark of multiple episodes of demyelination and remyelination, were also identified in the affected nerves. We further confirmed that this demyelination in Ngsk *Prnp*$^{0/0}$ mice could be successfully rescued by expressing transgenic mouse PrP^C [30]. The same demyelination was observed in the sciatic nerves of Zrch I *Prnp*$^{0/0}$ mice[30]. These results indicate that PrP^C is involved in the organization of a myelin sheath.

Oligodendrocytes and Schwann cells, professional cells forming myelin sheaths in the CNS and the peripheral nervous system, respectively, are known to express PrP^C on the surface[33,35]. It is therefore conceivable that PrP^C functions as an adhesion molecule within a myelin sheath and/or between a myelin sheath and an axon to form a tightly compacted myelin sheath. It is also possible that PrP^C is a trophic factor for oligodendrocytes and Schwann cells.

7.3. Identification of PrP-like protein and its ectopic expression in ataxic lines of mice devoid of PrP

7.3.1. Aberrant transcripts in ataxic lines of PrP-null mic

It is highly likely that the discrepant phenotypes observed among different lines of PrP-null mice, including ataxia, Purkinje cell degeneration, and gliosis, are attributable to the different targeting strategies used in these mutant mice. The *Prnp* promoter/enhancer region remained intact and therefore still active in all lines of $Prnp^{0/0}$ mice, indicating that this region is not associated with the discrepancy. However, we noticed that 0.27-kb of the 3' part of the *Prnp* intron 2 was commonly targeted in all ataxic lines of $Prnp^{0/0}$ mice (Figure 7.1). In contrast, this part was completely intact in non-ataxic lines of the mice (Figure 7.1). This part of the intron 2 possesses several specific elements that are important for splicing processes of pre-mRNA, such as a splicing acceptor signal, suggesting that the pre-mRNA transcribed from the residual *Prnp*promoter undergoes abnormal splicing in ataxic lines, but not in non-ataxic lines of $Prnp^{0/0}$ mice. Recent evidence shows that each of the pre-mRNA maturation processes, such as transcription, capping, splicing, and cleavage/polyadenylation, functionally interacts with the other processes to produce proper mRNAs[36]. It is conceivable, therefore, that the impaired splicing of the pre-mRNA transcribed from the *Prnp* promoter disturbs other normal processes of the pre-mRNA, leading to the generation of the aberrant transcripts etiologically relevant to the discrepancy.

To detect these predicted aberrant transcripts, we carried out Northern blotting of total RNAs extracted from the brains of ataxic Ngsk $Prnp^{0/0}$ mice with a cDNA probe consisting of the *Prnp* exons 1 and 2 (1/2), because these exons were intact in all lines of PrP-null mice[3]. In wild-type mice, the 2.2-kb authentic PrP mRNA was abundantly expressed in the brain (Figure 7.3). In Zrch I $Prnp^{0/0}$ mice, the 2.4-kb PrP-neo chimeric mRNA was detected as described previously (Figure 7.3). This chimeric mRNA also contained the residual sequence of the *Prnp* exon 3 (Figure 7.3). However, in the brains of Ngsk $Prnp^{0/0}$ mice,

Figure 7.3. Northern blotting of the brains of wild-type, Ngsk $Prnp^{0/0}$, Zrch I $Prnp^{0/0}$, Ngsk $Prnp^{0/+}$, and Ngsk/Zrch $Prnp^{0/0}$ mice using *Prnp* exons 1/2 and exon 3 probes. The 2.2- and 3.4-kb aberrant transcripts, which were hybridized to the exons 1/2 but not to the exon 3 probe, were specifically expressed from the Ngsk targeted *Prnp* allele. (Modified from Li et al.[3])

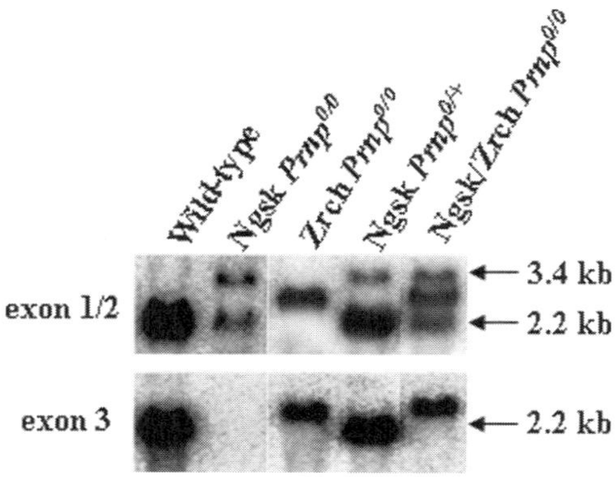

2.2- and 3.4-kb transcripts were specifically detectable (Figure 7.3). In contrast to the PrP-neo chimeric mRNA in Zrch I $Prnp^{0/0}$ mice, these transcripts could not be hybridized to an exon 3 probe (Figure 7.3). These aberrant transcripts were reported in the brains of another ataxic Rcm0 $Prnp^{0/0}$ mouse, but not in another non-ataxic Npu $Prnp^{0/0}$ mouse[2]. Taken together, these results are consistent with the earlier prediction that the abnormal transcripts including the *Prnp* exon 1/2 sequences but not the exon 3 are aberrantly expressed in ataxic lines of $Prnp^{0/0}$ mice.

We next examined whether these aberrant transcripts could be encoded on the targeted *Prnp* allele of ataxic $Prnp^{0/0}$ mice[3]. In the brains of Ngsk $Prnp^{0/+}$ mice, hemizygous for the targeted allele, the aberrant transcripts and the authentic PrP mRNA were expressed at similar levels (Figure 7.3). Moreover, Zrch I/Ngsk chimeric $Prnp^{0/0}$ mice expressed both the PrP-neo fused mRNA and the aberrant transcripts (Figure 7.3). These results clearly show that the expression of the aberrant transcripts is closely associated with the presence of the Ngsk targeted *Prnp* allele, but not with the Zrch I targeted allele, indicating that these aberrant transcripts are specifically encoded on the targeted *Prnp* alleles of ataxic lines of $Prnp^{0/0}$ mice.

7.3.2. The aberrant transcripts encode PrP-like protein

We cloned and sequenced the cDNAs of the aberrant 2.2- and 3.4-kb transcripts expressing in the brains of Ngsk $Prnp^{0/0}$ mice, and showed that they were polyadenylated mRNAs containing the conserved polyadenylation signal 20-bp upstream of the poly(A) start site[3]. The sequences for the *Prnp* exons 1/2 were present in these transcripts, consistent with the results of Northern blotting. Moreover, these sequences were located at the 5' terminus, indicating that the aberrant mRNAs were transcribed from the residual *Prnp* promoter. Novel sequences were identified downstream of the *Prnp* exon 1/2 sequences. According to a detailed sequence analysis, the 2.2- or 3.4-kb signal detected on Northern blotting was composed, not of a single mRNA species, but of at least four species of alternatively spliced mRNAs. Three of these had 103-, 163-, and 266 (103+163)-bp inserts just downstream of the *Prnp* exon 1/ 2 sequences, respectively, while the other contained no insert. The 103-bp inset was predominant among them. In addition to these alternatively spliced inserts, the longer insert of 1,237-bp locating in the center of the novel sequence of the 3.4-kb mRNAs was spliced out in the 2.2-kb mRNAs.

In the novel sequences of the aberrant transcripts, a common ORF encoding a putative protein with 179 amino acids was identified[3]. The deduced amino acid sequence of the protein was remarkably similar to that of PrP on a homology search, sharing ~23% identical amino acids with the C-terminal half of PrP. Thus, we named this putative protein PrP-like protein (PrPLP)[3]. Since Moore et al. independently identified the same sequences using a genomic walking technique and large scale sequencing, and termed it doppel (Dpl)[2], we will refer to this putative protein hereafter as PrPLP/Dpl. Subsequently, the expression of PrPLP/Dp in the brains of ataxic lines *Prnp*$^{0/0}$ mice was confirmed by immunoblotting using anti-PrPLP/Dpl antibodies[15,37].

7.3.3. PrP-like protein is encoded by an independent gene

Is PrPLP/Dpl expressed under physiological conditions? To address this question, we performed Northern blotting of various tissues of adult wild-type mice with the PrPLP/Dpl-coding sequence as a probe[38]. No substantial expression of PrPLP/Dpl mRNA could be detected in the brain where PrP was abundantly expressed[38]. In contrast, considerable amounts of 2.0- and 3.2-kb mRNAs were detectable in the testis, heart, kidney, and spleen[38]. Neither the *Prnp* exons 1/2 nor the exon 3 probes could hybridize to these mRNAs, suggesting that the PrPLP/Dpl expression is regulated by its own promoter, but not by the *Prnp* promoter, in wild-type mice. We also cloned and sequenced the cDNAs of these mRNAs expressing in the testis, and identified a 55-bp unique sequence at the 5' terminus instead of the *Prnp* exon 1/2 sequences, which was followed by the PrPLP/Dpl-coding sequence[3]. No inserts of the 103-, 163-, and 266-bp could be detected downstream of the 55-bp sequence. These results indicate that PrPLP/Dpl is encoded on an independent genomic gene. Moore et al. independently identified the same gene and named it *Prnd*[2].

7.3.4. PrP-like protein is ectopically expressed in neurons and non-neuronal cells of ataxic lines of PrP-null mice

We identified the cells expressing the aberrant PrPLP/Dpl mRNAs in the brain of ataxic Ngsk $Prnp^{0/0}$ mice by in situ hybridization with a probe for the PrPLP/Dpl-coding sequence[38]. Consistent with the results of Northern blotting, we could detect no expression of the PrPLP/Dpl-coding sequence in the brains of wild-type mice. In contrast, in Ngsk $Prnp^{0/0}$ mice, the mRNAs were abundantly expressed in almost all neurons and non-neuronal cells including glial cells throughout the brains, with the strongest expression in pyramidal cells of the hippocampus and in Purkinje cells[38]. The expression of the mRNAs in glial cells of Ngsk $Prnp^{0/0}$ mice was further confirmed by subjecting the primary cultured glial cells to Northern blotting analysis. The cell-expression profiles of the aberrant mRNAs in the brain of Ngsk $Prnp^{0/0}$ mice were nearly identical to those of PrP mRNA in wild-type mice, consistent with the finding that the aberrant mRNAs are abnormally expressed under the control of the *Prnp* promoter in ataxic lines of $Prnp^{0/0}$ mice.

7.3.5. Ectopic expression of PrP-like protein due to unusual intergenic splicing between *Prnp* and *Prnd*

Moore et al. successfully showed the complete genomic configuration of *Prnd* by sequencing of a cosmid clone, I/Ln-J-4[2]. *Prnd* was located 16-kb downstream of *Prnp*, and the 103-, 163-, and 266 (103 + 163)-bp inserts are derived from two exonic sequences present between *Prnp* and *Prnd*, indicating that the aberrant transcripts are generated as a result of an unusual intergenic splicing taking place between *Prnp* and *Prnd* on the targeted allele.

Pre-mRNAs usually undergo several modifications within the framework of a single gene, such as capping at the 5' end, splicing out intronic sequences, and cleavage and polyadenylation at the 3' end, and convert

Figure 7.4. Mechanisms of the generation of the aberrant chimeric mRNAs via intergenic splicing taking place between *Prnp* and *Prnd* in Ngsk *Prnp*$^{0/0}$ mice. A, wild-type mice; B, Ngsk *Prnp*$^{0/0}$ mice.

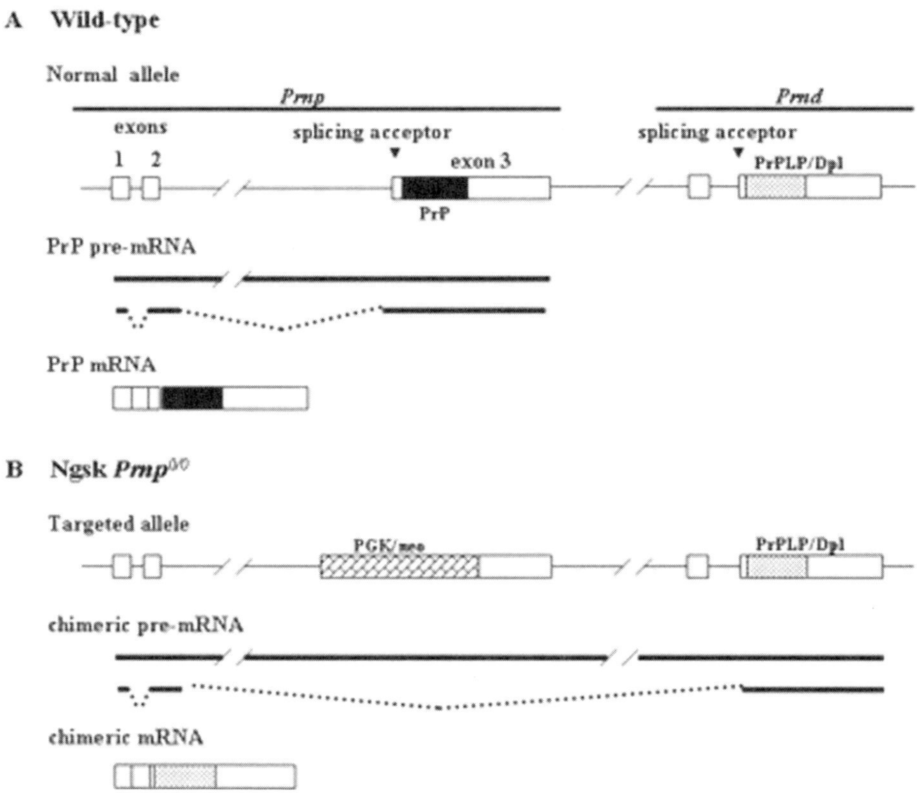

into mature mRNAs. In wild-type mice, the PrP pre-mRNA is normally cleaved and polyadenylated at the last exon of the *Prnp* (Figure 7.4A). However, in ataxic lines of *Prnp*$^{0/0}$ mice, probably due to the lack of the 3' part of intron 2, the pre-mRNA transcribed from the *Prnp* promoter could not efficiently undergo cleavage/polyadenylation at the last exon of *Prnp*[2,3]. Thereafter, it was further elongated until the last exon of *Prnd* and subjected to intergenic splicing between the residual *Prnp* exons 1/2 and the PrPLP/Dpl-coding exon (Figure 7.4B). As a result, PrPLP/Dpl became abnormally expressed under the control of the *Prnp* promoter, leading to the ectopic expression of PrPLP/Dpl in the brains of ataxic lines of *Prnp*$^{0/0}$ mice[2,3]. It is likely, therefore, that the aberrant expression of PrPLP/Dpl in the brains of ataxic *Prnp*$^{0/0}$ mice is attributable to the discrepant phenotypes observed among different lines of *Prnp*$^{0/0}$ mice.

7.4. Protein structure and normal function of PrP-like protein

7.4.1. PrP-like protein is a homologue of globular domain of PrP

PrPLP/Dpl is a GPI-anchored membrane glycoprotein[39]. On the biosynthesis of PrPLP/Dpl along the secretary pathway, the N-terminal 24 or 27 and the C-terminal 25 hydrophobic residues are removed as a signal peptide and a GPI-anchor signal peptide, respectively, and the two conserved asparagines are N-glycosylated (Figure 7.5). PrPLP/Dpl possesses two disulfide linkages (Figure 7.5). Disruption of these linkages by DTT was reported to predispose recombinant PrPLP/Dpl to be precipitatable, suggesting that these disulfide linkages are important for the proper folding of the protein[40].

PrPLP/Dpl resembles the C-terminal half of PrP^C in the amino acid sequence, lacking both the Cu^{2+}-binding octapeptide repeat and the hydrophobic domain present at the N-terminal half of PrP^C (Figure 7.5). Both proteins are anchored onto the cell membrane by a GPI moiety, indicating that these proteins are expressed on the same raft domain of the membrane. Indeed, Shaked et al. showed that PrPLP/Dpl and PrP^C are raft-associated proteins using flotation assays[41]. Moreover, Mo et al. analyzed a nuclear magnetic resonance (NMR) structure of recombinant PrPLP/Dpl and observed marked similarity in the protein structure between PrPLP/Dpl and the C-terminal half of PrP^C, both of which are composed of three α-helices and two short β-strands[40]. These stunning similarities in the cellular topology and protein structure of the

Figure 7.5. Protein structures of PrP and PrPLP/Dpl. α and β indicate α-helix and β-strand, respectively. S-S indicates a disulfide bond.

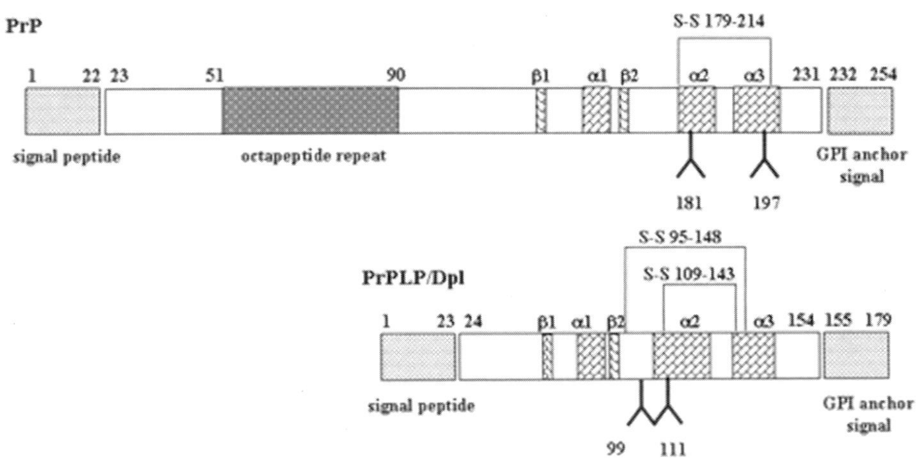

two proteins strongly suggest that PrPLP/Dpl is a homologue protein of the C-terminal globular domain of PrP^C.

7.4.2. Developmental expression of PrP-like protein in brain endothelial cells

PrPLP/Dpl was not expressed in the brains of adult mice[2,38]. However, we found substantial expression in the brains of neonatal mice[38]. The PrPLP/Dpl mRNA was already detected at 1 day after birth, peaked at about 1 week, and then gradually decreased to an undetectable level by at least 8 weeks. Using in situ hybridization together with immunohistochemistry, we found that PrPLP/Dpl was expressed specifically by brain endothelial cells[38]. This developmental expression of PrPLP/Dpl in the brain endothelial cells suggests that PrPLP/Dpl is involved in the development of brain blood vessels. It is also conceivable that PrPLP/Dpl is associated with the development of the blood-brain barrier.

7.4.3. Impaired spermatogenesis in mice devoid of PrP-like protein

To investigate the physiological functions of PrPLP/Dpl, Behrens et al. generated mice devoid of PrPLP/Dpl, designated $Prnd^{neo/neo}$ mice, using homologous recombination in ES cells[42]. The mutant mice were born normally and had no obvious defects, indicating that PrPLP/Dpl is not indispensable to embryonic development. Interestingly, female $Prnd^{neo/neo}$ mice were fertile, whereas male mutant mice showed sterility. The testes in these mice seemed developmentally normal, but both the number of spermatozoa and the motility of mutant sperm were significantly decreased. Moreover, the mutant sperm exhibited abnormal morphologies. The authors also showed that the acrosome function was impaired in the mutant spermatozoa. Immunohistochemical studies using anti-PrPLP/Dpl antibodies demonstrated the specific expression of PrPLP/Dpl in spermatids[42]. Also, Peoc'h et al. showed that, in human testes, spermatozoa and Sertoli cells expressed PrPLP/Dpl[43]. These results indicate that PrPLP/Dpl is involved in spermatogenesis.

7.5. Neurotoxic function of PrP-like protein

7.5.1. Neurotoxicity of ectopically expressed PrP-like protein in the absence of PrP

Ataxia and Purkinje cell degeneration have been shown to be closely associated with the ectopic expression of PrPLP/Dpl in the brains of ataxic *Prnp*$^{0/0}$ mice, strongly suggesting a causal relationship between the ectopically expressed PrPLP/Dpl and these discrepant phenotypes[2,3]. Moore et al. recently demonstrated the involvement of PrPLP/Dpl in these neurological abnormalities by showing that PrPLP/Dpl transgenically driven by the *Prnp* promoter rendered non-ataxic Zrch I *Prnp*$^{0/0}$ mice ataxic due to marked Purkinje cell loss[37]. This finding, taken with our previous results that these abnormal phenotypes in Ngsk *Prnp*$^{0/0}$ mice were successfully rescued by the introduction of a transgene encoding PrP^C [30], indicates that the ectopically expressed PrPLP/Dpl induces neurotoxicity on Purkinje cells in the absence of PrP^C, causing the ataxia and Purkinje cell degeneration as observed in ataxic lines of *Prnp*$^{0/0}$ mice.

7.5.2. PrP-like protein expressed on Purkinje cells is involved in neurodegeneration

We generated two different types of transgenic (tg) mice, designated tg(N-PrPLP/Dpl) and tg(P-PrPLP/Dpl)[44]. Three lines of tg(N-PrPLP/Dpl) mice, 25, 31, and 32, expressed PrPLP/Dpl specifically in nearly all neurons including Purkinje cells under the control of the neuron-specific enolase promoter, and four lines of tg(P-PrPLP/Dpl) mice, 18, 26, 27, and 48, expressed PrPLP/Dpl only in Purkinje cells under the control of the Purkinje cell protein-2 promoter. To investigate whether these cell-specific expressions of PrPLP/Dpl could induce ataxia and Purkinje cell degeneration in the absence of PrP^C, we backcrossed tg(N-PrPLP/Dpl)25 and 32 mice and tg(P-PrPLP/Dpl) 26 and 27 mice with non-ataxic Zrch I *Prnp*$^{0/0}$ mice to generate each tg line of mice carrying the Zrch I *Prnp*$^{0/0}$ alleles[44]. Neither ataxia nor Purkinje cell degeneration was detected in these tg mice on the wild-type genetic background (Figure 7.6). In contrast, all of the tg(N-PrPLP/Dpl)25 and 32 mice carrying the Zrch I *Prnp*$^{0/0}$ genotype began to exhibit ataxia at 359 ± 52 and 58 ± 15 days, respectively (Figure 7.6). Moreover, the tg(P-PrPLP/Dpl)26 and 27 mice with the Zrch I *Prnp*$^{0/0}$ background also developed similar ataxia at 268 ± 28 and 167 ± 13 days, respectively (Figure 7.6). These ataxic tg mice carrying the Zrch I *Prnp*$^{0/0}$ genotype showed marked degeneration of Purkinje cells throughout the cerebellar cortex. No pathological changes were detectable in other cells of these mice, including granule cells. Similarly, Anderson et al. recently reported that the targeted expression of PrPLP/Dpl to Purkinje cells of Zrch I *Prnp*$^{0/0}$ mice caused ataxia and Purkinje cell degeneration[45]. Taken together, these results indicate that PrPLP/Dpl ectopically overexpressed on Purkinje cells is itself neurotoxic enough to induce the degeneration of the cells in the absence of PrP^C.

Figure 7.6. The onset of ataxia in tg(N-PrPLP/Dpl)25 mice (A), tg(N-PrPLP/Dpl)32 mice (B), tg(P-PrPLP/Dpl)26 mice (C), and tg(P-PrPLP/Dpl)27 mice (D) on the genetic background of wild-type (*Prnp*$^{+/+}$), Zrch I *Prnp*$^{0/+}$, and Zrch I *Prnp*$^{0/0}$.

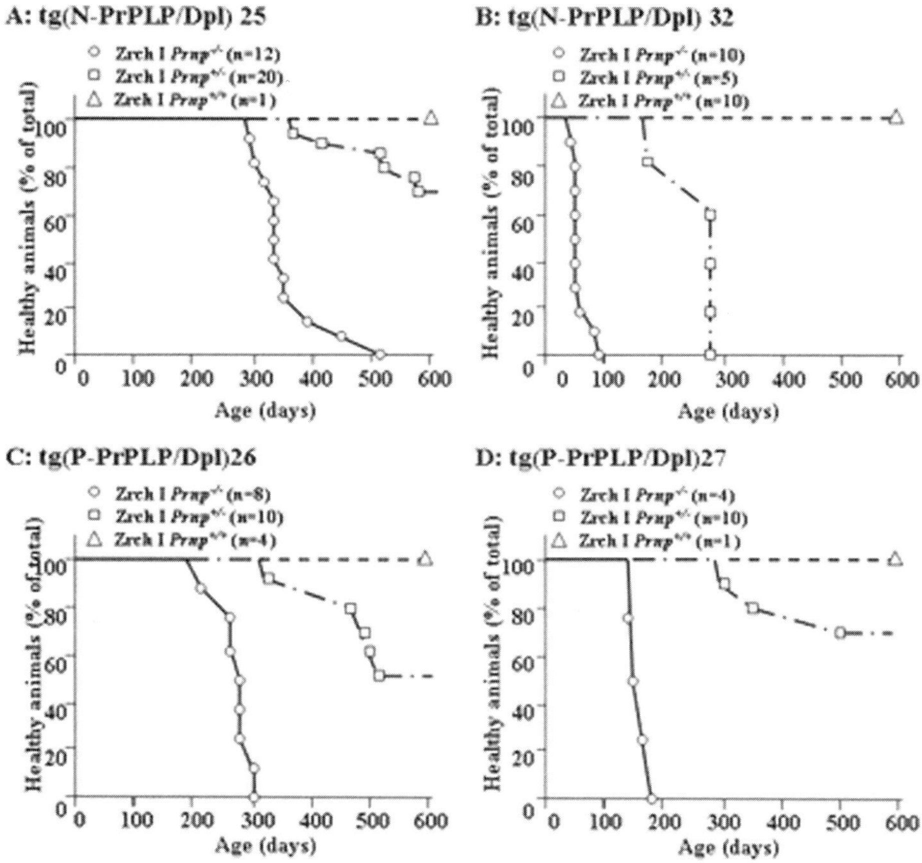

7.5.3. Neurotoxicity of PrP-like protein is stoichiometrically neutralized by PrP

Rossi et al. reported that the onset of ataxia and Purkinje cell degeneration occurred much earlier in Zrch II *Prnp*$^{0/0}$ mice than in Zrch I/Zrch II chimeric *Prnp*$^{0/0}$ mice hemizygous for the PrPLP/Dpl-encoding Zrch II targeted allele, and further showed that the onset of the ataxia in Zrch II *Prnp*$^{0/0}$ mice was hastened by the expression of additional transgenic PrPLP/Dpl[15]. Moore et al. also showed that Zrch I *Prnp*$^{0/0}$ mice transgenic for PrPLP/Dpl developed ataxia and Purkinje cell degeneration with incubation times inversely correlated to the expression levels of PrPLP/Dpl in the brain[37]. We reported similar results in both tg(N-PrPLP/Dpl) and tg(N-PrPLP/Dpl) mice[44]. Tg(N-PrPLP/Dpl)25 mice expressed PrPLP/Dpl in the cerebrum and cerebellum at a level less than a quarter that of Ngsk $Prnp^{0/0}$ mice on Western blotting, and developed ataxia at 359 ± 52 days on the Zrch I $Prnp^{0/0}$ background. On the other hand, tg(N-PrPLP/Dpl)32 mice expressed PrPLP/Dpl in the cerebrum and cerebellum at a level about 2–3 and 1–2 times more than that of Ngsk $Prnp^{0/0}$ mice, respectively, and succumbed to the disease much earlier at 58 ± 15 days on the Zrch I $Prnp^{0/0}$ background. Similarly, tg(P-PrPLP/Dpl)26 mice exhibiting weak signals in most Purkinje cells on in situ hybridization developed ataxia at 268 ± 28 days on the Zrch I $Prnp^{0/0}$ background, while tg(P-PrPLP/Dpl)27 mice with the heavily stained Purkinje cells evenly distributed throughout the cerebellar cortex got sick at 167 ± 13 days. Taken together, these results show that PrPLP/Dpl exerts its neurotoxicity on Purkinje cells in a dose-dependent manner.

We next examined the relationship between the onset of ataxia and the expression levels of PrP^C [44]. All tg(N-PrPLP/Dpl)32 mice hemizygous for Zrch I *Prnp* allele, $Prnp^{0/+}$, developed ataxia at 259 ± 48 days, and 6/20 tg(N-PrPLP/Dpl)25 mice exhibited similar symptoms at 495 ± 86 days on the Zrch I $Prnp^{0/+}$ background (Figure 7.6). Similarly, 5/10 tg(P-PrPLP/Dpl)26 mice and 3/10 tg(P-PrPLP/Dpl)27 mice became ataxic at 463 ± 81 and 391 ± 108 days, respectively, on the Zrch I $Prnp^{0/+}$ background (Figure 7.6). Pathological examinations revealed that all symptomatic tg mice with the Zrch I $Prnp^{0/+}$ genotype exhibited the degeneration of Purkinje cells. The onset of the ataxia in each line of tg mice with the Zrch I $Prnp^{0/+}$ background was greatly prolonged, compared with that of the same line of mice with the Zrch I $Prnp^{0/0}$ background, indicating that, in contrast to PrPLP/Dpl, the time to ataxia and Purkinje cell degeneration correlates with the expression levels of PrP^C in the brain. In other words, PrP^C neutralizes the neurotoxicity of PrPLP/Dpl in a stoichiometric manner.

7.5.4. Neutralization of PrP-like protein toxicity by PrP is mediated through N-terminal residues

Successful rescue of the ataxia and Purkinje cell degeneration in Ngsk $Prnp^{0/0}$ mice by introduction of a transgene encoding PrP^C prompted us to investigate the domain of PrP^C, which neutralizes the neurotoxicity of PrPLP/Dpl. We recently reported that hamster PrP^C could protect Ngsk $Prnp^{0/0}$ mice from these neurological abnormalities[46]. Human PrP^C was also shown to be capable of rescuing the electophysiological deficits in

Zrch *Prnp*$^{0/0}$ mice[25]. It is therefore conceivable that PrP^C is functionally equivalent in all mammalians.

We also showed that PrP with a familial prion disease-associated mutation (E199K) could fully neutralize the PrPLP/Dpl-induced neurotoxicity in Ngsk *Prnp*$^{0/0}$ mice[46]. The protein structure of recombinant human PrP 90-231 with the corresponding mutation (E200K) was shown to be almost identical to that of wild-type PrP[47]. Thus, the mutation (E200K) has little affect on the structure and function of PrP^C. This result indicates that other disease-associated mutant PrPs are also functionally competent. To promote understanding of the pathogenesis of familial prion diseases, it might be useful to examine whether such PrPs rescue Ngsk *Prnp*$^{0/0}$ mice from ataxia and Purkinje cell degeneration.

We also reported that Ngsk *Prnp*$^{0/0}$ mice expressing the PrP transgene with a deletion of the N-terminal residues 23–88 (MHM2.del23-88) developed ataxia and Purkinje cell degeneration on a time course identical to that of non-transgenic Ngsk *Prnp*$^{0/0}$ mice[46]. This result clearly indicates that PrP^C neutralizes the neurotoxicity of PrPLP/Dpl through its N-terminal residues containing the PrP-specific Cu^{2+}-binding octapeptide repeat region.

7.5.5. Possible mechanisms for Purkinje cell degeneration in PrP-null mice

Shmerling et al. previously reported that PrPΔ32-121 and PrPΔ32-134, which lack the N-terminal residues 32-121 and 32-134, respectively, caused ataxia and cerebellar degeneration characterized by marked granule cell death in Zrch I *Prnp*$^{0/0}$ mice[48]. They further showed that these neurological abnormalities could be rescued by full-length PrP^C[48]. Purkinje cells remained intact in these mice probably because the truncated PrPs were not expressed in Purkinje cells due to the insufficient activity of the PrP promoter they used[15]. Flechsig et al. subsequently demonstrated that ataxia and Purkinje cell loss could be induced by the targeted expression of PrPΔ32-134 to Purkinje cells of Zrch I *Prnp*$^{0/0}$ mice[49]. PrPΔ32-121 and PrPΔ32-134 lack the octapeptide repeat and central hydrophobic region, but preserve the C-terminal globular domain of PrP^C, the domain homologous to PrPLP/Dpl. It is therefore very likely that PrPLP/Dpl and the truncated PrPs utilize the same or at least a similar mechanism to exert their neurotoxicity on Purkinje cells.

Weissmann and colleagues have hypothesized that PrP^C and another conjectural PrP-like molecule, protein π, elicit a signal necessary for Purkinje cell survival upon interaction with a yet unidentified putative cognate ligand[48,50]. In non-ataxic *Prnp*$^{0/0}$ mice, the protein π signaling remains intact and therefore Purkinje cells are healthy. In ataxic *Prnp*$^{0/0}$ mice, PrPLP/Dpl and the N-terminally truncated PrPs compete with protein π for the ligand and disturb the signal, resulting in Purkinje cell death. However, this theory can be verified only when the putative molecules are identified. Interestingly, it was recently demonstrated that copper could bind not only to PrP^{C} but also to PrPLP/Dpl[51–53]. Hence, copper is a candidate for the hypothetical ligand. On the other hand, in contrast to PrPΔ32-121 and PrPΔ32-134, PrPΔ23-88 has never shown neurotoxicity in Zrch I *Prnp*$^{0/0}$ mice[54], indicating that the residues between 88 and 121 inhibit the neurotoxic potential of PrPLP/Dpl. This region, overlapping with the hydrophobic region, forms part of the binding sites for the heat shock protein, stress-inducible protein 1[55], and the extracellular matrix constituent, glycosaminoglycans[56]. It is therefore possible that these associating molecules are involved in the Purkinje cell degeneration in ataxic lines of *Prnp*$^{0/0}$ mice.

Wong et al. reported that oxidative stresses including radical oxygen species and nitric oxide were much more elevated in the brains of ataxic Rcm0 *Prnp*$^{0/0}$ mice than in those of non-ataxic Npu *Prnp*$^{0/0}$ mice[57]. Cui et al. showed that treatment of the cultured cerebellar neurons of non-ataxic Zrch I *Prnp*$^{0/0}$ mice with recombinant PrPLP/Dpl induced apoptotic cell death with concomitant increase in the expression of inducible and neuronal NO synthases and heme oxygenase-1, molecular markers for oxidative stresses[58]. They further demonstrated that a pharmacological inhibitor of NO synthases, L-*N*-acetyl methyl ester (NAME), could protect the neurons from apoptosis[58]. These results strongly suggest that PrPLP/Dpl is involved in the generation of oxidative stresses. There is an emerging hypothesis that PrP^{C} is associated with antioxidant activity by activating Cu^{2+}-dependent antioxidant enzymes such as superoxide dismutase via transfer of the octapeptide repeat-bound Cu^{2+}[59]. Indeed, the primary cultured cerebellar neurons from non-ataxic Npu *Prnp*$^{0/0}$ mice were shown to be more sensitive to oxidative stresses than those from wild-type mice[60]. Therefore, it is alternatively conceivable that PrPLP/Dpl produces oxidative stresses toxic to Purkinje neurons and that PrP^{C} detoxifies them.

Kuwahara et al. reported that hippocampal neuronal cells from ataxic *Prnp*$^{0/0}$ mice easily underwent apoptosis after withdrawal of serum, and that the apoptosis could be prevented by either re-introduction of PrP^{C} or by expressing an anti-apoptotic protein, Bcl-2, exogenously[61]. Moreover, Bounhar et al. showed that PrP^{C} protected human primary neurons from the apoptosis induced by the pro-apoptotic protein Bax[62]. Interestingly, they further showed that PrP lacking the octapeptide repeat region completely lost this neuroprotective potential[62], consistent with the unsuccessful rescue of the ataxia and Purkinje cell degeneration by PrP(MHM2.del23-88)[46]. These results suggest the further possibility that PrP^C is an anti-apoptotic protein while PrPLP/Dpl is pro-apoptotic.

7.6. Mice devoid of PrP shed light on the pathogenesis of prion diseases

There is a strong argument that the conformational conversion of PrP^C to PrP^{Sc} plays an essential role in the pathogenesis of prion diseases[7–9,14,63], but the exact nature of this role has not been elucidated. The constitutive conversion causes the accumulation of PrP^{Sc} in the affected brain. In contrast, PrP^C is reduced. Indeed, Yokoyama et al. showed that, in experimentally infected mice, the PrP^C-specific immunoreactivity was decreased in the brain regions where PrP^{Sc} had accumulated[64]. Therefore, it has been postulated that the functional loss of PrP^C induced by the conversion is involved in the pathogenesis. On the other hand, Forloni et al. showed that an amyloidgenic PrP peptide (PrP106-126) was highly toxic to primary cultured neurons, arguing that PrP^{Sc} could itself be neurotoxic[65]. Moreover, it is possible that both the functional loss of PrP^C and the neurotoxicity of PrP^{Sc} are required for the pathogenesis.

Non-ataxic $Prnp^{0/0}$ mice exhibited impaired LTP in the hippocampus CA1 region and altered circadian activity and sleep[24,29]. Demyelination was also observed in the spinal cord and peripheral nerves in $Prnp^{0/0}$ mice[30]. Prion diseases commonly present with memory loss or dementia[63]. Since LTP is a form of synaptic plasticity that is thought to underlie memory formation, the functional loss of PrP^C might be relevant to the dementia seen in prion diseases. Alteration in sleep and circadian rhythms is a characteristic symptom of the inherited human prion disease, fetal familiar insomnia[63]. Moreover, demyelinating peripheral neuropathy has been reported in some cases of inherited prion disease associated with the E200K mutation of PrP[66,67]. This remarkable similarity of the phenotypes of PrP-null mice to those of prion diseases strongly suggests that the neuropathological abnormalities in prion diseases are attributable to the functional loss of PrP^C.

It has been shown that, in the absence of functional PrP^C, the ectopic expression of PrPLP/Dpl, PrPΔ32-121, or PrPΔ32-134 caused ataxia, Purkinje cell degeneration, and gliosis in mice[37,49]. PrPΔ32-121 and PrPΔ32-134 were also highly toxic to cerebellar granule cells[48]. Ataxia is one of the major initial symptoms, and marked degeneration of both Purkinje cells and granule neurons is commonly noted in human prion diseases[63]. Gliosis is also invariably observed[63]. Therefore, it is conceivable that mechanisms involved in the neurodegeneration and gliosis of ataxic *Prnp*$^{0/0}$ mice could work in a similar manner in the neuropathologies of prion diseases. However, since no ectopic upregulation of PrPLP/Dpl has been reported in the brains affected by experimental prion diseases[68], it is possible that PrP^{Sc} accumulated in the neurons replaces PrPLP/Dpl in prion diseases, or that fragmented products derived from PrP^{Sc}, which are often observed in the affected brains[69], possess a neurotoxic potential equivalent to that of PrPLP/Dpl, PrPΔ32-121, or PrPΔ32-134.

7.7. Possible implication of ectopic expression of PrP-like protein in pathophysiolocal conditions

A body of evidence is emerging to show that pathophysiological stresses such as ischemia can disturb normal splicing processes and promote alternative splicing of a considerable number of genes expressed in the CNS[70–76]. Interestingly, Purkinje cells are highly vulnerable to ischemic stress[77]. It is possible that ischemic stress produces the chimeric *Prnp-Prnd* mRNA encoding for PrPLP/Dpl by up-regulation of alternative intergenic splicing in neurons, particularly in Purkinje cells. It will be intriguing, therefore, to examine the involvement of the PrPLP/Dpl-induced neurodegeneration in various neuropathological conditions.

7.8. Acknowledgments

I would like to thank Dr. S. Katamine, Dr. K. Shigematsu, Dr. N. Nishida, Dr. R. Atarashi, Dr. A. Li, Mr. N. Yamaguchi, and Mr. N. Okimura. This study is partly supported by a Research for Comprehensive Promotion of Study of Brain and a Grand-in-Aid for Scientific Research on Priority Areas-Advanced Brain Science Project- from the Ministry of Education, Culture, Sports, Science and Technology, Japan, and by a Research on Specific Diseases from the Ministry of Health Labour and Welfare, Japan.

7.9. References

1. E. N. Olson, H. H. Arnold, P. W. Rigby, and B. J. Wold, Know your neighbors: three phenotypes in null mutants of the myogenic bHLH gene MRF4. *Cell* **85**, 1–4 (1996).

2. R. C. Moore, I. Y. Lee, G. L. Silverman, P. M. Harrison, R. Strome, C. Heinrich, A. Karunaratne, S. H. Pasternak, M. A. Chishti, Y. Liang, P. Mastrangelo, K. Wang, A. F. Smit, S. Katamine, G. A. Carlson, F. E. Cohen, S. B. Prusiner, D. W. Melton,

P. Tremblay, L. E. Hood, and D. Westaway, Ataxia in prion protein (PrP)-deficient mice is associated with upregulation of the novel PrP-like protein doppel. *J Mol Biol* **292**, 797–817 (1999).

3. A. Li, S. Sakaguchi, R. Atarashi, B. C. Roy, R. Nakaoke, K. Arima, N. Okimura, J. Kopacek, and K. Shigematsu, Identification of a novel gene encoding a PrP-like protein expressed as chimeric transcripts fused to PrP exon 1/2 in ataxic mouse line with a disrupted PrP gene. *Cell Mol Neurobiol* **20**, 553–67 (2000).

4. B. Oesch, D. Westaway, M. Walchli, M. P. McKinley, S. B. Kent, R. Aebersold, R. A. Barry, P. Tempst, D. B. Teplow, L. E. Hood, and et al., A cellular gene encodes scrapie PrP 27–30 protein. *Cell* **40**, 735–46 (1985).

5. S. B. Prusiner, Molecular biology of prion diseases. *Science* **252**, 1515–22 (1991).

6. S. B. Prusiner, Prions. *Proc Natl Acad Sci U S A* **95**, 13363–83 (1998).

7. H. Bueler, A. Aguzzi, A. Sailer, R. A. Greiner, P. Autenried, M. Aguet, and C. Weissmann, Mice devoid of PrP are resistant to scrapie. *Cell* **73**, 1339–47 (1993).

8. S. B. Prusiner, D. Groth, A. Serban, R. Koehler, D. Foster, M. Torchia, D. Burton, S. L. Yang, and S. J. DeArmond, Ablation of the prion protein (PrP) gene in mice prevents scrapie and facilitates production of anti-PrP antibodies. *Proc Natl Acad Sci U S A* **90**, 10608–12 (1993).

9. J. C. Manson, A. R. Clarke, P. A. McBride, I. McConnell, and J. Hope, PrP gene dosage determines the timing but not the final intensity or distribution of lesions in scrapie pathology. *Neurodegeneration* **3**, 331–40 (1994).

10. S. Sakaguchi, S. Katamine, K. Yamanouchi, M. Kishikawa, R. Moriuchi, N. Yasukawa, T. Doi, and T. Miyamoto, Kinetics of infectivity are dissociated from PrP accumulation in salivary glands of Creutzfeldt-Jakob disease agent-inoculated mice. *J Gen Virol* **74 (Pt 10)**, 2117–23 (1993).

11. H. Bueler, M. Fischer, Y. Lang, H. Bluethmann, H. P. Lipp, S. J. DeArmond, S. B. Prusiner, M. Aguet, and C. Weissmann, Normal development and behaviour of mice lacking the neuronal cell-surface PrP protein. *Nature* **356**, 577–82 (1992).

12. J. C. Manson, A. R. Clarke, M. L. Hooper, L. Aitchison, I. McConnell, and J. Hope, 129/Ola mice carrying a null mutation in PrP that abolishes mRNA production are developmentally normal. *Mol Neurobiol* **8**, 121–7 (1994).

13. S. Sakaguchi, S. Katamine, N. Nishida, R. Moriuchi, K. Shigematsu, T. Sugimoto, A. Nakatani, Y. Kataoka, T. Houtani, S. Shirabe, H. Okada, S. Hasegawa, T. Miyamoto, and T. Noda, Loss of cerebellar Purkinje cells in aged mice homozygous for a disrupted PrP gene. *Nature* **380**, 528–31 (1996).

14. S. Sakaguchi, S. Katamine, K. Shigematsu, A. Nakatani, R. Moriuchi, N. Nishida, K. Kurokawa, R. Nakaoke, H. Sato, K. Jishage, and et al., Accumulation of proteinase K-resistant prion protein (PrP) is restricted by the expression level of normal PrP

in mice inoculated with a mouse-adapted strain of the Creutzfeldt-Jakob disease agent. *J Virol* **69**, 7586–92 (1995).

15. D. Rossi, A. Cozzio, E. Flechsig, M. A. Klein, T. Rulicke, A. Aguzzi, and C. Weissmann, Onset of ataxia and Purkinje cell loss in PrP null mice inversely correlated with Dpl level in brain. *Embo J* **20**, 694–702 (2001).

16. J. Manson, J. D. West, V. Thomson, P. McBride, M. H. Kaufman, and J. Hope, The prion protein gene: a role in mouse embryogenesis? *Development* **115**, 117–22 (1992).

17. S. J. DeArmond, W. C. Mobley, D. L. DeMott, R. A. Barry, J. H. Beckstead, and S. B. Prusiner, Changes in the localization of brain prion proteins during scrapie infection. *Neurology* **37**, 1271–80 (1987).

18. R. Morris, Developments of a water-maze procedure for studying spatial learning in the rat. *J Neurosci Methods* **11**, 47–60 (1984).

19. A. L. Phinney, M. E. Calhoun, D. P. Wolfer, H. P. Lipp, and H. Zheng, No hippocampal neuron or synaptic bouton loss in learningimpaired aged beta-amyloid precursor protein-null mice. *Neuroscience* **90**, 1207–1216 (1999).

20. H. P. Lipp, and H. Van der Loos, A computer-controlled Y-maze for the analysis of vibrissotactile discrimination learning in mice. *Behav Brain Res* **45**, 135–145 (1991).

21. N. Nishida, S. Katamine, K. Shigematsu, A. Nakatani, N. Sakamoto, S. Hasegawa, R. Nakaoke, R. Atarashi, Y. Kataoka, and T. Miyamoto, Prion protein is necessary for latent learning and long-term memory retention. *Cell Mol Neurobiol* **17**, 537–45 (1997).

22. A. Ettenberg, M. L. Moal, G. F. Koob, and F. E. Bloom, Vasopressin potentiation in the performance of a learned appetitive task: Reversal by a pressor antagonist analog of vasopressin. *Pharmacol Biochem Behav* **18**, 645–647 (1983).

23. H. P. Lipp, H. Schwegler, B. Heimrich, and P. Driscoll, Infrapyramidal mossy fibers and two-way avoidance learning: Developmental modification of hippocampal circuitry and adult behavior of rats and mice. *J Neurosci* **8**, 1905–1921 (1988).

24. J. Collinge, M. A. Whittington, K. C. Sidle, C. J. Smith, M. S. Palmer, A. R. Clarke, and J. G. Jefferys, Prion protein is necessary for normal synaptic function. *Nature* **370**, 295–7 (1994).

25. M. A. Whittington, K. C. Sidle, I. Gowland, J. Meads, A. F. Hill, M. S. Palmer, J. G. Jefferys, and J. Collinge, Rescue of neurophysiological phenotype seen in PrP null mice by transgene encoding human prion protein. *Nat Genet* **9**, 197–201 (1995).

26. J. Manson, J. Hope, A. R. Clarke, A. Johnston, C. Black, and N. MacLeod, PrP gene dosage and long term potentiation. *Neurodegeneration* **4**, 113–115 (1995).

27. J. W. Herms, H. A. Kretzchmar, S. Titz, and B. U. Keller, Patch-clamp analysis of synaptic transmission to cerebellar purkinje cells of prion protein knockout mice. *Eur J Neurosci* **7**, 2508–12 (1995).

28. P. M. Lledo, P. Tremblay, S. J. DeArmond, S. B. Prusiner, and R. A. Nicoll, Mice deficient for prion protein exhibit normal neuronal excitability and synaptic transmission in the hippocampus. *Proc Natl Acad Sci U S A* **93**, 2403–7 (1996).

29. I. Tobler, S. E. Gaus, T. Deboer, P. Achermann, M. Fischer, T. Rulicke, M. Moser, B. Oesch, P. A. McBride, and J. C. Manson, Altered circadian activity rhythms and sleep in mice devoid of prion protein. *Nature* **380**, 639–42 (1996).

30. N. Nishida, P. Tremblay, T. Sugimoto, K. Shigematsu, S. Shirabe, C. Petromilli, S. P. Erpel, R. Nakaoke, R. Atarashi, T. Houtani, M. Torchia, S. Sakaguchi, S. J. DeArmond, S. B. Prusiner, and S. Katamine, A mouse prion protein transgene rescues mice deficient for the prion protein gene from purkinje cell degeneration and demyelination. *Lab Invest* **79**, 689–97 (1999).

31. D. R. Borchelt, V. E. Koliatsos, M. Guarnieri, C. A. Pardo, S. S. Sisodia, and D. L. Price, Rapid anterograde axonal transport of the cellular prion glycoprotein in the peripheral and central nervous systems. *J Biol Chem* **269**, 14711–4 (1994).

32. R. Atarashi, S. Sakaguchi, K. Shigematsu, K. Arima, N. Okimura, N. Yamaguchi, A. Li, J. Kopacek, and S. Katamine, Abnormal activation of glial cells in the brains of prion protein-deficient mice ectopically expressing prion protein-like protein, PrPLP/Dpl. *Mol Med* **7**, 803–9 (2001).

33. M. Moser, R. J. Colello, U. Pott, and B. Oesch, Developmental expression of the prion protein gene in glial cells. *Neuron* **14**, 509–17 (1995).

34. D. R. Brown, A. Besinger, J. W. Herms, and H. A. Kretzschmar, Microglial expression of the prion protein. *Neuroreport* **9**, 1425–9 (1998).

35. J. Follet, C. Lemaire-Vieille, F. Blanquet-Grossard, V. Podevin-Dimster, S. Lehmann, J. P. Chauvin, J. P. Decavel, R. Varea, J. Grassi, M. Fontes, and J. Y. Cesbron, PrP expression and replication by Schwann cells: implications in prion spreading. *J Virol* **76**, 2434–9 (2002).

36. E. J. Steinmetz, Pre-mRNA processing and the CTD of RNA polymerase II: The tail that wags the dog. *Cell* **89**, 491–494 (1997).

37. R. C. Moore, P. Mastrangelo, E. Bouzamondo, C. Heinrich, G. Legname, S. B. Prusiner, L. Hood, D. Westaway, S. J. DeArmond, and P. Tremblay, Doppel-induced cerebellar degeneration in transgenic mice. *Proc Natl Acad Sci U S A* **98**, 15288–93 (2001).

38. A. Li, S. Sakaguchi, K. Shigematsu, R. Atarashi, B. C. Roy, R. Nakaoke, K. Arima, N. Okimura, J. Kopacek, and S. Katamine, Physiological expression of the gene for

PrP-like protein, PrPLP/Dpl, by brain endothelial cells and its ectopic expression in neurons of PrP-deficient mice ataxic due to Purkinje cell degeneration. *Am J Pathol* **157**, 1447–52 (2000).

39. G. L. Silverman, K. Qin, R. C. Moore, Y. Yang, P. Mastrangelo, P. Tremblay, S. B. Prusiner, F. E. Cohen, and D. Westaway, Doppel is an N-glycosylated, glycosylphosphatidylinositol-anchored protein. Expression in testis and ectopic production in the brains of Prnp(0/0) mice predisposed to Purkinje cell loss. *J Biol Chem* **275**, 26834–41 (2000).

40. H. Mo, R. C. Moore, F. E. Cohen, D. Westaway, S. B. Prusiner, P. E. Wright, and H. J. Dyson, Two different neurodegenerative diseases caused by proteins with similar structures. *Proc Natl Acad Sci U S A* **98**, 2352–7 (2001).

41. Y. Shaked, N. Hijazi, and R. Gabizon, Doppel and PrP(C) do not share the same membrane microenvironment. *FEBS Lett* **530**, 85–8 (2002).

42. A. Behrens, N. Genoud, H. Naumann, T. Rulicke, F. Janett, F. L. Heppner, B. Ledermann, and A. Aguzzi, Absence of the prion protein homologue Doppel causes male sterility. *Embo J* **21**, 3652–8 (2002).

43. K. Peoc'h, C. Serres, Y. Frobert, C. Martin, S. Lehmann, S. Chasseigneaux, V. Sazdovitch, J. Grassi, P. Jouannet, J. M. Launay, and J. L. Laplanche, The human "prion-like" protein Doppel is expressed in both Sertoli cells and spermatozoa. *J Biol Chem* **277**, 43071–8 (2002).

44. N. Yamaguchi, S. Sakaguchi, K. Shigematsu, N. Okimura, and S. Katamine, Doppel-induced Purkinje cell death is stoichiometrically abrogated by prion protein. *Biochem Biophys Res Commun* **319**, 1247–1252 (2004).

45. L. Anderson, D. Rossi, J. Linehan, S. Brandner, and C. Weissmann, Transgene-driven expression of the Doppel protein in Purkinje cells causes Purkinje cell degeneration and motor impairment. *Proc Natl Acad Sci U S A* **101**, 3644–3649 (2004).

46. R. Atarashi, N. Nishida, K. Shigematsu, S. Goto, T. Kondo, S. Sakaguchi, and S. Katamine, Deletion of N-terminal residues 23-88 from prion protein (PrP) abrogates the potential to rescue PrP-deficient mice from PrP-like protein/doppel-induced Neurodegeneration. *J Biol Chem* **278**, 28944–9 (2003).

47. Y. Zhang, W. Swietnicki, M. G. Zagorski, W. K. Surewicz, and F. D. Sonnichsen, Solution structure of the E200K variant of human prion protein. Implications for the mechanism of pathogenesis in familial prion diseases. *J Biol Chem* **275**, 33650–4 (2000).

48. D. Shmerling, I. Hegyi, M. Fischer, T. Blattler, S. Brandner, J. Gotz, T. Rulicke, E. Flechsig, A. Cozzio, C. von Mering, C. Hangartner, A. Aguzzi, and C. Weissmann, Expression of amino-terminally truncated PrP in the mouse leading to ataxia and specific cerebellar lesions. *Cell* **93**, 203–14 (1998).

49. E. Flechsig, I. Hegyi, R. Leimeroth, A. Zuniga, D. Rossi, A. Cozzio, P. Schwarz, T. Rulicke, J. Gotz, A. Aguzzi, and C. Weissmann, Expression of truncated PrP targeted to Purkinje cells of PrP knockout mice causes Purkinje cell death and ataxia. *Embo J* **22**, 3095–101 (2003).

50. C. Weissmann, and A. Aguzzi, Perspectives: neurobiology. PrP's double causes trouble. *Science* **286**, 914–5 (1999).

51. G. M. Cereghetti, A. Schweiger, R. Glockshuber, and S. Van Doorslaer, Electron paramagnetic resonance evidence for binding of Cu(2+) to the C-terminal domain of the murine prion protein. *Biophys J* **81**, 516–25 (2001).

52. G. S. Jackson, I. Murray, L. L. Hosszu, N. Gibbs, J. P. Waltho, A. R. Clarke, and J. Collinge, Location and properties of metal-binding sites on the human prion protein. *Proc Natl Acad Sci U S A* **98**, 8531–5 (2001).

53. K. Qin, J. Coomaraswamy, P. Mastrangelo, Y. Yang, S. Lugowski, C. Petromilli, S. B. Prusiner, P. E. Fraser, J. M. Goldberg, A. Chakrabartty, and D. Westaway, The PrP-like protein Doppel binds copper. *J Biol Chem* **278**, 8888–96 (2003).

54. T. Muramoto, S. J. DeArmond, M. Scott, G. C. Telling, F. E. Cohen, and S. B. Prusiner, Heritable disorder resembling neuronal storage disease in mice expressing prion protein with deletion of an alpha-helix. *Nat Med* **3**, 750–5 (1997).

55. S. M. Zanata, M. H. Lopes, A. F. Mercadante, G. N. Hajj, L. B. Chiarini, R. Nomizo, A. R. Freitas, A. L. Cabral, K. S. Lee, M. A. Juliano, E. de Oliveira, S. G. Jachieri, A. Burlingame, L. Huang, R. Linden, R. R. Brentani, and V. R. Martins, Stress-inducible protein 1 is a cell surface ligand for cellular prion that triggers neuroprotection. *EMBO J* **21**, 3307–16 (2002).

56. R. G. Warner, C. Hundt, S. Weiss, and J. E. Turnbull, Identification of the heparan sulfate binding sites in the cellular prion protein. *J Biol Chem* **277**, 18421–30 (2002).

57. B. S. Wong, T. Liu, D. Paisley, R. Li, T. Pan, S. G. Chen, G. Perry, R. B. Petersen, M. A. Smith, D. W. Melton, P. Gambetti, D. R. Brown, and M. S. Sy, Induction of HO-1 and NOS in doppel-expressing mice devoid of PrP: implications for doppel function. *Mol Cell Neurosci* **17**, 768–75 (2001).

58. T. Cui, A. Holme, J. Sassoon, and D. R. Brown, Analysis of doppel protein toxicity. *Mol Cell Neurosci* **23**, 144–55 (2003).

59. D. R. Brown, K. Qin, J. W. Herms, A. Madlung, J. Manson, R. Strome, P. E. Fraser, T. Kruck, A. von Bohlen, W. Schulz-Schaeffer, A. Giese, D. Westaway, and H. Kretzschmar, The cellular prion protein binds copper in vivo. *Nature* **390**, 684–7 (1997).

60. D. R. Brown, R. S. Nicholas, and L. Canevari, Lack of prion protein expression results in a neuronal phenotype sensitive to stress. *J Neurosci Res* **67**, 211–24 (2002).

61. C. Kuwahara, A. M. Takeuchi, T. Nishimura, K. Haraguchi, A. Kubosaki, Y. Matsumoto, K. Saeki, T. Yokoyama, S. Itohara, and T. Onodera, Prions prevent neuronal cell-line death. *Nature* **400**, 225–6 (1999).

62. Y. Bounhar, Y. Zhang, C. G. Goodyer, and A. LeBlanc, Prion protein protects human neurons against Bax-mediated apoptosis. *J Biol Chem* **276**, 39145–9 (2001).

63. S. J. DeArmond, and S. B. Prusiner, Etiology and pathogenesis of prion diseases. *Am J Pathol* **146**, 785–811 (1995).

64. T. Yokoyama, K. M. Kimura, Y. Ushiki, S. Yamada, A. Morooka, T. Nakashiba, T. Sassa, and S. Itohara, *In vivo* conversion of cellular prion protein to pathogenic isoforms, as monitored by conformation-specific antibodies. *J Biol Chem* **276**, 11265–71 (2001).

65. G. Forloni, N. Angeretti, R. Chiesa, E. Monzani, M. Salmona, O. Bugiani, and F. Tagliavini, Neurotoxicity of a prion protein fragment. *Nature* **362**, 543–6 (1993).

66. T. Kitamoto, N. Amano, Y. Terao, Y. Nakazato, T. Isshiki, T. Mizutani, and J. Tateishi, A new inherited prion disease (PrP-P105L mutation) showing spastic paraparesis. *Ann Neurol* **34**, 808–13 (1993).

67. M. Y. Neufeld, J. Josiphov, and A. D. Korczyn, Demyelinating peripheral neuropathy in Creutzfeldt-Jakob disease. *Muscle Nerve* **15**, 1234–9 (1992).

68. N. L. Tuzi, E. Gall, D. Melton, and J. C. Manson, Expression of doppel in the CNS of mice does not modulate transmissible spongiform encephalopathy disease. *J Gen Virol* **83**, 705–11 (2002).

69. S. G. Chen, D. B. Teplow, P. Parchi, J. K. Teller, P. Gambetti, and L. Autilio-Gambetti, Truncated forms of the human prion protein in normal brain and in prion diseases. *J Biol Chem* **270**, 19173–80 (1995).

70. R. Daoud, M. Da Penha Berzaghi, F. Siedler, M. Hubener, and S. Stamm, Activity-dependent regulation of alternative splicing patterns in the rat brain. *Eur J Neurosci* **11**, 788–802 (1999).

71. R. Daoud, G. Mies, A. Smialowska, L. Olah, K. A. Hossmann, and S. Stamm, Ischemia induces a translocation of the splicing factor tra2-beta 1 and changes alternative splicing patterns in the brain. *J Neurosci* **22**, 5889–99 (2002).

72. D. Grisaru, M. Sternfeld, A. Eldor, D. Glick, and H. Soreq, Structural roles of acetylcholinesterase variants in biology and pathology. *Eur J Biochem* **264**, 672–86 (1999).

73. Y. Imai, N. Matsuo, S. Ogawa, M. Tohyama, and T. Takagi, Cloning of a gene, YT521, for a novel RNA splicing-related protein induced by hypoxia/reoxygenation. *Brain Res Mol Brain Res* **53**, 33–40 (1998).

74. K. L. Jin, S. H. Graham, X. O. Mao, X. He, T. Nagayama, R. P. Simon, and D. A. Greenberg, Bax kappa, a novel Bax splice variant from ischemic rat brain lacking an ART domain, promotes neuronal cell death. *J Neurochem* **77**, 1508–19 (2001).

75. D. Kaufer, A. Friedman, S. Seidman, and H. Soreq, Acute stress facilitates long-lasting changes in cholinergic gene expression. *Nature* **393**, 373–7 (1998).

76. J. Xie, and D. L. Black, A CaMK IV responsive RNA element mediates depolarization-induced alternative splicing of ion channels. *Nature* **410**, 936–9 (2001).

77. J. R. Sarna, and R. Hawkes, Patterned Purkinje cell death in the cerebellum. *Prog Neurobiol* **70**, 473–507 (2003).

Chapter 8

OXIDATIVE STRESS AND MITOCHONDRIAL DYSFUNCTION IN NEURODEGENERATION OF TRANSMISSIBLE SPONGIFORM ENCEPHALOPATHIES (TSES)

Boe-Hyun Kim[1], Jae-Il Kim[2], Richard I. Carp[2], Yong-Sun Kim[1]

[1]*Ilsong Institute of Life Science and Department of Microbiology, College of Medicine, Hallym University, Anyang 431-060, South Korea*

[2]*New York State Institute for Basic Research in Developmental Disabilities, Staten Island, NY 10314, USA*

8.1. Abstract

Transmissible spongiform encephalopathies (TSEs) or prion diseases are neurodegenerative disorders that are invariably fatal in humans and animals. An important component of the infectious agent is a glycoprotein, termed PrP^{Sc}, which is derived from a normal cellular protein, termed PrP^{C}. The pathogenic mechanisms of TSEs are not clear, but several factors such as oxidative stress and mitochondrial dysfunction have been reported to be involved. In the current review, we will present data that supports a role for oxidative stress and mitochondrial dysfunction in the induction of these diseases. We will discuss the pathways whereby oxidative stress and mitochondrial dysfunction could lead to neuronal damage and the clinical manifestations of TSE diseases.

8.2. Introduction

Mitochondria are a major source and target of free radicals[1–3], and the dysfunction of mitochondria can initiate the signaling cascades involved in programmed cell death or apoptosis[4–7]. Mitochondria play a crucial role in the regeneration of antioxidants through the production of reducing equivalents[8–13]. The major role of mitochondria is producing the vast amounts of ATP within most cells and higher organisms through oxidative phosphorylation[3,14]. Functions of mitochondria also include regulation of intracellular calcium (Ca^{2+}) homeostasis and production of reactive oxygen species (ROS). Thus, mitochondria have been implicated as central executioners of cell death[3,14]. These functions of mitochondria mean that they play a major role in cell signaling, as well as in biosynthesis and degradation[15–20].

The endogenous production of ROS is thought to be a major limitation of cellular life span[1,21–25]. Mitochondria have received considerable attention as both a principal source and target of ROS. Mitochondrial oxidative stress has been implicated in heart diseases including myocardial preconditioning, ischemia/reperfusion, and other pathologies[1,2,4–7]. In addition, oxidative stress in the mitochondria is associated with the pathogenesis of Alzheimer's disease, Parkinson's disease, TSEs, and amyotrophic lateral sclerosis (ALS) as well as aging itself. Free radicals and ROS have been observed to influence molecular and biochemical processes and directly cause some of the changes observed in cells during differentiation, aging, and transformation[26]. Nearly half of the effects discussed involve members of the mitogen-activated protein (MAP) kinases and nuclear factor-κB (NF-κB) signaling pathways and genes regulated by these pathways.

Mitochondrial dysfunction associated with the loss of Ca^{2+} homeostasis and increased cellular oxidative stress have long been recognized to play a major role in cell damage[27].

In this chapter, we discuss the role of mitochondria in regulation of Ca^{2+} homeostasis and in oxidative stress. We relate these changes induced by mitochondria to general aspects of cell death and specifically relate these events to neurodegeneration in TSEs.

8.3. Putative relationship between prion protein (PrP) and oxidative stress

The deposition of abnormal protein fibrils is a prominent pathological feature of many different 'protein conformational' diseases, including some important neurodegenerative diseases such as Alzheimer disease (AD), Parkinson's disease (PD), motor neuron disease and the TSEs[28]. The underlying protein component of the pathological fibrils associated with TSEs almost invariably adopts, predominantly, an anti-parallel pattern. An important component of the protein fibrils is a glycoprotein, termed PrP^{Sc}, which is derived from a normal cellular protein termed

PrP^C. PrP^{Sc} molecules accumulate, sometimes in very large amounts, within the central nervous system (CNS), where they are thought to lead to neurodegeneration[28–30]. An important aspect of these proteins is that a significant number of aggregating proteins associated with neurodegeneration are known to lead to redox-activation. TSEs can exist in both non-inherited, apparently sporadic, infectious, and inherited forms[31]. Molecular genetics studies have revealed that the inherited forms of disease can often, but not always, be due to mutations in the gene encoding the actual fibril-forming polypeptide (PrP)[28,31]. These aberrant protein conformations may aggregate and form deposits which interfere with normal function, eventually leading to disease. This idea is supported by transgenic mouse models in which overexpression of mutant, fibril-forming protein leads to a pathological picture in the mice with many similarities to both human and animal TSEs[32,33].

Further evidence for a role of aggregating proteins in the initiation of disease are the findings that several fibril forming proteins or smaller fragments there of, have been shown to be toxic to cells in culture under certain conditions[30,34–36]. However, it is not clear how this relates to cell death *in vivo*. Nor is it clear if the mechanism of cell death is similar in all cases where toxicity has been observed. Although the precise molecular mechanism by which PrP^{Sc} mediates cell death has remained a matter of considerable dispute, it is possible that the production of ROS and the influx of calcium ions into cells are both involved in toxicity. One fascinating possibility is that the aggregating peptide itself may be able to produce ROS such as hydrogen peroxide (H_2O_2)[37,38].

8.4. The link between PrP and mitochondrial dysfunction

As noted above, it is postulated that PrP needs to be in a conformationally altered state, i.e., PrP^{Sc}, before it becomes toxic to cells. Several studies have demonstrated that there is a toxic form of the peptide[30,34–36]. In addition to the direct production of ROS from the peptide, various other hypotheses have been put forward to explain the cytotoxic effects of PrP^{Sc}. These include: (1) the induction by PrP^{Sc} or by PrP peptides (PrP 106-126 or PrP 118-135) of calcium influx, directly and/or indirectly, into cytosol and mitochondria[30,35,39,40]; (2) the alteration of superoxide dismutase (SOD) activities by PrP^{Sc} in mitochondria and cytosol[41,42]; (3) an interaction between PrP^{Sc} and intracellular signaling proteins, such as NF-κB[43]; and (4) effects on mitochondrial proteins such as Bax, Bcl-2 and cytochrome c[40]. The various possible modalities of mitochondrial dysfunction are not necessarily incompatible with the idea that an important aspect of the toxicity of PrP^{Sc} is due to its ability to generate ROS directly. The toxic process could have dual mechanisms involving binding or attachment of protein aggregates to cell components, followed by the direct and/or indirect induction of oxidative damage.

In the remainder of this article, we will review the evidence of recently published data that mitochondrial dysfunction can generate ROS and finally induce neuronal cell death, which is thought to be a major aspect of the pathogenesis of TSEs[39–43].

8.5. Mechanisms of mitochondrial dysfunction

8.5.1. Mitochondrial dysfunction in relation to alternation of calcium metabolism

A mitochondrial pathway is thought to be a major cause of activation of a cascade which leads to apoptosis[30,40]. Energy depletion and increased oxidative damage to several synaptic proteins, such as the N-methyl-D-aspartate (NMDA) receptor and the α-amino-3-hydroxy-5-methyl-4-isoxazolepropionate (AMPA) receptor lacking the glutamate receptor 2 (GluR2) subunit, may result in loss of local calcium (Ca^{2+}) homeostasis resulting in synaptic degeneration[14,39]. Ca^{2+} is known to activate several intracellular enzymes, such as phospholipase A_2, nitric oxide synthase (NOS), xanthine dehydrogenase, calcineurin, and endonucleases, many of which can elicit the generation of endogenous ROS[14]. Moreover, when taken up by mitochondria, an increase in mitochondrial Ca^{2+} can also promote ROS generation[44].

Recently it has been found that Ca^{2+} homeostasis is disrupted in cerebellar granule cells from prion-deficient mice and in a cell line which produces hypothalamic gonadotropin after treatment with either PrP 106-126 or with β-amyloid[45,46]. PrP 106-126 has also been found to exert effects on calcium channels[47–49]. Calcium is known as a mediator of apoptosis in response to many stimuli[50].

The synthetic fragments of PrP peptides 106-126 and 118-135 are partially resistant to proteinase-K, readily aggregate into amyloid fibrils and can be neurotoxic[30,34,39,51]. The toxic effect of these synthetic peptides and their relation to the activity of calcineurin were demonstrated by inhibition of calcineurin factor, FK506[39]. Calcineurin is Ca^{2+}/calmodulin-dependent phosphatase and when activated it dephosphorylates the proapoptotic Bad protein allowing its translocation to the mitochondria, where it binds to anti-apoptotic Bcl-2 and Bcl-xL family proteins; this results in breakdown of mitochondrial transmembrane potential, and induces cytochrome c (cyto c) release[40,52,53]. In the cytosol, cyto c binds apoptotic protease activating factor-1 (Apaf-1) and dATP to form a complex that activates caspase-9, which can activate the executioner caspase-3[54].

An interesting observation is that increased mitochondrial inner membrane permeability, defined by the formation of the permeability transition pore (PTP), has been suggested to be involved not only in apoptotic but also in necrotic cell death, and autophagy[55]. The occurrence of necrosis or apoptotic-type of cell death may depend upon the intracellular levels of ATP, as was noted some years ago: if the ATP levels fall, plasma membrane rupture occurs; if ATP levels are maintained, the caspase-dependent apoptotic cascade can proceed[14].

Proapoptotic Bax was previously reported to mediate the release of cyto c through the opening of the PTP[56]. However, it has also been shown that Bax triggers cyto c release from isolated mitochondria independent of Ca^{2+}-inducible PTP[57]. Activation of the receptors for neurotrophins plays an important role in regulating the apoptotic activity of the BH3-only proteins of the Bcl-2 protein family. One known example is the phosphorylation of Bad at Ser^{112} and Ser^{136} which mediates its sequestering by the cytosolic protein 14-3-3, avoiding its heterodimerization with Bcl-2 and Bcl-xL at the mitochondrial membrane. The signaling pathway involves phosphatidylinositol-3-kinase (PI3K)/Akt (protein kinase B). In contrast, if there is a Ca^{2+}-dependent apoptotic stimuli such as Ca^{2+}-calmodulin induced activation, dephosphorylated Bad is released from its 14-3-3 anchor and is translocated to the mitochondria[14]. It is then free to execute its proapoptotic activity. In TSEs, it is known that Bcl-2 and Bax proteins are differentially expressed in the brains of hamsters infected with 263K scrapie agent[58]. This study has shown that there is decreased expression level of Bcl-2 and significantly increased expression level of Bax in the scrapie-infected animal model. In a subsequent study, calcium/calmodulin-dependent protein kinase II (CaM kinase II) was markedly elevated in cerebral cortex and hippocampal CA1 regions of scrapie-infected mice[59]. In addition, it was shown that in a PrP-deficient cell model, the increased expression of Bax results in decreased expression of Bcl-2. These results are associated with release of cyto c and alteration of Ca^{2+} levels in mitochondria[40]. Taken together, these studies suggest that imbalance of calcium homeostasis may cause mitochondrial PTP formation; this, in turn, causes release of cyto c to the cytosol of the cell. The consequence of the above is that cytosolic protein Bad would be taken up by mitochondria membranes. These changes then proceed sequentially to caspase-dependent apoptosis by activation of caspase-3.

8.5.2. Activation of SOD and oxidative stimuli

The superoxide anion (O_2^-) is generated in a constant manner by normal mitochondria. An enhancement in O_2^- formation may be the basis for cyto c release, as reported in cerebellar granule neurons undergoing glutamate-induced excitotoxicity[60].

It has been reported that the levels of malondialdehyde (MDA) and heme oxygenase-1 (HO-1), which are oxidative stress markers, are increased in brains of scrapie-infected mice[41,61,62]. The activity of Cu/Zn-SOD, a cytosolic antioxidant enzyme which is responsible for scavenging ROS, was not affected by scrapie infection, whereas that of Mn-SOD, a mitochondrial O_2^- scavenging enzyme, was markedly decreased in scrapie-infected mice. The activity of SOD (Mn-SOD and/or Cu/Zn-SOD) controls the levels of O_2^- by producing hydrogen peroxide (H_2O_2). H_2O_2 is not very reactive, unless it encounters Fe^{2+} or Cu^+ to form the highly reactive and short-lived hydroxyl radical ($\cdot OH$) by the Fenton-Harber Weiss reaction[63,64]. Importantly, a decreased expression of Mn-SOD was shown to induce oxidative stress and intensify apoptotic cell death by increasing mitochondrial cyto c release and the activation of caspases[40–42].

8.5.3. Signal transduction pathways

ROS and antioxidants are known to influence the expression of a number of genes and signal transduction pathways[14,43,65]. Although many pathways are known to be redox sensitive, none have been more thoroughly examined than the MAP kinase and NF-κB signal transduction pathways. Many of the redox effects on cells are mediated either directly or indirectly through these pathways, which contain multiple steps sensitive to ROS. Because of the large number of studies of oxidants on the various components of the MAP kinase and NF-κB signal transduction pathways, we provide a brief overview of both pathways.

MAP kinases; Extracellular signaling molecules such as growth factors and cytokines induce changes in cell behavior via complex mechanisms that involve transmission of the signal from the plasma membrane to the nucleus, where gene expression is altered[66]. The first step of a signaling cascade usually involves the activation of receptors that either have protein kinase activity or activate protein kinases in the cytoplasm. The signal is eventually transmitted to the nucleus, where it activates the transcription factors that regulate gene expression. One of the most studied families of signal transduction pathways is the MAP kinase family. Four MAP kinase subfamilies have been identified to date; (1) extracellular regulated kinase (ERK)[67–71], (2) c-jun

NH_2-terminal kinase/stress activated protein (JNK/SAP) kinase[72–76], (3) p38 kinase[77–80], and (4) big MAP kinase (BMK/ERK5)[81,82]. All these pathways contain redox-sensitive sites[26].

In the ERK and JNK pathways, a guanine nucleotide exchange factor such as the mammalian homologue of Sos[83] activates Ras by conversion of Ras-GDP to Ras-GTP[84–86]. It is known that activated Ras stimulates production of copious amounts of O_2^- [87]. Activation of all MAP kinases depends on dual phosphorylation of specific tyrosine (T) and threonine (Y) residues, which are located on the TXY motif in the linker loop 12 (L12) region of kinase subdomain VIII[88]. These reactions are catalyzed by dual-specificity threonine and tyrosine kinases belonging to the MAP kinase kinase (MKK) family. Members of the ERK, JNK, and p38 MAP kinase subfamilies phosphorylate the COOH-terminal transcriptional activation domain of Elk-1 and SAP-1[89–93], which then associates with other nuclear proteins to form ternary complex factor (TCF).

Our recent data demonstrated that gene expression of the chemokine RANTES (regulated on activation normal T cell expressed and secreted) was detected in hippocampal region of scrapie-infected mice, mainly in reactive astrocytes and in PrP-amyloid deposits, whereas the chemokine was not found in hippocampus of controls. Furthermore the expression level of RANTES receptors, CCR1, CCR3 and CCR5, were markedly elevated in the hippocampal region of scrapie-infected group, especially in reactive astrocytes[94]. RANTES is a proinflammatory cytokine that mediates leukocyte migration and activation. The action of this signal molecule and the activation of its receptors (i.e. CCR1, CCR3, and CCR5) may induce changes in cell behavior, thus inducing the signaling cascade. In a subsequent study, we have shown that JNK, p38, and ERK are increased in scrapie-infected animals. Furthermore, the phosphorylated cAMP/calcium-responsive element binding protein (CREB) which is a downstream transcription factor of active ERK, was significantly increased in scrapie[95]. Indeed, NADPH oxidase, a major ROS generator in cells, and ERK1/2, two MAPKs, have been identified as targets of PrP^C-mediated signaling[96]. This study provides evidence of a PrP^C-Fyn coupling-mediated signal cascade in a neuroectodermal precursor cell line, 1C11. It was shown that the binding of anti-PrP antibody to PrP^C promotes NADPH oxidase activation and ERK1/2 phosphorylation in the 1C11 cell line and its neuronal derivatives, as well as in GT1-7 hypothalamic cells and BW5147 lymphoid cells. These results indicate that PrP^C signaling activity is not restricted to neuronal cells. Moreover, common intracellular targets are recruited by the PrP^C-mediated signals and the nature of the signaling intermediates involved in the PrP^C transduction cascade varies depending on the cell context. In addition, it has been demonstrated that PrP106-126, a synthetic peptide, maintains many PrP^{Sc} characteristics; it induces cell death and activates caspase-3 through the p38 MAP kinase pathway in the SH-SY5Y neuroblastoma cells and microglia cells[97,98]. Eventually, activation of the cytokine receptors to the MAP kinase family of proteins is stimulated in TSEs, suggesting that TSE agents are inducing many effects of redox-sensitive stress by the MAP kinase pathways. Thus, these redox changes exert downstream effects through increased ROS formation.

NF-κB; The NF-κB/Rel family of transcription factors is involved in the regulation of numerous genes, including acute phase proteins, cell surface receptors, and cytokines and they also regulate certain viral genes[14]. The activation of antioxidant responses is mediated partially through NF-κB, which has been found to be a factor involved in the transcriptional regulation of Mn-SOD[99,100]. In the unstimulated state, NF-κB is bound to inhibitory proteins (IκB), thus maintaining it in the cytosol[101]. ROS activate NF-κB by causing the release of the IκB from the NF-κB complex[102].

Previously, it has been reported that the activity of NF-κB is increased significantly in scrapie-infected mice; increased activity was most pronounced in hippocampus and thalamus[43]. It also has been shown that the staining of NF-κB is especially intense in reactive astrocytes and PrP-amyloid plaques[43]. Also, the expression level of cyclooxygenase-2 (COX-2) was co-localized with PrP^{Sc} and with NF-κB in scrapie-infected mice[103]. These results indicate that upregulation of COX-2 in reactive astrocytes may be related to accumulation of PrP^{Sc}. The proposal has been made that PrP^{Sc} accumulation in reactive astrocytes may activate NF-κB through increase of ROS production and in turn, lead to alterations of NF-κB-mediated gene expression. The result showing increased expression of heme oxygenase-1 (HO-1) in scrapie-infected rodents[62] combined with the fact that NF-κB is one of the transcription factors for HO-1 supports the correlation between increased oxidative stress and activation of NF-κB signaling pathway. Furthermore, gene expression of IL-6 and iNOS, representative target genes of NF-κB activation, was detected only in the scrapie-infected group[43,104].

8.5.4. Other factors inducing mitochondrial dysfunction by activating ROS and/or oxidative stress

There are a number of other factors that can cause ROS. In this section, we present studies on iron, copper, phospholipase D (PLD), and AGEs as potential activating factors of ROS and/or oxidative stress.

Iron; As another approach to the study of oxidative stress related to neurodegeneration, we examined the relationship between iron metabolism and scrapie infection in different scrapie strain-host combinations. Previously, we have reported that the levels of total and ferric (Fe^{3+}) iron were significantly increased and that the redox state of iron was changed in cerebral cortex, striatum and brainstem by scrapie infection[63]. The change of iron redox state in favor of Fe^{3+} is known to be a condition for iron to participate in the hydroxyl radical (·OH) formation and lipid peroxidation. In addition, the expression levels of iron regulatory proteins (IRPs; IRP1 and IRP2), which are known as iron sensing proteins, and its binding protein, iron response element (IRE), are altered in scrapie-infected rodents[105,106]. These results indicate that changes in iron content and its redox state may accompany the increase of ROS generation. These changes may cause mitochondria-dependent oxidative damage to neurons, leading to neurodegeneration in TSEs.

Copper; Copper is a transition metal ion that has intrinsic capability to cycle between oxidized Cu (II) and reduced Cu (I) forms. Copper acts as a cofactor in many enzymes that play key roles in cell metabolism. Indeed, in the presence of hydrogen peroxide (H_2O_2), trace amounts of Cu (I) may catalyze the production of hydroxyl radical (Fenton's reaction). Oxidative stress is amplified by copper reactivity, leading to the impairment of essential molecules such as lipids, proteins, and DNA[107]. The primary antioxidant defense of cells against ROS production is mediated by the Cu/Zn-SOD[108].

The mitochondrial matrix shows the presence of a Mn-SOD while the intermembrane space (IMS) has a small amount of Cu/Zn-SOD which is primarily a cytosolic enzyme[42]. The network of distribution of copper to the mitochondria is targeted by a chaperone called Cox17, localized both in the cytosol and in the IMS of mitochondria. Cox17 appears to be fundamental for the insertion of copper in the mitochondrial enzyme cytochrome c oxidase (Cytox). In its active assembled form, Cox17 contains three copper ions, and it is then capable of inducing mitochondrial respiration. Furthermore it has been shown that proper delivery of copper to mitochondria is critical for the maturation and assembly of the individual Cytox subunits into the functional holoenzymes[109]. The finding that Cu/Zn-SOD is located at the IMS[42,109] extends the role played by copper in mitochondria, from driving energy production by Cytox, to counteracting ROS by Cu/Zn-SOD. The presence of mechanisms against ROS in mitochondria suggest that oxidative damage to mitochondrial DNA, enzymes, and the electron transport complexes must be deleterious to mitochondrial function.

Phospholipase D (PLD); It has been reported that phospholipase D (PLD) can be induced by ROS and that breakdown of phospholipids by PLD can be recognized as an important signaling mechanism in the CNS[110]. Recently, we found that the expression level and enzyme activity of PLD1, which is an isozyme of PLD, were significantly increased in the brains of scrapie-infected group, especially in mitochondrial fraction[111]. In addition, PLD1 immunoreactivity was significantly increased in cerebral cortex and hippocampus of infected brain, especially in reactive astrocytes. These results suggest that PLD1 activation by scrapie infection may affect mitochondrial lipid metabolism and in turn, lead to cellular dysfunction during the pathogenesis of TSEs. In mitochondria of scrapie-infected mice, the level of the oxidized form of glutathione (GSSG) was markedly increased, whereas mitochondrial membrane potential and energy metabolisms (ATP/ADP ratio) were decreased. It is possible that alterations of mitochondrial lipid metabolism leads to alteration of mitochondrial permeability and of energy metabolism due to a disturbed mitochondrial respiratory system. These changes may be accompanied by abnormal calcium accumulation in the mitochondria of scrapie-infected rodents. Thus, mitochondrial dysfunction can be caused by oxidative damage, abnormal calcium accumulation and altered energy metabolism. Mitochondrial dysfunction can certainly contribute to neurodegeneration in TSEs.

Advanced glycation end products (AGEs); AGEs are one of the carbohydrate modifications associated with oxidative stress. AGEs have been reported to be associated with the pathogenesis of vascular disease, AD and PD, suggesting that AGEs may contribute to the progressive deterioration associated with all chronic diseases. Tau is a protein associated with paired helical filaments (PHFs), which play an important role in AD pathology. Tau is advanced-glycated in AD and the deleterious effects of AGEs may be associated with oxidative stress[112–114]. We have shown that AGEs develop during TSEs[115]. We reported that one or more lysines at residues, 23, 24 and 27 of PrP^{Sc} are covalently modified with AGEs, yielding carboxymethyl-lysine (CML), one of the chemical varieties of AGEs. The nonenzymatic glycation of amino groups of proteins, termed the Maillard reaction, produces reversible Schiff bases and Amadori products. These early glycated products then undergo advanced glycation and oxidation (glycoxidation), which elicits irreversible modification, to form heteromorphic and fluorescent derivatives of AGEs[115]. It is possible that post-translational AGE-modification of PrP^{Sc} might play a role in protein stabilization and thereby be responsible, in part, for the accumulation of the aberrant protein in the brain. In summary, our results with AGE modification indicate that nonenzymatic glycation plays a role in the post-translational processing of PrP^{Sc}. Furthermore, the deleterious effects of AGEs may be associated with oxidative stress[113,114,116]. These studies support the concept that the

Figure 8.1. Possible oxidative stress mechanisms and signaling pathways in prion diseases. PrP^C, cellular normal form of host prion protein; PrP^{Sc}, scrapie or abnormal form of prion protein; AGE, advanced glycation end product; ROS, reactive oxygen species; PLD, phospholipase D; IRPs, iron regulatory proteins; NF-κB, nuclear factor-κB; ATM, ataxia-telangiectasia mutated kinase; ERK, extracellular regulated kinase; JNK, c-*jun* NH_2-terminal kinase; p38, MAP kinase protein; JAK, Janus kinase; STAT, signal transducers and activators of transcription.

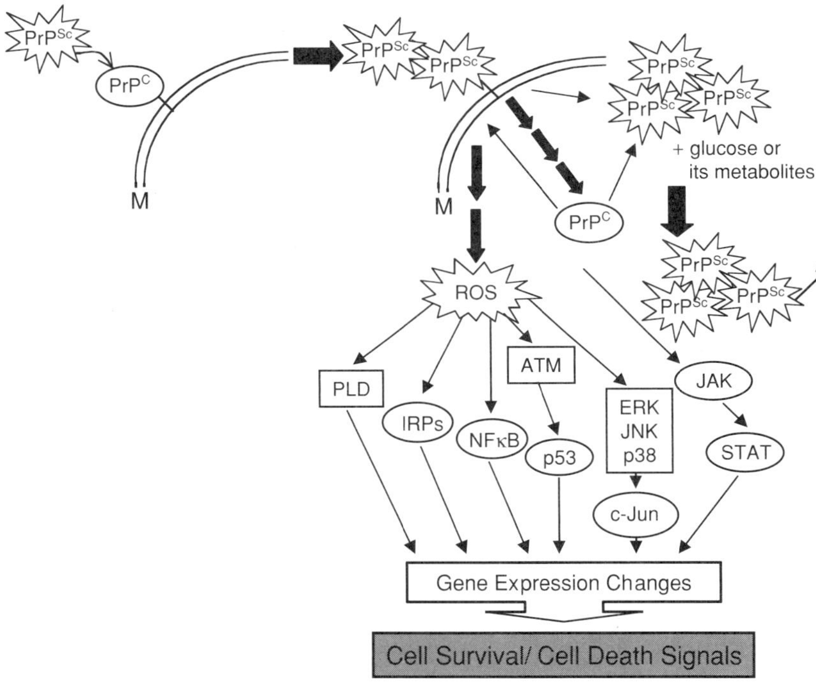

conformational changes of PrP^{Sc} protein caused by glycation may be involved in the activation of ROS signaling pathway, cause mitochondrial dysfunction and thereby may contribute to neurodegeneration in TSEs (Figure 8.1).

8.6. Conclusion

Here, we briefly reviewed and discussed the current research on TSEs, particularly focused on demonstrating that increased oxidative stress and mitochondrial dysfunction can contribute to neurodegenerative processes. In this context, alteration of calcium metabolism, regulation of SOD activities, activation of ROS-mediated signal transduction pathways, disturbances in regulation of iron and copper,

Figure 8.2. Possible cell death mechanisms of neurodegeneration via mitochondria in prion diseases by oxidative stress. Bax, Bcl-2-associated X protein; PARP, poly(ADP-ribose) polymerase.

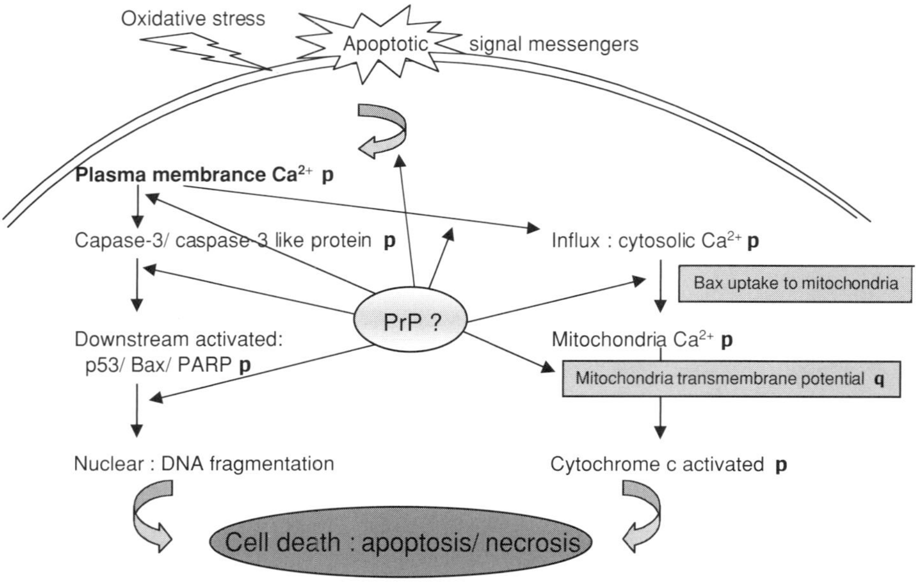

PLD, and formation of AGEs may be closely related to mitochondrial damage and pathogenic mechanisms of neurodegeneration in TSEs.

Neurons are highly dependent on glucose for ATP generation that is necessary for many biochemical processes. Neurons also produce ROS as by-products of oxidative phosphorylation within their mitochondria. If the amounts of ROS produced overcome the antioxidants, oxidative stress occurs, often followed by neuronal damage. The CNS is particularly susceptible to ROS-induced damage[117] because, (1) it has a high consumption of oxygen; (2) it contains high levels of membrane polyunsaturated fatty acids susceptible to free radical attack; (3) it is relatively deficient in oxidative defenses (poor catalase activity, and moderate SOD and glutathione peroxidase activities); and (4) a high content of iron and ascorbate can be found in some regions of the CNS, enabling the generation of more ROS through the Fenton-Haber Weiss reaction.

The mode of cell death by mitochondrial dysfunction, necrosis vs. apoptosis, in the pathogenic process of TSEs still remains to be elucidated. Part of the uncertainty is related to the fact that increased levels of calcium in mitochondria is a characteristic feature of both necrosis and apoptosis. The observation that altered calcium metabolism is related to mitochondrial dysfunction in scrapie-infected animals argues in favor of the concept that these changes contribute to necrotic cell death. However, results in experiments with neuronal cells under oxidative stress conditions, which activate ROS signaling, support the apoptotic pathway of neurodegeneration. Therefore, neuronal cell death, which gives rise to neurodegeneration in TSEs, can occur either by apoptosis or necrosis via mitochondrial dysfunction. Our understanding of these two cell death mechanisms in TSEs is summarized in Figure 8.2, which shows the necrotic pathway on the right and the apoptotic pathway on the left.

Thus, it remains to be established if the oxidative stress effect on mitochondria plays a causative role in the neurodegenerative process in TSEs or is a consequence of the disease. Nevertheless, we are confident that PrP^{Sc} replication affects mitochondria and that this organelle plays an important role in oxidative stress and in apoptotic cell death in TSEs.

8.7. Acknowledgments

We gratefully acknowledge documentary segmentation assistance from Mrs Eun-Ju Choi. This work was supported by a grant of the Korea Health 21 R & D Project, Ministry of Health and Welfare, Republic of Korea (02-PJ10-PG6-AG01-0003).

8.8. References

1. S. Melove, Mitochondrial oxidative stress: physiologic consequences and potential for a role in aging, *Ann. N. Y. Acad. Sci.* **908**, 219–225 (2000).

2. S. Melove, P. Coskun, M. Patel, R. Tuinistra, B. Cottrell, A. S. Jun, T. H. Zastawny, M. Dizdaroglu, S. I. Goodman, T. Huang, H. Miziorko, C. J. Epstein, D. C. Wallace, Mitochondrial disease in superoxide dismutase 2 mutant mice, *Proc. Natl. Acad. Sci. USA.* **96**, 846–851 (1999).

3. M. F. Beal, N. Howell, I. Bodis Wallner, eds., Mitochondria and free radicals in neurodegenerative diseases. New York, N. Y., Wiley-Liss (1997).

4. D. R. Green, J. C. Reed, Mitochondria and apoptosis, *Science* **281**, 1309–1312 (1998)

5. D. G. Nicholls, S.L. Budd, Mitochondria and neuronal survival, *Physiol. Rev.* **80**, 315–360 (2000).

6. T. Kuwana, J. J. Smith, M. Muzio, V. Dixit, D. D. Newmeyer, S. Kornbluth, Apoptosis induction by caspase-8 is amplified through the mitochondrial release of cytochrome c, *J. Biol. Chem.* **273**, 16589–16594 (1998).

7. S. A. Susin, H. K. Lorenzo, N. I. Zamzami, A. Marzo, B. E. Snow, G. M. Brothers, J. Mangion, E. Jacotot, P. Costantini, M. Loeffler, N. Larochette, D.R. Goodlett, R. Aebersold, D.P. Siderovski, J.M. Penninger, G. Kroemer, Molecular characterization of mitochondiral apoptosis-inducing factor, *Nature* **397**, 441–446 (1999).

8. V. E. Kagan, E. Serbinova, L. Packer, Antioxidant effects of ubiquinones in microsomes and mitochondria are mediated by tocopherol recycling, *Biochem. Biophys. Res. Commun.* **169**, 851–857 (1990).

9. V. E. Kagan, A. Shvedova, E. Serbinova, S. Khan, C. Swanson, R. Powell, L. Packer, Dihydrolipoic acid-a universal antioxidant both in the membrane and in the aqueous phase: reduction of peroxyl, ascorbyl and chromanoxyl radicals, *Biochem. Pharmacol.* **44**, 1637–1649 (1992).

10. V. E. Kagan, Y. Y. Tyurina, Recycling and redox cycling of phenolic antioxidant, *Ann. N. Y. Acad. Sci.* **854**, 425–434 (1998).

11. J. J. Maguire, V. Kagan, B. A. C. Ackrell, E. Serbinova, L.Packer, Succinate-ubiquinone reductase linked recycling of α-tocopherol in reconstituted systems and mitochondria: requirement for reduced ubiquinone, *Arch. Biochem. Biophys.* **292**, 47–53 (1992).

12. D. A. Stoyanovsky, A. N. Osipov, P. J. Quinn, V. E. Kagan, Ubiquinone-dependent recycling of vitamin E radicals by superoxide, *Arch. Biochem. Biophys.* **323**, 343–351 (1995).

13. M. Hiramatsu, R. D. Velasco, D. S. Wilson, L. Packer, Ubiquinone protects against loss of tocopherol in rat liver microsomes and mitochondrial membranes, *Res. Commun. Chem. Phaol. Pharmacol.* **72**, 231–241 (1991).

14. A. C. Rego, C. R. Oliveira, Mitochondrial dysfunction and reactive oxygen species in excitotoxicity and apoptosis: Implications for the pathogenesis of neurodegenerative diseases, *Neurochem. Res.* **28**, 1563–1574 (2003).

15. J. G. McCormack, A. P. Halestrap, R. M. Denton, Role of calcium ions in regulation of mammalian intramitochondrial metabolism, *Physiol. Rev.* **70**, 391–425 (1990).

16. T. E. Gunter, K. K. Gunter, S. S. Sheu, C. E. Gavin, Mitochondrial calcium transport: physiological and pathological evidence, *Am. J. Physiol.* **267**, C313–C339 (1994).

17. T. E. Gunter, L. Buntinas, G. C. Sparagna, K. K. Gunter, The Ca^{2+} transport mechanisms of mitochondria and Ca^{2+} uptake from physiological-type Ca^{2+} transients, *Biochim. Biophys. Acta.* **1366**, 5–15 (1998).

18. G. Hajnóczky, L. D. Robb-Gaspers, M. B. Seitz, A. P. Thomas, Decoding the cytosolic calcium oscillations in the mitochondria, *Cell* **82**, 415–424 (1995).

19. R. Rizzuto, A. W. M. Simpson, M. Brini, T. Pozzan, Rapid changes of mitochondrial Ca^{2+} revealed by specifically targeted recombinant aequorin, *Nature* **358**, 325–327 (1992).

20. L. D. Robb-Gaspers, G. A. Rutter, P. G. Burnett, P. Hajnóczky, R. M. Denton, A. P. Thomas, Coupling between cytosolic and mitochondrial calcium oscillations: role in the regulation of hepatic metabolism, *Biochim. Biophys. Acta.* **1366**, 17–22 (1998).

21. J. V. Leonard, A. H. Schapira, Mitochondrial respiratory chain disorders, I: mitochondrial DNA defects, *Lancet* **355**, 299–304 (2000).

22. A. H. Schapira, Mitochondrial disorders, *Biochim. Biophys. Acta.* **1410**, 99–102 (1999).

23. M. Hirano, M. Davidson, S. DiMauro, Mitochondria and the heart, *Curr. Opin. Cardiol.* **16**, 201–210 (2001).

24. H. H. M. Dahl, Getting to the nucleus of mitochondrial disorders: identification of respiratory chain-enzyme genes causing Leigh syndrome, *Am. J. Hum. Genet.* **63**, 1594–1597 (1998).

25. M. B. Delatycki, R. Williamson, S. M. Forrest, Freidreich ataxia: an overview, *J. Med. Genet.* **37**, 1–8 (2000).

26. R. G. Allen, M. Tresini, Oxidative stress and gene regulation, *Free. Radic. Biol. Med.* **28**, 463–499 (2000).

27. A. Frandsen, A. Schousboe, Excitoatory amino acid-mediated cytotoxicity and calcium homeostasis in cultured neurons, *J. Neurochem.* **60**, 1202–1211 (1993).

28. B. J. Tabner, S. Turnbull, O. M. A. El-Agnaf, D. Allsop, Production of reactive oxygen species from aggregating proteins implicated in Alzheimer'a disease, Parkinson's disease and other neurodegenerative diseases, *Curr. Top. Med. Chem.* **1**, 507–517 (2001).

29. D. A. Harris, Cellular biology of prion diseases, *Clin. Microbiol. Rev.* **12**, 429–444 (1999).

30. C. N. O'Donovan, D. Tobin, T. G. Cotterm, Prion protein fragment PrP-(106-126) induces apoptosis via mitochondrial disruption in human neuronal SH-SY5Y cells, *J. Biol. Chem.* **276**, 43516–43523 (2001).

31. S. B. Prusiner, Prions, *Proc. Natl. Acad. Sci. USA* **95**, 13363–13383 (1998).

32. M. R. Scott, S. Supattapone, H. O. Nguyen, S. J. DeArmond, S. B. Prusiner, Transgenic models of prion disease, *Arch. Virol. Suppl.* **16**, 113–124 (2000).

33. C. Weissmann, E. Flechsig, PrP knock-out and PrP transgenic mice in prion research, *Br. Med. Bull.* **66**, 43–60 (2003).

34. T. Pillot, L. Lins, M. Goethals, B. Vanloo, J. Baert, J. Vandekerckhove, M. Rosseneu, R. Brasseur, The 118–135 peptide of the human prion protein forms amyloid fibrils and induces liposome fusion, *J. Mol. Biol.* **274**, 381–393 (1997).

35. M. Pérez, F. Wandosell, C. Colaco, J. Avila, Sulphated glycosaminoglycans prevent the neurotoxicity of a human prion protein fragment, *Biochem. J.* **335**, 369–374 (1998).

36. G. Forloni, N. Angeretti, R. Chiesa, E. Monzani, M. Salmona, O. Bugiani, F. Taliavinim, Neurotoxicity of a prion protein fragment, *Nature* **362**, 543–546 (1993).

37. X. Huang, C. S. Atwood, M. A. Hartshorn, G. Multhaup, L. E. Goldstein, R. C. Scarpa, M. P. Caujungco, D. N. Gray, J. Lim, R. D. Moir, R. E. Tanzi, A. I. Bush, The Aβ peptide of Alzheimer's disease directly produces hydrogen peroxide through metal iron reduction, *Biochemistry* **38**, 7609–7616 (1999).

38. K. Hensley, J. M. Carney, M. P. Mattson, M. Aksenova, M. Harris, J. F. Wu, R. A. Floyd, D. A. Butterfield, A model for β-amyloid aggregation and neurotoxicity based on free radical generation by the peptide: Relevance to Alzheimer disease, *Proc. Natl. Acad. Sci. USA* **91**, 3270–3274 (1994).

39. P. Agostinho, C. R. Oliveira, Involvement of calcineurin in the neurotoxic effects induced by amyloid-beta and prion peptides, *Eur, J. Neurosci.* **17**, 1189–1196 (2003).

40. B. H. Kim, H. G. Lee, J. K. Choi, J. I. Kim, E. K. Choi, R. I. Carp, Y. S. Kim, The cellular prion protein (PrP^C) prevents apoptotic neuronal cell death and mitochondrial dysfunction induced by serum deprivation, *Brain Res. Mol. Brain Res.* **124**, 40–50 (2004).

41. S. I. Choi, W. K. Ju, E. K. Choi, J. Kim, H. Z. Lea, R. I. Carp, H. M. Wisniewski, Y. S. Kim, Mitochondrial dysfunction induced by oxidative stress in the brains of hamsters infected with the 263K scrapie agent, *Acta Neuropathol. (Berl.)* **96**, 279–286 (1998).

42. D. R. Brown, R. St. J. Nicholas, L. Canevari, Lack of prion protein expression results in a neuronal phenotype sensitive to stress, *J. Neurosci. Res.* **67**, 211–224 (2002).

43. J. I. Kim, W. K. Ju, J. H. Choi, J. Kim, E. K. Choi, R. I. Carp, H. M. Wisniewski, Y. S. Kim, Expression of cytokine genes and increased nuclear factor-kappa B activity in the brains of scrapie-infected mice, *Brain Res. Mol. Brain. Res.* **73**, 17–27 (1999).

44. A. J. Kowaltowski, R. F. Castilho, A. E. Vercesi, Ca^{2+}-induced mitochondrial membrane permeabilization: Role of coenzyme Q redox state, *Am. J. Physiol.* **269**, 141–147 (1995).

45. J. W. Herms, S. Korte, S. Gall, I. Schneider, S. Dunker, H.A. Kretzschmar, Altered intracellular calcium homeostasis in cerebellar granule cells of prion protein-deficient mice, *J. Neurochem.* **75**, 1487–1492 (2000).

46. M. Kuwahara, Y. Kuroda, N. Arispe, E. Rojas, Alzheimer's β-amyloid, human islet amylin, and prion protein fragment evoke intracellular free calcium elevations by a common mechanism in a hypothalamic GnRH neuronal cell line, *J. Biol. Chem.* **275**, 14077–14083 (2000).

47. T. Forio, S. Thellung, C. Amico, M. Robello, M. Salmona, Prion protein fragment 106-126 induces apoptotic cell death and impairment of L-type voltage-sensitive calcium channel activity in the GH3 cell line, *J. Neurosci. Res.* **54**, 341–352 (1998).

48. S. Thellung, T. Florio, V. Villa, A. Corsaro, S. Arena, F. Tagliavini, G. Schettini, Apoptotic cell death and impairment of L-type voltage-sensitive calcium channel activity in rat cerebellar granule cells treated with the prion protein fragment 106–126, *Neurobiol. Dis.* **7**, 299–309 (2000).

49. V. Silei, C. Fabrizi, G. Venturini, M. Salmona, O. Bugiani, F. Taliavini, G.M. Lauro, Activation of microglial cells by PrP and β-amyloid fragments raises intracellular calcium through L-type voltage sensitive calcium channels , *Brain Res.* **818**, 168–170 (1999).

50. P. D. Guta, K. Pushkala, Importance of the role of calcium in programmed cell death: a review, *Cytobios.* **99**, 83–95 (1999).

51. D. R. Brown, J. W. Herms, B. Schmidt, H. A. Kretzschmar, PrP and β-amyloid fragments activate different neurotoxic mechanisms in cultured mouse cells, *Eur. J. Neurosci.* **9**, 1162–1169 (1997).

52. L. Zhang, J. Chen, I. I. Fu, Suppression of apoptosis signal regulating kinase I-induced cell death by 14-3-3 proteins, *Proc. Natl. Acad. Sci. USA* **96**, 8511–8515 (1999).

53. J. Springer, R. D. Azbill, S. A. Nottingham, S. E. Kennedy, Clacineurin-mediated BAD dephosphorylation activated the caspase-3 apoptotic cascade in traumatic spinal cord injury, *J. Neurosci.* **20**, 7246–7251 (2000).

54. S. Desagher, J. C. Martinou, Mitochondria as the central control point of apoptosis, *Trends. Cell. Biol.* **10**, 369–377 (2000).

55. J. J. Lemasters, T. Qian, S. P. Elmore, L. C. Brenner, A. L. Nieminen, Confocal microscopy of the mitochondrial permeability transition in necrotic cell killing, apoptosis and autophagy, *Biofactors* **8**, 283–285 (1998).

56. J. G. Pastorino, M. Tafani, R. J. Rothman, A. Marcineviniute, J. B. Hoek, J. L. Farboer, Functional consequences of the sustained or transition pore, *J. Biol. Chem.* **274**, 31734–31739 (1999).

57. R. Eskes, B. Antonsson, A. Osen-Sand, S. Montessuit, C. Richter, R. Sadoul, G. Mazzei, A. Nichols, J.C. Martinou, Bax-induced cytochrome c release from mitochondria is independent of the permeability transition pore but highly dependent on Mg^{2+} ions, *J. Cell. Biol.* **143**, 217–224 (1998).

58. S. K. Park, S. I. Choi, J. K. Jin, E. K. Choi, J. I. Kim, R. I. Carp, Y. S. Kim, Differential expression of Bax and Bcl-2 in the brains of hamsters infected with 263K scrapie agent, *Neuroreport* **11**, 1677–1682 (2000).

59. J. K. Jin, J. K. Choi, H. G. Lee, Y. S. Kim, R. I. Carp, E. K. Choi, Increased expression of CaM kinase II alpha in the brains of scrapie-infected mice, *Neurosci. Lett.* **273**, 37–40 (1999).

60. A. Atlante, P. Calissano, A. Bobba, A. Azzariti, E. Marra, S. Passarella, Cytochrome c is released from mitochondria in a reactive oxygen species (ROS)-dependent fashion and can operate as a ROS scavenger and as a respiratory substrate in cerebellar neurons undergoing exitotoxic death, *J. Biol. Chem.* **275**, 37159–37166 (2000).

61. H. G. Lee, S. I. Choi, S. K. Park, N. H. Kim, E. K. Choi, R. I. Carp, Y. S. Kim, Alternation of glutathoine metabolism and abnormal calcium accumulation in the mitochondria of hamster brain infected with scrapie agent, *Neurobiol. Aging* **21**, S151 (2000).

62. Y. G. Choi, J. I. Kim, H. P. Lee, J. K. Jin, E. K. Choi, R. I. Carp, Y. S. Kim, Induction of heme oxygenase-1 in the brains of scrapie-infected mice, *Neurosci. Lett.* **289**, 173–176 (2000).

63. N. H. Kim, S. J. Park, J. K. Jin, M. S. Kwon, E. K. Choi, R. I. Carp, Y. S. Kim, Increased ferric iron content and iron-induced oxidative stress in the brains of scrpie-infected mice, *Brain Res.* **884**, 98–103 (2000).

64. D. R. Brown, K. F. Qin, J. W. Herms, A. Madlung, J. Manson, R. Strome, P. E. Frase, T. Kruck, A. Bonbohlen, W. Schulzschaeffer, A. Giese, D. Westaway, H. Kretzschmar, The cellular prion protein binds copper in vivo, *Nature* **390**, 684–687 (1997).

65. R. G. Allen, Oxidative stress and supeoxide dismutase in development, aging and gene regulation, *Age* **21**, 47–76 (1998).

66. M. Karin, T. Hunter, Transcriptional control by protein phosphorylation: signal transmission from the cell surface to the nucleus, *Curr. Biol.* **5**, 747–757 (1995).

67. A. Baba, E. Lee, T. Matsuda, T. Kihara, H. Iwata, Reversible inhibition of adenylate cyclase activity of rat brain caudate nucleus by oxidized glutathione. *Biochim. Biophys. Acta.* **85**, 1204–1210 (1978).

68. T. G. Boulton, S. H. Nye, D. J. Robbins, N. Y. Ip, E. Radziejewska, S. D. Morgenbesser, R. A. DePinho, N. Panayotatos, M. H. Cobb, G. D. Yancopoulos, ERKs: a family of protein-serine/threonine kinases that are activated and tyrosine phosphorylated in response to insulin and NGF, *Cell* **65**, 663–675 (1991).

69. D. L. Charest, G. Mordret, K. W. Harder, F. Jirik, S. L. Pelech, Molecular cloning, expression, and characterization of the human mitogen-activated protein kinase $p44^{ERK1}$, *Mol. Cell. Biol.* **13**, 4679–4690 (1993).

70. C. Lechner, M. A. Zahalka, J. F. Giot, N. P. Moller, A. Ullrich, ERK6, a mitogen-activated protein kinase involved in C2C12 myoblast differentiation, *Proc. Natl. Acad. Sci.* USA **93**, 4355–4359 (1996).

71. A. X. Zhu, Y. Zhao, D. E. Moller, J. S. Flier, Cloning and characterization of p97^{MAPK}, a novel human homolog of rat ERK-3, *Mol. Cell, Biol.* **14**, 8202–8211 (1994).

72. R. Carletti, S. Tacconi, E. Bettini, F. Ferraguti, Stress activated protein kinases, a novel family of mitogen-activated protein kinases, are heterogeneously expressed in the adult rat brain and differentially distributed from extracellular-signal-regulated protein kinases, *Neuroscience* **69**, 1103–1110 (1995).

73. R. J. Davis, MAPKs: new JNK expands the group, *Trends Biochem. Sci.* **19**, 470–473 (1994).

74. J. M. Kyriakis, P. Banerjee, E. Nikolakaki, T. Dai, E. A. Rubie, M. F. Ahmad, J. Avruch, J. R. Woodgett, The stress-activated protein kinase subfamily of c-jun kinases, *Nature* **369**, 156–160 (1994).

75. S. Mertens, M. Craxton, M. Goedert, SAP kinase-3, a new member of the family mammalian stress-activated protein kinases, *FEBS Lett.* **383**, 273–276 (1996).

76. J. R. Woodgett, J. Avruch, J. Kyriakis, The stress activated protein kinase pathway, *Cancer Surv.* **27**, 127–138 (1996).

77. Y. Jiang, C. Chen, Z. Li, W. Guo, J. A. Gegner, S. Lin, J. Han, Characterization of the structure and function of a new mitogen-activated protein kinase (p38β), *J. Biol. Chem.* **271**, 17920–17926 (1996).

78. Y. Jiang, H. Gram, M. Zhao, L. New, J. Gu, L. Feng, F. Di Padova, R. J. Ulevitch, J. Han, Characterization of the structure and function of the fourth member of p38 group mitogen-activated protein kinases, p38δ, *J. Biol. Chem.* **272**, 30122–30128 (1997).

79. Z. Li, Y. Jiang, R. J. Ulevitch, J. Han, The primary structure of p38γ: a new member of p38 group of MAP kinases, *Biochem. Biophys. Res. Commun.* **228**, 334–340 (1996).

80. B. Stein, M. X. Yang, D. B. Young, R. Janknecht, T. Hunter, B. W. Murray, M. S. Barbosa, p38-2, a novel mitogen-activated protein kinase with distinct properties, *J. Biol. Chem.* **272**, 19509–19517 (1997).

81. J. D. Lee, R. J. Ulevitch, J. Han, Primary structure of BMK1: a new mammalian map kinase, *Biochem. Biophys. Res. Commun.* **213**, 715–724 (1995).

82. G. Zhou, Z. Q. Bao, J. E. Dixon, Components of a new human protein kinase signal transduction pathway, *J. Biol. Chem.* **270**, 12665–12669 (1995).

83. P. Chardin, J. H. Camonis, N. W. Gale, L. van Aelst, J. Schlessinger, M. H. Wigler, D. Bar-Sagi, Human Sos1: a guanine nucleotide exchange factor for Ras that binds to GRB2, *Science* **260**, 1338–1343 (1993).

84. R. J. Grand, D. Owen, The biochemistry of ras p21, *Biochem. J.* **279**, 609–631 (1991).

85. I. G. Macara, K. M. Lounsbury, S. A. Richards, C. McKiernan, D. Bar-Sagi, The Ras superfamily of GTPases, *FASEB J.* **10**, 625–630 (1996).

86. S. J. Taylor, D. Shalloway, Cell cycle-dependent activation of Ras, *Curr. Biol.* **6**, 1621–1627 (1996).

87. K. Irani, Y. Xia, J. L. Zweier, S. J. Sollott, C. J. Der, E. R. Fearson, M. Sundaresan, T. Finkel, P.J. Goldschmidt-Clemont, Mitogenic signaling mediated by oxidants in Ras-transformed fibroblasts, *Science* **275**, 1649–1652 (1997).

88. F. Zhang, A. Strand, D. Robbins, M. H. Cobb, E. J. Goldsmith, Atomic structure of the MAP kinase ERK2 at 2.3 A resolution, *Nature* **367**, 704–711 (1994).

89. R. Janknecht, W. H. Ernst, V. Pingoud, A. Nordheim, Activation of ternary complex factor Elk-1 by MAP kinase, *EMBO J.* **12**, 5097–5104 (1993).

90. R. Janknecht, W. H. Ernst, A. Nordheim, SAP1a is a nuclear target of signaling cascades involving ERKs, *Oncogene* **10**, 1209–1216 (1995).

91. A. J. Whitmarsh, P. Shore, A. D. Sharrocks, R. J. Davis, Integration of MAP kinase signal transduction pathways at the serum responsive element, *Science* **269**, 403–407 (1995).

92. R. Janknecht, T. Hunter, Convergence of MAP kinase pathways on the ternary complex factor Sap-1α, *EMBO J.* **16**, 1620–1627 (1997).

93. M. A. Price, C. Hill, R. Treisman, Integration of growth factor signals at the c-fos serum response element, *Philos. Trans. R. Soc. Lond. B. Biol. Sci.* **351**, 551–559 (1996).

94. H. P. Lee, Y. C. Jeon, J. K. Choi, J. I. Kim, R. I. Carp, Y. S. Kim, The expression of RANTES and chemokine receptors in the brains of scrapie-infected mice, *J. Neuroimmunol* **158**, 26–33 (2005).

95. Y. S. Kim, H. P. Lee, Y. C. Jun, J. K. Choi, J. I. Kim, R. I. Carp, Activation of mitogen-activated protein kinases (MAPKS) in hamster brains infected with 263K scrapie agent, *Neurobiol. Aging Abstr.* 25, S457 (2004).

96. B. Schneider, V. Mutel, M. Pietri, M. Ermonval, S. Mouillet-Richard, O. Kellermann, NADPH oxidase and extracelular regulated kinases 1/2 are targets of prion protein signaling in neuronal and nonneuronal cells, *Proc. Natl. Acad. Sci.* USA **100**, 13326–13331 (2003).

97. S. Thellung, V. Villa, A. Corsaro, S. Arena, E. Millo, G. Damonte, U. Benatti, F. Tagliavini, T. Florio, G. Schettini, p38 MAP kinase mediates the cell death induced by PrP106-126 in the SH-SY5Y neuroblastoma cells, *Neurobiol. Dis.* **9**, 69–81 (2002).

98. F. B. Hafiz, D. R. Brown, A model for the mechanism of astrogliosis in prion disease, *Mol. Cell. Neurosci.* **16**, 221–232 (2000).

99. B. Meyrick, M. A. Magnusin, Identification and functional characterization of the bovine manganous superoxide dismutase gene, *Am. J. Respir. Cell Mol. Biol.* **10**, 113–121 (1994).

100. X. S. Wan, M. N. Devalaraja, D. K. St. Clair, Molecular structure and organization of the human manganese superoxide dismutase gene, *DNA Cell Biol.* **13**, 1127–1136 (1994).

101. A. A. Beg, S. M. Ruben, R. I. Scheinman, S. Haskill, C. A. Rosen, A. S. Baldwin Jr., IκB interacts with the nuclear localization sequences of the subunits of NF-κB: a mechanism for cytoplasmic retention, *Genes Dev.* **6**, 1899–1913 (1992).

102. R. Schreck, P. Rieber, P. A. Baeuerle, Reactive oxygen intermediates as apparently widely used messengers in the activation of the NF-κB transcription factor and HIV-1, *EMBO J.* **10**, 2247–2258 (1991).

103. J. I. Kim, J. K. Jin, E. K. Choi, S. I. Choi, R. I. Carp, Y. S. Kim, Overexpression of cyclooxygenase-2 in astrocytes of scrapie-infected mice, *Neurobiol. Aging* **21**, S87 (2000).

104. W. K. Ju, K. J. Park, E. K. Choi, J. Kim, R. I. Carp, H. M. Wisniewski, Y. S. Kim, Expression of inducible nitric oxide synthase in the brains of scrapie-infected mice, *J. Neurovirol.* **4**, 445–450 (1998).

105. Y. S. Kim, K, Hur, Y. C. Jun, E. A. Leibold, J. R. Connor, R. I. Carp, Altered expression of iron regulatory proteins (IRP1 and IRP2) and ferritin in the brains of scrapie-infected mice, *2002 Abstrac Viewer/Itinerary Planner. Washington, DC: Soc. Neurosci.* Program No. 692.12 (2002), Online.

106. K. Hur, J. I. Kim, S. I. Choi, E. K. Choi, R. I. Carp, Y. S. Kim, The pathogenic mechanisms of prion diseases, *Mech. Ageing. Dev.* **123**, 1637–1647 (2002).

107. G. Rotilio, L. Rossi, A. De Martino, A. M. Da Costa Ferreira, M. R. Ciriolo, Free radical, metal ions and oxidative stress: Chemical mechanisms of damage and protection in living systems, *J. Braz. Chem. Soc.* **6**, 221–227 (1995).

108. L. Rossi, M. F. Lombardo, M. R. Cirriolo, G. Rotilio, Mitochondrial dysfunction in neurodegenerative diseases associated with copper imbalance, *Neurochem. Res.* **29**, 493–504 (2004).

109. M. Glerum, A. Shtanko, A. Tzagoloff, Characterization of Cox17, a yeast gene involved in copper metabolism and assembly of cytochrome oxidase, *J. Biol. Chem.* **271**, 14504–14509 (1996).

110. J. Klein, V. Chalifa, M. Liscovitch, K. Loffelholz, Role of phospholipase D activation in nervous system physiology and pathophysiology, *J. Neurochem.* **65**, 1445–1455 (1995).

111. J. K. Jin, N. H. Kim, D. S. Min, J. I. Kim, B. H. Jeong, S. I. Choi, E. K. Choi, R. I. Carp, Y. S. Kim, Increased expression of phospholipase D1 in the brains of scrapie-infected mice, *J. Neurochem*, [in press] (2005).

112. S. D. Yan, X. Chen, A. M. Schmidt, J. Brett, G. Godman, Y. S. Zou, C. W. Scott, C.Caputo, T. Frappier, M. A. Smith, G. Perry, S. H. Yen, D. Stern, Glycated Tau protein in Alzheimer disease: A mechanism for induction of oxidant stress, *Proc. Natl. Acad. Sci.* USA **91**, 7787–7791 (1994).

113. M. Brownlee, A. Cerami, and H. Vlassara, Advanced glycosylation end products in tissue and the biochemical basis of diabetic complications, *N. Engl. J. Med.***318**, 1315–1321 (1988).

114. R. Castellani, M. A. Smith, G. L. Richey and G. Perry, Glycoxidation and oxidative stress in Parkinson disease and diffuse Lewy body disease, *Brain Res.* **737**, 195–200 (1996).

115. Y. G. Choi, J. I. Kim, Y. C. Jeon, S. J. Park, E. K. Choi, R. Rubenstein, R. J. Kascsak, R. I. Carp, Y. S. Kim, Nonenzymatic glycation at the N-terminus of pathogenic prion protein in transmissible spongiform encephalopathies, *J. Biol. Chem.* **279**, 30402–30409 (2004).

116. M. A. Smith, S. Taneda, P. L. Richey, S. Miyata, S. Yan, D. Stern, L. M. Sayre, V. M. Monnier, and G. Perry, Advanced Maillard Reaction End Products are Associated with Alzheimer Disease Pathology, *Proc. Natl. Acad. Sci.* USA **91**, 5710–5714 (1994).

117. B. Halliwell, Reactive oxygen species and the central nervous system, *J. Neurochem.* **59**, 1609–1623 (1992).

Chapter 9

MECHANISMS OF PRION TOXICITY AND THEIR RELATIONSHIP TO PRION INFECTIVITY

Laura Vella[1,2], Andrew F. Hill[1,2,3] and Roberto Cappai[2,3]

[1]*Department of Biochemistry and Molecular Biology and* [2]*Department of Pathology, and* [3]*The Centre for Neuroscience, The University of Melbourne, Melbourne, Victoria 3010, Australia*

9.1. Introduction

According to the protein only hypothesis, an abnormal isoform of the host encoded prion protein (PrP^C), referred to as PrP^{Sc}, is the sole or major component of the infectious agent the "prion" causing transmissible spongiform encephalopathies[1]. PrP^C expression is required for PrP^{Sc} propagation and was definitively established by the resistance of PrP null mice to infectious prions[2]. PrP^C is absolutely required for not only infectivity, but also PrP associated neurotoxicity[3]. This was demonstrated by grafting of embryonic telencephalic tissue from transgenic mice overexpressing PrP into the brains of PrP null mice and inoculating the PrP expressing graft with mice prions. The PrP expressing grafts accumulated high levels of infectious PrP^{Sc}, and developed severe histopathological changes characteristic of spongiform encephalopathy. Surprisingly, the graft derived PrP^{Sc} which migrated to areas of the PrP null host brain had no neuropathology, suggesting toxicity requires PrP^C expression. Whilst these studies highlight the absolute requirement of neuronal PrP^C expression for prion infection and toxicity, numerous studies to be discussed in this review illustrate a demarcation between prion infectivity and toxicity. Interestingly, infectivity can be present in wild-type brains in the absence of neurotoxicity and conversely neuropathology can be present in the absence or low levels of PrP^{Sc}. Collectively this data question PrP^{Sc} as the sole neurotoxic molecule, the nature of the toxic entity and its relationship to prion infectivity.

PrP^{Sc} is derived from PrP^{C} by a post-translational mechanism[4], suggesting that *in vitro* generation of prions from highly purified recombinant PrP should be possible. Conversion reactions using purified hamster PrP^{C} and a large excess of hamster PrP^{Sc} [5] can be stimulated by chaotropes, detergents or chaperone proteins[5–7] to generate protease resistant PrP. However the converted PrP is not infectious and is referred to as PrP^{res} [8]. These data strongly suggest that cellular factors, other than those added *in vitro*, are necessary for the acquisition of infectious properties. Due to the difficulty in isolating pure infectious PrP^{Sc} and detecting the toxic and or transmissible agent at the molecular level, synthetic PrP peptides are utilised to model prion disease *in vivo* and unravel the enigma associated with PrP^{Sc} infectivity and toxicity. This review will detail some of the neurotoxic mechanisms revealed by synthetic PrP peptides and PrP^{Sc} and provide a summary of the literature suggesting an uncoupling between prion infectivity and toxicity.

9.2. Synthetic PrP peptides as models to study PrP^{Sc} neurotoxicity

Neurodegenerative diseases aside from prion disease are associated with protein misfolding and concomitant neurotoxicity. Although these diseases are not transmissible, the toxic mechanisms associated with disease are similar. An example of this is the neurotoxic amyloid-β (Aβ) peptide which is the principal constituent of the extracellular amyloid deposits in Alzheimer's disease[9]. Aβ is a 42–43 amino acid peptide, formed from amyloid precursor protein by proteolytic processing (Figure 9.1). The neurotoxic effect of Aβ was explored initially using rat PC12 cells transfected with Aβ. The viability of these cells was decreased significantly and treatment with an antibody against Aβ reversed this deficit[10]. Subsequently, primary rat hippocampal cells were treated with synthetic peptide or fragments of synthetic peptide corresponding to the Aβ sequence, and exhibited reduced neuronal survival indicating a direct neurotoxic effect of Aβ[11].

On the basis of these observations Forloni *et al* (1993) created synthetic peptides corresponding to the PrP present in amyloid deposits from the cerebral cortex of patients affected by the human prion disease, Gerstmann-Stäussler-Scheinker Syndrome (GSS). The effects of synthetic PrP peptides on the viability of primary rat hippocampal cultures was tested and a peptide homologous to a highly conserved region of the prion protein, residues 106-126 (PrP106-126), exhibited physiochemical

Figure 9.1. Schematic of (a) PrP^C, PrP^{Sc} and PrP106-126 (b) Amyloid precursor protein and Amyloid-β peptide. Bold residues represent metal binding site ligands. Hydrophobic sequences are underlined.

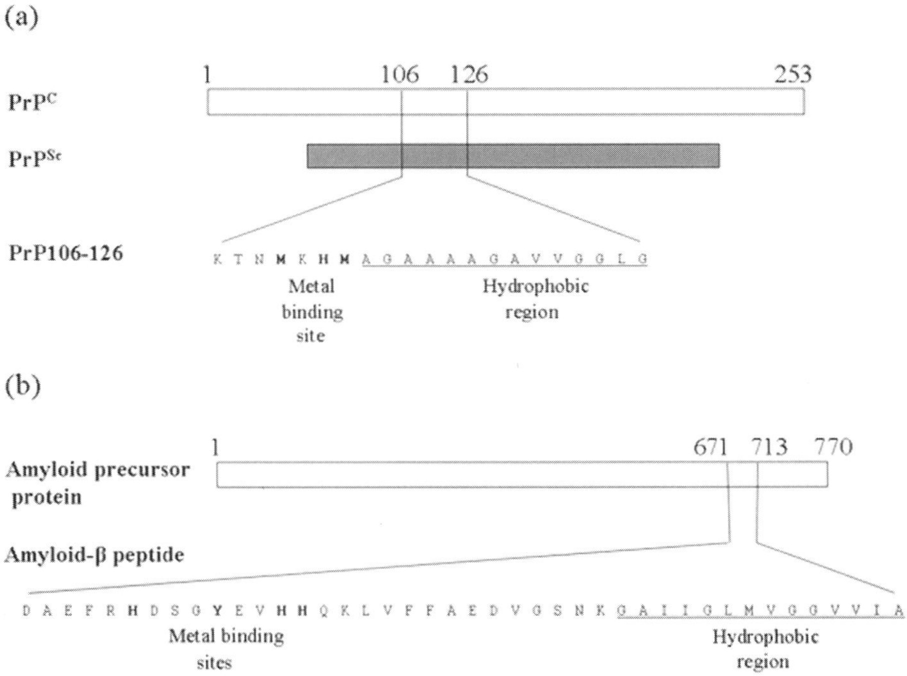

properties resembling PrP^{Sc} (Table 9.1). Significantly, PrP106-126 had a neurotoxic effect when applied to primary rat hippocampal cells, triggering neuronal death by apoptosis[12]. In contrast, other PrP peptides such as PrP57-64, PrP89-106, PrP106-114, PrP127-135, PrP127-147, did not significantly reduce cell viability, further suggesting that PrP106-126 represents a principal toxic component in prion disease. Unlike the Aβ peptide utilised in Alzheimer's research, PrP106-126 is not a naturally

Table 9.1. Similarities and differences between PrP^C, PrP^{Sc} and PrP106-126.

PrP^C	PrP^{Sc}	PrP106-126
• normal cellular form	• abnormal isoform	• synthetic peptide
• proteinase K sensitive	• proteinase K resistant	• proteinase K resistant
• alpha helical structure	• beta sheet structure	• beta sheet structure
• binds copper	• binds copper	• binds copper
	• forms amyloid	• forms amyloid
	• toxic to neurons	• toxic to primary neurons

occurring peptide, questioning the physiological relevance of these experiments. However, the absolute requirement of PrP^C expression for PrP106-126 toxicity validates its use as a model peptide for PrP^{Sc}, with PrP null neurons showing no alteration in cell viability when exposed to the peptide[13]. Furthermore, a longer synthetic PrP peptide, PrP82-146, identical to a natural PrP peptide fragment found in plaques from patients with GSS also forms amyloid fibrils and is toxic to neurons in culture, suggesting physiological relevance. Due to the difficulty in isolating infectious PrP^{Sc} to homogeneity *in vitro*, PrP106-126 is the most widely used synthetic peptide to model prion associated toxicity.

9.2.1. Physiochemical and toxic properties of PrP106-126

Prion disease in animals and humans is characterised by accumulation of PrP^{Sc} (sometimes as amyloid deposits[14]), astrogliosis and neuronal loss, resulting in spongiform vacuolation[15]. Homologous to PrP, the structure of PrP106-126 is highly hydrophobic and exhibits structural plasticity which is modulated by pH, ionic strength or a membrane like environment[16,17] (Table 9.1). PrP106-126 has a marked tendency to form stable β-sheet structures at acidic pH and in the presence of artificial membrane liposomes PrP106-126 secondary structure changes from a random coil to a predominantly β-sheet arrangement (16). The hydrophobic core sequence of PrP106-126 modulates formation of its stable β-sheet secondary structure[18] and like PrP^{Sc} is partially resistant to proteinase K digestion. Mutations that lower the hydrophobicity of the hydrophobic core ablate PrP106-126 neurotoxic activity, highlighting the importance of β-sheet formation[18].

PrP106-126 is highly amyloidogenic and readily forms amyloid fibrils in solution, which exhibit morphological, tinctorial, optical and conformational properties similar to those found *in vivo*[12,16,17,19]. There is conflicting data regarding the necessity of an amyloidogenic state for neurotoxicity as non-amyloidogenic neurotoxic variants of PrP106-126 can precede the formation of mature amyloid[20]. Furthermore, in a non-amyloidogenic state PrP106-126 can cause cell death, suggesting amyloid formation is not necessary for the toxic activity of the peptide[19,21]. It is likely that both highly aggregated insoluble material with high β-sheet content and insoluble non-aggregated material with low β-sheet content are capable of mediating PrP106-126 neurotoxicity[22]. This has similarity to recent studies in Alzheimer's disease where soluble oligomers, rather than the protease-resistant plaque, are the neurotoxic species[23,24] (Table 9.2).

Table 9.2. Similarities between the natural Amyloid-β peptide and synthetic PrP106-126 peptide.

- Forms amyloid
- Hydrophobic and copper binding regions
- Soluble & insoluble form are toxic to primary neurons
- Activates microglia
- Triggers primary neuronal cell death via apoptosis
- Upregulation of arachidonic acid pathway through the 5-lox pathway
- Causes oxidative stress
- Changes Ca^{2+} homeostasis
- Interaction with Cu^{2+} ions
- Can be of sporadic or genetic origin

9.2.2. Requirement of PrP^C for PrP^{Sc} and PrP106-126 toxicity

Neuronal cultures from PrP knockout mice incubated with PrP106-126 show survival rates equal to or greater than that of untreated wild-type cultures, demonstrating the necessity of PrP^C expression for PrP106-126 toxicity[13]. This mimics the requirement of PrP^C for prion infectivity *in vivo*[3]. Cultured neurons from transgenic mice overexpressing PrP^C 8-10 fold more than endogenous levels (Tg35 mice), have increased susceptibility to PrP106-126 toxicity[25], however transgenic mice which over express PrP^C 12-14 fold (Tg20 mice) are no more sensitive to the peptide than Tg35 mice. Microglia express PrP^C [26–28] and interestingly the microglia in Tg35 mice express more PrP^C than wild-type or Tg20 mice microglia[25], suggesting that in addition to neuronal PrP^C expression, the level of microglial PrP^C expression influences the degree of PrP106-126 toxicity.

9.2.3. The involvement of microglial PrP^C expression in PrP^{Sc} and PrP106-126 toxicity

Microglia function as macrophages in the central nervous system and upon activation by inflammatory stimuli migrate to sites of injury releasing a variety of factors including cytokines and acute phase proteins[29]. When microglia are activated due to a certain pathological imbalance, the inflammatory factors and reactive oxygen species they produce contribute to neuronal degeneration, as observed in Alzheimer's disease[30]. A role for microglia in prion disease has been confirmed *in vivo*[28,31–36]. Mice with clinical signs of prion disease demonstrate microglial activation in brain areas with PrP^{Sc} deposition[34–36], neuronal apoptosis and vacuolation[32,33]. Microglial activation precedes apoptotic neuronal death[37] and both the time course and the spatial distribution of microglial activation closely resembles the pattern of PrP^{Sc} deposition *in vivo*, suggesting microglial activation is involved in the neurotoxicity of PrP^{Sc}[28]. PrP^{C} expressing microglia affect the neurotoxicity of PrP106-126 *in vitro*, with PrP106-126 toxicity dependent upon both neuronal and microglial PrP^{C} expression[2,25]. PrP106-126 activates PrP^{C} expressing microglia *in vitro*[38,39] and it is hypothesised that the degree of activation is dependent upon the level of microglial PrP^{C} expressed[40]. Microglia appear to be activated by PrP106-126 or PrP^{Sc}, in a similar manner to that which occurs with Aβ[41,42]; an inflammatory response is initiated via a tyrosine kinase signaling cascade[43] which induces inflammatory gene expression, increasing reactive oxygen production, inflammatory cytokines and other potential neurotoxins[38]. In addition, PrP106-126 induces microglial DNA synthesis, thereby increasing microglia numbers and enhancing the subsequent inflammatory response, however this effect appears to be independent of microglial PrP^{C} levels[44].

9.2.4. Subcellular investigation of PrP^{Sc} and PrP 106-126 toxicity

Astrogliosis is one of the hallmarks of prion disease[14] and occurs late in the course of prion disease, following accumulation of PrP^{Sc}[45,46] and microglial activation[31]. The major astrocytic function is to support neurons by regulating extracellular glutamate levels[47]. Glutamate is an excitatory neurotransmitter, responsible for neuronal N-methyl-D-aspartate (NMDA) receptor activation. Regulation of extracellular glutamate is critical, as high glutamate concentrations are toxic to neurons[48].

PrP^{Sc} and PrP106-126 neurotoxicity involves activation of NMDA receptors in neuronal cells[49,50] as NMDA receptor antagonists produce a neuroprotective effect against prion toxicity[50–52]. In response to NMDA receptor stimulation, intracellular Ca^{2+} is increased[53], resulting in upregulation of arachidonic acid (AA)[54,55] presumably through the 5-lipoxygenase pathway (5-lox)[56]. Quinacrine, which inhibits PrP106-126 toxicity[56] is proposed as a potential therapeutic based on its ability to cross the blood-brain barrier and prevent the conversion of PrP into PrP^{Sc} in cell culture models[57]. However an *in vivo* study has showed no efficacy[58].

High levels of AA have a positive feedback response on NMDA receptors, potentiating further Ca^{2+} influx and subsequently up-regulating AA catabolism. The AA generates oxidised metabolites which are toxic to neuronal cells and triggers the synaptic release of glutamate which promotes further NMDA activation[52] and inhibits glutamate uptake by astrocytes[54,59]. Factors released from activated microglia following PrP106-126 treatment include Interleukin-1 (IL-1) and Interleukin-6 (IL-6)[38,60]. These factors not only stimulate astrocyte proliferation *in vitro*[61] but also up-regulate AA pathways[62,63], abetting the neurotoxic effect of glutamate. Interestingly, mice expressing PrP^C exclusively in astrocytes[64] are susceptible to prion disease. This was shown by application of PrP106-126 in cell culture systems, whereby PrP106-126 caused the death of PrP null neurons when PrP expressing astrocytes were present in the cultures[65]. The PrP null neuronal cell death is believed to occur because of increased astrocytic glutamate release[66] and inhibition of clearance of extracellular glutamate[67], leading to increased activation of neuronal NMDA receptors. Collectively this data suggests a role for PrP^C in glutamate uptake[50].

9.2.5. The role of copper in PrP^{Sc} and PrP106-126 toxicity

Physiologically, copper is necessary for synaptic transmission and is an integral component of multiple cellular enzymes, including cytochromes and superoxide dismutase[68]. The prion protein is highly concentrated at the synaptic cleft of neuronal cells and selectively binds Cu^{2+} ions *in vivo* and *in vitro*[69–72]. A role for copper in prion disease is suggested from studies in mice inoculated with mouse prions or humans with Creutzfeldt-Jakob disease. In both cases there is a reduction in brain copper levels[73,74]. The N-terminal octameric repeat region of PrP is highly conserved in mammals[75,76]. It has been established by several laboratories that copper binds and interacts with PrP^C at this hydrophobic region, inducing conformational change[71,77,78], raising the possibility that copper may trigger some functional alteration in PrP^C[79]. Interestingly, the N-terminal polar region of PrP106-126 contains a high affinity binding site for copper and mutagenesis of the metal binding site ablates the neurotoxic activity of PrP106-126, demonstrating the requirement of metal binding for PrP106-126 toxicity[80]. At physiological pH, aggregation of PrP106–126 is critically dependent on copper and zinc binding[80], further, altering the levels of copper in the culture media modulates PrP peptide toxicity as does addition of a copper chelator (50).

PrP null mice have reduced superoxide dismutase activity and are more vulnerable to oxidative stress[81–83], suggesting PrP^C can protect cells against oxidative stress and copper toxicity via metabolism or transport of copper at the synaptic cleft[84,85]. Whilst PrP^C binds copper at physiological concentrations, it does not appear to transport copper from the extracellular medium to the cytoplasm[86]. A role for astrocytes in the uptake and clearance of copper released by neurons has recently been suggested. It is hypothesised that PrP^C collects copper from the extracellular environment and shuttles it to astrocytes, protecting neurons from copper toxicity[87].

9.2.6. The role of calcium in PrP^{Sc} and PrP106-126 toxicity

In comparison to wild-type mice, neuronal preparations from PrP null mice have disrupted synaptic transmission and show weakened γ-aminobutyric acid type A ($GABA_A$) receptor-mediated fast inhibition and impaired long term potentiation[88]. This deficit can be reversed by reintroducing PrP expression[89].

Neuronal cells from PrP^C null mice appear to have abnormal electrophysiological properties such as disrupted Ca^{2+} activated K+ currents in particular in outward late after-hyperpolarization currents[90,91]. Studies with cultured prion infected cells indicated PrP^{Sc} can alter receptor mediated calcium response[92,93] and these alterations in excitability were also demonstrated in electrophysiological studies on prion infected mice[94] and hamsters[95] suggesting neuronal cell loss might be due to decreased Ca^{2+} entry and reduced intracellular free Ca^{2+} levels. This was confirmed in PrP null mice which had reduced Ca^{2+} -activated potassium currents due to an alteration in intracellular calcium homeostasis (48). PrP106-126 can alter Ca^{2+} levels in neuronal and glial cells[39,96,97]. The neurotoxic effect of the peptide was found to be reduced by compounds which block voltage-gated Ca^{2+} channels (VGCCs)[49], and further studies gave evidence of a direct alteration of L-type VGCCs by PrP106-126 suggesting this phenomenon is related to cell death induced by PrP106-126[98].

9.3. Other neurotoxic PrP peptides

Whilst PrP106-126 is the most studied of the PrP toxic peptides, other peptides have been identified as neurotoxic. The PrP178-193 peptide which corresponds to a PrP α-helical region, promotes Cu^{2+} -induced lipid peroxidation and cytotoxicity in primary neuronal cultures[99]. Furthermore, several point mutations of the PrP gene have been associated with prion disease. A mutation at residue 102 altered the biological activity of the PrP89-106 peptide causing it to became neurotoxic without changing its fibrillogenic capacity[100]. A mutation at residue 178 in the PrP169-185 peptide strongly increased the neurotoxic activity of the peptide with no clear alteration of its structural conformation[100].

The majority of neurotoxic PrP peptides tested include the 106-126 sequence. Four predicted α-helical regions which comprise the PrP106-126 sequence, when synthesised as individual peptides such as, PrP109-122[101] and PrP109-141[102] form β-sheets and PrP106-147 forms amyloid[103].

The ability to reconstitute *in vivo*, susceptibility to prion infection in PrP null mice has delineated PrP^C regions dispensable for susceptibility to prion disease. PrP null mice overexpressing the PrP gene with deletions from either residues 32-80[104] thus maintaining one octarepeat or 32-93, devoid of all octarepeats are susceptible to prion disease[105,106]. A truncated PrP gene of 106 amino acids with two large deletions (residues 23-88 and 141-176) expressed in PrP null mice confers susceptibility to prion disease[107] and the corresponding synthetic peptide has similar neurotoxic effects *in vitro* as PrP106-126[108]. These studies indicate that at least 60 residues of the N-terminal region of mature PrP^C are expendable for prion propagation, but the critical residues spanning 106-126 comprise part of the crucial PrP sequence necessary for susceptibility to prion disease.

9.4. Is prion infectivity the same as prion toxicity?

The studies detailed above clearly demonstrate fragments of PrP can be neurotoxic but how this relates to disease transmission is unclear. Examples exist where PrP toxicity can be uncoupled from transmission, thus questioning our understanding of prion disease.

9.4.1. Infectious PrP^{Sc} in the absence of neurotoxicity

Prion infected animals have been traditionally diagnosed based on clinical symptoms and neuropathological examination after death. The lack of clinical signs was assumed to reflect a failure of prion transmission[109]. However, the prion disease incubation period can exceed an animals natural life span and tissue from asymptomatic infected mice is capable of causing clinical disease when transmitted into healthy mice[110]. This is referred to as subclinical prion disease and applies to animals harbouring high levels of infectivity without exhibiting any clinical signs of disease during their normal lifespan[111]. The subclinical models discussed below suggest that the mere presence of PrP^{Sc} is not enough to cause neurotoxicity.

The most widely used model examining transmission of prion disease between different species, utilizes mice inoculated with hamster scrapie strain 263K. The 263K hamster strain had been reported to be non-pathogenic to mice[109,112], based on the absence of clinical symptoms and terminal disease within the normal lifespan of the infected animal. However evidence of detectable PrP^{Sc} propagation was not investigated. Whilst the transmission of disease was noted in hamsters inoculated with extracts from clinically normal 263K infected mice, it was concluded the infectivity was due to the presence of the original inoculum and that *de novo* replication had not occurred[113]. This view has been challenged with recent experiments which utilised antibodies to distinguish between mice and hamster 263K PrP^{Sc} and identified high levels of mouse PrP^{Sc} (and not hamster) in mice with no clinical signs of prion disease[111]. Moreover, the transmission of these latent prions into further mice caused terminal prion disease[114].

Subclinical disease can also occur when prions are passaged within the same species. Frigg *et al* (1999) investigated PrP expression in B-lymphocytes and discovered that mice lacking differentiated B-lymphocytes can harbor PrP^{Sc} in their brains following peripheral challenge with prion infection, yet do not develop clinical symptoms[115]. Brain material from these mice resulted in transmission of clinical disease. Interestingly, the immunodeficient mice harbored infectious titers equaling or even exceeding those of terminally sick wild-type controls. Typically, studies such as those discussed above, have utilised intracerebral injected high dose inoculum to produce subclinical infection. However, mice injected with dilutions of inoculum below end point titration can harbor similar levels of infectivity as terminally ill mice without exhibiting clinical signs of disease[116]. Moreover, the oral route, which is considered the most likely mode of transmission in natural prion diseases, can induce subclinical prion disease in wild-type mice[117].

Evidence of a disassociation between PrP^{Sc} and neurotoxicity is also observed in mice with clinical disease[3,118]. Mice with only one functional PrP allele express half the level of PrP as wild-type mice and accumulate infectivity levels as high as wild-type mice, but develop terminal disease only after an extended incubation period[118]. The presence of PrP^{Sc} in the absence of neuropathology or clinical symptoms seriously questions whether infectious PrP^{Sc} is by itself highly neurotoxic.

9.4.2. Neurotoxicity in the absence or low levels of PrP^{Sc}

Evidence that PrP^{Sc} is the toxic entity in prion disease arises from the correlation that is often observed between the accumulation of PrP^{Sc} and the appearance of neuropathology and clinical symptoms[45,119]. Whilst PrP^{Sc} copurifies with infectivity[120], neuropathology can develop in the absence or very low levels of PrP^{Sc}.

Fatal familial insomnia (FFI) is a human prion disease, presenting severe neuropathological features, yet PrP^{Sc} levels are low or even undetectable[121,122]. In some cases fatal familial insomnia can be transmitted to susceptible animals, confirming the presence of infectious prions[123,124]. Likewise, bovine spongiform encephalopathy (BSE) can be transmitted to mice in the absence of detectable PrP^{Sc}. Mice infected with homogenate from BSE-infected cattle, exhibited neurological symptoms and neuronal death, however more than 55% had no detectable PrP^{Sc}[125].

Mice expressing PrP^{C} lacking residues 32-93 are susceptible to prion disease[105,106]. Interestingly, there is no neuropathology typical for mice prion disease and terminally ill animals have infectivity levels 10–30 times lower and prion titers 30–50 times lower than their wild-type counterparts, indicating a dissociation between PrP^{Sc} and clinical disease[105]. Mouse cell lines infected with a mouse adapted human strain of CJD can propagate PrP^{Sc} and provides further evidence of uncoupling between PrP^{Sc} and infectivity[126]. After several passages in culture following the initial infection a 650-fold increase in infectivity bioassays was noted. However the level of PrP^{Sc}, as detected by Western blot, increased only 2-fold. Whereas PrP^{Sc} indicated persistent infection, the amount was not quantitatively related to infectivity.

9.5. Conclusion

Similar to the Aβ peptide in Alzheimer's disease, a synthetic PrP peptide with a hydrophobic C-terminal sequence can induce neurotoxicity (Table 9.2). PrP106-126 in contrast with other PrP fragments, is fibrillogenic, has proteinase resistant properties and induces astrogliosis in primary neuronal cultures. Although this fragment has not been detected in amyloid deposits of prion disease brain, the 106-126 sequence is an integral part of abnormal PrP isoforms and a crucial sequence for susceptibility to prion disease. PrP106-126 has proven to be a useful model for studying neurotoxicity in prion disease due to the difficulty in isolating pure PrP^{Sc}.

An unresolved question is how neurotoxicity relates to infectivity? Numerous studies have shown that irrespective of whether subclinical or terminal disease occurs following prion inoculation, there is a lack of correlation between PrP^{Sc} levels, prion infectivity and prion neurotoxicity. While PrP^{Sc} may directly cause some neurodegeneration, these data question PrP^{Sc} as the sole neurotoxic molecule in prion disease and the nature of the neurotoxic entity.

The absolute requirement of PrP^{C} expression for infection and toxicity, suggested prion disease may be due, at least in part, to the loss of PrP^{C} function. However, the generation of "healthy" PrP knockout mice, argued against this[127]. Furthermore, complete ablation of PrP expression in adult mice using conditional gene expression does not cause disease thus excluding the possibility that PrP null mice do not develop neurodegeneration because of compensatory adaptations during neurodevelopment[91]. One hypothesis is that the principal factor determining the clinical outcome of prion infection is the conversion rate of PrP^{C} to PrP^{Sc}. Tg20 mice, which express 12–14 fold more PrP^{C} than wild-type, succumb to terminal disease more rapidly than wild-type mice and terminal disease is associated with low levels of PrP^{Sc}[104]. Conversly, mice that express half the wildtype level of PrP have long disease incubation periods, associated with high levels of PrP^{Sc}[118].

It is possible that a neurotoxic intermediate species, designated PrP^{L}, whereby 'L' stands for lethal, is produced in the conversion of PrP^{C} to PrP^{Sc} whereby PrP^{Sc} is a highly aggregated material that could be a relatively inert end-product[124] (Figure 9.2). Different forms of PrP, such as an alternative topological variant called ^{Ctm}PrP has been proposed as the neurotoxic intermediate[128]. It is suggested that the amount of PrP^{Sc} that accumulates is inversely related to the amount of ^{Ctm}PrP present, indicating that ^{Ctm}PrP rather than PrP^{Sc} may be the proximate cause of neurodegeneration[129,130]. Whether toxic but non-infectious forms of PrP are generated as intermediates and in addition to PrP^{Sc} are the primary neurotoxic species is unknown.

The rate at which PrP^{L} is formed could determine the severity of neurodegeneration[124]. Such intermediates may form slowly and be cleared or tolerated by the host at low levels, providing some explanation for the long incubation periods and progressive accumulation of the PrP^{Sc} end product (Figure 9.2). Over time the intermediate accumulates until a threshold is reached resulting in neurotoxicity. In Tg20 mice, the high level of PrP expression may hasten the conversion reaction in inoculated animals, yielding more toxic intermediate product at a faster rate compared to wild-type mice. Consequently there is no extended incubation period and PrP^{Sc} has limited time to form. It is plausible that

Figure 9.2. **PrP^C**; normal cellular form. **PrP^{Sc}**; abnormal partially toxic and infectious form. **PrP^L**; abnormal toxic form. PrP^C expression or suboptimal conditions may determine the fate of PrP^L. PrP^L may be cleared or tolerated by the host at low levels, hence the end-product, PrP^{Sc}, accumulates overtime (Subclinical disease). Conversly, conditions that quicken the formation of PrP^L, may overwhelm clearance mechanisms, leading to accumulation of PrP^L to neurotoxic levels, consequently PrP^{Sc} has limited time to form. Accumulation of PrP^L may lead to NMDA receptor activation, increase in Ca^{2+} activity, upregulation of AA (5-lox pathway) and consequently neurotoxicity leading to death (FFI).

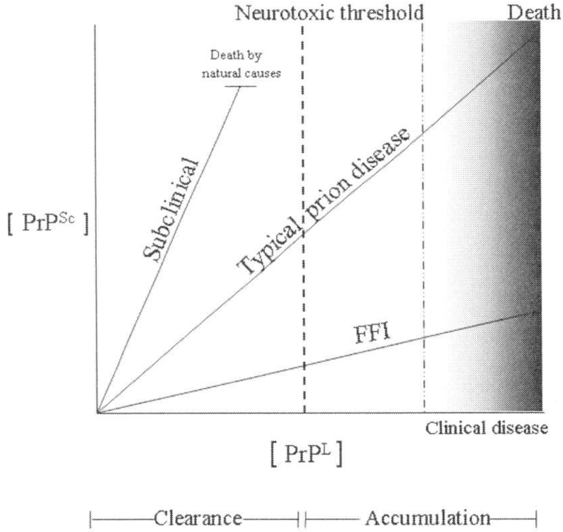

subclinical disease is the result of suboptimal conditions and the critical threshold for PrP^L may not be reached or maintained.

The literature reviewed in this chapter collectively raise questions on the relationship between PrP^{Sc}, infectivity and toxicity. What is the nature of the infectious agent? What is the neurotoxic molecule associated with prion disease? Which cellular proteins assist in PrP^C conversion into toxic PrP? How can we interfere with this conversion? The answers to these questions will ultimately lead to early diagnosis and treatments as the accumulation of PrP^{Sc} is currently used for biochemical diagnosis of TSEs. We need other markers of prion disease, that allow us to test for prion infectivity in the absence of PrP^{Sc} or vice-versa. The studies discussed here suggest a demarcation between infectious prions and neurotoxic prions, whereby infectious PrP^{Sc} can be present in the brain without causing neurotoxicity and conversely neuropathology can develop in the absence or low levels of PrP^{Sc}. These discrepancies indicate that alternative forms of PrP, distinct from infectious PrP and/or unidentified cellular factors, may be involved in the mechanisms of prion toxicity.

9.6. References

1. S. B. Prusiner, Novel proteinaceous infectious particles cause scrapie. *Science* **216**, 136–44 (1982).

2. H. Bueler, A. Aguzzi, A. Sailer, R. A. Greiner, P. Autenried, M. Aguet and C. Weissmann, Mice devoid of PrP are resistant to scrapie. *Cell* **73**, 1339–47 (1993).

3. S. Brandner, S. Isenmann, A. Raeber, M. Fischer, A. Sailer, Y. Kobayashi, S. Marino, C. Weissmann and A. Aguzzi, Normal host prion protein necessary for scrapie-induced neurotoxicity. *Nature* **379**, 339–43 (1996).

4. D. R. Borchelt, M. Scott, A. Taraboulos, N. Stahl and S. B. Prusiner, Scrapie and cellular prion proteins differ in their kinetics of synthesis and topology in cultured cells. *J Cell Biol* **110**, 743–52 (1990).

5. D. A. Kocisko, J. H. Come, S. A. Priola, B. Chesebro, G. J. Raymond, P. T. Lansbury and B. Caughey, Cell-free formation of protease-resistant prion protein. *Nature* **370**, 471–4 (1994).

6. S. K. DebBurman, G. J. Raymond, B. Caughey and S. Lindquist, Chaperone-supervised conversion of prion protein to its protease-resistant form. *Proc Natl Acad Sci USA* **94**, 13938–43 (1997).

7. J. Chabry, S. A. Priola, K. Wehrly, J. Nishio, J. Hope and B. Chesebro, Species-independent inhibition of abnormal prion protein (PrP) formation by a peptide containing a conserved PrP sequence. *J Virol* **73**, 6245–50 (1999).

8. A. F. Hill, M. Antoniou and J. Collinge, Protease-resistant prion protein produced in vitro lacks detectable infectivity. *J Gen Virol* **80 (Pt 1)**, 11–4 (1999).

9. C. L. Masters, G. Simms, N. A. Weinman, G. Multhaup, B. L. McDonald and K. Beyreuther, Amyloid plaque core protein in Alzheimer disease and Down syndrome. *Proc Natl Acad Sci USA* **82**, 4245–9 (1985).

10. B. A. Yankner, L. R. Dawes, S. Fisher, L. Villa-Komaroff, M. L. Oster-Granite and R. L. Neve, Neurotoxicity of a fragment of the amyloid precursor associated with Alzheimer's disease. *Science* **245**, 417–20 (1989).

11. R. L. Neve, L. R. Dawes, B. A. Yankner, L. I. Benowitz, W. Rodriguez and G. A. Higgins, Genetics and biology of the Alzheimer amyloid precursor. *Prog Brain Res* **86**, 257–67 (1990).

12. G. Forloni, N. Angeretti, R. Chiesa, E. Monzani, M. Salmona, O. Bugiani and F. Tagliavini, Neurotoxicity of a prion protein fragment. *Nature* **362**, 543–6 (1993).

13. D. R. Brown, J. Herms and H. A. Kretzschmar, Mouse cortical cells lacking cellular PrP survive in culture with a neurotoxic PrP fragment. *Neuroreport* **5**, 2057–60 (1994).

14. S. B. Prusiner, Prions. *Proc Natl Acad Sci USA* **95**, 13363–83 (1998).

15. H. Budka, A. Aguzzi, P. Brown, J. M. Brucher, O. Bugiani, F. Gullotta, M. Haltia, J. J. Hauw, J. W. Ironside, K. Jellinger and et al., Neuropathological diagnostic criteria for Creutzfeldt-Jakob disease (CJD) and other human spongiform encephalopathies (prion diseases). *Brain Pathol* **5**, 459–66 (1995).

16. C. Selvaggini, L. De Gioia, L. Cantu, E. Ghibaudi, L. Diomede, F. Passerini, G. Forloni, O. Bugiani, F. Tagliavini and M. Salmona, Molecular characteristics of a protease-resistant, amyloidogenic and neurotoxic peptide homologous to residues 106–126 of the prion protein. *Biochem Biophys Res Commun* **194**, 1380–6 (1993).

17. L. De Gioia, C. Selvaggini, E. Ghibaudi, L. Diomede, O. Bugiani, G. Forloni, F. Tagliavini and M. Salmona, Conformational polymorphism of the amyloidogenic and neurotoxic peptide homologous to residues 106-126 of the prion protein. *J Biol Chem* **269**, 7859–62 (1994).

18. M. F. Jobling, L. R. Stewart, A. R. White, C. McLean, A. Friedhuber, F. Maher, K. Beyreuther, C. L. Masters, C. J. Barrow, S. J. Collins and R. Cappai, The hydrophobic core sequence modulates the neurotoxic and secondary structure properties of the prion peptide 106-126. *J Neurochem* **73**, 1557–65 (1999).

19. M. Salmona, P. Malesani, L. De Gioia, S. Gorla, M. Bruschi, A. Molinari, F. Della Vedova, B. Pedrotti, M. A. Marrari, T. Awan, O. Bugiani, G. Forloni and F. Tagliavini, Molecular determinants of the physicochemical properties of a critical prion protein region comprising residues 106–126. *Biochem J* **342 (Pt 1)**, 207–14 (1999).

20. M. Bucciantini, E. Giannoni, F. Chiti, F. Baroni, L. Formigli, J. Zurdo, N. Taddei, G. Ramponi, C. M. Dobson and M. Stefani, Inherent toxicity of aggregates implies a common mechanism for protein misfolding diseases. *Nature* **416**, 507–11 (2002).

21. A. Corsaro, S. Thellung, V. Villa, D. R. Principe, D. Paludi, S. Arena, E. Millo, D. Schettini, G. Damonte, A. Aceto, G. Schettini and T. Florio, Prion protein fragment 106-126 induces a p38 MAP kinase-dependent apoptosis in SH-SY5Y neuroblastoma cells independently from the amyloid fibril formation. *Ann NY Acad Sci* **1010**, 610–22 (2003).

22. D. R. Brown, M. Pitschke, D. Riesner and H. A. Kretzschmar, Cellular effects of a neurotoxic prion protein are related to its β-sheet content. *Neurosci. Res. Comm.* **23**, 119–128 (1998).

23. T. Oda, P. Wals, H. H. Osterburg, S. A. Johnson, G. M. Pasinetti, T. E. Morgan, I. Rozovsky, W. B. Stine, S. W. Snyder, T. F. Holzman and et al., Clusterin (apoJ) alters the aggregation of amyloid beta-peptide (A beta 1–42) and forms slowly sedimenting A beta complexes that cause oxidative stress. *Exp Neurol* **136**, 22–31 (1995).

24. M. P. Lambert, A. K. Barlow, B. A. Chromy, C. Edwards, R. Freed, M. Liosatos, T. E. Morgan, I. Rozovsky, B. Trommer, K. L. Viola, P. Wals, C. Zhang, C. E. Finch, G. A. Krafft and W. L. Klein, Diffusible, nonfibrillar ligands derived from Abeta 1-42 are potent central nervous system neurotoxins. *PNAS* **95**, 6448–6453 (1998).

25. D. R. Brown, Prion protein-overexpressing cells show altered response to a neurotoxic prion protein peptide. *J Neurosci Res* **54**, 331–40 (1998).

26. D. R. Brown, A. Besinger, J. W. Herms and H. A. Kretzschmar, Microglial expression of the prion protein. *Neuroreport* **9**, 1425–9 (1998).

27. D. R. Brown, B. Schmidt and H. A. Kretzschmar, A prion protein fragment primes type 1 astrocytes to proliferation signals from microglia. *Neurobiol Dis* **4**, 410–22 (1998).

28. A. Giese, D. R. Brown, M. H. Groschup, C. Feldmann, I. Haist and H. A. Kretzschmar, Role of microglia in neuronal cell death in prion disease. *Brain Pathol* **8**, 449–57 (1998).

29. R. N. Kalaria, Microglia and Alzheimer's disease. *Curr Opin Hematol* **6**, 15–24 (1999).

30. S. Itagaki, P. L. McGeer, H. Akiyama, S. Zhu and D. Selkoe, Relationship of microglia and astrocytes to amyloid deposits of Alzheimer disease. *J Neuroimmunol* **24**, 173–82 (1989).

31. A. Williams, P. J. Lucassen, D. Ritchie and M. Bruce, PrP deposition, microglial activation, and neuronal apoptosis in murine scrapie. *Exp Neurol* **144**, 433–8 (1997).

32. P. J. Lucassen, A. Williams, W. C. Chung and H. Fraser, Detection of apoptosis in murine scrapie. *Neurosci Lett* **198**, 185–8 (1995).

33. A. E. Williams, L. J. Lawson, V. H. Perry and H. Fraser, Characterization of the microglial response in murine scrapie. *Neuropathol Appl Neurobiol* **20**, 47–55 (1994).

34. D. C. Guiroy, I. Wakayama, P. P. Liberski and D. C. Gajdusek, Relationship of microglia and scrapie amyloid-immunoreactive plaques in kuru, Creutzfeldt-Jakob disease and Gerstmann-Straussler syndrome. *Acta Neuropathol (Berl)* **87**, 526–30 (1994).

35. J. W. Ironside, L. McCardle, P. A. Hayward and J. E. Bell, Ubiquitin immunocytochemistry in human spongiform encephalopathies. *Neuropathol Appl Neurobiol* **19**, 134–40 (1993).

36. M. Miyazono, T. Iwaki, T. Kitamoto, Y. Kaneko, K. Doh-ura and J. Tateishi, A comparative immunohistochemical study of Kuru and senile plaques with a special reference to glial reactions at various stages of amyloid plaque formation. *Am J Pathol* **139**, 589–98 (1991).

37. S. Betmouni, V. H. Perry and J. L. Gordon, Evidence for an early inflammatory response in the central nervous system of mice with scrapie. *Neuroscience* **74**, 1–5 (1996).

38. J. M. Peyrin, C. I. Lasmezas, S. Haik, F. Tagliavini, M. Salmona, A. Williams, D. Richie, J. P. Deslys and D. Dormont, Microglial cells respond to amyloidogenic

PrP peptide by the production of inflammatory cytokines. *Neuroreport* **10**, 723–9 (1999).

39. V. Silei, C. Fabrizi, G. Venturini, M. Salmona, O. Bugiani, F. Tagliavini and G. M. Lauro, Activation of microglial cells by PrP and beta-amyloid fragments raises intracellular calcium through L-type voltage sensitive calcium channels. *Brain Res* **818**, 168–70 (1999).

40. D. R. Brown and H. A. Kretzschmar, Microglia and prion disease: a review. *Histol Histopathol* **12**, 883–92 (1997).

41. D. Lorton, J. M. Kocsis, L. King, K. Madden and K. R. Brunden, beta-Amyloid induces increased release of interleukin-1 beta from lipopolysaccharide-activated human monocytes. *J Neuroimmunol* **67**, 21–9 (1996).

42. S. L. Yates, L. H. Burgess, J. Kocsis-Angle, J. M. Antal, M. D. Dority, P. B. Embury, A. M. Piotrkowski and K. R. Brunden, Amyloid beta and amylin fibrils induce increases in proinflammatory cytokine and chemokine production by THP-1 cells and murine microglia. *J Neurochem* **74**, 1017–25 (2000).

43. C. K. Combs, D. E. Johnson, S. B. Cannady, T. M. Lehman and G. E. Landreth, Identification of microglial signal transduction pathways mediating a neurotoxic response to amyloidogenic fragments of beta-amyloid and prion proteins. *J Neurosci* **19**, 928–39 (1999).

44. J. A. Hamilton, G. Whitty, A. R. White, M. F. Jobling, A. Thompson, C. J. Barrow, R. Cappai, K. Beyreuther and C. L. Masters, Alzheimer's disease amyloid beta and prion protein amyloidogenic peptides promote macrophage survival, DNA synthesis and enhanced proliferative response to CSF-1 (M-CSF). *Brain Res* **940**, 49–54 (2002).

45. K. Jendroska, F. P. Heinzel, M. Torchia, L. Stowring, H. A. Kretzschmar, A. Kon, A. Stern, S. B. Prusiner and S. J. DeArmond, Proteinase-resistant prion protein accumulation in Syrian hamster brain correlates with regional pathology and scrapie infectivity. *Neurology* **41**, 1482–90 (1991).

46. S. J. DeArmond, S. L. Yang, A. Lee, R. Bowler, A. Taraboulos, D. Groth and S. B. Prusiner, Three scrapie prion isolates exhibit different accumulation patterns of the prion protein scrapie isoform. *Proc Natl Acad Sci USA* **90**, 6449–53 (1993).

47. H. McLennan and H. V. Wheal, The interaction of glutamic and aspartic acids with excitatory amino acid receptors in the mammalian central nervous system. *Can J Physiol Pharmacol* **54**, 70–2 (1976).

48. G. Garthwaite, G. D. Williams and J. Garthwaite, Glutamate Toxicity: An Experimental and Theoretical Analysis. *Eur J Neurosci* **4**, 353–360 (1992).

49. D. R. Brown, J. W. Herms, B. Schmidt and H. A. Kretzschmar, PrP and beta-amyloid fragments activate different neurotoxic mechanisms in cultured mouse cells. *Eur J Neurosci* **9**, 1162–9 (1997).

50. W. E. Muller, H. Ushijima, H. C. Schroder, J. M. Forrest, W. F. Schatton, P. G. Rytik and M. Heffner-Lauc, Cytoprotective effect of NMDA receptor antagonists on prion protein (PrionSc)-induced toxicity in rat cortical cell cultures. *Eur J Pharmacol* **246**, 261–7 (1993).

51. S. Perovic, G. Pergande, H. Ushijima, M. Kelve, J. Forrest and W. E. Muller, Flupirtine partially prevents neuronal injury induced by prion protein fragment and lead acetate. *Neurodegeneration* **4**, 369–74 (1995).

52. H. C. Schroder, S. Perovic, V. Kavsan, H. Ushijima and W. E. Muller, Mechanisms of prionSc- and HIV-1 gp120 induced neuronal cell death. *Neurotoxicology* **19**, 683–8 (1998).

53. B. Miller, M. Sarantis, S. F. Traynelis and D. Attwell, Potentiation of NMDA receptor currents by arachidonic acid. *Nature* **355**, 722–5 (1992).

54. B. Barbour, M. Szatkowski, N. Ingledew and D. Attwell, Arachidonic acid induces a prolonged inhibition of glutamate uptake into glial cells. *Nature* **342**, 918–20 (1989).

55. A. Dumuis, M. Sebben, L. Haynes, J. P. Pin and J. Bockaert, NMDA receptors activate the arachidonic acid cascade system in striatal neurons. *Nature* **336**, 68–70 (1988).

56. L. R. Stewart, A. R. White, M. F. Jobling, B. E. Needham, F. Maher, J. Thyer, K. Beyreuther, C. L. Masters, S. J. Collins and R. Cappai, Involvement of the 5-lipoxygenase pathway in the neurotoxicity of the prion peptide PrP106-126. *J Neurosci Res* **65**, 565–72 (2001).

57. C. Korth, B. C. May, F. E. Cohen and S. B. Prusiner, Acridine and phenothiazine derivatives as pharmacotherapeutics for prion disease. *Proc Natl Acad Sci USA* **98**, 9836–41 (2001).

58. A. Barret, F. Tagliavini, G. Forloni, C. Bate, M. Salmona, L. Colombo, A. De Luigi, L. Limido, S. Suardi, G. Rossi, F. Auvre, K. T. Adjou, N. Sales, A. Williams, C. Lasmezas and J. P. Deslys, Evaluation of quinacrine treatment for prion diseases. *J Virol* **77**, 8462–9 (2003).

59. M. Daniels, G. M. Cereghetti and D. R. Brown, Toxicity of novel C-terminal prion protein fragments and peptides harbouring disease-related C-terminal mutations. *Eur J Biochem* **268**, 6155–64 (2001).

60. F. B. Hafiz and D. R. Brown, A model for the mechanism of astrogliosis in prion disease. *Mol Cell Neurosci* **16**, 221–32 (2000).

61. D. R. Brown, Prion protein peptide neurotoxicity can be mediated by astrocytes. *J Neurochem* **73**, 1105–13 (1999).

62. E. T. Dayton and E. O. Major, Recombinant human interleukin 1 beta induces production of prostaglandins in primary human fetal astrocytes and immortalized human fetal astrocyte cultures. *J Neuroimmunol* **71**, 11–8 (1996).

63. N. Stella, A. Estelles, J. Siciliano, M. Tence, S. Desagher, D. Piomelli, J. Glowinski and J. Premont, Interleukin-1 enhances the ATP-evoked release of arachidonic acid from mouse astrocytes. *J Neurosci* **17**, 2939–46 (1997).

64. M. Moser, R. J. Colello, U. Pott and B. Oesch, Developmental expression of the prion protein gene in glial cells. *Neuron* **14**, 509–17 (1995).

65. A. J. Raeber, R. E. Race, S. Brandner, S. A. Priola, A. Sailer, R. A. Bessen, L. Mucke, J. Manson, A. Aguzzi, M. B. Oldstone, C. Weissmann and B. Chesebro, Astrocyte-specific expression of hamster prion protein (PrP) renders PrP knockout mice susceptible to hamster scrapie. *EMBO J* **16**, 6057–65 (1997).

66. J. Sassoon, M. Daniels and D. R. Brown, Astrocytic regulation of NMDA receptor subunit composition modulates the toxicity of prion peptide PrP106-126. *Mol Cell Neurosci* **25**, 181–91 (2004).

67. D. R. Brown and C. M. Mohn, Astrocytic glutamate uptake and prion protein expression. *Glia* **25**, 282–92 (1999).

68. N. Vassallo and J. Herms, Cellular prion protein function in copper homeostasis and redox signalling at the synapse. *J Neurochem* **86**, 538–44 (2003).

69. D. R. Brown, K. Qin, J. W. Herms, A. Madlung, J. Manson, R. Strome, P. E. Fraser, T. Kruck, A. von Bohlen, W. Schulz-Schaeffer, A. Giese, D. Westaway and H. Kretzschmar, The cellular prion protein binds copper *in vivo*. *Nature* **390**, 684–7 (1997).

70. J. Stockel, J. Safar, A. C. Wallace, F. E. Cohen and S. B. Prusiner, Prion protein selectively binds copper(II) ions. *Biochemistry* **37**, 7185–93 (1998).

71. M. P. Hornshaw, J. R. McDermott, J. M. Candy and J. H. Lakey, Copper binding to the N-terminal tandem repeat region of mammalian and avian prion protein: structural studies using synthetic peptides. *Biochem Biophys Res Commun* **214**, 993–9 (1995).

72. D. R. Brown, F. Hafiz, L. L. Glasssmith, B. S. Wong, I. M. Jones, C. Clive and S. J. Haswell, Consequences of manganese replacement of copper for prion protein function and proteinase resistance. *EMBO J* **19**, 1180–6 (2000).

73. B. S. Wong, D. R. Brown, T. Pan, M. Whiteman, T. Liu, X. Bu, R. Li, P. Gambetti, J. Olesik, R. Rubenstein and M. S. Sy, Oxidative impairment in scrapie-infected mice is associated with brain metals perturbations and altered antioxidant activities. *J Neurochem* **79**, 689–98 (2001).

74. A. M. Thackray, R. Knight, S. J. Haswell, R. Bujdoso and D. R. Brown, Metal imbalance and compromised antioxidant function are early changes in prion disease. *Biochem J* **362**, 253–8 (2002).

75. H. A. Kretzschmar, L. E. Stowring, D. Westaway, W. H. Stubblebine, S. B. Prusiner and S. J. Dearmond, Molecular cloning of a human prion protein cDNA. *DNA* **5**, 315–24 (1986).

76. J. M. Gabriel, B. Oesch, H. Kretzschmar, M. Scott and S. B. Prusiner, Molecular cloning of a candidate chicken prion protein. *Proc Natl Acad Sci USA* **89**, 9097–101 (1992).

77. B. Belosi, E. Gaggelli, R. Guerrini, H. Kozlowski, M. Luczkowski, F. M. Mancini, M. Remelli, D. Valensin and G. Valensin, Copper binding to the neurotoxic peptide PrP106-126: thermodynamic and structural studies. *Chembiochem* **5**, 349–59 (2004).

78. E. Wong, A. M. Thackray and R. Bujdoso, Copper induces increased beta-sheet content in the scrapie-susceptible ovine prion protein PrPVRQ compared with the resistant allelic variant PrPARR. *Biochem J* **380**, 273–82 (2004).

79. T. Miura, A. Hori-i and H. Takeuchi, Metal-dependent alpha-helix formation promoted by the glycine-rich octapeptide region of prion protein. *FEBS Lett* **396**, 248–52 (1996).

80. M. F. Jobling, X. Huang, L. R. Stewart, K. J. Barnham, C. Curtain, I. Volitakis, M. Perugini, A. R. White, R. A. Cherny, C. L. Masters, C. J. Barrow, S. J. Collins, A. I. Bush and R. Cappai, Copper and zinc binding modulates the aggregation and neurotoxic properties of the prion peptide PrP106-126. *Biochemistry* **40**, 8073–84 (2001).

81. D. R. Brown and A. Besinger, Prion protein expression and superoxide dismutase activity. *Biochem J* **334 (Pt 2)**, 423–9 (1998).

82. F. Klamt, F. Dal-Pizzol, M. J. Conte da Frota, R. Walz, M. E. Andrades, E. G. da Silva, R. R. Brentani, I. Izquierdo and J. C. Fonseca Moreira, Imbalance of antioxidant defense in mice lacking cellular prion protein. *Free Radic Biol Med* **30**, 1137–44 (2001).

83. D. R. Brown, R. S. Nicholas and L. Canevari, Lack of prion protein expression results in a neuronal phenotype sensitive to stress. *J Neurosci Res* **67**, 211–24 (2002).

84. D. R. Brown and J. Sassoon, Copper-dependent functions for the prion protein. *Mol Biotechnol* **22**, 165–78 (2002).

85. P. C. Pauly and D. A. Harris, Copper stimulates endocytosis of the prion protein. *J Biol Chem* **273**, 33107–10 (1998).

86. W. Rachidi, A. Mange, A. Senator, P. Guiraud, J. Riondel, M. Benboubetra, A. Favier and S. Lehmann, Prion infection impairs copper binding of cultured cells. *J Biol Chem* **278**, 14595–8 (2003).

87. D. R. Brown, Role of the prion protein in copper turnover in astrocytes. *Neurobiol Dis* **15**, 534–43 (2004).

88. J. Collinge, M. A. Whittington, K. C. Sidle, C. J. Smith, M. S. Palmer, A. R. Clarke and J. G. Jefferys, Prion protein is necessary for normal synaptic function. *Nature* **370**, 295–7 (1994).

89. M. A. Whittington, K. C. Sidle, I. Gowland, J. Meads, A. F. Hill, M. S. Palmer, J. G. Jefferys and J. Collinge, Rescue of neurophysiological phenotype seen in PrP null mice by transgene encoding human prion protein. *Nat Genet* **9**, 197–201 (1995).

90. S. B. Colling, J. Collinge and J. G. Jefferys, Hippocampal slices from prion protein null mice: disrupted Ca(2+)-activated K+ currents. *Neurosci Lett* **209**, 49–52 (1996).

91. G. R. Mallucci, S. Ratte, E. A. Asante, J. Linehan, I. Gowland, J. G. Jefferys and J. Collinge, Post-natal knockout of prion protein alters hippocampal CA1 properties, but does not result in neurodegeneration. *EMBO J* **21**, 202–10 (2002).

92. K. Kristensson, B. Feuerstein, A. Taraboulos, W. C. Hyun, S. B. Prusiner and S. J. DeArmond, Scrapie prions alter receptor-mediated calcium responses in cultured cells. *Neurology* **43**, 2335–2341 (1993).

93. K. Wong, Y. Qiu, W. Hyun, R. Nixon, J. VanCleff, J. Sanchez-Salazar, S. B. Prusiner and S. J. DeArmond, Decreased receptor-mediated calcium response in prion-infected cells correlates with decreased membrane fluidity and IP3 release. *Neurology* **47**, 741–750 (1996).

94. A. R. Johnston, J. R. Fraser, M. Jeffrey and N. MacLeod, Synaptic plasticity in the CA1 area of the hippocampus of scrapie-infected mice. *Neurobiol Dis* **5**, 188–95 (1998).

95. P. A. Barrow, C. D. Holmgren, A. J. Tapper and J. G. Jefferys, Intrinsic physiological and morphological properties of principal cells of the hippocampus and neocortex in hamsters infected with scrapie. *Neurobiol Dis* **6**, 406–23 (1999).

96. T. Florio, S. Thellung, C. Amico, M. Robello, M. Salmona, O. Bugiani, F. Tagliavini, G. Forloni and G. Schettini, Prion protein fragment 106-126 induces apoptotic cell death and impairment of L-type voltage-sensitive calcium channel activity in the GH3 cell line. *J Neurosci Res* **54**, 341–52 (1998).

97. T. Florio, M. Grimaldi, A. Scorziello, M. Salmona, O. Bugiani, F. Tagliavini, G. Forloni and G. Schettini, Intracellular calcium rise through L-type calcium channels, as molecular mechanism for prion protein fragment 106–126-induced astroglial proliferation. *Biochem Biophys Res Commun* **228**, 397–405 (1996).

98. S. Thellung, T. Florio, V. Villa, A. Corsaro, S. Arena, C. Amico, M. Robello, M. Salmona, G. Forloni, O. Bugiani, F. Tagliavini and G. Schettini, Apoptotic cell death and impairment of L-type voltage-sensitive calcium channel activity in rat cerebellar granule cells treated with the prion protein fragment 106–126. *Neurobiol Dis* **7**, 299–309 (2000).

99. A. Thompson, A. R. White, C. McLean, C. L. Masters, R. Cappai and C. J. Barrow, Amyloidogenicity and neurotoxicity of peptides corresponding to the helical regions of PrP(C). *J Neurosci Res* **62**, 293–301 (2000).

100. G. Forloni, N. Angeretti, P. Malesani, E. Peressini, T. Rodriguez Martin, P. Della Torre and M. Salmona, Influence of mutations associated with familial prion-related

encephalopathies on biological activity of prion protein peptides. *Ann Neurol* **45**, 489–94 (1999).

101. J. Nguyen, M. A. Baldwin, F. E. Cohen and S. B. Prusiner, Prion protein peptides induce alpha-helix to beta-sheet conformational transitions. *Biochemistry* **34**, 4186–92 (1995).

102. H. Zhang, K. Kaneko, J. T. Nguyen, T. L. Livshits, M. A. Baldwin, F. E. Cohen, T. L. James and S. B. Prusiner, Conformational transitions in peptides containing two putative alpha-helices of the prion protein. *J Mol Biol* **250**, 514–26 (1995).

103. F. Tagliavini, F. Prelli, L. Verga, G. Giaccone, R. Sarma, P. Gorevic, B.Ghetti, F. Passerini, E. Ghibaudi, G. Forloni and et al., Synthetic peptides homologous to prion protein residues 106–147 form amyloid-like fibrils in vitro. *Proc Natl Acad Sci USA* **90**, 9678–82 (1993).

104. M. Fischer, T. Rulicke, A. Raeber, A. Sailer, M. Moser, B. Oesch, S. Brandner, A. Aguzzi and C. Weissmann, Prion protein (PrP) with amino-proximal deletions restoring susceptibility of PrP knockout mice to scrapie. *EMBO J* **15**, 1255–64 (1996).

105. E. Flechsig, D. Shmerling, I. Hegyi, A. J. Raeber, M. Fischer, A. Cozzio, C. von Mering, A. Aguzzi and C. Weissmann, Prion protein devoid of the octapeptide repeat region restores susceptibility to scrapie in PrP knockout mice. *Neuron* **27**, 399–408 (2000).

106. D. Shmerling, I. Hegyi, M. Fischer, T. Blattler, S. Brandner, J. Gotz, T. Rulicke, E. Flechsig, A. Cozzio, C. von Mering, C. Hangartner, A. Aguzzi and C. Weissmann, Expression of amino-terminally truncated PrP in the mouse leading to ataxia and specific cerebellar lesions. *Cell* **93**, 203–14 (1998).

107. S. Supattapone, P. Bosque, T. Muramoto, H. Wille, C. Aagaard, D. Peretz, H. O. Nguyen, C. Heinrich, M. Torchia, J. Safar, F. E. Cohen, S. J. DeArmond, S. B. Prusiner and M. Scott, Prion protein of 106 residues creates an artifical transmission barrier for prion replication in transgenic mice. *Cell* **96**, 869–78 (1999).

108. V. Bonetto, T. Massignan, R. Chiesa, M. Morbin, G. Mazzoleni, L. Diomede, N. Angeretti, L. Colombo, G. Forloni, F. Tagliavini and M. Salmona, Synthetic miniprion PrP106. *J Biol Chem* **277**, 31327–34 (2002).

109. R. H. Kimberlin and C. A. Walker, Pathogenesis of scrapie: agent multiplication in brain at the first and second passage of hamster scrapie in mice. *J Gen Virol* **42**, 107–17 (1979).

110. A. G. Dickinson, H. Fraser and G. W. Outram, Scrapie incubation time can exceed natural lifespan. *Nature* **256**, 732–3 (1975).

111. A. F. Hill, S. Joiner, J. Linehan, M. Desbruslais, P. L. Lantos and J. Collinge, Species-barrier-independent prion replication in apparently resistant species. *Proc Natl Acad Sci USA* **97**, 10248–53 (2000).

112. R. Race and B. Chesebro, Scrapie infectivity found in resistant species. *Nature* **392**, 770 (1998).

113. R. H. Kimberlin, C. A. Walker and H. Fraser, The genomic identity of different strains of mouse scrapie is expressed in hamsters and preserved on reisolation in mice. *J Gen Virol* **70 (Pt 8)**, 2017–25 (1989).

114. R. Race, A. Raines, G. J. Raymond, B. Caughey and B. Chesebro, Long-Term Subclinical Carrier State Precedes Scrapie Replication and Adaptation in a Resistant Species: Analogies to Bovine Spongiform Encephalopathy and Variant Creutzfeldt-Jakob Disease in Humans. *J. Virol.* **75**, 10106–10112 (2001).

115. R. Frigg, M. A. Klein, I. Hegyi, R. M. Zinkernagel and A. Aguzzi, Scrapie pathogenesis in subclinically infected B-cell-deficient mice. *J Virol* **73**, 9584–8 (1999).

116. A. M. Thackray, M. A. Klein, A. Aguzzi and R. Bujdoso, Chronic Subclinical Prion Disease Induced by Low-Dose Inoculum. *J. Virol.* **76**, 2510–2517 (2002).

117. A. M. Thackray, M. A. Klein and R. Bujdoso, Subclinical Prion Disease Induced by Oral Inoculation. *J. Virol.* **77**, 7991–7998 (2003).

118. H. Bueler, A. Raeber, A. Sailer, M. Fischer, A. Aguzzi and C. Weissmann, High prion and PrPSc levels but delayed onset of disease in scrapie-inoculated mice heterozygous for a disrupted PrP gene. *Mol Med* **1**, 19–30 (1994).

119. S. J. DeArmond, W. C. Mobley, D. L. DeMott, R. A. Barry, J. H. Beckstead and S. B. Prusiner, Changes in the localization of brain prion proteins during scrapie infection. *Neurology* **37**, 1271–80 (1987).

120. R. Gabizon, M. P. McKinley, D. Groth and S. B. Prusiner, Immunoaffinity purification and neutralization of scrapie prion infectivity. *Proc Natl Acad Sci USA* **85**, 6617–21 (1988).

121. R. Medori, P. Montagna, H. J. Tritschler, A. LeBlanc, P. Cortelli, P. Tinuper, E. Lugaresi and P. Gambetti, Fatal familial insomnia: a second kindred with mutation of prion protein gene at codon 178. *Neurology* **42**, 669–70 (1992).

122. J. Collinge, F. Owen, M. Poulter, M. Leach, T. J. Crow, M. N. Rossor, J. Hardy, M. J. Mullan, I. Janota and P. L. Lantos, Prion dementia without characteristic pathology. *Lancet* **336**, 7–9 (1990).

123. J. Collinge, M. S. Palmer, K. C. Sidle, I. Gowland, R. Medori, J. Ironside and P. Lantos, Transmission of fatal familial insomnia to laboratory animals. *Lancet* **346**, 569–70 (1995).

124. A. F. Hill and J. Collinge, Subclinical prion infection. *Trends in Microbiology* **11**, 578–584 (2003).

125. C. I. Lasmezas, J. P. Deslys, O. Robain, A. Jaegly, V. Beringue, J. M. Peyrin, J. G. Fournier, J. J. Hauw, J. Rossier and D. Dormont, Transmission of the BSE agent

to mice in the absence of detectable abnormal prion protein. *Science* **275**, 402–5 (1997).

126. A. Arjona, L. Simarro, F. Islinger, N. Nishida and L. Manuelidis, Two Creutzfeldt-Jakob disease agents reproduce prion protein-independent identities in cell cultures. *Proc Natl Acad Sci U S A* **101**, 8768–73 (2004).

127. H. Bueler, M. Fischer, Y. Lang, H. Bluethmann, H. P. Lipp, S. J. DeArmond, S. B. Prusiner, M. Aguet and C. Weissmann, Normal development and behaviour of mice lacking the neuronal cell-surface PrP protein. *Nature* **356**, 577–82 (1992).

128. R. S. Hegde, J. A. Mastrianni, M. R. Scott, K. A. DeFea, P. Tremblay, M. Torchia, S. J. DeArmond, S. B. Prusiner and V. R. Lingappa, A transmembrane form of the prion protein in neurodegenerative disease. *Science* **279**, 827–34 (1998).

129. R. S. Hegde, P. Tremblay, D. Groth, S. J. DeArmond, S. B. Prusiner and V. R. Lingappa, Transmissible and genetic prion diseases share a common pathway of neurodegeneration. *Nature* **402**, 822–6 (1999).

130. I. Sponne, A. Fifre, V. Koziel, B. Kriem, T. Oster, J. L. Olivier and T. Pillot, Oligodendrocytes are susceptible to apoptotic cell death induced by prion protein-derived peptides. *Glia* **47**, 1–8 (2004).

Chapter 10

A STONE GUEST ON THE BRAIN: DEATH AS A PRION

David R. Brown

Department of Biology and Biochemistry,
University of Bath, Bath BA2 7AY, UK

10.1. Introduction

There has been considerable interest in determining the mechanism of conversion of the normal cellular form of the prion protein (PrP^{c}) into a protease resistant form (PrP^{Sc}). This is because the protease resistant form of the prion protein can be detected in animals or humans with prion diseases. There is a clear link between the prion protein and prion diseases, such as scrapie, because mice where the gene for PrP (*prnp*) is ablated are resistant to infection with mouse passaged scrapie[1]. Recently, it has been shown, with a conditional knockout mouse strain, that switching off PrP expression during onset of mouse scrapie results in an arrest in the disease progress[2]. Despite this, the most common feature of prion disease is neurodegeneration. The "progression" in these diseases relates to the gradual loss of neurones of various brain regions, which results in dementia and failure in locomotion. Therefore, progress in dealing with or understanding the mechanism of this disease requires knowledge of the mechanism leading to this neurodegeneration. There is a clear link between the expression of PrP^{c}, deposition of PrP^{Sc} and neuronal death. In contrast, there remains a body of researches that remain unconvinced about the central nature of PrP^{Sc} to prion disease causality. Since there are a number of reports that scrapie can be transmitted in the absence of detectable PrP^{Sc} [3,4,5], then it is possible that PrP^{Sc} is little more than a tombstone of a pathological process. It is also possible that the true culprit is another form of PrP or something else.

It remains irrefutable that PrP^{c} expression is essential for prion disease to develop because of the failure of the disease to be transmitted to PrP-knockout mice[1] and the lack of neurodegeneration in PrP-deficient tissue, even in the presence of PrP^{Sc} deposition[6].

At a basic level, the possibilities for the induction of cell death are limited. One or a combination of these possibilities is likely to be the cause in initiating cell death. The first of these possibilities is that PrP^{Sc} is neurotoxic directly by a physical interaction with neurones. Alternatively, PrP^{Sc} could induce secondary effects such as oxidative stress, which then act to trigger cell death. Another possibility is that neuronal death is triggered not by the production of PrP^{Sc}, but by the loss of functional PrP^{c}. There is now considerable evidence that PrP^{c} acts either as an antioxidant or a Cu transporting protein or both. Therefore, the potential loss of antioxidant activity could potentially be responsible for cell death given the clear presence of oxidative stress in the brains of animals and patients with prion diseases. Opposing this is the finding that mice deficient in the expression of PrP are healthy and do not show signs of neuronal loss[7]. There could be compensations that lead to functional replacement in the PrP-knockout mouse. However, given that inactivation of expression during prion disease causes disease arrest[2] then it is unlikely that functional replacement is the key to the resistance of PrP-knockout mice to disease transmission. Nevertheless, as switching off PrP^{c} expression reverses prion disease, there is a clear link between protein conversion and maintaining prion disease progress. In the absence of PrP^{c} expression there is clearly the potential for the defensive mechanism to become active. The implication is that inactivation of PrP^{c} is necessary for disease progress. This means that loss of PrP^{c} function is likely to be involved in the mechanism of neuronal death.

A further possibility as to the cause of neuronal death in prion disease is that another form of PrP is causal or another protein altogether. The doppel protein, a homologue of PrP is neurotoxic[8]. However, despite its possible neurodegenerative effects in some transgenic mice, there is currently no evidence that doppel has a direct or even auxiliary effect in prion disease. Another possibility has been transmembrane forms of PrP. Although there is some evidence for the existence of transmembrane forms of PrP in *in vitro* translation systems[9] and some cell culture models of inherited mutations of the prion protein, there is scant evidence that such forms of the protein exist *in vivo*. Indeed, one study reports that mutations supposed to generate transmembrane PrP actually prevent PrP from being exported from the cell and do not result in transmembrane protein at all[10]. Most of the evidence suggests that wild-type PrP^{c} is linked to the extracellular side of the plasma membrane by a GPI anchor. It has been suggested that some PrP mutations increase membrane association, but this on its own is not the cause of cell death.

Furthermore, mice expressing a mutant form of PrP, supposedly only as a transmembrane form, are resistant to experimental transmission of prion disease[11].

Another variation, potentially more plausible than transmembrane PrP is the potential export of abnormal PrP into the cytoplasm[12,13] with or without failure of the proteosome to degrade this misdirected PrP[14]. Although there is some evidence this occurs, the current literature suggests that this cytoplasmic PrP is not toxic[15].

Many recent models of neurodegeneration in prion disease suggest that neurodegeneration results from the cells response to the presence of mis-folded protein prior to export from the cell or because of failure of such protein to be exported. Such possibilities hold a high level of credibility. A recent paper suggests that abnormal PrP folding results in endoplasmic reticulum (ER) stress[16]. This then leads to the activation of cell death via the ER specific caspase 12. The only problem with this suggestion is that the study was carried out on human cells, which do not express functional caspase 12. Neither the universality nor the veracity of this finding has been confirmed.

Currently, the majority of research suggests neurodegeneration results from the innate toxicity of PrP^{Sc}. As this toxicity cannot occur without neuronal production of PrP^{c}, then clearly the neurotoxic mechanism requires two species of PrP. PrP^{Sc} is potentially necessary, but not sufficient for neuronal death. Similarly, loss of PrP^{c} function is necessary but not sufficient. The majority of experimental evidence suggests that PrP^{c} is a copper binding protein[17] that can function to transport copper[18] and/or PrP^{c} is an antioxidant requiring copper binding for its activity[19]. Further consistent evidence from research suggests that oxidative damage is a feature of prion diseases[20] implying that breakdown of oxidative defence might contribute to neuronal death in prion disease. Generation of PrP by the brain might be sufficient to inactivate compensatory mechanisms that might come into play when PrP expression is impossible because of genetic ablation of the gene or temporal inactivation of gene expression. Therefore, expression of PrP that is either functionally inactive or can be converted to PrP^{Sc} is necessary for the mechanism of neurodegeneration. The rest of this review will consider research related to this model.

10.2. Peptides as paradigms

The majority of research examining the mechanism of neurodegeneration that might be relevant to prion diseases has been carried out using cell culture techniques. The assumption used in these studies is that exposure of pure or mixed cell populations to PrP^{Sc} results in the induction of cell death, either by apoptosis or some other mechanism.

Figure 10.1. Toxicity of PrP106-126
The synthetic peptide is toxic to neurones by an apoptotic mechanism. Cerebellar cells were from wild-type mice were prepared in culture. These cells were treated with PrP106-126 at various concentrations for five days (O). Further cultures were treated with PrP106-126 and 10 μM $CuSO_4$ (Δ) or 1 mM bathocuproine sulfonate. (•). After 5 days, the survival of the cells was assessed using an MTT assay. Values were compared to those of untreated controls as a percentage. * indicated the lowest concentration where there are significant differences (Student's t test, $p < 0.05$) between the treatments. As can be seen, PrP106-126 on its own is highly toxic reducing neuronal survival by more than 70%. Addition of Cu increases the toxicity further. 10 μM $CuSO_4$ on its own had not toxic effect. The cell impermeable Cu chelator, bathocuproine sulfonate greatly inhibited PrP106-126 toxicity suggesting this toxicity is dependent on Cu present in the medium.

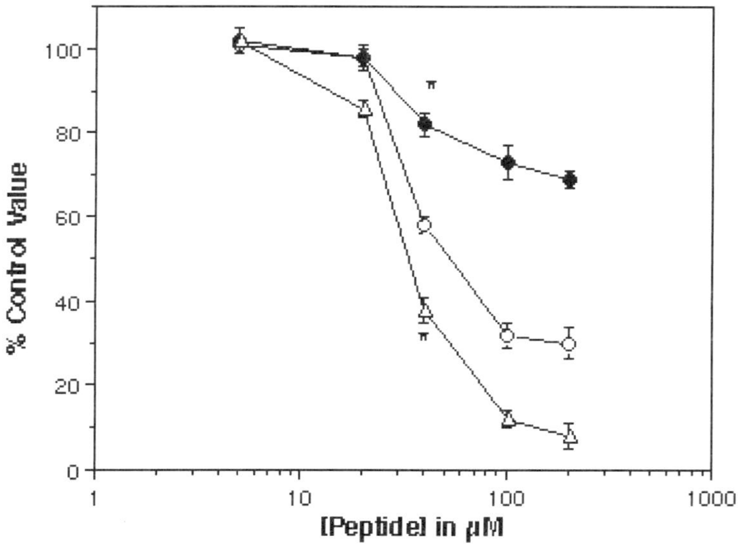

A considerable percentage of this research has focused on a single peptide fragment known as PrP106-126 because it is based on amino acid residues 106–126 of the human PrP sequence. The first study using this peptide[21] established that this fragment is more toxic than other fragments and induces apoptosis in neurones. Since then, several other studies have examined different fragments either overlapping this region[22] or from elsewhere in the protein sequence[23]. Some of these fragments have the advantage of being more soluble than PrP106-126. However, the relative insolubility of PrP106-126 can be overcome by preparing the peptide in a different solvent such as DMSO and adding it to the culture in this form[24]. Whereas some PrP peptides are toxic, others have been found to be protective. In particular, peptides related to the octameric repeat region of the N-terminus have been shown to protect neurones from copper toxicity by their potential to bind copper[25,26]. This is in contrast to PrP106-126 (Figure 10.1) where there is some evidence that the toxic effects of the peptide are mediated by or require an interaction with copper ions[27,28,29].

Mice have been generated that express truncated versions of PrP^c. In particular, a mouse expressing a prion protein of 106 amino acids[30] has been used to identify those parts of the protein necessary for both infection and neurodegeneration in prion disease. The truncated version of PrP^c can be converted into a truncated PrP^{Sc} capable of infecting the same transgenic mice and inducing neurodegeneration. The 106 amino acids comprise the amino acid residues 89–140 and from 177 to the C-terminus. As neuronal death also occurs in these mice, then these regions must also contain the main neurotoxic domain of PrP. There is general agreement that the hydrophobic domain of PrP is the principal toxic domain. This implies that the toxic domain lies within amino acid 89–140. A number of studies have looked at the toxicity of this fragment or the so-called mini-prion[31]. Apart from confirming the toxicity of the fragment and providing structural information[32], these studies have done little to advance the understanding of toxicity beyond that of studies using the peptide fragment PrP106-126.

Although some mice expressing truncated forms of PrP^c do not develop spontaneous disease[30,33], others show spontaneous disease[34,35]. Of particular interest are mice with N-terminal deletions lacking amino-residues up to 121 or 134 of the mouse sequence. These mice express truncated protein at the cell surface and develop degeneration in the cerebellum shortly after birth. This disease is not prion disease, but represents a novel form of cerebellum specific degeneration. Indeed, like doppel (see below) the toxicity of this fragment is inhibited by the co-expression of wild-type PrP. Analyses using peptides and primary neurones in culture showed that the fragment 121–231 is toxic to neurones (Figure 10.2)[23]. Smaller peptides consisting of individual helices within this region are also toxic. Two peptides, PrP163-184 and PrP196-220, appeared to be the most toxic. As these fragments are also toxic to neurones lacking PrP-expression, this toxicity is not related to that produced by PrP^{Sc}. However, it is quite possible that these fragments could produce some cell death or exacerbate neuronal death induced by PrP^{Sc}. Mutations equivalent to those that are causal to inherited forms of prion disease in humans enhance the toxicity of the PrP121-231 fragment[23]. This suggests that perhaps this fragment could play a role in cell death observed in inherited prion diseases such as Gerstmann-Sträussler-Scheinker syndrome. This could possibly be due to the generation of hydrogen peroxide by the interaction of the C-terminal fragment with metals[36]. Interestingly, the toxic hydrophobic domain of PrP (i.e. the region including PrP106-126) is able to neutralise the toxicity of this domain[23]. Perhaps the gain of toxicity of PrP^{Sc} is a result of two

Figure 10.2. PrP106-126 interacts with PrP^c. Peptides were applied to cerebellar neuronal cultures for four days. These peptides included PrP106-126, a peptide with the same composite amino acids as PrP106-126 but in a random order (scrambled) and a beta-amyloid peptide (βA25-35). Peptides applied to cells were collected from the supernatant by centrifugation. The pellets were dissolved in urea and electrophoresed on a PAGE gel. The resulting gel was blotted and PrP detected with and antibody to PrP (DR1) that does not recognise PrP106-126. 1 = no peptide, 2 = 100 μM scrambled peptide, 3 = 20 μM PrP106-126, 4 = 100 μM PrP106-126, 5 = 100 μM βA25-35. These results show that full length PrP becomes bound to PrP106-126 when it is applied to cultures. In addition, a truncated fragment could also be detected. Further analysis has shown that this fragment is approximately 20 kD, whereas the normal protein is 25 kD. These bands could be seen on a coomassie stained gel of the same material suggesting there was very little other protein bound to PrP106-126. This and other data (Brown, 2000b) suggest that there is a specific interaction between PrP106-126 and PrP^c.

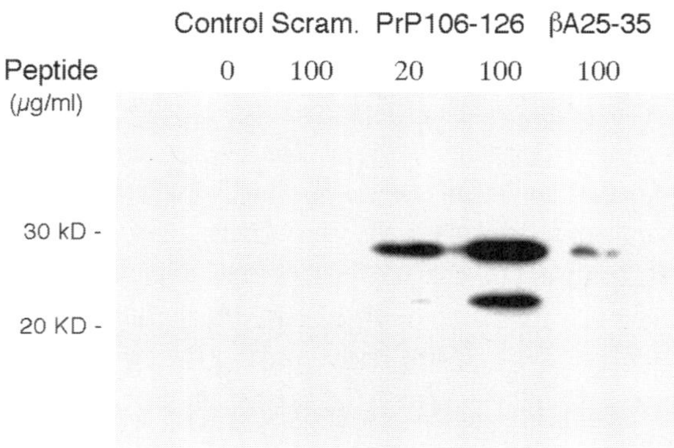

potentially toxic domains within the protein being unable to interact due to changes in protein conformation. This fits with the generally supported hypothesis that change in protein conformation makes PrP toxic.

At present only PrP106-126 and PrP118-135 have been shown to be toxic *in vivo*[37,38]. Although study of PrP118-135 is interesting, its toxicity to PrP-knockout cells makes it an inferior model to that of PrP106-126. PrP118-135 is similar to another peptide PrP127-136. This peptide was first reported to have very little neurotoxicity[21] but was later shown to be toxic to PrP-knockout neurones[24]. Attempts to generate the smallest toxic fragment of PrP led to the finding that a peptide 113–126, just 14 amino acid residues in length, is the smallest toxic molecule with the same properties as PrP^{Sc} [24]. This fragment formed fibrils and was high in beta-sheet and resisted proteinase K digestion. The biophysical characteristics of PrP106-126 and other fragments have been looked at in detail by many authors[39,40,41], but the fibrillar nature of the peptide depends largely on an eight amino acid residue palendrome AGAAAAGA[24]. This is possibly the smallest fragment of protein able to generate amyloid fibrils.

PrP106-126 has many cellular effects and binds to a multitude of other proteins. This includes effects on membrane micro-viscosity and the activity of calcium channels. These studies are dealt with in detail elsewhere in this book and have also been reviewed[42]. In the current chapter the mechanism of toxicity will be looked at from the point of view of our laboratory.

Our analyses of the toxicity of PrP106-126 showed that the mechanism involves two basic effects. These effects will be discussed in detail. Both effects are necessary for the toxic effect.

1. **Direct Effects:** PrP106-126 interacts directly with cells and alters the cells' response to toxic substances produced by other cells or are present in the extracellular environment. The peptide also changes the cellular response to non-toxic signals received in chemical form from other cells. These effects require the neurone to express PrP^{c}. As mentioned above, without continued cellular expression of PrP^{c}, neither PrP^{Sc} nor PrP106-126 has a toxic effect. Possibly, these effects require a direct interaction between PrP106-126 and PrP^{c}. There could be other direct effects PrP106-126 has on cells that exacerbate the toxic effects, but these are secondary for the requirement of PrP^{c} expression.
2. **Indirect Effects:** PrP106-126 can also interact with other cells besides neurones. As glia vastly outnumber neurones in the brain, it is quite likely that PrP^{Sc} will encounter glia in an animal with prion disease. Glial responses to either PrP106-126 or PrP^{Sc} have been assessed. Activation of astrocytes result in reduced activity of glutamate and/or copper clearance by uptake mechanisms. This can result in increased levels of toxic substances in the vicinity of the neurones. Similarly, microglia activated by PrP106-126 or PrP^{Sc} result in the production of radicals, especially superoxide and possibly nitrogen radicals but also inflammatory cytokines. In this way, neurones are exposed to increases in stress agents or potentially toxic molecules.
3. **Combined Effects:** On their own, these changes are probably insufficient to initiate cell death. However, reduced neuronal resistance to oxidative stress combined with an increase in oxidative stress in the vicinity are potentially the two factors necessary to activate apoptosis.

4. **Age Effects:** One of the central curiosities of human prion diseases is the age of onset. In particular, both GSS and sCJD have a clear age dependence. GSS, although an inherited disease, remains dormant until the patients reach their 50's. This clearly implicates age dependent changes in the nervous system. One possibility is the increased inability of the ageing nervous system to deal with oxidative stress. As oxidative stress is already implicated in prion diseases, then this might be the deciding factor in what initiates sporadic diseases such as CJD.

Some of these issues will be discussed in detail.

10.3. Interactions as initiators

There is now a long list of proteins that bind to PrP. Some of these have been identified as binding partners specifically for PrP106-126[43,44], while others bind PrP^c [45]. It has been known for some time that PrP can interact with itself. In particular, peptides can bind to PrP^c and cause it to gain protease resistance[46]. Whether PrP106-126 interacts with PrP^c or another protein, this interaction is likely to be necessary to cause changes to intracellular signalling. The nature of the intracellular signalling resulting from this interaction and potentially leading to apoptosis remains elusive. Although caspases and calcium are involved this is extremely non-specific and superficial.

Our findings suggest that all direct effects related to the toxicity of PrP106-126 are mediated through direct interaction with PrP^c. That PrP106-126 does form fibrils clearly indicates that it can interact with itself. It has already been suggested that the form of PrP peptides that interact with the full-length protein is soluble[47]. Furthermore, the neuronal component of the cell death mechanism is optimal for more soluble PrP106-126 as the use of membrane filters indicate that the peptide must be able to pass through a 0.4 μM pore to initiate neuronal death[48]. Further research has shown that prevention of the interaction of PrP106-126 and PrP^c blocks toxicity[27]. Toxicity can be prevented by using the peptide AGAAAAGA[24]. Analysis of the binding of PrP106-126 and PrP^{Sc} to recombinant wild type and mutant PrP suggest that this region of the protein is the site at which the interaction occurs[27].

Binding of PrP106-126 has several effects. It causes direct inhibition of the antioxidant activity of the protein[27]. Treatment of neurones with PrP106-126 causes a marked reduction in the resistance of neurones to oxidative stress[49]. There is also a decrease in the activity of other antioxidant enzymes such as Cu/Zn superoxide dismutase[49,50]. In addition, the peptide causes a decrease in the uptake of Cu by neuronal cells[27] and causes a decrease of Cu incorporation into Cu/Zn superoxide dismutase[51]. It is unclear how these changes are brought about. However, there is evidence that PrP106-126 can enter cells[52] and might interact with intracellular proteins such as microtubules[44]. There is also evidence that aggregates of PrP106-126 can cause the shedding of PrP^c (Figure 10.3), which then becomes trapped in the aggregates[27]. Interestingly, a subset of antibodies to PrP can also cause apoptosis suggesting that possibly inhibiting either the breakdown of the protein or its ability to bind Cu or its interaction with other proteins might be sufficient to trigger cell death[25]. Recently it has been confirmed that similar antibodies injected *in vivo* can also cause neuronal apoptosis[53]. As antibodies to PrP increase the toxicity of Cu to cells, then it is quite possible that interfering in the protein's role in Cu metabolism might be central to the ability of PrP106-126 to initiate cell death. However, there is sufficient evidence to suggest that direct effects of PrP106-126 on neurones, mediated through PrP^c might be necessary to induce cell death, but are insufficient for its execution. PrP106-126 has the effect of compromising the neurones ability to deal with stressful conditions[54] and in this way gives neurones a phenotype like that observed for PrP-deficient neurones[55]. Execution of cell death then comes about as a result of this compromised phenotype and one of a number of different stress events such as the production of superoxide[54].

10.4. Microglia as mediators

A microglial response during the course of prion disease is now well documented. As part of gliosis there is widespread activation of microglia that precedes the onset of neuronal death[56,57,58]. Additionally, microglia probably proliferate and alter the kinds of molecules they release[59,60]. There is association between microglial response and deposits of abnormal PrP[61,62,63]. Also, there are reports that microglia are damaged and possibly die during the disease progress[64]. There is evidence that microglia may be associated with PrP^{Sc} plaques in human disease[61,65,66]. A clearer picture comes from an animal model of prion disease: scrapie infected mice. Mice infected with scrapie show significant activation of microglia[67]. There is also evidence that this microglial activation precedes neuronal death. Studies of apoptosis using *in situ* end-labelling with terminal transferase have lead to evidence for the time course of apoptosis in mice infected with several different strains of scrapie[57]. This apoptosis was found to follow activation of microglia sequentially as indicated by immunohistochemistry for markers of activated microglia[68]. There is other evidence that microglia

Figure 10.3. The C-terminal fragment of PrP, PrP121-231 is highly toxic to neurones. Cerebellar neruones from wild-type mice were treated with either full-length recombinant mouse PrP (PrP23-231) (A) or the C-terminal fragment PrP121-231 (B). The cells were stained with the Hoechst reagent to detect apoptotic nuclei. An increase in aberrant nuclei was only detected for the PrP121-231 treatment. Scale bar = 50 μm. C shows the quantitation of this result for four separate experiments.

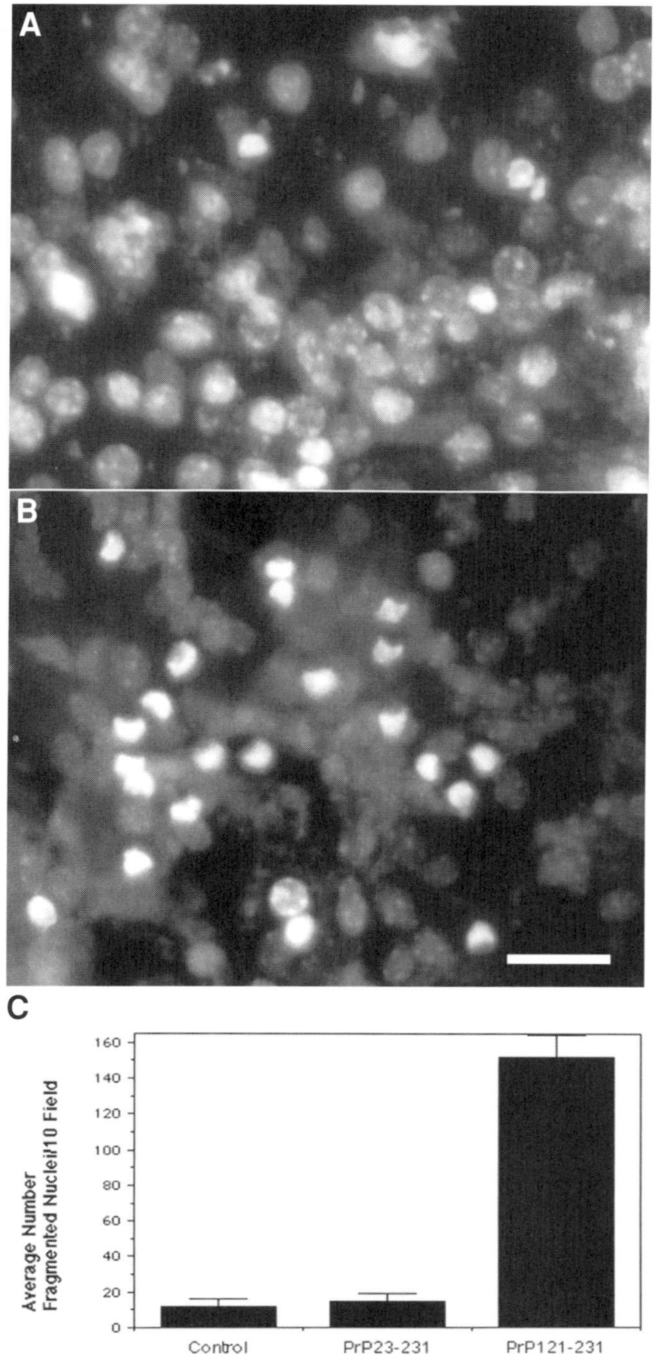

activation precedes neuronal death in general and especially precedes clinical symptoms of scrapie[67] and that large numbers of microglia are present in the scrapie infected mouse brain throughout the disease[56]. This suggests that microglial activation precedes neuronal death in mouse scrapie. Unfortunately, there is no proof that all the neuronal death in prion disease is apoptotic and similarly, there remains no proof that microglial activation is causative as regards apoptosis. Recent work suggests that microglial changes are accompanied by both changes in synapses and minor behavioural changes not normally described as a response to scrapie infection[69]. These changes have been seen as early as 13 weeks post experimental inoculation of mice with scrapie.

There is more evidence for the production of microglia products and especially cytokines. Increased expression of interleukin-1 and interleukin-6 as well as the tumor necrosis factor α has been detected in the brains of hamsters[70] and mice[71]. As indicated above, the expression of both these cytokines is increased when microglia are treated with PrP106-126. Also these cytokines are involved in PrP106-126 induced proliferation of astrocytes. Astrocytes in scrapie infected hamsters show increased levels of the transcription factor NF-KB[70]. It is known that interleukin causes increased nuclear entry of NF-KB in astrocytes[72]. Increased levels of NF-KB can lead to increased expression of cytokines by astrocytes.

Several studies have now identified markers of oxidative stress in the brains of rodents with prion disease. There are increased levels of oxidised lipids in the brains of scrapie infected hamsters[73]. Another study[20] has shown increased levels of nitrotyrosine and hemeoxygenase-1 in the brains of scrapie infected mice. These observations imply that significant free radical damage is being generated in the brains of scrapie infected mice. Additionally, there is evidence for mitochondrial damage in cells from brains of scrapie infected hamsters and mice[74,75]. These changes include reduction in the activity of mitochondrial enzymes and structural abnormalities in the mitochondria. Other enzymes known to be associated with resistance to oxidative stress, such as catalase and glutathione-S-transferase, show increased expression[75]. Together, these results suggest that oxidative stress is involved in the pathology of prion diseases. It is possible that the oxidative damage to the brain in scrapie might be a result of the damage to mitochondria, which can generate superoxide. However, the measured level of oxygen radicals detected with dichlorofluorescein in mitochondrial fractions from the brains of scrapie infected mice was not greatly increased above that of controls[70]. Reactive oxygen species such as superoxide are generated by microglia and the implication of this is that microglia cause damage in the brain of scrapie infected mice. However, microglia activation has also been postulated to be a response to neuronal damage rather than the cause of it. Even if neuronal damage was the sole cause of the microglial response (which is unlikely given the complexity of cell-cell interactions), then the microglial activation is still likely to cause significant production of toxic substances to trigger neuronal apoptosis.

Studies with the synthetic peptide PrP106-126 have shown that depletion of microglia from neuronal cultures by the use of microglia specific toxins results in a significant reduction in peptide toxicity[54]. Similarly, addition of microglia to neuronal cultures can increase the toxicity of both PrP106-126 and PrP^{Sc} [76]. Also, microglia respond to signals from peptide-treated neurones stimulating them to produce cytokines and possibly other agents that might alter neuronal survival, such as interleukin-6[77]. However, the majority of changes to microglial activity induced by the peptide occur in the absence of other cells[78]. These changes include proliferation[79] superoxide production[54], activation of nitric oxide release[78] and the release of cytokines[80]. As well as releasing neurotoxic substances in response to PrP106-126, microglia might also mediate the astrocytic response to PrP106-126. PrP106-126 causes the release of cytokines from microglia such as interleukins 1 and 6[80]. It has been shown that, in an astrocyte-microglia co-culture, microglia response to PrP106-126 stimulate astrocyte proliferation[79]. Similar to microglia-neurone interactions leading to death, direct effects of PrP106-126 on astrocytes are necessary to stimulate astrocytic responses to the mitogens released by microglia[81]. Thus, PrP106-126 causes the mitogen release from microglia and primes astrocytes to respond to the microglial signals and it is possible that the interleukins released from microglia in response to PrP106-126 trigger the proliferation of PrP106-126-primed astrocytes[82]. In this way, microglia could be mediators of both neuronal apoptosis and astrogliosis as seen in prion disease.

10.5. Astrocytes as accelerators

Astrocytosis is also a very typical reaction that occurs in prion diseases. Increased GFAP (glial fibrillary acidic protein) was an early molecular marker associated with prion disease because of the strong astrocytic reaction. Like microglial responses, astrocytic activation is considered a response to the presence of damaged neurones. However, advances in the understanding of signalling between neurones, astrocytes and other cells are shining a new light into this assumption, suggesting it is probably false. A recent paper by Marella and Chabry[83] suggest that prion disease causes a change in the signalling between neurones and astrocytes, which in turn alters signalling to microglia via chemokines. Also, older evidence suggests that astrocyte activation in prion disease occurs prior to onset of symptoms of the disease[84].

Previous studies have shown that mouse cerebellar neurones become dependent on astrocytes for protection from glutamate toxicity[85,86]. Cerebellar neurones co-cultured with astrocytes show an increased sensitivity to the toxicity of glutamate as compared to cerebellar neurones not co-cultured. This effect is not dependent on regional origin of the astrocytes used for co-culture[86]. Region specific effects of astrocytes have been suggested to be contact-mediated and changes to glutamate sensitivity are related to diffusable factors, some of which have been defined[86]. In the presence of astrocytes, this increased sensitivity to the toxic effects of glutamate is normally not noticeable because of the survival promoting effects of astrocytes, such as clearance of glutamate and release of protective factors. Increased sensitivity to glutamate toxicity is only of consequence if the protectiveness of astrocytes is compromised. Several mechanisms causing this compromise, which lead to glutamate induced neuronal death, have been found; (1) physical removal of astrocytes, (2) addition of substances which inhibit astrocytic glutamate uptake[86], (3) substances which activate astrocytes (eg. TGF-β)[85] or (4) inactivation of protective factors such as interleukin-6 (IL-6)[85]. This increased sensitivity to glutamate is induced in neurones by a factor released by astrocytes. Neurones treated with conditioned medium (NAM) from astrocytes exposed to medium from neuronal cultures showed decreased survival. This decreased survival was a result of glutamate toxicity, but not because of increased glutamate concentration[86]. Thus, some additional factor released by astrocytes into NAM makes neurones more sensitive to glutamate toxicity. Release of this factor is enhanced by the presence of neurones themselves or by conditioned medium from neurones[85]. Clearly, identification of this factor and its mechanism of action are of great importance because of their possible role in regulating neuronal sensitivity to excitotoxic death.

Following from this, work changes in NMDA receptor subunit subtype composition have been identified which parallel increased sensitivity to glutamate toxicity. Changes in the NMDA receptors NR1 and NR2 were found to occur. In particular, NAM caused increase in the subtypes NR1b and NR2a subtypes[87]. It has been suggested that the subunit NR1b is associated with increased sensitivity to glutamate toxicity[88,89]. In our experiments, we observed that both co-culture with astrocytes and treatment with NAM resulted in an increase in NR1b. Anti-NR1b oligonucleotide inhibited NAM toxicity implying that this subunit is involved in the increased toxicity of NAM. As cerebellar neurones are protected by astrocytic clearance of glutamate when co-cultured, then this would explain why the expression of this subunit in co-cultured cerebellar neurones does not result in increased cell death. Despite its role in sensitivity to glutamate, the NR1b isoform has been shown to be necessary for normal regeneration in certain systems such as the retina[90].

During normal granule cell development, there is a known progression in the expression of the NR2 subunit subtypes. Early in development, the predominant subtype is NR2b, but as development progresses, there is a switch to NR2a and then an increase in the levels of NR2c[91,92]. Change in the subunit composition of NMDA receptors alters electrophysiological responses of cerebellar neurones[93,94]. Co-cultured cerebellar neurones demonstrated a NR2 subtype profile more like that of a maturer cerebellar neurone. The cerebellar neurones in NAM showed an increase in type NR2a, but this seemed to be associated with increased sensitivity to NAM whereas NR2c appeared to be protective. Therefore, increased sensitivity to glutamate in the NAM treated cerebellar neurones might represent the inadequacies of a transition phase in development. Oligonucleotide knockdown of expression of the subunit subtypes NR1b and NR2a inhibited glutamate toxicity markedly[87]. Oligonucleotide therapy based on these observations might provide an effective way to combat excitotoxic neurone death in a variety of diseases (Figure 10.4).

The importance of these findings to prion disease are highlighted in analyses of survival of cerebellar granule cells co-cultured with large numbers of astrocytes when treated with the PrP106-126 peptide. Cerebellar cell cultures contain a mixture of cells including microglia, astrocytes and neurones. Specifically, killing astrocytes in such cultures was shown to reduce the toxicity of PrP106-126[95]. PrP106-126 also causes inhibition of glutamate uptake[96]. In order to test whether PrP106-126 can influence the toxicity of glutamate to neurones via its interaction with astrocytes, the following culture system was devised. Neurones from PrP-knockout mice were cultured with wild-type neurones and treated with PrP106-126. Neurones from PrP-knockout mice are resistant to the toxicity of PrP106-126[97]. PrP106-126 has been shown to have no effect on the toxicity of glutamate to neurones when cultured alone. The effects of PrP106-126 on astrocytes requires their expression of PrP^{c}[81]. Under these conditions PrP106-126 was found to be toxic to the co-cultured PrP-knockout neurones[96]. This toxicity could be inhibited by MK801, suggesting that the toxicity is mediated by glutamate. In such a culture system, wild-type astrocytes were found to greatly block the toxicity of added glutamate to the PrP-knockout neurones. This protection was blocked if PrP106-126 was added to the co-culture. As described above, co-culture with astrocytes makes neurones more sensitive to glutamate. In the presence of neurones, astrocytes are able to rapidly clear

Figure 10.4. Knockdown of specific NMDA receptor subunits can block toxicity of PrP106-126. Cerebellar neurones were grown in culture and treated with one of five different antisense oligonucleotides at 100 nM or combinations (replenished daily) for five days. These antisense DNA oligos were designed to bind to the DNA for one of five NMDA receptor subuniut subtypes. These were NR1a, NR1b, NR2a, NR2b or NR2c. Either total NR1 or NR2 was detected using RT-PCR. The upper panel shows treatment with NR1 oligonucleotides either on their own (NR1a = a, NR1b = b) or combined (a + b). The lower panel shows treatment with NR2 oligonucleotides either on their own (NR2a = a, NR2b = b, NR2c = c) or combined (a + b + c). These results show that the oligonucleotides effectively blocked expression of the NMDA receptor subunits. In the presence of astrocytes and Cu PrP106-126 is toxic at very low concentrations and this toxicity is mediated by glutamate (Sassoon et al., 2004). Cerebellar neurones were co-cultured with astrocytes. Parallel co-cultured cerebellar cell cultures were either treated with 20 μM PrP106-126 (open bars) or 20 μM PrP106-126 and 10 μM $CuSO_4$ (black bars). Cultures were co-treated with one of the five antisense oligonucleotides. Relative survival was determined with an MTT assay. The bar graph shows the toxicity of PrP106-126 under these conditions to the cells with or without addition of the oligonucleotides. One oligonucleotide in particular inhibited the toxicity of glutamate-mediated PrP106-126 toxicity. This suggests that the receptor's subunit subtype NR2a is involved in transducing the toxic effect of PrP106-126.

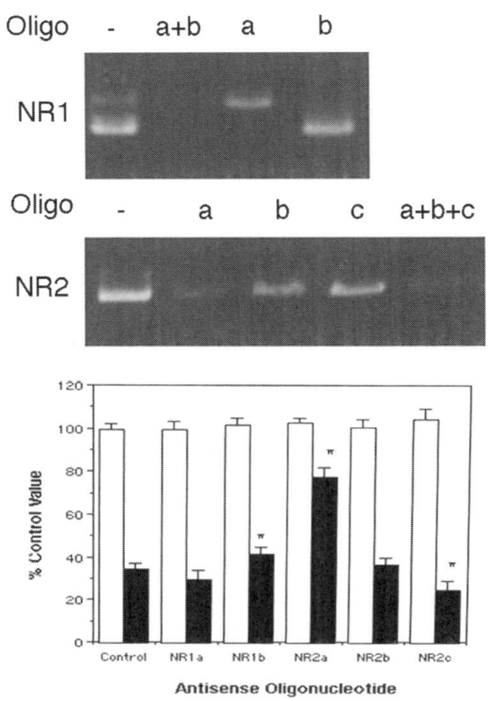

glutamate and secrete neuroprotective factors. This increased neuronal sensitivity to glutamate usually has no impact on neuronal survival. However, when astrocytic protection is inhibited by PrP106-126, then the increased sensitivity to glutamate makes the level of glutamate in

Figure 10.5. Mechanism of PrP106-126 toxicity—interplay of several cell types. This summary figure shows the details of the theoretical toxic mechanism of PrP106-126 to neurones as described in the text.

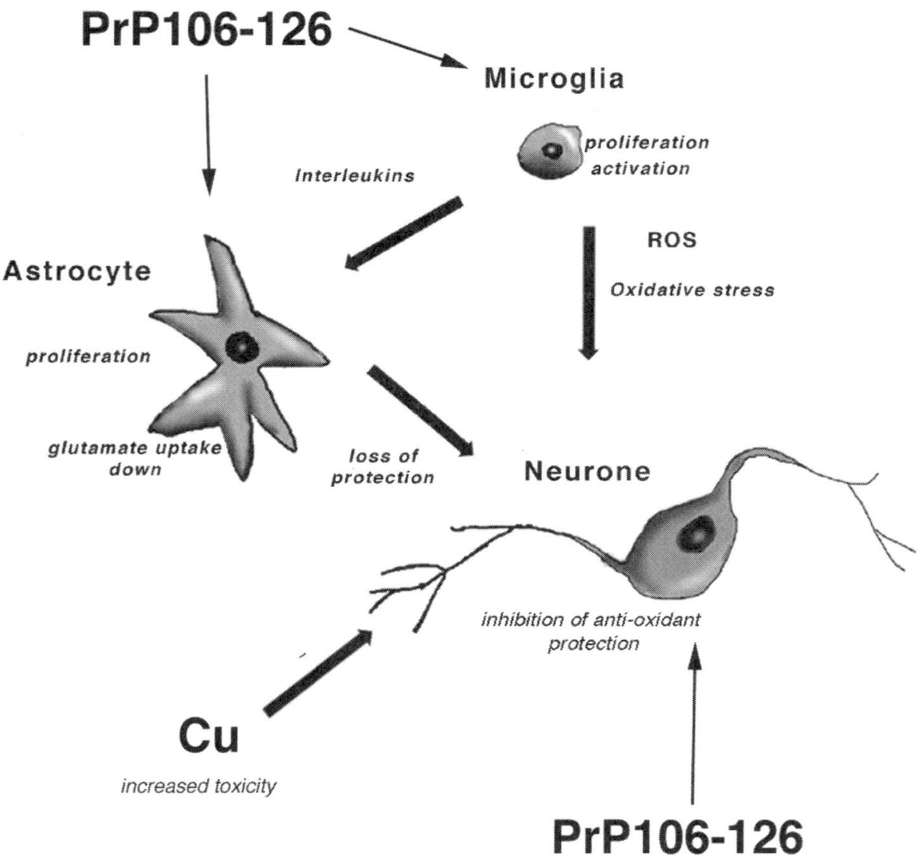

the culture medium potentially more toxic. Thus, in this culture system, glutamate mediated toxicity emerges without any substantial increase in glutamate concentration in the culture[96]. New data also suggest that PrP106-126 alters NMDA receptor composition in neurones which would also increase the toxicity of glutamate to these cells[98]. Using antisense DNA it was shown that specifically inhibiting the expression of NMDA receptor subunit subtype NR2a effectively blocked the toxicity of PrP106-126 mediated via astrocytes.

This astrocyte model has the potential to greatly enhance neuronal death *in vitro*. Rapid neuronal loss is seen in the late stages of prion disease following astrogliosis. Thus, while microglial effects are likely to initiate the pathological changes, astrocytes could greatly accelerate the rate of neuronal loss by the mechanism described above (Figure 10.5).

10.6. Doppel induced death

Two separately derived "strains" of mice (Npu and Zrk1) in which protein expression had been knocked out were examined for gross disturbances in behaviour and development[7,99]. None were found and on the basis of this, some experts suggested that the protein had a redundant function or no function at all. If this were the case, why would the sequence of the PrP^c be so highly conserved from turtle to humans? Possibly its function is so essential, that like many important proteins normal metabolism has mechanisms to compensate for its loss.

Yet, even this picture was blurred when another strain of PrP^c deficient mice (Ngs) was found to develop late onset motor disturbances and the loss of Purkinje cells in the cerebellum[100]. A recent paper has suggested that this and two other strains of PrP^c deficient mice (Rcm0, Zrk2) become ill because another protein, termed Doppel (Dpl), with a small degree of homology ($\sim$25%) to PrP^c, is highly expressed in these mice[101]. This expression is possibly driven by the PrP^c promoter running directly into the Dpl reading frame, which is directly in tandem with that of PrP^c. Whatever the role of Dpl in causing the phenotype of these PrP^c-deficient mice, the late onset pathology is abrogated by re-introducing PrP^c expression. It is possible that renewed PrP^c expression has a negative feedback effect on the PrP^c promoter inhibiting Dpl expression. Whatever the explanation, the implication is that PrP^c expression has a function in preventing disease.

The discovery of Dpl has generated a great interest because of its possible role in prion disease. Its structure has been analysed by NMR spectroscopy and compared to that of PrP^c [102,103]. Dpl contains a very similar secondary structure with three helices forming a globular domain. However, unlike PrP^c, Dpl will not fold into intermediates that would result in a beta-sheet rich form of the protein. Thus, Dpl is unlikely to form the kind of misfolded protein found in prion disease[104]. Polymorphisms in the protein were investigated to determine if any were associated with disease[105,106]. No relationship between gene sequence and prion disease was found. Altering the expression of Dpl does not alter the susceptibility of transplanted brain tissue to neurodegeneration caused by infection with scrapie[107]. Increasing Dpl expression in mouse brain does not alter the incubation time for prion disease[108]. Dpl is expressed in both Sertoli cells and spermatozoa[109] and plays a role in the late stage of spermatogenesis. Although Dpl clearly plays a role in male gametogenesis and sperm-egg interactions[110], its actual function remains unknown. Furthermore, what its relation is to PrP^c beyond structural homology is also purely speculative at present.

Unlike PrP-knockout mice, Dpl-knockout mice have a distinct phenotype which is an altered fertility[7,110]. This phenotype shows that the main role of Dpl is in the regulation of male fertility especially in regard to sperm maturation and the ability of sperm to penetrate the zona pellucida of oocytes[110]. Expression of Dpl in testes has been shown to be much higher than in the brain[111]. Thus, when considering the possible role of Dpl in prion disease, evidence needs to be provided that Dpl expression actually reaches levels where questions of Dpl toxicity might be relevant. However, at present no such evidence exists. Nevertheless, it is clear that expression of Dpl can be toxic to neurones. This has sparked interest in determining the possible neurotoxic mechanism of Dpl. The evidence from the study of RcmO and other Dpl expressing PrP-knockout mice is that Dpl is toxic in the absence of PrP^c expression.

Initial studies of Dpl expressing PrP-knockout mice to try and determine a mechanism for this cell death suggested a possible involvement of hemoxygenase and nitric oxide systems due to alterations of the relevant proteins in Rcm0 mice[112]. In addition, these mice showed increased evidence of oxidative damage. The use of animal models for mechanistic dissection of toxic pathways is rather cumbersome and time consuming. Therefore further study used culture models to provide a clearer insight into Dpl toxicity[8].

Dpl toxicity can occur in the presence of PrP^c expression (Figure 10.6). However, blocking toxicity of Dpl is clearly PrP^c dose dependent[8]. Dpl toxicity to neurones overexpressing PrP^c ten fold is abolished. Additionally co-application of recombinant PrP with Dpl also inhibits its toxicity. This inhibitory effect is greater if the protein has Cu bound suggesting that the Cu-dependent function of PrP^c is protective. Nevertheless, we also have evidence that a fragment of PrP can interact directly with Dpl, altering its structure and inhibiting its toxicity. Thus, it is also possible that PrP with Cu bound is more able to interact with Dpl than Cu-free PrP. Behrens and Aguzzi[113] have described three possible mechanisms by which Dpl might be toxic in the absence of PrP^c. The first, initially proposed by Weissmann and Aguzzi[114], suggests that Dpl and PrP^c both bind a common ligand. This ligand binds preferentially to PrP^c, but in its absence the ligand binds to Dpl and initiates cell death. There is currently no evidence to support this model. The second proposed mechanism is that initially proposed by Wong et al.[112], which suggested that Dpl and PrP^c have complementary activities in the brain. It has been shown that Dpl activates the NOS system[8] thus providing some evidence for this hypothesis. The third proposed mechanism is that Dpl and PrP^c can both form some type of multimeric complex and

Figure 10.6. Doppel is toxic to neurones, but toxicity is inhibited by PrP^c. A Culture of cerebellar neurones was prepared from newborn mice. These mice were either wild-type expressing PrP^c (•), PrP-knockout mice (O) or mice overexpressing PrP^c(■). The cells were treated for 4 days with increasing concentration of recombinant mouse Dpl. At the end of the time, the survival of the neurones was determined relative to untreated control culture. The MTT assay values indicated that Dpl is toxic to neuronal cultures at concentrations above 5 μg/ml. The toxicity was decreased by the level of expression of PrP^c. Dpl was not toxic at all to overexpressing neurones. B Cerebellar neurones from PrP-knockout mice were treated with Dpl protein at 50 μg/ml. Concurrently, the same cultures were treated with recombinant mouse PrP. The protein used was either the apo (O) or holo (•) form of PrP. In other words, holo-PrP had Cu bound but apo-PrP did not. Binding of Cu to PrP endows it with antioxidant activity (Brown et al., 1999). After 4 days, survival of the cultures was determined. PrP inhibited the toxicity of Dpl in a dose dependent manner. The inhibition was greater for holo-PrP. This suggests that functional PrP might be better at protecting against Dpl toxicity. It is possible that apo-PrP was protective because it bound Cu present in the cell culture medium. However, it remains possible that PrP could inhibit Dpl toxicity without Cu being bound.

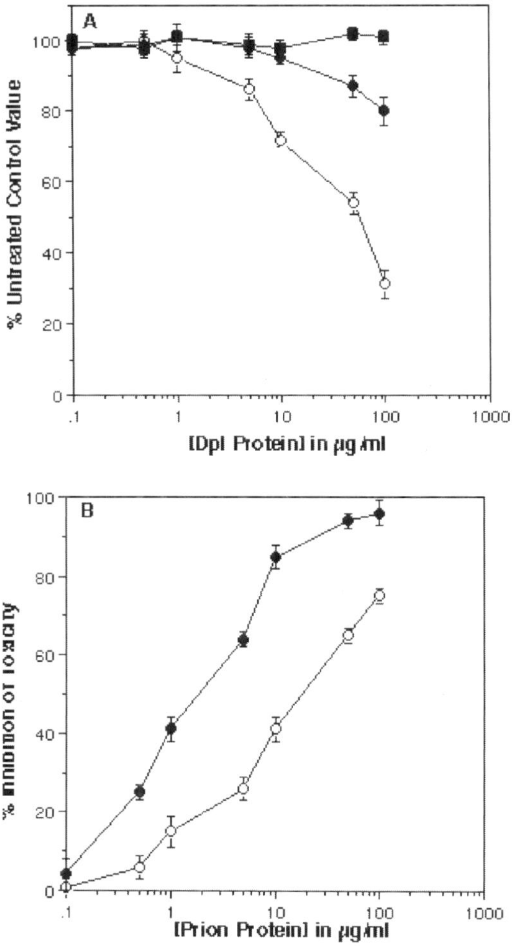

the one composed only of Dpl would lead to toxicity. Again, there is no direct evidence. However, on the other hand, there is evidence for a model like this in that Dpl toxicity might be inhibited by direct interaction of Dpl and PrP^{c}. Therefore a mixture of physical interaction with PrP^{c} and the protective activity of PrP^{c} might explain the anti-Dpl effects of PrP^{c}. As PrP^{c} is an antioxidant that requires copper for its activity[115], this then suggests that Dpl toxicity involves an oxidative process. In support of the notion that the copper dependent activity of PrP^{c} protects cells from Dpl toxicity, a recent paper by Atarashi et al.[115] examined co-expression of a truncated form of PrP within Ngsk PrP-knockout mice that over expressed Dpl. Unlike wild-type PrP^{c}, which protected the mice from Dpl induced neurodegeneration, the truncated PrP, did not. The implication is that the N-terminal domain is key in the ability of PrPc to protect against Dpl toxicity. This could either be due to its copper dependent function or because this domain interacts physically with Dpl and renders it non-toxic.

Exposure of neurones derived from PrP-knockout mice to recombinant Dpl showed that Dpl toxicity is related to the generation of NO (nitric oxide)[8]. NOS (NO synthase) inhibitors inhibit Dpl toxicity, but this toxicity did not require the involvement of glia suggesting the toxic NO is generated from nNOS in neurones. Another study of PrP-knockout mice that expressed Dpl suggested that microglial and astrocyte activation occurs before the onset of Purkinje cell death in the brains of the mice[116]. It is quite possible that in the brains of mice expressing Dpl that glial activation could exacerbate cell death. However, the glial activation observed by Atarashi et al.[116] could have been as a result of damaged neurones or the production of NO by neurones. Also, glial activation occurred in areas of the brain where neuronal loss did not occur. This suggests that the glial response was a response to brain changes rather than a direct cause of them.

Despite the wide interest in Dpl, there is currently little reason to think that Dpl has any relevance to the nature of neurodegeneration cause by prion disease.

10.7. Conclusion

Research into the mechanism of neurodegeneration in prion diseases has made considerable advances in the last ten years. Ten years ago reviews into "neurodegeneration and prion disease"[117] did little more than mention that neuronal death occurred. Now, as this text book verifies, there is a wide range of theories and research avenues that are being heavily investigated. In particular, new transgenic models are being used to make bold insights into the role of PrP in the cell death mechanism. However, it is perhaps highly important to keep a keen eye on previous and on going research using cell culture based models. Although often condemned as artificial, the majority of the significant results from cell culture models of cell death in prion diseases, have been reproduced by the in vivo models (e.g. the need for continued PrP^{c} expression). It is therefore more important than ever for researchers to acknowledge the insights of such work or they will be forced to "reinvent the wheel" at every turn. Nevertheless, as research continues in the prion field in its seemingly haphazard manner, we are gradually getting to know the "stone guest" of prion disease much better. Eventually as research leads to treatment, the unwanted "stone guest" will be asked to leave.

10.8. References

1. H. Büeler, A. Aguzzi, A. Sailer, R. A. Greiner, P. Autenried, M. Aguet and C.Weissmann, Mice devoid of PrP are resistant to scrapie. *Cell* **73**, 1339–1347 (1993).

2. G. Mallucci, A. Dickinson, J. Linehan, P. C. Klohn, S. Brandner and J. Collinge, Depleting neuronal PrP in prion infection prevents disease and reverses spongiosis. *Science.* **302**, 871–874 (2003).

3. L. Manuelidis, T. Sklaviadis and E. E. Manuelidis, Evidence suggesting that PrP is not the infectious agent in Creutzfeldt-Jakob disease. *EMBO J.* **6**, 341–347 (1987).

4. R. Race, K. Meade-White, A. Raines, G. J. Raymond, B. Caughey and B. Chesebro Subclinical scrapie infection in a resistant species: persistence, replication, and adaptation of infectivity during four passages. *J. Infect. Dis.* **186 Suppl 2**, S166–70 (2002).

5. G. M. Shaked, G. Fridlander, Z. Meiner, A. Taraboulos and R. Gabizon, Protease-resistant and detergent-insoluble prion protein is not necessarily associated with prion infectivity. *J. Biol. Chem.* **274**, 17981–17986 (1999).

6. S. Brandner, S. Isenmann , A. Raeber, M. Fischer, A. Sailer, Y. Kobayashi, S. Marino, C. Weissmann and A. Aguzzi, Normal host prion protein necessary for scrapie-induced neurotoxicity. *Nature,* **379**, 339–343, (1996).

7. H. Büeler, M. Fischer, Y. Lang, H. Bluethmann, H.-P. Lipp, S. J. DeArmond, S. B. Prusiner, M. Aguet and C. Weissmann, Normal development and behaviour of mice lacking the neuronal cell-surface PrP protein. *Nature*, **356**, 577–582, (1992).

8. T. Cui, A. Holme, J. Sassoon, and D. R. Brown, (2003) Analysis of doppel protein toxicity. *Mol. Cell Neurosci.* **23**, 144–145, (2003).

9. R. S. Hegde, J. A. Mastrianni, M. R. Scott, K. D. Defea, P. Tremblay, M. Torchia, S. J. DeArmond, S.B. Prusiner and V. R. Lingappa, A transmembrane form of the prion protein in neurodegenerative disease. *Science* **279**, 827–834 (1998).

10. A. Holme, M. Daniels, J. Sassoon, and D. R. Brown, A novel method of generating neuronal cell lines from gene-knockout mice to study prion protein membrane orientation. *Eur. J. Neurosci.* **18**, 571–579 (2003).

11. R. S. Hegde, P. Tremblay, D. Groth, S. J. DeArmond, S. B. Prusiner, and V. R. Lingappa, Transmissible and genetic prion diseases share a common pathway of neurodegeneration. *Nature* **402**, 822–826 (1999).

12. J. Ma and S. Lindquist S, Conversion of PrP to a self-perpetuating PrPSc-like conformation in the cytosol. *Science* **298**, 1785–1788 (2002).

13. J. Ma, R. Wollmann, S. Lindquist, Neurotoxicity and neurodegeneration when PrP accumulates in the cytosol. *Science*. **298**, 1781–1785 (2002).

14. B. Drisaldi, R. S. Stewart, C. Adles, L. R. Stewart, E. Quaglio, E. Biasini, L. Fioriti, R. Chiesa and D. A. Harris, Mutant PrP is delayed in its exit from the endoplasmic reticulum, but neither wild-type nor mutant PrP undergoes retrotranslocation prior to proteasomal degradation. *J. Biol. Chem*. **278**, 21732–21743(2003).

15. X. Roucou, Q. Guo, Y. Zhang, C.G. Goodyer and A. C. LeBlanc, Cytosolic prion protein is not toxic and protects against Bax-mediated cell death in human primary neurons. *J. Biol. Chem.* **278**, 40877–40881(2003).

16. C. Hetz, M. Russelakis-Carneiro, K. Maundrell, J. Castilla and C. Soto, Caspase-12 and endoplasmic reticulum stress mediate neurotoxicity of pathological prion protein. *EMBO J.* **22**, 5435–5445 (2003).

17. D. R Brown, K. Qin, J. W. Herms, A. Madlung, J. Manson, R. Strome, P. E. Fraser, T. Kruck, A. von Bohlen, W. Schulz-Schaeffer, A. Giese, D. Westaway and H. Kretzschmar, The cellular prion protein binds copper *in vivo*. *Nature* **390**, 684–687(1997).

18. D. R. Brown, Prion protein expression aids cellular uptake and veratridine-induced release of copper. *J. Neurosci. Res*. 58, 717–725 (1999).

19. D. R. Brown, B. S. Wong, F. Hafiz, C. Clive, S. Haswell and I. M. Jones, Normal prion protein has an activity like that of superoxide dismutase. *Biochem. J.* **344**, 1–5 (1999).

20. M. Guentchev T. Voigtländer, C. Haberler, M. H. Groschup, and H. Budka, Evidence for oxidative stress in experimental prion disease. *Neurobiol. Dis*. **7**, 270–273 (2000).

21. G. Forloni, N. Angeretti, R. Chiesa, E. Monzani, M. Salmona, O. Bugiani and F. Tagliavini, Neurotoxicity of a prion protein fragment. *Nature* **362**, 543–546 (1993).

22. T. Pillot, B. Drouet, M. Pincon-Raymond, J. Vandekerckhove, M. Rosseneu, J. Chambaz, A nonfibrillar form of the fusogenic prion protein fragment [118–135] induces apoptotic cell death in rat cortical neurons. *J. Neurochem.* **75**, 2298–2308 (2000).

23. M. Daniels, G. M. Cereghetti, and D. R. Brown, Toxicity of novel C-terminal prion protein fragments and peptides harbouring disease-related C-terminal mutations. *Eur. J. Biochem.* **268**, 6155–6164 (2001).

24. D. R. Brown, Prion protein peptides: Optimal toxicity and peptide blockade of toxicity. *Mol. Cell Neurosci.* **15**, 66–78 (2000).

25. D. R. Brown, B. Schmidt and H. A. Kretzschmar, Effects of copper on survival of prion protein knockout neurones and glia. *J. Neurochem.* **70**, 1686–1693(1998).

26. M. A. Chacon, M. I. Barria, R. Lorca , J. P. Huidobro-Toro and N. C. Inestrosa, A human prion protein peptide (PrP(59-91)) protects against copper neurotoxicity. Mol *Psychiatry*. **8**, 853–862 (2003).

27. D. R. Brown, PrPSc-like prion protein peptide inhibits the function of cellular prion protein. *Biochem. J.* **352**, 511–518(2000).

28. M. F. Jobling, X. Huang, L. R. Stewart, K. J. Barnham, C. Curtain, I. Volitakis, M. Perugini, A. R. White, A. R. Cherny, R. A., C. L. Masters, C. J. Barrow, S. J. Collins, A. I. Bush and R. Cappai, Copper and zinc binding modulates the aggregation and neurotoxic properties of the prion peptide PrP106-126. *Biochemistry* **40**, 8073–8084 (2001).

29. S. Turnbull, B. J. Tabner, D. R. Brown and D. Allsop, Copper-dependent generation of hydrogen peroxide from the toxic prion protein fragment PrP106-126. *Neurosci. Lett.* **336**, 159–162 (2003).

30. S. Supattapone, P. Bosque, T. Muramoto, H. Wille, C. Aagaard, D. Peretz, H. O. Nguyen, C. Heinrich, M. Torchia, J. Safar, F. E. Cohen, S. J. DeArmond, S. B. Prusiner and M. Scott, Prion protein of 106 residues creates an artificial transmission barrier for prion replication in transgenic mice. *Cell*, **96**, 869–878 (1999).

31. V. Bonetto, T. Massignan, R. Chiesa, M. Morbin, G. Mazzoleni, L. Diomede, N. Angeretti, L. Colombo, G. Forloni, F. Tagliavini, M. Salmona, Synthetic miniprion PrP106. *J. Biol. Chem.* **277**, 31327–31334 (2002).

32. M. Salmona, M. Morbin, T. Massignan, L. Colombo, G. Mazzoleni, R. Capobianco, L. Diomede, F. Thaler, L. Mollica, G. Musco, J. J. Kourie, O. Bugiani, D. Sharma, H. Inouye, D. A. Kirschner, G. Forloni and F. Tagliavini. Structural properties of

Gerstmann-Straussler-Scheinker disease amyloid protein. *J. Biol. Chem.* **278**, 48146–48153 (2003).

33. M. Fischer, T. Rülicke, A. Raeber, A. Sailer, M. Moser, B. Oesch, S. Brandner, A. Aguzzi and C. Weissmann, Prion protein (PrP) with amino-proximal deletions restoring suceptability of PrP knockout mice to scrapie. *EMBO J.* **15**, 1255–1264 (1996).

34. T. Muramoto, S. J. DeArmond, M. Scott G. C. Telling, F. Cohen and S. B. Prusiner, Heritable disorder resembling neuronal storage disease in mice expressing prion protein with deletion of an alpha-helix. *Nature Med.***3,** 750–755 (1997).

35. D. Shmerling, I. Hegyi, M. Fischer, T. Blattler, S. Brandner, J. Gotz, T. Rulicke, E. Flechsig, A. Cozzio, C. von Mering, C. Hangartner, A. Aguzzi and C. Weissmann, Expression of amino-terminally truncated PrP in the mouse leading to ataxia and specific cerebellar lesions. *Cell.* **93** 203–214 (1998).

36. S. Turnbull, B. J. Tabner, D. R. Brown and D. Allsop, Generation of hydrogen peroxide from mutant forms of the prion protein fragment PrP121-231 *Biochemistry* **42**, 7675–7681 (2003).

37. M. Ettaiche, R. Pichot, J. P. Vincent and J. Chabry, In vivo cytotoxicity of the prion protein fragment 106-126. *J. Biol. Chem.* **275**, 36487–36490 (2000).

38. J. Chabry, C. Ratsimanohatra, I. Sponne, P. P. Elena, J. P. Vincent and T. Pillot, In vivo and in vitro neurotoxicity of the human prion protein (PrP) fragment P118-135 independently of PrP expression. *J. Neurosci.* **23**, 462–469 (2003).

39. L. De Gioia, C. Selvaggini, E. Ghibaudi, L. Diomede, O. Bugiani, G. Forloni, F. Tagliavini and M. Salmona, Conformational polymorphism of the amyloidogenic and neurotoxic peptide homologous to residues 106-126 of the prion protein. *J. Biol. Chem.* **269**, 7859–7862 (1994).

40. M. Salmona, P. Malesani, L. De Gioia, S. Gorla, M. Bruschi, A. Molinari, F. Della Vedova, B. Pedrotti, M. A. Marrari, T. Awan, O. Bugiani, G. Forloni and F. Tagliavini, Molecular determinants of the physicochemical properties of a critical prion protein region comprising residues 106-126. *Biochem. J.* **342,** 207–214 (1999).

41. K. Kuwata, T. Matumoto, H. Cheng, K. Nagayama, T. L. James and H. Roder, NMR-detected hydrogen exchange and molecular dynamics simulations provide structural insight into fibril formation of prion protein fragment 106–126. *Proc. Natl. Acad. Sci. USA* **100,** 14790–14795 (2003).

42. D. R. Brown, Mayhem of the multiple mechanisms: Modelling neurodegeneration in prion disease. *J. Neurochem.* **82**, 209–215 (2002).

43. V. R. Martins, E. Graner, J. Garcia-Abreu, S. J. de Souza, A. F. Mercadante, S. S. Veiga, S. M. Zanata, V. M. Neto and R. R. Brentani Complementary hydropathy identifies a cellular prion protein receptor. *Nature Med.* **3**, 1376–1382 (1997).

44. D. R. Brown, B. Schmidt and H. A. Kretzschmar, A prion protein fragment interacts with PrP-deficient cells. *J. Neurosci. Res.* **52**, 260–267 (1998).

45. R. Rieger, F. Edenhofer, C. I. Lasmezas and S. Weiss, The human 37-kDa laminin receptor precursor interacts with the prion protein in eukaryotic cells. *Nature Med.* **3**, 1383–1388 (1997).

46. K. Kaneko, D. Peretz, K. M. Pan, T. C. Blochberger, H. Wille, R. Gabizon, O. H. Griffith, F. E. Cohen, M. A. Baldwin and S. B. Prusiner, Prion protein (PrP) synthetic peptides induce cellular PrP to acquire properties of the scrapie isoform. *Proc. Natl. Acad. Sci. USA* **92**, 11160–11164 (1995).

47. K. Kaneko, H. Wille, I. Mehlhorn, H. Zhang H. Ball, F. E. Cohen, M. A. Baldwin and S. B. Prusiner, Molecular properties of complexes formed between the prion protein and synthetic peptides. *J. Mol. Biol.* **270**, 574–586 (1997).

48. D. R. Brown, M. Pitschke, D. Riesner and H. A. Kretzschmar, Cellular effects of a neurotoxic prion protein peptide are related to its β-sheet content. *Neurosci. Res. Comm.* **23**, 119–128 (1998).

49. D. R. Brown, W. J. Schultz-Schaeffer, B. Schmidt and H. A Kretzschmar, Prion protein-deficient cells show altered response to oxidative stress due to decreased SOD-1 activity. *Exp. Neurol.* **146**, 104–112 (1997).

50. A. R. White, S. J. Collins, F. Maher, M. F. Jobling, L. R. Stewart, J. M. Thyer, K. Beyreuther, C. L. Masters and R. Cappai Prion protein-deficient neurons reveal lower glutathione reductase activity and increased susceptibility to hydrogen peroxide toxicity. *Am. J. Pathol.* **155**, 1723–1730 (1999).

51. D. R. Brown and A. Besinger, Prion protein expression and superoxide dismutase activity. *Biochem. J.* **334**, 423–429 (1998).

52. S. J. McHattie, D. R. Brown and M. M. Bird, Cellular uptake of the prion protein fragment PrP106-126 *in vitro*. *J. Neurocytol.***28**, 145–155 (1999).

53. L. Solforosi, J. R. Criado, D. B. McGavern, S. Wirz, M. Sanchez-Alavez, S. Sugama, L. A. DeGiorgio, B. T. Volpe, E. Wiseman, G. Abalos, E. Masliah, D. Gilden, M. B. Oldstone, B. Conti and R. A. Williamson, Cross-linking cellular prion protein triggers neuronal apoptosis *in vivo*. *Science*. **303**, 1514–1516 (2004).

54. D. R. Brown, B. Schmidt and H. A. Kretzschmar, Role of microglia and host prion protein in neurotoxicity of a prion protein fragment. *Nature* **380**, 345–347 (1996)

55. D. R. Brown, St. J. R. Nicholas and L. Canevari, L. Lack of prion protein expression results in a neuronal phenotype sensitive to stress. *J. Neurosci. Res.* **67**, 211–224 (2002).

56. A. E. Williams, L. J. Lawson, V. H. Perry and H. Faser, Characterization of microglial response in murine scrapie. *Neuropathol. Appl. Neurobiol.*, **20**, 47–55 (1994).

57. A. Giese, M. H. Groschup, B. Hess and H. A. Kretzschmar, Neuronal c ell death in scrapie-infected mice is due to apoptosis. *Brain Pathol.* **5**, 213–221 (1995).

58. B. Van Everbroeck, E. Dewulf, P. Pals, U. Lubke, J. J. Martin and P. Cras, The role of cytokines, astrocytes, microglia and apoptosis in Creutzfeldt-Jakob disease. *Neurobiol. Aging.* **23**, 59–64 (2002).

59. C. A. Baker, D. Martin and L. Manuelidis,. Microglia from Creutzfeldt-Jakob disease-infected brains are infectious and show specific mRNA activation profiles. *J. Virol.* **76**, 10905–10913 (2001)

60. C. A. Baker and L. Manuelidis, Unique inflammatory RNA profiles of microglia in Creutzfeldt-Jakob disease. *Proc. Natl. Acad. Sci. USA* **100**, 675–679 (2003).

61. D. C. Guiroy, I. Wakayama, P. P. Liberski and D. C. Gajdusek, Relationship of microglia and scrapie amyloid-immunoreactive plaques in kuru,Creutzfeldt-Jakob disease and Gerstmann-Straussler syndrome. *Acta Neuropathol.* **87**, 526–530 (1994).

62. P. P. Liberski, J. Ironside, L. McCardle and A. Sherring, Ultrastructural analysis of the florid plaque in variant Creutzfeldt-Jakob disease. *Folia Neuropathol.* **38**, 167–170 (2000).

63. P. P. Liberski, B. Sikorska, J. Bratosiewicz-Wlsik, A. Walic, P. Brown and D. R. Brown, Exuberant cellular reaction of the optic nerves in experimental Creutzfeldt-Jakob disease in rodents: an ultrastructural study. *Acta Neurbiol. Exp.* **63**, 309–318 (2003).

64. U. v Eitzen, R. Egensperger, S. Kosel, E. M. Grasbon-Frodl, Y.Imai, K. Bise, S. Kohsaka, P. Mehraein and M. B. Graeber, Microglia and the development of spongiform change in Creutzfeldt-Jakob disease. *J. Neuropathol. Exp. Neurol.* **57**, 246–256 (1998).

65. M. Miyazono, T. Iwaki, T. Kitamoto, Y. Kaneko, K. Doh-ura and J. Tateishi, A comparative immunohistochemical study of Kuru and senile plaques with a special reference to glial reactions at various stages of amyloid plaque formation. *Am. J. Pathol.***139**, 589–598 (1991).

66. J. W. Ironside, L. McCardle, P. A. Hayward and J. E. Bell, Ubiquitin immunocytochemistry in human spongiform encephalopathies. *Neuropathol. Appl. Neurobiol.* **19**, 134–40 (1993).

67. S. Betmouni, V. H. Perry and J. L. Gordon, Evidence for an early inflammatory response in the central nervous system of mice with scrapie. *Neuroscience*, **74**, 1–5 (1996).

68. A. Giese, D. R. Brown, M. H. Groschup, C. Feldmann, I. Haist and H. A. Kretzschmar, Role of microglia in neuronal cell death in prion disease. *Brain Pathol.* **8**, 449–57 (1998).

69. C. Cunningham, R. Deacon, H. Wells, D. Boche, S. Waters, C. P. Diniz, H. Scott, J. N. Rawlins and V. H. Perry, Synaptic changes characterize early behavioural signs in the ME7 model of murine prion disease. *Eur J Neurosci.* **17**, 2147–2155 (2003).

70. J. I. Kim, W.-K. Ju, J.-H. Choi, J. Kim, E.-K. Choi, R. I. Carp, H. M. Wisniewski and Y.-S. Kim, Expression of cytokine genes and increased nuclear factor-kappa B activity in the brains of scrapie-infected mice. *Mol. Brain Res.* **73**, 17–27 (1999).

71. A. E. Williams, A.-M. van Dam, W. K. H. Man-A-Hing, F. Berkenbosch, F., P. Eikelenboom and H. Fraser, Cytokines, prostaglandins and lipocortin-1 are present in the brains of scrapie-infected mice. *Brain Res.* **654**, 200–206 (1994).

72. P. N. Moynagh, D. C. Williams and L. A. J. O'Niell, Interleukin-1 activates transcriptional factor NF-KB in glial cells. *Biochem. J.* **294**, 343–347 (1993).

73. Z. Guan, M. Soderberg, P. Sindelar, S. B. Prusiner, K. Kristensson and G. Dallner, Lipid composition in scrapie-infected mouse brain: prion infection increases the levels of dolichyl phosphate and ubiquinone. *J. Neurochem.***66**, 277–285 (1996).

74. S. I. Choi, W. K. Ju, E. K. Choi, J. Kim, H. Z. Lea, R. I. Carp, H. M. Wisniewski and Y. S. Kim, Mitochondrial dysfunction induced by oxidative stress in the brains of hamsters infected with the 263 K scrapie agent. *Acta Neuropathol.* **96**, 279–286 (1998).

75. D. W. Lee, H. O. Sohn, H. B. Lim, Y. G. Lee, Y.-S. Kim, R. I. Carp and H. M. Wisniewski, Alteration of free radical metabolism in the brain of mice infected with scrapie agent. *Free Radic. Res.* **30**, 499–507 (1999).

76. C. Bate, R. S. Boshuizen, J. P. Langeveld and A. Williams, Temporal and spatial relationship between the death of PrP-damaged neurones and microglial activation. *Neuroreport.* **13**, 1695–700 (2002).

77. C. Bate, S. Reid and A. Williams, Killing of prion-damaged neurones by microglia. *Neuroreport.* **12**, 2589–2594 (2001).

78. C. Fabrizi, V. Silei, M. Menegazzi, M. Salmona, O. Bugiani, F. Tagliavini, H. Suzuki and G. M. Lauro, The stimulation of inducible nitric-oxide synthase by the prion protein fragment 106–126 in human microglia is tumor necrosis factor-alpha-dependent and involves p38 mitogen-activated protein kinase. *J. Biol. Chem.* **276**, 25692–25696 (2001).

79. D. R. Brown, B. Schmidt and H. A. Kretzschmar, A neurotoxic prion protein fragment enhances proliferation of microglia but not astrocytes in culture. *Glia* **18,** 59–67 (1996).

80. J.-M. Peyrin, C. I. Lasmezas, S. Haïk, F. Tagliavini, M. Salmona, A. Williams, D. Richie, J.-P. Deslys. and D. Dormont, Microglial cells respond to amyloido-

genic PrP peptide by the production of inflammatory cytokines. *Neuroreport* **10**, 723–729 (1999).

81. D. R. Brown, B. Schmidt and H. A. Kretzschmar, A prion protein fragment primes type 1 astrocytes to proliferation signals from microglia. *Neurobiol. Disease* **4**, 410–422 (1998).

82. F. Hafiz and D. R. Brown, A model for the mechanism of astrogliosis in prion disease. *Mol. Cell Neurosci.* **16,** 221–232 (2000).

83. M. Marella and J. Chabry, Neurons and astrocytes respond to prion infection by inducing microglia recruitment. *J. Neurosci.* **24**, 620–627 (2004).

84. K. Jendroska, F. P. Heinzel, M. Torchia, L. Stowring, H. A. Kretzschmar, A. Kon, A. Stern, S. B. Prusiner and S. J. DeArmond, Proteinase-resistant prion protein accumulation in Syrian hamster brain correlates with regional pathology and scrapie infectivity. *Neurology* **41**, 1482–1490 (1991).

85. D. R. Brown, Dependence of neurones on astrocytes in a co-culture system renders neurones sensitive to TGF-β1 induced glutamate toxicity. *J. Neurochem.* **72,** 943–953 (1999).

86. D. R. Brown, Neurones depend on astrocytes in a co-culture system for protection from glutamate toxicity. *Mol. Cell. Neurosci.* **13**, 379–389 (1999).

87. M. Daniels and D. R. Brown Astrocytes regulate *N*-methyl-D-aspartate receptor subunit composition increasing neuronal sensitivity to excitotoxicity. *J. Biol. Chem.* **276**, 22446–22452 (2001).

88. G. M. Durand, P. Gregor, X. Zheng, M. R. V. Bennett, G. R. Uhu, and R. S. Zukin, Cloning of an apparent splice variant of the N-methyl-D-aspartate receptor NMDAR1 with altered sensitivity to polyamines and activators of protein kinase C. *Proc. Natl. Acad. Sci. USA* **89**, 9359–9363 (1992).

89. N. Nakanishi, R. Alex and N. A. Schneider, Alternative splicing generates functionally distinct N-methyl-D-aspartate receptors. *Proc. Natl. Acad. Sci. USA* **89**, 8552–8556 (1992).

90. M. R. Kreutz, T. M. Böckers, J. Bockmannm, C. I. Seidenbecher, B. Kracht, C. K. Vorwerk, J. Weise and B. A. Sabel, Axonal injury alters alternative splicing of the retinal NR1 receptor: the preferential expression of the NR1b isoform is crucial for retinal ganglion cell survival. *J. Neurosci.* **18**, 8278–8291 (1998).

91. H. Monyer, N. Burnashev, D. J. Laurie, B. Sakmann and P. H. Seeburg, Developmental and regional expression in the rat brain and functional properties of four NMDA receptors. *Neuron*, **12**, 529–540 (1994).

92. M. Watanabe, M. Mishina, and Y. Inoue, Distinct spatiotemporal expression of five NMDA receptor channel subunit mRNAs in the cerebellum. *J. Comp. Neurol.* **343**, 513–519 (1994).

93. E. Audinat, B. Lambolez, J. Rossier and F. Crepel, Activity dependent regulation of N-methyl-D-aspartate receptor subunits expression in cerebellar granule cells. *Eur. J. Neurosci.* **6**, 1792–1800 (1994).

94. M. Farrant, D. Feldmeyer, T. Takahashi, and S. G. Cull-Candy, NMDA-receptor channel diversity in the developing cerebellum. *Nature* **368**, 335–339 (1994).

95. D. R. Brown, Prion protein peptide neurotoxicity can be mediated by astrocytes. *J. Neurochem.* **73**, 1105–1113 (1999).

96. D. R. Brown and C. M. Mohn Astrocytic glutamate uptake and prion protein expression. *Glia* **25**, 282–292 (1999).

97. D. R. Brown, J. Herms, and H. A. Kretzschmar, Mouse cortical cells lacking cellular PrP survive in culture with a neurotoxic PrP fragment. *Neuroreport* **5**, 2057–60 (1994).

98. J. Sassoon, M. Daniels and D. R. Brown, Astrocytic regulation of NMDA receptor subunit composition modulates the toxicity of prion peptide PrP106-126. *Mol. Cell Neurosci.* **25**, 181–191 (2004).

99. J. C. Manson, A. R. Clarke, M. L. Hooper, L. Aitchison, I. McConnell and J. Hope, 129/Ola mice carring a null mutation in PrP that abolishes mRNA production are developmentally normal. *Mol. Neurobiol.* **8**, 121–127 (1994).

100. S. Sakaguchi, S. Katamine, N. Nishida, R. Moriuchi, K. Shigematsu, T. Sugimoto, A. Nakatani, Y. Kataoka, T. Houtani, S. Shirabe, H. Okada, S. Hasegawa, T. Miyamoto and T. Noda Loss of cerebellar Purkinje cells in aged mice homozygous for a disrupted PrP gene. *Nature*, **380**, 528–531 (1996).

101. R. C. Moore, I. Y. Lee, G. L. Silverman, P. M. Harrison, R. Strome, C. Heinrich, A. Karunaratne, S. H. Pasternak, M. A. Chishti, Y. Liang, P. Mastrangelo, K. Wang, A. F. A. Smit, S. Katamine, G. A. Carlson, F. E. Cohen, S. B. Prusiner, D. W. Melton, P., Tremblay, L. E. Hood, and D. Westaway, Ataxia in prion protein (PrP)-deficient mice is associated with upregulation of the novel PrP-like protein doppel. *J. Mol. Biol.*, **292,** 797–817 (1999).

102. K. Lu, W. Wang, Z. Xie, B. S. Wong, R. Li, R. B. Petersen, M. S. Sy. and S. G. Chen, Expression and structural characterization of the recombinant human doppel protein. *Biochemistry,* **39**, 13575–13583 (2000).

103. H. Mo, R. C. Moore, F. E. Cohen, D. Westaway, S. B. Prusiner and P. E. Wright, Two different neurodegenerative diseases caused by proteins with similar structures. *Proc. Natl. Acad. Sci. USA* **98,** 2352–2357 (2001).

104. E. M. Nicholson, H. Mo, S. B. Prusiner, F. E. Cohen, and S. Marqusee, Differences between the prion protein and its homolog Doppel: a partially structured stat with implications for scrapie formation. *J. Mol. Biol.* **316**, 807–815 (2002).

105. K. Poec'h, C. Guérin, J.-P. Brandel, J-M. Launay and J.-L. Laplanche, First report of polymorphisms in the prion-like protein gene (PRND): implications for human prion disease. *Neurosci. Lett.* **286**, 144–148 (2000).

106. S. Mead, J. Beck, A. Dickinson, E. M. Fisher and J. Collinge, Examination of the human prion protein-like gene doppel for geenetic susceptibility to sporadic and variant Creutzfeldt-Jakob disease disease. *Neurosci. Lett.* **290**, 117–120 (2000).

107. A. Behrens, S. Brandner, N. Genoud and A. Aguzzi Normal neurogenesis and scrapie pathogenesis in neural grafts lacking the prion protein homologue Doppel. *EMBO Rep.* **2**, 347–352 (2001).

108. N. L. Tuzi, E. Gall, D. Melton and J. C. Manson, (2002) Expression of doppel in the CNS of mice does not modulate transmissible spongiform encephalopathy disease. *J. Gen. Virol.* **83**, 705–711.

109. K. Poec'h, C. Serres, Y. Frobert, C. Martin, S. Lehmann, S. Chasseigneaux, V. Sazdovitch, J. Grassi, P. Jouannet, J. M. Launay and J. L. Laplanche, The human "prion-like" protein Doppel is expressed in both Sertoli cells and spermatozoa. *J. Biol. Chem.* **277**, 43071–43078 (2002).

110. A. Behrens, N. Genoud, H. Neumann, T. Rülicke, F. Janett, F. L. Heppner, B. Ledermann and A. Aguzzi, Absence of the prion protein homologue Doppel causes male sterility. *EMBO J.* **21**, 3652–3658 (2002).

111. G. L. Silverman, K. Qin, R. C. Moore, Y. Yang, P. Mastrangelo, P. Tremblay, S. B. Prusiner, F. E. Cohen and D. Westaway, Doppel is an N-glycosylated, glycosylphosphatidylinositol-anchored protein. Expression in testis and ectopic production in the brains of Prnpo/o mice predisposed to Purkinje cell loss. *J. Biol. Chem.* **275**, 26834–26841 (2000).

112. B.-S. Wong, T. Liu, D. Paisley, R. Li, T. Pan, S. G. Chen, G. Perry, R. B. Petersen, M. A. Smith, D. W. Melton, P. Gambetti, D. R. Brown and M.-S. Sy, Induction of HO-1 and NOS in doppel-expressing mice devoid of PrP: Implications for Doppel function. *Mol. Cell Neurosci.* **17**, 768–775 (2001).

113. A Behrens and A. Aguzzi, Small is not beautiful: antagonizing functions for the prion protein PrP^c and its homologue Dpl. *Trends Neurosci.* **25**, 150–154 (2002).

114. C. Weissmann and A. Aguzzi PrP's double causes trouble. *Science* **286** 914–915 (1999).

115. D. R. Brown, Prion and prejudice: normal protein at the synapse. *Trends Neurosci.* **24**, 85–90 (2001).

116. R. Atarashi, N. Nishida, K. Shigematsu, S. Gtot, T. Kondo, S. Sakaguchi and S. Katamine, Deletion of N-terminal residues 23–88 from prion protein (PrP) abrogates the potential to rescue PrP-deficient mice from PrP-like protein/doppel-induced neurodegeneration. *J. Biol. Chem.* **278**, 28944–28949 (2003).

117. R. Atarashi, S. Sakagushi, K. Shigematsu, K. Arima, N. Okimura, N., Yamaguchi, A. Li, J. Kopacek and S. Katamine, Abnormal activation of glial cells in the brains of prion protein-deficient mice ectopically expression prion protein-like, PrPLP/Dpl. *Mol. Med.* **7**, 803–809 (2000).

118. S. B. Prusiner and S. J. DeArmond, Prion diseases and neurodegeneration. *Ann. Rev. Neurosci.* **17**, 311–39 (1994).

Chapter 11

MOLECULAR MECHANISMS MEDIATING NEURONAL CELL DEATH IN EXPERIMENTAL MODELS OF PRION DISEASES, *IN VITRO*

Tullio Florio[1,2], Stefano Thellung[2] and Gennaro Schettini[1,2]

[1]*Department of Oncology, Biology and Genetics, University of Genova and*
[2]*Pharmacology and Neuroscience, National Institute for Cancer Research, Genova, Italy*

11.1. Summary

In this chapter we will review the growing bulk of experimental observations regarding the modulation of intracellular pathways mediating the neuronal cell death in *in vitro* models of prion diseases. In particular the effects of the prion-derived peptide PrP106-126 in different cell culture models will be described. PrP106-126 represents the first experimental approach to the prion-dependent neurotoxicity that has now been largely accepted by the scientific community.

This peptide, although incompletely, reproduces *in vitro* many of the biochemical and pathological characteristics of the PrP^{Sc}, allowing a detailed analysis of the signal transduction mediating the prion toxicity.

The analysis will be focused on the different intracellular pathways modulated by PrP106-126, or related peptides, to induce neuronal cell death in both primary neuronal cultures or neuronal-like cell lines, in relation with its the structural characteristics. In particular the effects of this peptide on the intracellular ion concentration, MAP kinase cascades and radical oxygen species generation will be discussed.

11.2. Introduction

One of the main limitations in the developing effective drugs for prion diseases, is represented by the lack of validated experimental models to study the neuropathological effects of pathogenic prion protein (PrP) molecules. Indeed, while several studies shed some light on the mechanisms of prion replication and on the biochemical features of the "infective" PrP species, only recently some information has been made available on the mechanisms of prion toxicity. These data were generated mainly using a 21 amino acid peptide (PrP106-126) derived from the PrP sequence, that Forloni et al.[1] proposed as the neurotoxic core of PrP^{Sc}. This peptide was mainly tested in cell culture studies where it was reported to induce neurotoxic[1] and gliotrophic[2,3] effects. However, neurotoxic effects of PrP106-126 were also reported in few *in vivo* studies[4].

In all these studies, the experimental approach to test the toxicity of PrP106-126 (or similar peptides, see below) is based on its administration externally to the cells in the culture medium. This protocol, although still not unanimously accepted, is based on the emerging concept that the PrP replication and toxicity are distinct phenomena[5]. Indeed, although not yet definitively proved, a growing number of evidence supports this notion.

First, prion replication process has been deeply studied in PrP^{Sc} infected cells, (mainly neuroblastoma cells) in which, although high level of protease K-resistant intracellular PrP is generated, no direct cell toxicity is observed; second, clinical studies showed that in the brain of some patients affected by prion diseases were identified large areas where PrP^{Sc} accumulated in high concentrations with minimal neuropathological changes[6]. Conversely, although PrP^{Sc} was temporally and spatially associated with development of the lesions[7] in several experimental models[8–11] and some clinical cases[12], it was reported that neurodegeneration may occur in the absence of detectable PrP^{Sc}. Moreover, it was reported that most of the familial prion diseases are much more difficult to be transmitted to mice then the acquired or sporadic forms[13], raising the hypothesis that some mutant PrP may be toxic without being infective[5]. All these evidence support the notion that PrP^{Sc}, although important for the prion replication, could not be responsible of the prion-induced neurodegeneration. In principle, the latter process may derive from toxic species generated as intermediates along the prion replication process that may or may not be themselves infectious or a by-product of PrP^{Sc} conversion[5]. Thus, the prion-dependent neurotoxic activity may reside in the activation of specific "death signals" via the interaction of toxic PrP species, experimentally represented by PrP106-126 or related peptides, with specific cell structures. Thus, either via the interaction with yet unknown membrane receptors or through the internalization into the endoplasmic reticulum, these peptides may trigger neuronal cell death, activating specific intracellular pathways[14–16]. On this premise, the use of synthetic peptides to delve deeper into the cellular mechanisms of prion toxicity, may allow the identification of valuable specific targets to the possible development of novel therapeutic strategies.

Importantly, recent data further validated this experimental approach. In fact, it was reported that the same mechanisms mediating neuronal death by extracellular application of PrP-derived peptides were observed both in vivo and *in vitro* and were reproduced using different synthetic PrP peptides[1,16], recombinant longer fragments[17], as well as using purified PrP^{Sc}[15,18–20].

Thus, all these observations strongly support the relevance, for PrP diseases, of the information that can be obtained using the above described peptide approach.

In this chapter we will critically analyze the available data on the biochemical characteristics and the neurotoxic effects of synthetic peptides derived from the PrP sequence, mainly focusing on the intracellular mechanisms regulated to induce neuronal cell death in relation to their structural characteristics.

11.3. The identification of PrP106-126

In 1993, Forloni and coll., analyzed the *in vitro* toxicity of several synthetic peptides derived from the sequence of an amyloid protein purified from the cerebral cortex isolated from a patients deceased from Gerstman-Straussler-Sheinker disease (GSS)[1]. In particular, in this seminal study, the Authors isolated a protein spanning the amino acids 57-150 of the PrP, that was divided in 6 sequences (amino acids 57-64, 89-106, 106-114, 106-126, 127-135 and 127-147) that were used to synthesize corresponding peptides. The rationale from this study was the hypothesis that, similarly to Alzheimer disease, β-structured peptides accumulating into the brain of prion patients as amyloid aggregates, may induce neuronal death. The results obtained were very surprising, since only the peptide corresponding to the amino acids 106-126 was able to elicit significant neurotoxicity via the triggering of the apoptotic pathway[1]. Subsequently, these results were repeated by a number of different laboratories and a wide consensus was reached in the scientific community for the use of this peptide to study the mechanisms of neuronal cell death.

Importantly, the characterization of the physical-chemical and biological characteristics of PrP106-126, beside neurotoxicity, led to the identification of several characteristics already identified in the PrP^{Sc} (high β-sheet content, protease K resistance, direct activation of astrocytes and microglia)[2,3,21–23]. All these data further support the relevance of this peptide as an *in vitro* model of PrP^{Sc}-induced neurotoxicity.

During the years, other extended peptides were analyzed for their structural and biological properties but, interestingly all of them included, at least partially the 106-126 sequence (i.e. PrP118-135, PrP82-146 and, more recently, a long peptide of 106 amino acids encompassing the amino acids 89-231 of the full length PrP with the deletion of the amino acids 141-176 [named PrP_{106}]).

Thus, all these studies allowed the establishment of a role for the amino acid 106-126 as the "death motif" internal to the PrP sequence.

11.4. Structural characteristics of PrP106-126

The analysis of the three-dimensional characteristics of the PrP106-126 peptide showed a high tendency to adopt a β-sheet configuration, when analyzed in phosphate buffer at pH 5.0[21]. In contrast, at pH 7.2 a rapid aggregation of the peptide is observed by laser light scatter analysis[21] and circular dichroism experiments, after the removal of the precipitated material, showed mainly a random coiled structure[24]. The aggregates show the features of amyloid fibrils as measured by electron microscopy and Congo red staining[24,25]; the generation of PrP106-126 macroaggregates (up to 6,000 molecules[26]) was reported to be dependent on the copper binding (at His_{111}, Met_{102} or Met_{112})[27,28] and, finally, a partial resistance of the peptide to protease K digestion was observed by reverse phase HPLC[26].

All these characteristics resemble the features identified in the pathogenic PrP^{Sc} isoform, strongly supporting the view that PrP106-126 represents an useful model to study the toxic pathways activated in prion diseases.

More recently, possible molecular determinants for PrP106-126 aggregation were identified in mutational experiments. It was observed that PrP106-126, when dissolved in viscous solvent (i.e. 10% glycerol), assumed a clear β-sheet configuration also at pH 7.2. Thus, on the basis of a computerized analysis of prediction of structure (AGADIR), Gly_{114} and Gly_{119} in the PrP106-126 sequence were identified as possible determinants of the aggregation process. Indeed it was hypothesized that, in force of their intrinsic flexibility, these residues may prevent that the peptide assuming a structured stable conformation favoring its aggregation. This prediction was experimentally confirmed analyzing the structural features of a mutant peptide in which the two glycines were

Figure 11.1. Diagramatic representation of the most commonly used PrP-derived peptides and their main characteristics (aggregation and *in vitro* toxicity)

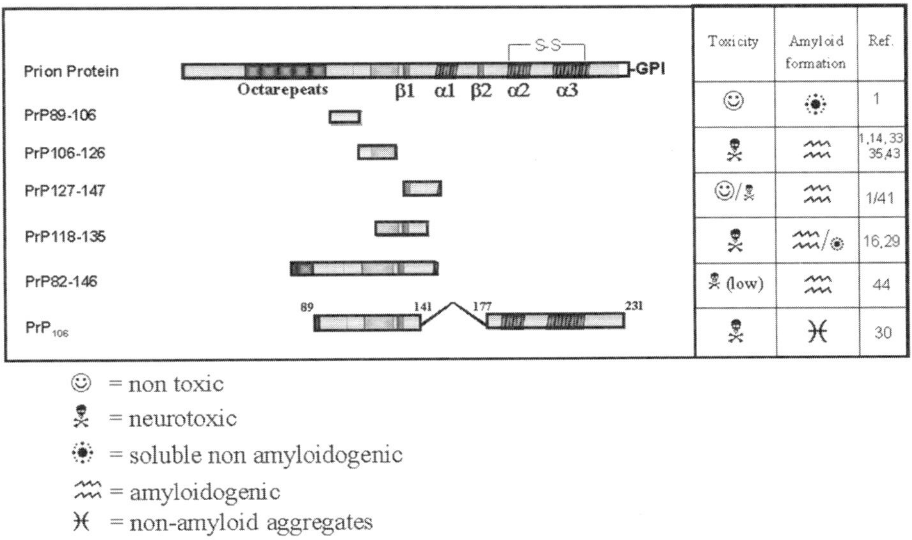

substituted with alanines[24]. This peptide was indeed completely soluble and β-structured although non amyloidogenic, thus suggesting that the steric restriction introduced in the 106-126 prion sequence upon the double Gly/Ala mutation, allowed the formation of a soluble state of the peptide that has a predominant propensity towards the β-structure. All these results indicate that Gly_{114} and Gly_{119} residues are very critical for the peptide flexibility as well as for its high tendency to form amyloid fibrils[24].

It is important to remark, that not all the PrP-derived peptides, able to form amyloid fibrils, are neurotoxic and, conversely, that not all the neurotoxic peptides are fibrillogenic (Figure 11.1).

In particular, in the original paper, beside PrP106-126, also the peptide corresponding to the amino acids 127-146 was able to generate amyloid fibrils, although non neurotoxic[1] (see below). Conversely, different peptides (i.e. PrP118-135, PrP_{106} or mutants of PrP106-126), used in experimental conditions in which they are not fibrillogenic, retain their toxicity[24,27,29,30].

This apparent discrepancy is however in line with many different studies about different amyloidogenic neurodegenerative diseases (for example Alzheimer's disease) in which the neuronal cell death is not related to the amyloid deposition but depends on protofibrillar intermediates[31]. Again this observation, seems to validate the data obtained using PrP-derived peptides in the context of the prion diseases.

Important information, on the determinants of the structural organization of PrP, using the peptide model, was provided by Salmona et al.[32] analyzing the peptide PrP82-146. Indeed, these Authors analyzed the structural characteristics of a peptide with a longer sequence that included both the fragments, previously reported to be amyloidogenic when analyzed as isolated fragments (amino acids 106-126 and 127-146)[1]. Moreover, by independently scrambling these regions, they tried to determine the individual contribution of these sequences to the conformational characteristics of the peptide.

Interestingly, their data show that the sequence 127-146, but not the sequence 106-126, is extremely relevant for the protease K resistance of the peptide and its aggregability, while the amyloid formation was affected by altering both structures, although a more pronounced inhibition occurred in the PrP82-146 scrambled in the 127-146 region[32].

Again, these data, considering the high toxicity of the PrP106-126, suggest that the amyloid aggregation properties, may not be relevant for the toxicity of the PrP peptides.

11.5. Toxicity of PrP106-126

To date PrP106-126 has been reported to induce cell death on a wide range of neuronal primary cultures of rodent hippocampal, cortical and cerebellar neurons[1,33–38] as well as in established human or murine neuroendocrine cell lines[14,39–43]. Similar toxic activity was reported also for the other PrP-derived peptides more commonly studied: PrP118-135[29], PrP$_{106}$[30] and, although with minor efficiency, PrP82-147[44].

Interestingly, in few studies it was demonstrated that these peptides may also induce neurodegeneration after administration in vivo. Indeed, the intravitreal injection of PrP106-126[4], as well as PrP118-135[16], caused a significant degeneration of retinal neurons, that, belonging to the CNS, represent a close system, with low content of proteolytic activity, allowing in an easily way the reaching of toxic concentration of the peptide compared to the brain injection.

Although the PrP106-126-induced cell death is now a well-recognized model to study the toxic effects of prions, some objective practical and theoretical limitations have to be considered when analyzing these kind of studies: first, in almost all the studies, independently from the cell model utilized, the cell death occurred only using extremely high concentrations of peptide (80–100 μM), and, second, the toxic effects were extremely delayed requiring from 3 to 10 days of treatment. It is interestingly to note that the similar concentrations (10 μM) and time of incubation (7 days) was also observed for PrP$_{106}$, the synthetic 106 amino acid long peptide although it was proposed, in light of its physico-chemical characteristics, to represent a "miniprion"[30].

In particular, when primary cultures are used, such prolonged *in vitro* culture may *per se* induce a reduction in cell viability or increase the susceptibility to external *noxae* that, in turn, may affect the final results observed. Thus, although numerous convincing studies have already addressed these aspects, it is always important to consider these issues when analyzing this kind of papers.

In both *in vitro* and *in vivo* studies, the cell death induced by PrP106-126 showed almost invariably the typical features of the programmed cell death or apoptosis (nuclear condensation, DNA internucleosomal cleavage, annexin V staining, caspase activation, etc.), although, in few cases, the appearance of necrosis (i.e. LDH release) was also observed[39,45]. Recently, it has been reported that purified PrP^{Sc} induces a caspase-dependent apoptosis in a mouse neuroblastoma cell line[15], further showing a relationship between the effects observed using PrP106-126 and PrP^{Sc}.

Using the human neuroblastoma cell line 5H-SY-5Y, O'Donovan and Coll. characterized in details the initial events by which PrP106-126 triggers apoptosis: the first step is a rapid mitochondrial membrane depolarization, followed by cytochrome C release into the cytosol and caspase activation[43]. However, the full comprehension of the interaction of the enzymatic steps caused by PrP106-126 is still elusive. In particular, it was reported by different groups that the exposure of neurons and neuronal like cells to PrP106-126 induces the activation of caspases, but the blockade of proapoptotic enzymes does not inhibit cell death induced by the peptide[14,43,46]. Thus, it has been proposed that other mechanisms, such as free radical production, calpains and MAP kinase activation (see below), could provide an alternative pathway to execute the neuron[14,43,46].

A precise correlation between the three-dimensional structure of PrP and its neurotoxic effects has been identified in all prion diseases. Thus it is extremely relevant, when using experimental models such the PrP-derived peptides, to verify whether such relationship is preserved. Numerous studies addressed this issue using mutant or modified PrP106-126[24,27,35,41], and a general consensus was identified on the absolute requirement of β-structured peptides to induce neurotoxic effects. A more complex scenario involved the analysis of the role of the generation of amyloid in such toxic effects. Indeed while it was reported that β-structured PrP106-126 mutants, unable to aggregate in classical amyloid fibrils, retain the same neurotoxic activity and intracellular mechanism activation like the PrP106-126 w.t.[24,27,47] and similar data were also reported using the PrP118-135[29] and PrP_{106}[30], in other reports the fibrillogenesis was considered a prerequisite for the capability of PrP106-126 to kill cells[35,41]. Indeed, in these reports introducing mutation that abolished the amyloid aggregation of PrP106-126 was observed also a lack of neurotoxic activity. This apparent discordance of results may be resolved considering that in these reports the relative participation to the toxic effects of the soluble β-structured peptide and the amyloid fibrils was not evaluated. Indeed in the previously cited papers, the soluble component of the PrP106-126, representing protofibrillar intermediates, likely present in all the β-structured peptides, was reported to be neurotoxic. A similar concept was recently developed for all the amylodogenic diseases, including Alzheimer's and Parkinson, as well as for fibrillogenic proteins unrelated to human pathologies[31].

To identify the molecular pathways involved in the apoptotic activity of PrP106-126, a great effort was dedicated to the search for possible cellular proteins interacting with the peptide able to trigger biochemical signals inside the cell. Indeed, the peptide through a direct interaction, was reported to alter some properties of the cell membrane such as the increase in its microviscosity[48]. Numerous candidates were proposed, in different studies, as PrP106-126 receptors, including the p75 NGF receptor[49], the formyl peptide receptor-like 1[50] and the L-type voltage sensitive calcium channels[2], while other authors identified a direct interaction with DNA[51]. Such a various array of possible transducer of PrP106-126 effects reflects mainly the different cell types analyzed but does not allow the identification of a unifying theory. Another research line proposes that the interaction of PrP106-126 with the target cells is dependent on the binding to the cellular PrP (PrP^C). The first evidence in this direction was based on experiments using neuronal cells derived from PrP-knockout mice. Using these cells, Brown et al.[52] reported that in the absence of PrP^C no signs of PrP106-126 (but not PrP118-135[16])-dependent toxicity are observed. This observation, that was subsequently replicated in few reports [33,35,53], represents another analogy between the cellular effects of PrP106-126 and PrP^{Sc}, since the latter protein require PrP^C to replicate itself and, in turn, cause neurodegeneration.

The interaction of PrP106-126 with PrP^C was shown to involve the amino acids 112-119[54] and the preincubation with antibody directed against such amino acids was able to rescue neuronal cells from the PrP106-126 neurotoxicity. Moreover, the relevance of these amino acids for the interaction between PrP molecules was further demonstrated in a report showing that a peptide encompassing the amino acids 113-120 of PrP was able to prevent the *in vitro* conversion of PrP^C in PrP^{Sc}[55]. A still open issue in these studies is represented by the biological effects induced by the interaction between the PrP106-126 and PrP^C. A first paper, using immunoprecipitation protocols showed that PrP106-126 was able to strip the PrP^C from the cell membrane and, thus, it was speculated that the toxicity of the peptide relay in a loss of function of survival signals activated by PrP^{C}, such as its proposed superoxide dismutase activity[54,56].

A completely different scenario was proposed by Singh and Coll. that showed that chronic incubation of neuroblastoma cells with PrP106-126 resulted in an accumulation of the peptide inside the cells causing, through the interaction with PrP^{C}, the accumulation of PrP^{Sc}-like isoforms into the lysosomes favoring the generation of PrP truncated isoforms (^{Ctm}PrP) that were responsible of the cell death[57,58]. This is a very suggestive hypothesis since ^{Ctm}PrP was identified also in the brain of some prion diseased patients, although the incredibly (at least for cell culture studies) long incubation necessary to obtain this effect (1 to 4 months) raise some questions on the real relevance of this phenomenon.

An alternative mechanism by which the interaction with the cells may result in a toxic effect was proposed following the observation that PrP106-126[59], as well as PrP82-146[60] caused the formation of cation channels in planar lipid membranes. Thus, it was suggested that the alteration of the intracellular cation concentration via the peptide-formed channels may represent another possible mechanisms of toxicity, although this activity was not unanimously confirmed[61].

Finally, co-culture studies showed that, beside a direct toxicity, PrP-derived peptides may also affect cell viability through an indirect mechanism involving the generation of oxygen radical species (ROS) production by the microglia. Indeed it was reported that co-culturing cerebellar neurons in the presence of purified microglial cells the toxicity of PrP106-126 was greatly increased[33]. Moreover, these Authors reported that PrP106-126 was able to directly induce microglia to release ROS and that the pretreatment with antioxidants prevented the neurotoxic effects of the peptide[33]. This observation, altogether with other reports showing a modulation of astrocyte and microglia activity by PrP106-126[2,3,22,23,34,62–64], is particularly relevant since in prion diseases affected brain the accumulation of prions starts in glial cells, gliosis is a precocious hallmark of these diseases[7,65] and microglia were reported to retain the same infectivity as whole brain homogenate[66].

Thus, glia activation is now considered an early effect in prion-dependent neurodegeneration and the *in vitro* reproduction of such effect by PrP106-126 further reinforce the significance of this peptide as a model to address the pathological changes during prion diseases. Interestingly, microglial cells derived from $PrP^{o/o}$ mice are still able to respond to PrP106-126 treatment with the release of radical species, in a way resembling that observed in microglial cultures derived from *wild type* animals[33]. From these evidences emerge that glial activation by PrP106-126 may involve completely different mechanisms than

Figure 11.2. Examples of PrP106-126 neurotoxicity *in vitro*.
A. Cerebellar granule cells cultured for 14 DIV (days *in vitro*) and treated for the last 3 and 7 days with PrP106-126 100 μM. Peptide neurotoxicity is evidenced by the progressive fragmentation of neurite network and condensation of cell bodies.
B. Human neuroblastoma SH-SY5Y cells have been differentiated with cis-retinoic acid and treated for 5 days with PrP106-126 (100 μM). The exposure to the peptide induces cell shrinkage, cell body fragmentation and their progressive detachment from the substrate.

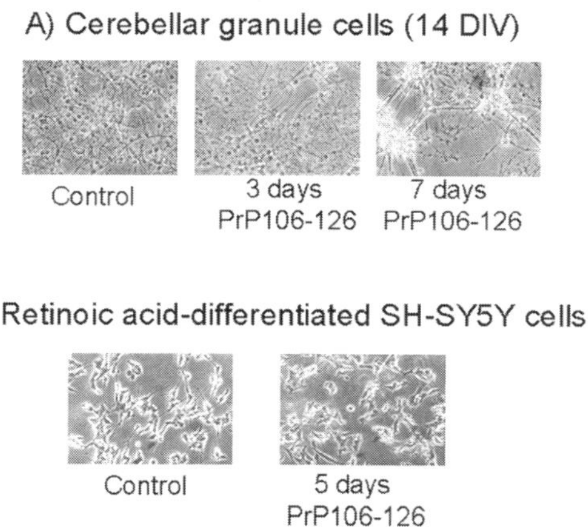

the direct toxicity induced by the peptide, since the latter phenomenon requires the expression of PrP^C.

11.6. Signal transduction of PrP106-126

To date, a large number of the studies about the neurotoxic effects of PrP-derived synthetic peptides were aimed to identify specific intracellular pathways activated by PrP106-126 or similar peptides.

However, although an impressive number of enzymatic systems have been reported to be activated or inhibited by these peptides (see Table 11.1), a clear and exhaustive identification of an intracellular cascade that specifically mediates prion peptides neurotoxicity still need to be provided[67].

For example, although it is generally accepted that PrP106-126 and other prion related peptides can activate the apoptotic program, some uncertainty exists about the chronology or the events that follow the interaction of the cell with the peptide and how they contribute to the apoptotic cascade.

Late apoptosis features, including annexin-V binding, chromatin condensation and DNA cleavage have been widely reported using prion derived peptides[1,14,29,37,40,43,52], but less consistent are the reports on

Table 11.1. Intracellular signalling induced by PrP-derived peptides.

Proposed signalling	Experimental models	Peptide	References
Caspase activation	CGC, CN	PrP106-126	14,43,46,68
	CN	PrP118-135	16,29
NMDA receptor over-stimulation	CN, CGC	PrP106-126	69
VSSC channel blockade	CGC, GH3	PrP106-126	36, 40
Mitochondrial depolarization, cytochrome C translocation to cytosol	SH-SY5Y	PrP106-126	43
p38 MAP kinase activation	SH-SY5Y	PrP106-126	14, 24, 47
	microglia		64
Calcineurin activation	CN	PrP106-126	45
AA release, 5-LOX activation	CGC	PrP106-126	38
GSH reduction, bcl-2 reduction	CN	PrP106-126	79
Hydrogen peroxide generation	Cell free reaction*	PrP106-126	82
Free radicals and proinflammatory cytokines production	microglia	PrP106-126	33, 63, 64
Plasmamembrane destabilization	CN	PrP118-135	29
Plasmamembrane microviscosity	CGC and glia	PrP106-126	48

CGC: cerebellar granule cells, CN: cortical neurons, GH3: rat pituitary adenoma cells, SH-SY5Y: human neuroblastoma cells. *PrP106-126 is incubated with Cu^{2+} before the addition of Fe^{2+}

the early events. In some studies, mitochondrial depolarization, with the consequent release of cytochrome C, has been shown to be an early event following cell exposure to PrP106-126, causing the subsequent activation of caspases[43]. More recently, PrP106-126, along with the Alzheimer's disease-related peptide Aβ25-35, has been reported to induce mitochondrial cytochrome C release and caspase activation via the activation of calcineurin and BAD[45]. However, although PrP106-126 and other amyloidogenic peptides have been commonly reported to elicit the activation of caspases, their role in the execution of the apoptotic program is still poorly understood. Indeed, studies using different cell models have shown that either the direct or indirect blockade of these proapoptotic enzymes does not inhibits cell death[14,43,46,68]. It has therefore been proposed that other events may act in concert with caspases to kill neurons.

Unfortunately the studies that tried to investigate this aspect led to the identification of very different intracellular players, although, in spite of these discrepancies, likely dependent on the cell model used, it is still possible to identify few key mediators of the proapoptotic effects of PrP106-126. Among them, a higher consensus was identified on perturbation of Ca^{++} homeostasis, activation of p38 MAP kinase, and generation of free oxygen radicals.

Alterations in the intracellular Ca^{++} concentrations represents one of the more commonly identified PrP106-126-mediated cytotoxic signal. For example in both the previously mentioned papers[43,45], as well as in experiments using purified PrP^{Sc}[15], a release of Ca^{++} from the endoplasmic reticulum precedes the activation of caspases. In particular Agostinho et al. correlate this event with the activation of calcineurin while O'Donovan et al. identified the activation of the lysosomial enzyme calpain as the target for the increased $[Ca^{++}]_i$. Interestingly, in both cases the inhibition of calcineurin or calpain was required to prevent the PrP106-126 toxicity, being the simple caspase blockade ineffective.

Several other laboratories have also investigated whether $[Ca^{++}]_i$ variation may play a role in the prion related cell death using cultured neurons or neuronal-like cell lines as experimental models[36,40,42,69]. For example, the exposure of cortical neurons to PrP106-126 evokes an increase of intracellular $[Ca^{++}]$ through the activation of both the voltage dependent Ca^{++}channels and the NMDA gutamate receptors[69]. On the other hand, using electrophysiological measurements, it was reported that the blockade of the L-type voltage sensitive calcium channels by PrP106-126 is responsible for the apoptosis in cerebellar granule cells[36]. This discrepancy could be explained considering the peculiarity of the cell system studied being the cerebellar granule cells, differently from cortical neurons, absolutely dependent for their survival from the chronic activity of the voltage dependent Ca^{++} channels[70]. Moreover, a similar inhibitory effect was obtained using longer recombinant PrP fragments[71]. Thus, the experimental model used seems to deeply influence the intracellular signalling activated by PrP106-126 to induce apoptosis. In any case studies on glial cells (both astrocytes and microglia), in which PrP106-126 exerts a trophic action (proliferation, stimulation of the release of cytokines or radical species) rather than cell death, perturbation of the $[Ca^{++}]_i$ represent also one of the main intracellular systems regulated by PrP peptides[2,72,73], further supporting the relevance of this signalling in the prion related cellular effects.

The human neuroblastoma cell lines, such as SH-SY5Y cells, are responsive to the apoptotic cell death induced by PrP106-126. These cells have been extensively studied to dissect the intracellular pathways occurring during neuronal death and therefore have been also used to discover the intracellular pathways activated by PrP-derived peptides.

Interestingly studies on these cells allowed the identification of the MAP kinase p38, but not ERK1/2 or JNK, as one of the mediator of the PrP106-126-dependent apoptosis[14,47]. Among the member of the MAP kinase family, p38 was reported to induce apoptotic cell death in response to a variety of external stimuli, being often involved as mediator of cell death in different neurodegenerative disorders[74]. This observation suggests that also in prion diseases p38 may play a significant role in the induction of neuronal death. Again, in this study was reported that although both caspases 3/7 and p38 were activated after PrP106-126 treatment, the inhibition of caspase activity alone did not prevent the cells to undergo to apoptosis but was also necessary the blockade of p38[14]. In this context it is important to observe that a caspase independent[75] and p38 dependent[76] neuronal death has been reported.

It has been also proposed that PrP106-126 induces neuronal death inducing oxidative stress through the generation of free radical production[67]. In particular, the exposure of primary neurons to the peptide causes a reduction of Cu^{++}/Zn^{++} superoxide dismutase (SOD) and glutathione reductase activities[77–79]. Interestingly, PrP^{C} has been reported to bind copper via its octarepeats domain and has been suggested to function as a superoxide dismutase[56]. Thus, it was proposed that the toxicity of PrP106-126 is dependent on its interaction to the copper binding region of PrP^{C} that in turn may impair the antioxidant properties of the cellular protein[54]. The decrease in GSH content caused by the exposure of cortical neurons to PrP106-126 was also coupled to a decrease of the expression of the antiapoptotic protein Bcl-2 and both mechanisms are believed to cooperate to kill neurons[78,79].

The contribution of metal (mainly Cu^{++} and Zn^{++}) to PrP106-126 induced cell death was also supported by the identification of metal binding sites on the peptide[28] and by the observation that mutations of these residues (His_{111}, Met_{102}) or metal chelation, reverted PrP106-126 toxicity.

It is important to note that most evidences about the involvement of oxidative stress and copper imbalance in the neuronal death by PrP106-126 have been obtained by *in vitro* studies, although metal perturbation and oxidative damage actually occur during prion diseases *in vivo*. In fact, immunochemical studies reported that cortices from scrapie-infected mice show positive immunostaining for nitrosylated tyrosine residues, a marker for peroxynitrite generation[80]. In mice, intracerebrally transmitted scrapie causes perturbation of brain metal homeostasis that is characterized by a decrease in copper content and an increase in manganese, that correlate with the onset of neurodegeneration[81]. Along with Cu^{++}/Mn^{++} imbalance, a progressive reduction of Cu^{++}/Zn^{++} SOD was also reported[81]. However, to date, a causal relationship between copper chelating property of prion protein and its antioxidant activity is not completely characterized, from a mechanistic point of view. In fact, $PrP^{0/0}$ neurons are in the meantime more sensitive to oxidative stress[78] and insensitive to PrP106-126 toxicity[52]: the combination of these findings contrasts with the idea that the oxidative stress may represent a pivotal and specific mechanism in the neurodegeneration induced by the peptide. Nevertheless, it is possible that the free radical generation and the consequent oxidative stress may represent a sort of intermediate executioner necessary to induce neuronal death by the PrP106-126. In fact, in cell free systems, PrP106-126[82] as well as recombinant PrP fragments[83], has been also reported to generate free radicals that may participate in render cells more vulnerable to oxidative stress. In these studies, by electrospin resonance spectroscopy (ESR), it was observed that PrP106-126, in the presence of reduced copper ions (Cu^{++}), catalyzes the formation of peroxide hydrogen that can be further converted into hydroxyl radical by the fenton reaction.

It is thereafter conceivable that the oxidative damage evoked by neuronal exposure to PrP106-126 may arises from both a direct generation of radical species by the peptide and a reduction of the free radical scavenging systems of the neuron.

Finally, PrP106-126 has been also used to investigate the role of arachidonic acid (AA) metabolism in prion encephalopathies. In particular, Cappai and Coll. demonstrated that the toxicity of the peptide in primary cultures of cerebellar granule cells is mediated by the activation of NMDA receptors with consequent elevation of $[Ca^{++}]_i$ that in turn regulates the release of AA and the activation of the 5'-lipoxygenase pathway[38].

The wide range of different intracellular systems involved in the neurotoxic effects of PrP106-126 coming out from all these data, although likely reflecting the real complexity of the scenario occurring during the neuronal death in prion diseases, has not allowed the identification of specific targets for the possible development of new drugs. However, these data may represent a starting point for a more stringent analysis using more specific experimental models.

11.7. PrP106-126 as a tool to screen for anti-neurodegenerative drugs effective in prion diseases

As detailed in the previous paragraph, an impressive amount of studies tried to identify possible therapeutical target for the prion-dependent neurodegeneration. Indeed, the development of *in vitro* models to study the neuronal cell death during prion diseases, represents the main goal of all the researches using PrP106-126 or other PrP-derived synthetic peptides, although their usefulness to develop therapeutic strategies for prion diseases has been questioned since these peptides are not infectious and do not cause conversion of PrP^{C} to PrP^{Sc}. Nevertheless, PrP106-126 sharing with PrP^{Sc} neurotoxic/gliotrophic properties, according to the hypothesis that infectivity and cell death effects are distinct phenomena, may represent the first *in vitro* approach to evaluate the potential effectiveness of compounds able to prevent the neuronal death occurring during prion diseases. In particular the research was focused mainly analyzing the effects of drugs affecting the most reproduced cell alterations induced by PrP106-126[84]. Moreover in few cases the same protective effects were observed using both synthetic peptides and purified PrP^{Sc} (for example antagonism of NMDA receptors)[85].

To date, the mechanisms of protection for the most promising drugs, involving both direct and indirect effects are:

1. antioxidant activity both direct by free radicals scavengers or indirect targeted to increase cell resistance to oxidative stress;
2. impairment of prion peptides ability to acquire toxic properties (aggregation and PK resistance);
3. down-stream effects (prevention of apoptosis).

The first possible therapeutical approach was proposed by Brown and Coll.[33] that identified an indirect mechanisms for the PrP106-126 mediated cell death, via the microglia production of free radicals. They demonstrated that the co-incubation of vitamin E and N-acetyl cysteine prevented the cerebellar neuron death induced by the peptide.

The NMDA antagonist flupirtine was reported to inhibit neuronal death induced by PrP106-126 via the interference with radical oxygen generation. However, in these studies the protective effect was induced by preventing the inhibitory effect of PrP106-126 on the intracellular content of reduced glutathione (GSH) resulting in an increased expression of the antiapoptotic protein Bcl-2[79,85]. The same neuroprotective activity of flupirtine was also observed against PrP^{Sc}-dependent neuronal death and was mimicked by the other NMDA antagonist memantine[18].

The anti-malaric drug quinacrine represents the first therapeutical approach in patients affected by Creutzfeldt-Jacob disease. Indeed it was reported that quinacrine (as well as chorpromazine, a phenothiazine derivative) is able to abolish scrapie infectivity *in vitro*, by preventing PrP^{C}-PrP^{Sc} conversion[86]. Recently, quinacrine has been observed to reduce the neuronal death *in vitro* induced by PrP106-126. This protective effect was reported to be dependent on the ability of quinacrine to block hydrogen peroxide generation and scavenge the hydroxyl radicals produced upon neuronal exposure to PrP106-126[87]. Moreover, quinacrine has been reported to inhibit the activity of the cation channel formed by PrP106-126 in synthetic lipid bilayers[88]. Conversely, quinacrine (differently from tetracyclines, see below) was reported to be ineffective in revering the partial protease resistance of PrP106-126, PrP82-146 or purified PrPSc[89].

Good therapeutic expectations derived from the study of the potential anti-prion activity of tetracyclines. Indeed, compounds of this class of antibiotics showed, *in vivo*, the ability to prolong the incubation time of the disease after experimental scrapie transmission[90]. The possible molecular mechanism for such effect was also partially identified. Tetracycline, by virtue of its hydrophobicity, was report to bind to PrP106-126 and PrP82-146 inhibiting their ability to aggregate and their PK resistance[91]. Importantly these physico-chemical changes induced by tetracycline treatment, were correlated with the abolishment of both neuronal death and glial proliferation caused by PrP106-126[91].

The compound squalestatin (that inhibits the conversion of squalene to cholesterol) has been reported to protect neurons against the toxicity induced by the prion peptides PrP106-126, PrP82-146 and $sPrP_{106,}$ as well as partially purified PrPSc. It has been proposed that squalestatin interferes with the peptide induced increase in cellular cholesterol and prostaglandin E_2(PGE_2) release, events related to neuronal injury in prion diseases[44].

Finally, the neuronal cell death induced by the peptide PrP118-135 was reported to be inhibited by humanin, a peptide that has already shown to protect neurons from the Alzheimer's amyloid-β peptide neurotoxicity, presumably through the blockade of a still unknown apoptotic early event[92]. In Table 11.2 are summarized the most studied compounds that have been demonstrated to interfere with the toxicity of prion-related synthetic peptides, *in vitro*, and the proposed mechanisms for their activity.

11.8. Conclusions and future perspectives

The identification of effective drugs for prion diseases represents a goal still very far to be achieved. The peculiarity of the transmission of these diseases, being infective, sporadic or familial, does not allow even the identification of specific targets for pharmacological intervention. At least three possible targets may be proposed for an ideal antiprion drug: 1) inhibition of the prion replication process; 2) inhibition of the generation of toxic isoforms; 3) inhibition of specific neurotoxic intracellular pathways responsible for the neuronal death. In particular the studies performed with PrP-derived synthetic peptides have mainly addressed

Table 11.2. Mechanism of action of potential drugs, reported to be effective as neuroprotective agents in experimental models of prion diseases, *in vitro*

Drug	Possible Mechanism	Peptide	References
Vitamin E, N-acetyl cysteine	Radical scavenger activity	PrP106-126	33
Flupirtine	Inhibition of GSH reduction, increase of bcl-2 expression	PrP106-126	79,85
Quinacrine	Inhibition of H_2O_2 generation, hydroxyl radical scavenging Blockade of the channels formed by the peptide.	PrP106-126	87 88
Tetracyclines	Blockade of aggregation, disruption of preformed fibrils, inhibition of PK resistance.	PrP106-126 PrP82-146	91
Squalestatin	Cholesterol depletion, decrease of PGE_2 release	PrP106-126 PrP82-146 PrP_{106}	44

the latter issue. Unfortunately, although they significantly increased our knowledge about some biophysical, biochemical and pathological characteristics of PrP^{Sc}, due to the intrinsic limitations of the experimental model did not completely accomplished the duty of identify specific molecular targets for antiprion drugs. Indeed, being intrinsically toxic, these peptides were able to interfere with proapoptotic intracellular pathways specific for the neuronal cell model used rather than with "prion specific" mechanisms. This limitation caused the generation of an incredible high amount of data without the definition of potential drug targets. Thus, it is mandatory to evolve this experimental model. In particular, in the past few years, using the recombinant DNA techniques have been generated a number of extended PrP fragment (i.e. corresponding to the a.a. 90-231 or 121-231) that seem to represent a more faithful experimental model for prion disease. In particular for their longer sequence, the use of this fragment may conjugate the three-dimensional features of PrP with its toxic properties. In this way these recombinant PrP molecules may provide information on the relationship between the structural characteristics of PrP, its molecular determinants of toxicity and the final cellular effects. Hopefully the development and validation of these new experimental models may favour the identification of really effective therapies at least to control the progression of prion diseases.

11.9. Acknowledgements

This work has been supported by grants from Telethon Italy, MIUR FIRB and MIUR PRIN to TF and GS.

11.10. References

1. G. Forloni, N. Angeretti, R. Chiesa, E. Monzani, M. Salmona, O. Bugiani and F. Tagliavini. Neurotoxicity of a prion protein fragment. *Nature*. 362, 543–546. (1993)

2. T. Florio, M. Grimaldi, A. Scorziello, M. Salmona, O. Bugiani, F. Tagliavini, G. Forloni and G. Schettini. Intracellular calcium rise through L-type calcium channels, as molecular mechanism for prion protein fragment 106–126-induced astroglial proliferation. *Biochem Biophys Res Commun*. 228, 397–405. (1996)

3. D. R. Brown, B. Schmidt and H. A. Kretzschmar. A prion protein fragment primes type 1 astrocytes to proliferation signals from microglia. *Neurobiol Dis*. 4, 410–422. (1998)

4. M. Ettaiche, R. Pichot, J. P. Vincent and J. Chabry. In vivo cytotoxicity of the prion protein fragment 106-126. *J Biol Chem*. 275, 36487–36490. (2000)

5. R. Chiesa and D. A. Harris. Prion diseases: what is the neurotoxic molecule? *Neurobiol Dis*. 8, 743–763. (2001)

6. P. Parchi, R. Castellani, P. Cortelli, P. Montagna, S. G. Chen, R. B. Petersen, V. Manetto, C. L. Vnencak-Jones, M. J. McLean, J. R. Sheller and et al. Regional distribution of protease-resistant prion protein in fatal familial insomnia. *Ann Neurol*. 38, 21–29. (1995)

7. S. J. DeArmond, W. C. Mobley, D. L. DeMott, R. A. Barry, J. H. Beckstead and S. B. Prusiner. Changes in the localization of brain prion proteins during scrapie infection. *Neurology*. 37, 1271–1280. (1987)

8. J. Collinge, M. S. Palmer, K. C. Sidle, I. Gowland, R. Medori, J. Ironside and P. Lantos. Transmission of fatal familial insomnia to laboratory animals. *Lancet*. 346, 569–570. (1995)

9. M. Fischer, T. Rulicke, A. Raeber, A. Sailer, M. Moser, B. Oesch, S. Brandner, A. Aguzzi and C. Weissmann. Prion protein (PrP) with amino-proximal deletions restoring susceptibility of PrP knockout mice to scrapie. *EMBO J*. 15, 1255–1264. (1996)

10. J. C. Manson, E. Jamieson, H. Baybutt, N. L. Tuzi, R. Barron, I. McConnell, R. Somerville, J. Ironside, R. Will, M. S. Sy, D. W. Melton, J. Hope and C. Bostock. A single amino acid alteration (101L) introduced into murine PrP dramatically alters incubation time of transmissible spongiform encephalopathy. *EMBO J*. 18, 6855–6864. (1999)

11. E. Flechsig, D. Shmerling, I. Hegyi, A. J. Raeber, M. Fischer, A. Cozzio, C. von Mering, A. Aguzzi and C. Weissmann. Prion protein devoid of the octapeptide repeat region restores susceptibility to scrapie in PrP knockout mice. *Neuron*. 27, 399–408. (2000)

12. R. Medori, P. Montagna, H. J. Tritschler, A. LeBlanc, P. Cortelli, P. Tinuper, E. Lugaresi and P. Gambetti. Fatal familial insomnia: a second kindred with mutation of prion protein gene at codon 178. *Neurology*. 42, 669–670. (1992)

13. P. Brown, C. J. Gibbs, Jr., P. Rodgers-Johnson, D. M. Asher, M. P. Sulima, A. Bacote, L. G. Goldfarb and D. C. Gajdusek. Human spongiform encephalopathy: the National Institutes of Health series of 300 cases of experimentally transmitted disease. *Ann Neurol*. 35, 513–529. (1994)

14. S. Thellung, V. Villa, A. Corsaro, S. Arena, E. Millo, G. Damonte, U. Benatti, F. Tagliavini, T. Florio and G. Schettini. p38 MAP kinase mediates the cell death induced by PrP106-126 in the SH-SY5Y neuroblastoma cells. *Neurobiol Dis*. 9, 69–81. (2002)

15. C. Hetz, M. Russelakis-Carneiro, K. Maundrell, J. Castilla and C. Soto. Caspase-12 and endoplasmic reticulum stress mediate neurotoxicity of pathological prion protein. *EMBO J*. 22, 5435–5445. (2003)

16. J. Chabry, C. Ratsimanohatra, I. Sponne, P. P. Elena, J. P. Vincent, T. Pillot, B. Drouet, M. Pincon-Raymond, J. Vandekerckhove, M. Rosseneu and J. Chambaz. In vivo and *in vitro* neurotoxicity of the human prion protein (PrP) fragment P118-135 independently of PrP expression. A nonfibrillar form of the fusogenic prion protein fragment [118-135] induces apoptotic cell death in rat cortical neurons. *J Neurosci*. 23, 462–469. (2003)

17. M. Daniels, G. M. Cereghetti and D. R. Brown. Toxicity of novel C-terminal prion protein fragments and peptides harbouring disease-related C-terminal mutations. *Eur J Biochem*. 268, 6155–6164. (2001)

18. W. E. Muller, H. Ushijima, H. C. Schroder, J. M. Forrest, W. F. Schatton, P. G. Rytik and M. Heffner-Lauc. Cytoprotective effect of NMDA receptor antagonists on prion protein (PrionSc)-induced toxicity in rat cortical cell cultures. *Eur J Pharmacol*. 246, 261–267. (1993)

19. A. Giese, D. R. Brown, M. H. Groschup, C. Feldmann, I. Haist and H. A. Kretzschmar. Role of microglia in neuronal cell death in prion disease. *Brain Pathol*. 8, 449–457. (1998)

20. K. Post, D. R. Brown, M. Groschup, H. A. Kretzschmar and D. Riesner. Neurotoxicity but not infectivity of prion proteins can be induced reversibly *in vitro*. *Arch Virol Suppl*. 265–273. (2000)

21. L. De Gioia, C. Selvaggini, E. Ghibaudi, L. Diomede, O. Bugiani, G. Forloni, F. Tagliavini and M. Salmona. Conformational polymorphism of the amyloidogenic and neurotoxic peptide homologous to residues 106-126 of the prion protein. *J Biol Chem*. 269, 7859–7862. (1994)

22. C. K. Combs, D. E. Johnson, S. B. Cannady, T. M. Lehman and G. E. Landreth. Identification of microglial signal transduction pathways mediating a neurotoxic response to amyloidogenic fragments of beta-amyloid and prion proteins. *J Neurosci*. 19, 928–939. (1999)

23. F. B. Hafiz and D. R. Brown. A model for the mechanism of astrogliosis in prion disease. *Mol Cell Neurosci*. 16, 221–232. (2000)

24. T. Florio, D. Paludi, V. Villa, D. R. Principe, A. Corsaro, E. Millo, G. Damonte, C. D'Arrigo, C. Russo, G. Schettini and A. Aceto. Contribution of two conserved glycine residues to fibrillogenesis of the 106-126 prion protein fragment. Evidence that a soluble variant of the 106-126 peptide is neurotoxic. *J Neurochem*. 85, 62–72. (2003)

25. F. Tagliavini, F. Prelli, L. Verga, G. Giaccone, R. Sarma, P. Gorevic, B. Ghetti, F. Passerini, E. Ghibaudi, G. Forloni and et al. Synthetic peptides homologous to prion protein residues 106-147 form amyloid-like fibrils *in vitro*. *Proc Natl Acad Sci U S A*. 90, 9678–9682. (1993)

26. C. Selvaggini, L. De Gioia, L. Cantu, E. Ghibaudi, L. Diomede, F. Passerini, G. Forloni, O. Bugiani, F. Tagliavini and M. Salmona. Molecular characteristics of a protease-resistant, amyloidogenic and neurotoxic peptide homologous to residues 106-126 of the prion protein. *Biochem Biophys Res Commun*. 194, 1380–1386. (1993)

27. M. Salmona, P. Malesani, L. De Gioia, S. Gorla, M. Bruschi, A. Molinari, F. Della Vedova, B. Pedrotti, M. A. Marrari, T. Awan, O. Bugiani, G. Forloni and F. Tagliavini. Molecular determinants of the physicochemical properties of a critical prion protein region comprising residues 106-126. *Biochem J*. 342, 207–214. (1999)

28. M. F. Jobling, X. Huang, L. R. Stewart, K. J. Barnham, C. Curtain, I. Volitakis, M. Perugini, A. R. White, R. A. Cherny, C. L. Masters, C. J. Barrow, S. J. Collins, A. I. Bush and R. Cappai. Copper and zinc binding modulates the aggregation and neurotoxic properties of the prion peptide PrP106-126. *Biochemistry*. 40, 8073–8084. (2001)

29. T. Pillot, B. Drouet, M. Pincon-Raymond, J. Vandekerckhove, M. Rosseneu and J. Chambaz. A nonfibrillar form of the fusogenic prion protein fragment [118-135] induces apoptotic cell death in rat cortical neurons. *J Neurochem*. 75, 2298–2308. (2000)

30. V. Bonetto, T. Massignan, R. Chiesa, M. Morbin, G. Mazzoleni, L. Diomede, N. Angeretti, L. Colombo, G. Forloni, F. Tagliavini and M. Salmona. Synthetic miniprion PrP106. *J Biol Chem*. 277, 31327–31334. (2002)

31. M. Bucciantini, E. Giannoni, F. Chiti, F. Baroni, L. Formigli, J. Zurdo, N. Taddei, G. Ramponi, C. M. Dobson and M. Stefani. Inherent toxicity of aggregates implies a common mechanism for protein misfolding diseases. *Nature*. 416, 507–511. (2002)

32. M. Salmona, M. Morbin, T. Massignan, L. Colombo, G. Mazzoleni, R. Capobianco, L. Diomede, F. Thaler, L. Mollica, G. Musco, J. J. Kourie, O. Bugiani, D. Sharma, H. Inouye, D. A. Kirschner, G. Forloni and F. Tagliavini. Structural properties of Gerstmann-Straussler-Scheinker disease amyloid protein. *J Biol Chem*. 278, 48146–48153. (2003)

33. D. R. Brown, B. Schmidt and H. A. Kretzschmar. Role of microglia and host prion protein in neurotoxicity of a prion protein fragment. *Nature*. 380, 345–347. (1996)

34. D. R. Brown. Prion protein peptide neurotoxicity can be mediated by astrocytes. *J Neurochem*. 73, 1105–1113. (1999)

35. M. F. Jobling, L. R. Stewart, A. R. White, C. McLean, A. Friedhuber, F. Maher, K. Beyreuther, C. L. Masters, C. J. Barrow, S. J. Collins and R. Cappai. The hydrophobic core sequence modulates the neurotoxic and secondary structure properties of the prion peptide 106-126. *J Neurochem*. 73, 1557–1565. (1999)

36. S. Thellung, T. Florio, V. Villa, A. Corsaro, S. Arena, C. Amico, M. Robello, M. Salmona, G. Forloni, O. Bugiani, F. Tagliavini and G. Schettini. Apoptotic cell death and impairment of L-type voltage-sensitive calcium channel activity in rat cerebellar granule cells treated with the prion protein fragment 106-126. *Neurobiol Dis*. 7, 299–309. (2000)

37. S. Thellung, T. Florio, A. Corsaro, S. Arena, M. Merlino, M. Salmona, F. Tagliavini, O. Bugiani, G. Forloni and G. Schettini. Intracellular mechanisms mediating the neuronal death and astrogliosis induced by the prion protein fragment 106-126. *Int J Dev Neurosci*. 18, 481–492. (2000)

38. L. R. Stewart, A. R. White, M. F. Jobling, B. E. Needham, F. Maher, J. Thyer, K. Beyreuther, C. L. Masters, S. J. Collins and R. Cappai. Involvement of the 5-lipoxygenase pathway in the neurotoxicity of the prion peptide PrP106-126. *J Neurosci Res*. 65, 565–572. (2001)

39. J. Hope, M. S. Shearman, H. C. Baxter, A. Chong, S. M. Kelly and N. C. Price. Cytotoxicity of prion protein peptide (PrP106-126) differs in mechanism from the cytotoxic activity of the Alzheimer's disease amyloid peptide, A beta 25-35. *Neurodegeneration*. 5, 1–11. (1996)

40. T. Florio, S. Thellung, C. Amico, M. Robello, M. Salmona, O. Bugiani, F. Tagliavini, G. Forloni and G. Schettini. Prion protein fragment 106-126 induces apoptotic cell death and impairment of L-type voltage-sensitive calcium channel activity in the GH3 cell line. *J Neurosci Res*. 54, 341–352. (1998)

41. D. L. Rymer and T. A. Good. The role of prion peptide structure and aggregation in toxicity and membrane binding. *J Neurochem*. 75, 2536–2545. (2000)

42. M. Kawahara, Y. Kuroda, N. Arispe and E. Rojas. Alzheimer's beta-amyloid, human islet amylin, and prion protein fragment evoke intracellular free calcium elevations by a common mechanism in a hypothalamic GnRH neuronal cell line. *J Biol Chem*. 275, 14077–14083. (2000)

43. C. N. O'Donovan, D. Tobin and T. G. Cotter. Prion protein fragment PrP-(106-126) induces apoptosis via mitochondrial disruption in human neuronal SH-SY5Y cells. *J Biol Chem*. 276, 43516–43523. (2001)

44. C. Bate, M. Salmona, L. Diomede and A. Williams. Squalestatin cures prion-infected neurons and protects against prion neurotoxicity. *J Biol Chem*. 279, 14983–14990. (2004)

45. P. Agostinho and C. R. Oliveira. Involvement of calcineurin in the neurotoxic effects induced by amyloid-beta and prion peptides. *Eur J Neurosci*. 17, 1189–1196. (2003)

46. A. R. White, R. Guirguis, M. W. Brazier, M. F. Jobling, A. F. Hill, K. Beyreuther, C. J. Barrow, C. L. Masters, S. J. Collins and R. Cappai. Sublethal concentrations of prion peptide PrP106-126 or the amyloid beta peptide of Alzheimer's disease activates expression of proapoptotic markers in primary cortical neurons. *Neurobiol Dis*. 8, 299–316. (2001)

47. A. Corsaro, S. Thellung, V. Villa, D. R. Principe, D. Paludi, S. Arena, E. Millo, D. Schettini, G. Damonte, A. Aceto, G. Schettini and T. Florio. Prion protein fragment 106–126 induces a p38 MAP kinase-dependent apoptosis in SH-SY5Y neuroblastoma cells independently from the amyloid fibril formation. *Ann N Y Acad Sci*. 1010, 610–622. (2003)

48. M. Salmona, G. Forloni, L. Diomede, M. Algeri, L. De Gioia, N. Angeretti, G. Giaccone, F. Tagliavini and O. Bugiani. A neurotoxic and gliotrophic fragment of the prion protein increases plasma membrane microviscosity. *Neurobiol Dis*. 4, 47–57. (1997)

49. V. Della-Bianca, F. Rossi, U. Armato, I. Dal-Pra, C. Costantini, G. Perini, V. Politi and G. Della Valle. Neurotrophin p75 receptor is involved in neuronal damage by prion peptide-(106-126). *J Biol Chem*. 276, 38929–38933. (2001)

50. Y. Le, H. Yazawa, W. Gong, Z. Yu, V. J. Ferrans, P. M. Murphy and J. M. Wang. The neurotoxic prion peptide fragment PrP(106-126) is a chemotactic agonist for the G protein-coupled receptor formyl peptide receptor-like 1. *J Immunol*. 166, 1448–1451. (2001)

51. P. K. Nandi. Interaction of prion peptide HuPrP106-126 with nucleic acid. *Arch Virol*. 142, 2537–2545. (1997)

52. D. R. Brown, J. Herms and H. A. Kretzschmar. Mouse cortical cells lacking cellular PrP survive in culture with a neurotoxic PrP fragment. *Neuroreport*. 5, 2057–2060. (1994)

53. D. R. Brown. Prion protein peptides: optimal toxicity and peptide blockade of toxicity. *Mol Cell Neurosci*. 15, 66–78. (2000)

54. D. R. Brown. PrPSc-like prion protein peptide inhibits the function of cellular prion protein. *Biochem J*. 352 Pt 2, 511–518. (2000)

55. J. Chabry, B. Caughey and B. Chesebro. Specific inhibition of *in vitro* formation of protease-resistant prion protein by synthetic peptides. *J Biol Chem*. 273, 13203–13207. (1998)

56. D. R. Brown, B. S. Wong, F. Hafiz, C. Clive, S. J. Haswell and I. M. Jones. Normal prion protein has an activity like that of superoxide dismutase. *Biochem J*. 344 Pt 1, 1–5. (1999)

57. Y. Gu, H. Fujioka, R. S. Mishra, R. Li and N. Singh. Prion peptide 106-126 modulates the aggregation of cellular prion protein and induces the synthesis of potentially neurotoxic transmembrane PrP. *J Biol Chem*. 277, 2275–2286. (2002)

58. N. Singh, Y. Gu, S. Bose, S. Kalepu, R. S. Mishra and S. Verghese. Prion peptide 106-126 as a model for prion replication and neurotoxicity. *Front Biosci*. 7, a60–71. (2002)

59. M. C. Lin, T. Mirzabekov and B. L. Kagan. Channel formation by a neurotoxic prion protein fragment. *J Biol Chem*. 272, 44–47. (1997)

60. R. Bahadi, P. V. Farrelly, B. L. Kenna, J. I. Kourie, F. Tagliavini, G. Forloni and M. Salmona. Channels formed with a mutant prion protein PrP(82-146) homologous to a 7-kDa fragment in diseased brain of GSS patients. *Am J Physiol Cell Physiol*. 285, C862–872. (2003)

61. M. Manunta, B. Kunz, E. Sandmeier, P. Christen and H. Schindler. Reported channel formation by prion protein fragment 106-126 in planar lipid bilayers cannot be reproduced. *FEBS Lett*. 474, 255–256. (2000)

62. G. Forloni, R. Del Bo, N. Angeretti, R. Chiesa, S. Smiroldo, R. Doni, E. Ghibaudi, M. Salmona, M. Porro, L. Verga and et al. A neurotoxic prion protein fragment induces rat astroglial proliferation and hypertrophy. *Eur J Neurosci*. 6, 1415–1422. (1994)

63. J. M. Peyrin, C. I. Lasmezas, S. Haik, F. Tagliavini, M. Salmona, A. Williams, D. Richie, J. P. Deslys and D. Dormont. Microglial cells respond to amyloidogenic PrP peptide by the production of inflammatory cytokines. *Neuroreport*. 10, 723–729. (1999)

64. C. Fabrizi, V. Silei, M. Menegazzi, M. Salmona, O. Bugiani, F. Tagliavini, H. Suzuki and G. M. Lauro. The stimulation of inducible nitric-oxide synthase by the prion protein fragment 106-126 in human microglia is tumor necrosis factor-alpha-dependent and involves p38 mitogen-activated protein kinase. *J Biol Chem*. 276, 25692–25696 (2001)

65. J. F. Diedrich, P. E. Bendheim, Y. S. Kim, R. I. Carp and A. T. Haase. Scrapie-associated prion protein accumulates in astrocytes during scrapie infection. *Proc Natl Acad Sci U S A*. 88, 375–379. (1991)

66. C. A. Baker, D. Martin and L. Manuelidis. Microglia from Creutzfeldt-Jakob disease-infected brains are infectious and show specific mRNA activation profiles. *J Virol*. 76, 10905–10913. (2002)

67. D. R. Brown. Mayhem of the multiple mechanisms: modelling neurodegeneration in prion disease. *J Neurochem*. 82, 209–215. (2002)

68. J. Saez-Valero, N. Angeretti and G. Forloni. Caspase-3 activation by beta-amyloid and prion protein peptides is independent from their neurotoxic effect. *Neurosci Lett*. 293, 207–210. (2000)

69. D. R. Brown, J. W. Herms, B. Schmidt and H. A. Kretzschmar. PrP and beta-amyloid fragments activate different neurotoxic mechanisms in cultured mouse cells. *Eur J Neurosci*. 9, 1162–1169. (1997)

70. C. Galli, O. Meucci, A. Scorziello, T. M. Werge, P. Calissano and G. Schettini. Apoptosis in cerebellar granule cells is blocked by high KCl, forskolin, and IGF-1 through distinct mechanisms of action: the involvement of intracellular calcium and RNA synthesis. *J Neurosci*. 15, 1172–1179. (1995)

71. S. Korte, N. Vassallo, M. L. Kramer, H. A. Kretzschmar and J. Herms. Modulation of L-type voltage-gated calcium channels by recombinant prion protein. *J Neurochem*. 87, 1037–1042. (2003)

72. V. Silei, C. Fabrizi, G. Venturini, M. Salmona, O. Bugiani, F. Tagliavini and G. M. Lauro. Activation of microglial cells by PrP and beta-amyloid fragments raises intracellular calcium through L-type voltage sensitive calcium channels. *Brain Res*. 818, 168–170. (1999)

73. J. W. Herms, A. Madlung, D. R. Brown and H. A. Kretzschmar. Increase of intracellular free Ca2+ in microglia activated by prion protein fragment. *Glia*. 21, 253–257. (1997)

74. K. Hensley, R. A. Floyd, N. Y. Zheng, R. Nael, K. A. Robinson, X. Nguyen, Q. N. Pye, C. A. Stewart, J. Geddes, W. R. Markesbery, E. Patel, G. V. Johnson and G. Bing. p38 kinase is activated in the Alzheimer's disease brain. *J Neurochem*. 72, 2053–2058. (1999)

75. R. A. Lockshin and Z. Zakeri. Caspase-independent cell deaths. *Curr Opin Cell Biol*. 14, 727–733. (2002)

76. T. Wada and J. M. Penninger. Mitogen-activated protein kinases in apoptosis regulation. *Oncogene*. 23, 2838–2849. (2004)

77. D. R. Brown, W. J. Schulz-Schaeffer, B. Schmidt and H. A. Kretzschmar. Prion protein-deficient cells show altered response to oxidative stress due to decreased SOD-1 activity. *Exp Neurol*. 146, 104–112. (1997)

78. A. R. White, S. J. Collins, F. Maher, M. F. Jobling, L. R. Stewart, J. M. Thyer, K. Beyreuther, C. L. Masters and R. Cappai. Prion protein-deficient neurons reveal lower glutathione reductase activity and increased susceptibility to hydrogen peroxide toxicity. *Am J Pathol*. 155, 1723–1730. (1999)

79. S. Perovic, H. C. Schroder, G. Pergande, H. Ushijima and W. E. Muller. Effect of flupirtine on Bcl-2 and glutathione level in neuronal cells treated *in vitro* with the prion protein fragment (PrP106-126). *Exp Neurol*. 147, 518–524. (1997)

80. M. Guentchev, T. Voigtlander, C. Haberler, M. H. Groschup and H. Budka. Evidence for oxidative stress in experimental prion disease. *Neurobiol Dis*. 7, 270–273. (2000)

81. A. M. Thackray, R. Knight, S. J. Haswell, R. Bujdoso and D. R. Brown. Metal imbalance and compromised antioxidant function are early changes in prion disease. *Biochem J*. 362, 253–258. (2002)

82. S. Turnbull, B. J. Tabner, D. R. Brown and D. Allsop. Copper-dependent generation of hydrogen peroxide from the toxic prion protein fragment PrP106-126. *Neurosci Lett*. 336, 159–162. (2003)

83. S. Turnbull, B. J. Tabner, D. R. Brown and D. Allsop. Generation of hydrogen peroxide from mutant forms of the prion protein fragment PrP121-231. *Biochemistry*. 42, 7675–7681. (2003)

84. T. Koster, K. Singh, M. Zimmermann and E. Gruys. Emerging therapeutic agents for transmissible spongiform encephalopathies: a review. *J Vet Pharmacol Ther*. 26, 315–326. (2003)

85. H. C. Schroder and W. E. Muller. Neuroprotective effect of flupirtine in prion disease. *Drugs Today (Barc)*. 38, 49–58. (2002)

86. C. Korth, B. C. May, F. E. Cohen and S. B. Prusiner. Acridine and phenothiazine derivatives as pharmacotherapeutics for prion disease. *Proc Natl Acad Sci U S A*. 98, 9836–9841. (2001)

87. S. Turnbull, B. J. Tabner, D. R. Brown and D. Allsop. Quinacrine acts as an antioxidant and reduces the toxicity of the prion peptide PrP106-126. *Neuroreport*. 14, 1743–1745. (2003)

88. P. V. Farrelly, B. L. Kenna, K. L. Laohachai, R. Bahadi, M. Salmona, G. Forloni and J. I. Kourie. Quinacrine blocks PrP (106-126)-formed channels. *J Neurosci Res*. 74, 934–941. (2003)

89. A. Barret, F. Tagliavini, G. Forloni, C. Bate, M. Salmona, L. Colombo, A. De Luigi, L. Limido, S. Suardi, G. Rossi, F. Auvre, K. T. Adjou, N. Sales, A. Williams, C. Lasmezas and J. P. Deslys. Evaluation of quinacrine treatment for prion diseases. *J Virol*. 77, 8462–8469. (2003)

90. G. Forloni, S. Iussich, T. Awan, L. Colombo, N. Angeretti, L. Girola, I. Bertani, G. Poli, M. Caramelli, M. Grazia Bruzzone, L. Farina, L. Limido, G. Rossi, G. Giaccone, J. W. Ironside, O. Bugiani, M. Salmona and F. Tagliavini. Tetracyclines affect prion infectivity. *Proc Natl Acad Sci U S A*. 99, 10849–10854. (2002)

91. F. Tagliavini, G. Forloni, L. Colombo, G. Rossi, L. Girola, B. Canciani, N. Angeretti, L. Giampaolo, E. Peressini, T. Awan, L. De Gioia, E. Ragg, O. Bugiani and M. Salmona. Tetracycline affects abnormal properties of synthetic PrP peptides and PrP(Sc) *in vitro*. *J Mol Biol*. 300, 1309–1322. (2000)

92. I. Sponne, A. Fifre, V. Koziel, B. Kriem, T. Oster, T. Pillot. Humanin rescues cortical neurons from prion-peptide-induced apoptosis. *Mol Cell Neurosci*. 25, 95–102. (2004)

Chapter 12

PROCESSING AND MIS-PROCESSING OF THE PRION PROTEIN: INSIGHTS INTO THE PATHOGENESIS OF FAMILIAL PRION DISORDERS

Neena Singh[1], Yaping Gu[1], Sharmila Bose[2], Subhabrata Basu[1], Xiu Luo[1], Richa Mishra[1], and Oscar Kuruvilla[1]

[1]*Institute of Pathology, Case Western Reserve University, Cleveland, OH USA*
[2]*Department of Molecular, Cellular and Developmental Biology, The University of Michigan, Ann Arbor, MI USA*

12.1. Summary

The cellular prion protein (PrP^C), though apparently innocuous, is the main agent responsible for infectious, familial, and sporadic prion disorders. Through its remarkable ability to undergo a change in conformation from a mainly α-helical to a β-sheet rich conformation commonly referred to as PrP-scrapie (PrP^{Sc}), PrP^C becomes infectious and pathogenic, a feature that is unique to this glycoprotein[1–4]. Over the years, most studies have focused on the mechanism of PrP^C to PrP^{Sc} conversion and its subsequent transmission to susceptible hosts, ignoring the less common familial prion disorders that result from point mutations in the prion protein gene (*PRNP*). In these disorders, mutant PrP (PrP^M) is presumed to undergo a spontaneous change in conformation to PrP^{Sc} without participation from an exogenous source of infectious PrP^{Sc}. Once initiated, the process proceeds exponentially, and deposits of mutant PrP^{Sc} are believed to result in the neurotoxicity observed in familial cases of prion disorders[5,6]. However, prion-specific neuropathology is often observed in the absence of detectable PrP^{Sc}, indicating the presence of alternative pathways of neurotoxicity in certain cases of prion disorders[7]. Recent studies on the processing of normal and mutant PrP underscore the importance of abnormal metabolism and various topological forms of PrP in prion disease pathogenesis[8]. In this chapter, we will review information on the complex pathways of intracellular trafficking and metabolism of normal and various mutant PrP forms, and highlight some of the abnormal pathways that may contribute to neurotoxicity in familial prion disorders.

12.2. Introduction

Familial disorders result from point mutations in the prion protein gene, and account for ~15% of all reported cases of prion disorders. Based on the phenotypic presentation, these disorders have been classified as Gerstmann Straussler Scheinker disease (GSS), Creutzfeldt Jakob disease (CJD), and fatal familial insomnia (FFI). These disorders are inherited as autosomal dominant traits with a penetrance of ~90%. More than 25 different point mutations and octapeptide repeat insertions in *PRNP* have been described, with little correlation between the mutation and the associated phenotype. Point mutations at codons P102L, P105L, A117V, G131V, Y145stop, F198S, D202N, Q217R, M232T, and insertions of octapeptide repeat sequences segregate with GSS, whereas mutations at codons D178N, V180N/M232R, T183A, E200K, R208H, V210I, and M232R are associated with CJD. The neuropathology in most familial cases has been attributed to a spontaneous change in the conformation of PrP^{M} to PrP^{Sc}, resulting in intra- or extracellular accumulation of aggregated and partially protease-resistant PrP^{Sc} (2,3,5). However, reports indicating the presence of prion-specific neuropathology in the absence of detectable PrP^{Sc} leave open the question whether PrP^{Sc} deposition is an obligatory step for the development of neurotoxicity[9,10]. Recent evidence indicates that mis-processing of PrP^{C} and PrP^{M}, and in some cases alternative topological forms of PrP may be the underlying cause of neurotoxicity in several familial prion disorders[11–16]. A diagrammatic representation of *PRNP* mutations segregating with familial disorders is shown in Figure 12.1.

12.3. Is PrP^{Sc} the key pathogenic agent in familial prion disorders?

The high correlation between PrP^{Sc} deposits and neurodegeneration has led to the belief that accumulation of aggregated and protease-resistant PrP^{Sc} is an obligatory step in the pathogenesis of prion

Figure 12.1. Point mutations in PrP associated with disease. Precursor PrP polypeptide that has not undergone processing in the ER is shown to represent the mutations in the GPI-SP. Following translocation into the ER, both the N-SP and the GPI-SP are cleaved, and a GPI anchor is added to residue 231 of PrP.

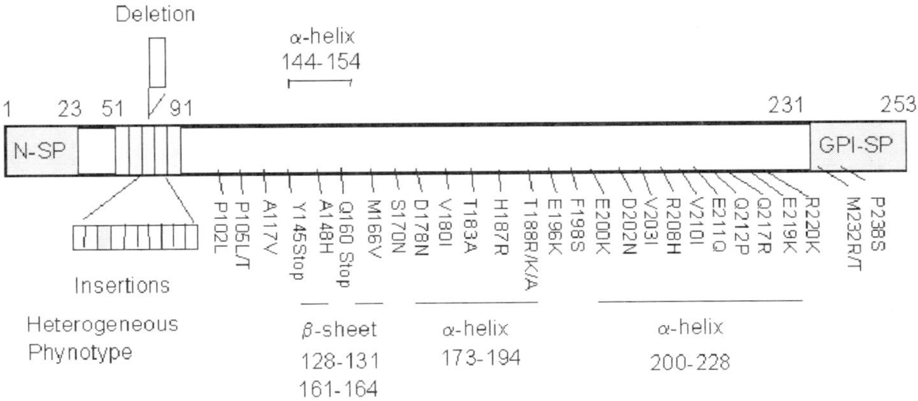

disorders[7]. Immunohistochemical staining of diseased brains shows a high correlation between the rate and extent of PrP^{Sc} accumulation and the incubation period and severity of the central nervous system (CNS) vacuolation, lending credence to the above belief. In most cases PrP^{Sc} deposition precedes vacuolation, suggesting a causal relationship with the observed neuropathology. Nevertheless, as in other amyloid-related diseases, neurodegeneration does not always show the same topographic distribution as amyloid plaques. In fact, prion disease with all its pathological hallmarks has been reported without any detectable PK-resistant PrP^{Sc}, and PK-resistant PrP^{Sc} has been generated without infectivity[10]. In GSS resulting from PrP^{102}, brain pathology and transmissibility to animals has been demonstrated without any PK-resistant PrP^{Sc}. Similar observations have been made in transgenic mice expressing PrP^{102}. Brain extracts from these animals transmit disease to recipient animals with consequent spongiform degeneration and amyloid plaques with little or no PrP^{Sc} (9). In GSS due to PrP^{145}, the conventional PrP^{Sc} fragment does not even exist due to a stop codon mutation at residue 145 of PrP, resulting in a C-terminally truncated protein. Yet the brain shows numerous PrP plaques and Tau positive neurofibrillary tangles[17].

In a recent report, cytosolic expression of even small amounts of PrP led to neurotoxicity in transgenic mice, resulting in typical features of prion disease without significant PrP^{Sc} accumulation[13–16]. Increasing amounts of cytosolic PrP led to the formation of PrP^{Sc} and its further propagation[16]. Subsequent studies have challenged these results, suggesting that the cytosolic presence of PrP is probably an experimental artifact[18]. Another report indicates a protective role of cytosolic PrP in human neurons rather than toxicity[20]. Although the role of cytosolic PrP is controversial, retro-translocation of wild type and mutant PrP forms from the ER to the cytosol has been reported in several instances, and could potentially lead to significant accumulations if proteasomal function is sub-optimal[21,22]. Further studies are needed to clarify the mechanism of toxicity by cytosolic PrP and improve our understanding of this intriguing phenomenon. Additional observations demonstrating prion-mediated neurotoxicity in the absence of significant PrP^{Sc} deposits include transgenic mouse models with astrocyte-specific PrP expression. In one mouse model, neuropathological changes induced by inoculation of infectious prions were reversed by turning off PrP expression in the neuronal population[23]. Although the authors ascribe this effect to a lack of PrP substrate in the neurons, expression of PrP only in astrocytes has been shown to induce neuronal death regardless of neuronal PrP expression[24]. A similar observation has been reported in PrP^{null} mice with only astrocyte-specific expression of PrP[25]. Likewise, exposure of astrocytes to PrP^{Sc} brain homogenate has been shown to induce neuronal death even when separated from the neuronal culture by a membrane[26]. Together, these observations suggest an indirect role of PrP^{Sc} in the generation of neurotoxicity, perhaps through permeable factors generated by astrocytes or other supporting cells in the brain parenchyma. However, the nature of cells that mediate this toxicity is unclear at present. A similar complexity surrounds the nature of the neurotoxic signal in familial prion disorders, some of which are not infectious in nature. Although PrP^{Sc} deposits are observed in some familial cases, the correlation is not persuasive, suggesting the contribution of alternative or additional metabolic processes in the initiation of neuronal death in such cases.

12.4. Processing of cellular PrP

In human neuroblastoma cells, PrP^C is synthesized in the endoplasmic reticulum (ER) in three topological forms; secretory PrP (${}^{sec}PrP$), and two transmembrane forms with either the carboxyl terminus or the amino terminus in the ER (${}^{Ctm}PrP$ and ${}^{Ntm}PrP$ respectively)[11,12]. ${}^{Sec}PrP$, synonymous with PrP^C, is by far the most abundant. Like other membrane proteins, the N-terminal signal peptide (N-SP) of ${}^{sec}PrP$ is cleaved co-translationally, and the C-terminal GPI signal peptide (GPI-SP) is removed in a transamidation reaction with the concomitant addition of a pre-assembled GPI anchor in the ER. Simultaneously or in a subsequent reaction, immature high mannose glycans and the disulfide bond are added to the nascent polypeptide, which is then transported along the secretory path. During its transit through the Golgi, the glycans are modified further, resulting in three distinct PrP forms; the unglycosylated form that does not acquire high mannose glycans in the ER and thus maintains its unglycosylated nature, mono-glycosylated form, and the di-glycosylated, sialylated forms. All three glycoforms are expressed on the cell surface, linked to the outer leaflet of the plasma membrane by the GPI anchor[27–29]. Thus, PrP^C or ^{Sec}PrP belongs to the distinct subset of 'GPI-linked' proteins that exhibit several special characteristics conferred by the GPI moiety[30,31]. In addition to serving as an elegant anchoring device, the GPI anchor influences the targeting and functioning of PrP in important ways. Like most GPI-linked proteins, PrP resides in cholesterol and sphingolipid rich membrane domains on the cell surface[32]. It recycles constitutively between the plasma membrane and the endocytic compartment, where 1–5% of the total pool is proteolytically cleaved, and the C-terminal 18kDa fragment is cycled back to the cell surface[33]. This fragment has a long half-life on the plasma membrane, and it is unclear if it undergoes any further recycling. Convincing evidence indicates that the conformational change of PrP^C to PrP^{Sc} occurs in these lipid-rich membrane domains at the cell surface or in an endocytic compartment[32]. Since β-sheet rich proteins have a propensity to intercalate within cholesterol-rich domains, these lipid microdomains serve as an optimal site for the conversion of PrP^C to PrP^{Sc}. Moreover, PrP^{Sc} and PrP^C would reside longer in these domains since GPI-linked proteins are endocytosed three-times slower than other membrane components, allowing more time for interaction. Depletion of cellular cholesterol or replacement of the GPI anchor of PrP with a transmembrane anchor diminishes the efficiency of PrP^{Sc} formation[32], probably since such a treatment enhances the rate of recycling of GPI linked proteins or displaces them from lipid rafts[30]. Thus, the GPI anchor facilitates PrP^{Sc} propagation because of its special characteristics.

However, conflicting reports suggest that PrP undergoes clathrin-mediated endocytosis despite the presence of a GPI anchor, perhaps through interaction with a neighboring transmembrane protein that leads PrP to the classical endocytic pathway[34]. The N-terminal domain of PrP is believed to mediate this interaction, since a series of deletions in this region reduce the internalization of PrP[35]. Interestingly, binding of copper ions to this same region enhances the endocytosis of PrP, an observation that has been explained by hypothesizing that perhaps copper increases the interaction of the putative transmembrane protein with PrP[36]. Sulfated glycans have also been proposed to stimulate the endocytosis of PrP selectively, and consequently reduce the formation of PrP^{Sc} in infected cells[37]. However, the mechanistic details of either phenomenon are unclear, and raise interesting questions regarding the endocytosis and recycling of PrP^C, and its conversion to PrP^{Sc}.

At steady state, majority of PrP on the cell surface is represented by the proteolytically processed, C-terminal 18kDa fragment of PrP linked by the GPI anchor, and not by full-length PrP. This fragment arises from PrP by cleavage at residues 111/112 in an endocytic compartment during its normal recycling from the plasma membrane[8,33]. In addition, a C-terminal fragment of 20kDa is also expressed on the cell surface, linked by the GPI anchor. This fragment appears to be a metabolic product of transmembrane PrP (^{Ctm}PrP), derived from proteolytic cleavage of ^{Ctm}PrP at the ER membrane[40,41]. Full-length PrP forms have a half-life of about 6 hours, whereas the truncated 18kDa fragment has a much slower rate of turnover[28,29]. Lysosomes are the major compartment of PrP turnover. However, a recent report indicates that a small amount of normal PrP (10%) undergoes degradation by the proteasomal pathway[21]. The molecular details of this process are presently unclear.

The transmembrane ^{Ctm}PrP and ^{Ntm}PrP forms span the lipid bilayer through a conserved, hydrophobic segment of PrP including residues 111-134. Normally, these forms represent ~2–10% of the total PrP synthesized in a given cell. However, point mutations in or close to the hydrophobic domain or infection by exogenous PrP^{Sc} upregulate the synthesis of ^{Ctm}PrP to as much as 20–30% of the total PrP, implicating this transmembrane form as an important mediator of cytotoxicity in prion disorders[11,12]. The ^{Ctm}PrP is glycosylated, GPI-linked, and is believed to maintain an uncleaved N-terminal signal peptide in the cytosol[38,39]. Details about the transport of ^{Ctm}PrP from the ER are presently unclear. Conflicting reports indicate its retention in the ER, acquisition of endoglycosidase-resistant mature glycans indicating transport to the Golgi, or processing in the ER to generate a C-terminal fragment that is transported to the cell surface[11,12,38–41]. Information about any post-translational modifications of ^{Ntm}PrP is presently lacking. A diagrammatic representation of the processing and turnover of PrP^C is shown in Figure 12.2.

12.5. Mis-processing of mutant PrP

Mis-processing of normal and mutant PrP has the potential to initiate neurotoxicity by a variety of mechanisms. Aberrant processing may occur due to point mutations in PrP, or as a sporadic event. The percentage of PrP that is degraded by the proteasomes increases substantially in the presence of point mutations, which confer distinct post-translational anomalies depending on the location of a particular mutation in the PrP

Figure 12.2. Processing of PrP^C. PrP^C is synthesized in three distinct topological forms at the ER. PrP^C or ^{Sec}PrP is transported and linked to the plasma membrane by a GPI anchor. During constitutive rounds of recycling through lipid rafts or clathrin-coated pits, ^{Sec}PrP is truncated at residue 111/112, and the C-terminal 18kDa fragment is transported back to the plasma membrane. ^{Ctm}PrP is either retained in the ER (1), truncated at ~residue 90 and the C-terminal fragment is transported to the plasma membrane (2), or is transported to the Golgi apparatus (3), beyond which its fate is uncertain. The fate of ^{Ntm}PrP is unclear. PM: plasma membrane; E: endosome; L: lysosome; ER: endoplasmic reticulum; N: nucleus.

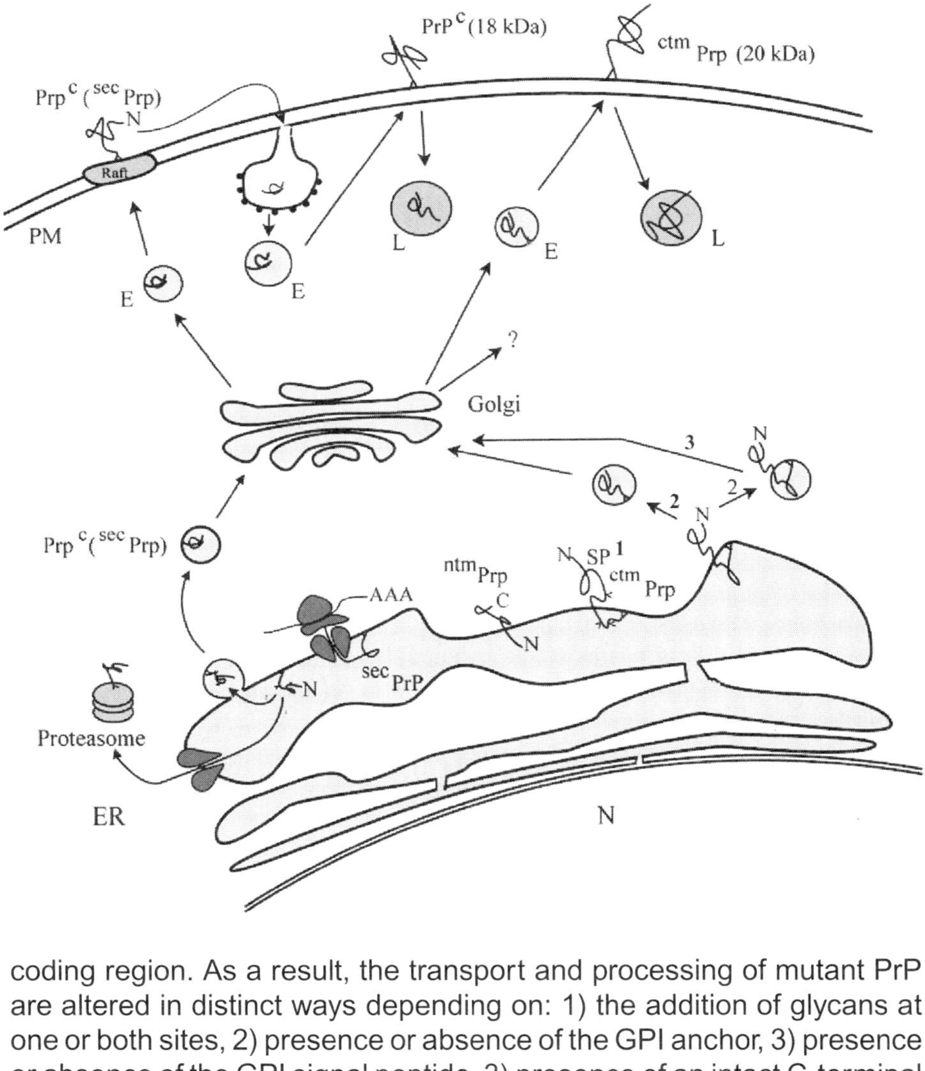

coding region. As a result, the transport and processing of mutant PrP are altered in distinct ways depending on: 1) the addition of glycans at one or both sites, 2) presence or absence of the GPI anchor, 3) presence or absence of the GPI signal peptide, 3) presence of an intact C-terminal domain which is eliminated in certain stop codon mutations, and 4) the overall folding state of the polypeptide. In each case, abnormal PrP forms are marked by the ER quality control[42]. However, unlike other misfolded proteins, PrP may mediate cytotoxicity by undergoing a conformational change to PrP^{Sc} in the ER, or by alternative, presently poorly understood pathways.

To avoid accumulation of such misfolded and potentially toxic protein aggregates, cells have developed sophisticated mechanisms to ensure correct folding and targeting of polypeptides, and expend a considerable amount of basal metabolic energy in eliminating improperly folded proteins[43]. Several checkpoints have been developed along the secretory path to ensure the highly selective nature of this process, beginning with the ER[19]. Examples of stringent quality control are evident in several cases of mutant PrP processing. Thus, abnormally glycosylated mutant PrP^{183A} associated with CJD is retained in the ER[44], whereas PrP^{200K} carrying abnormal glycans exits the ER but accumulates in the lysosomal compartment[45]. In other cases, aberrantly glycosylated mutant PrP forms have been reported to assume a PrP^{Sc}-like conformation[46,47]. Mutant PrP in GSS Q217R is diverted to the lysosomes, where it assumes a partially protease-resistant conformation[55]. The GPI anchor, in addition to its function as an anchor, plays a critical role in promoting PrP folding and transport[48,49]. Although a deficiency of the anchor by itself does not cause a significant disruption in the normal processing and transport of PrP[50], its absence due to persistence of the GPI signal peptide disrupts PrP transport. PrP with the uncleaved GPI signal peptide is retained in the ER in association with the chaperone BiP, and is targeted for proteasomal degradation. Although the efforts at refolding this form are ultimately futile, the association with BiP keeps this abnormal PrP in a relatively soluble state in the ER until its final delivery to the proteasomes through BiP-mediated retrotranslocation to the cytosol[51]. If a conservative serine to threonine mutation is introduced at codon 231 of PrP, the GPI-SP is cleaved without the addition of an anchor, and the association with BiP is lost (Figure 12.3A). As a result, a small but significant amount of $PrP^{217R/231T}$ is transported out of the ER (Figure 12.3B). However, $\sim$80% still remains aggregated in the ER[52]. Surprisingly, in the absence of the Q217R mutation, the S231T mutation causes complete retention of the mutant protein in the ER[52].

Thus, a combination of direct and indirect effects of the mutation in *PRNP* determines post-translational processing and transport, and its cellular site of accumulation. Depending on the milieu at that site, mutant PrP either undergoes aggregation and conversion to PrP^{Sc} with consequent cellular toxicity, or initiates intracellular signals that result in cell death. Thus, in CJD and FFI associated with 178N, retro-translocated mutant PrP accumulates in the cytosol of COS cells in aggresome-like structures[14]. In addition, interference with disulfide bond formation or glycan addition to normal PrP induces its transformation to a PrP^{Sc}-like form, suggesting that a reducing and deglycosylating environment in the cytosol may be conducive to PrP aggregation[13]. On the other hand,

Figure 12.3. Processing of PrP^{217R} and $PrP^{217R/231T}$. (A) Following a pulse of 2 hours with ^{35}S-mthionine, PrP^{217R} and $PrP^{217R/231T}$ cells were lysed and processed for immunoprecipitation with anti-PrP (lanes 1 and 2), or anti-KDEL antibodies (lanes 3 and 4), and analyzed by SDS-PAGE-fluorography. Lysates of PrP^{217R} immunoprecipitated with anti-PrP antibody 3F4 show the expected glycoforms of PrP, and an additional band migrating at 32 kDa that contains the uncleaved GPI-SP (lane 1) (55). The 32 kDa band also immunoprecipitates with anti-KDEL antibody, suggesting an association with BiP (lane 3) (51). Similar processing of $PrP^{217R/231T}$ with 3F4 demonstrates the disappearance of 32 kDa band (lane 2), and absence of co-immunoprecipitation of any PrP bands with anti-KDEL (lane 4). (B) Pulse chase analysis followed by immunoprecipitation of PrP^{Q217R} lysates with 3F4 shows rescue of the 32 kDa band in the presence of ALLN. A similar analysis of $PrP^{Q217R/S231T}$ lysates shows absence of the 32 kDa band, and rescue of the mono- and diglycosylated forms of PrP in the presence of ALLN.

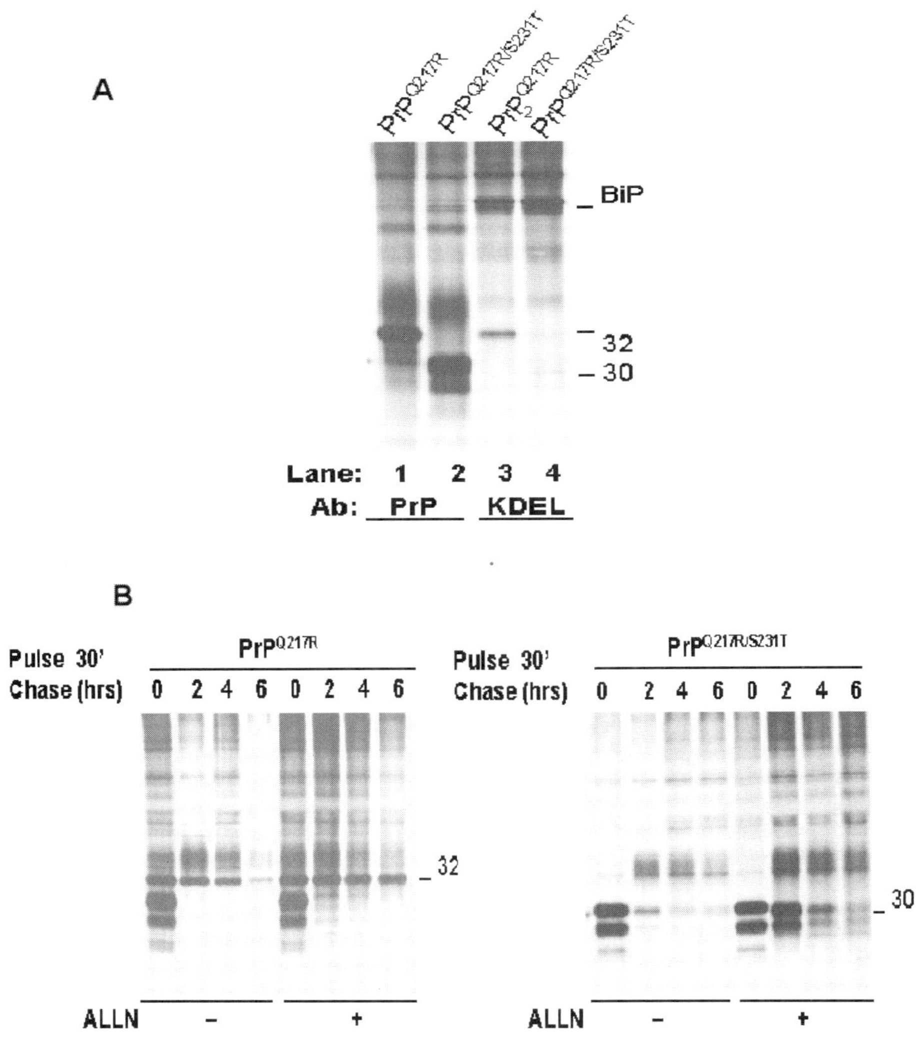

C-terminally truncated PrP forms associated with GSS carrying stop codon mutations at residues 145 or 160, although retro-translocated to the cytosol, do not aggregate. Instead, these forms accumulate in the nuclei of transfected cells[53,54]. Thus, a decline in the ER quality control with advancing age could initiate or enhance toxicity by PrP^{M} through several potential mechanisms: 1) in the cytosol, PrP could form aggregates or cause direct cytotoxicity, 2) in the lysosomes, the low pH could induce a change in conformation to PrP^{Sc} with consequent cellular toxicity, 3) in the nucleus, binding of PrP to nucleic acids could induce or suppress the transcription of a variety of genes that may initiate cell death by different mechanisms. The potential transport pathways of mutant PrP forms are depicted in Figure 12.4.

Figure 12.4. Processing of PrP^{M}. Inhibition of proteasomal function results in the accumulation of mutant PrP^{178D}, PrP^{203I}, PrP^{211Q}, and PrP^{212P} in aggresome-like structures in the cytosol, and of $PrP^{145stop}$ and $PrP^{160stop}$ in the nucleus (1). PrP32 with an intact GPI signal peptide is retained in the ER (2). PrP^{217R} and PrP^{200K} are targeted mainly to the lysosomes (3), whereas PrP^{102L} shows inefficient re-cycling from the PM (4). Several other mutant PrP forms acquire resistance to PI-PLC cleavage at the plasma membrane (5). PM: plasma membrane; E: endosome; L: lysosome; ER: endoplasmic reticulum; N: nucleus.

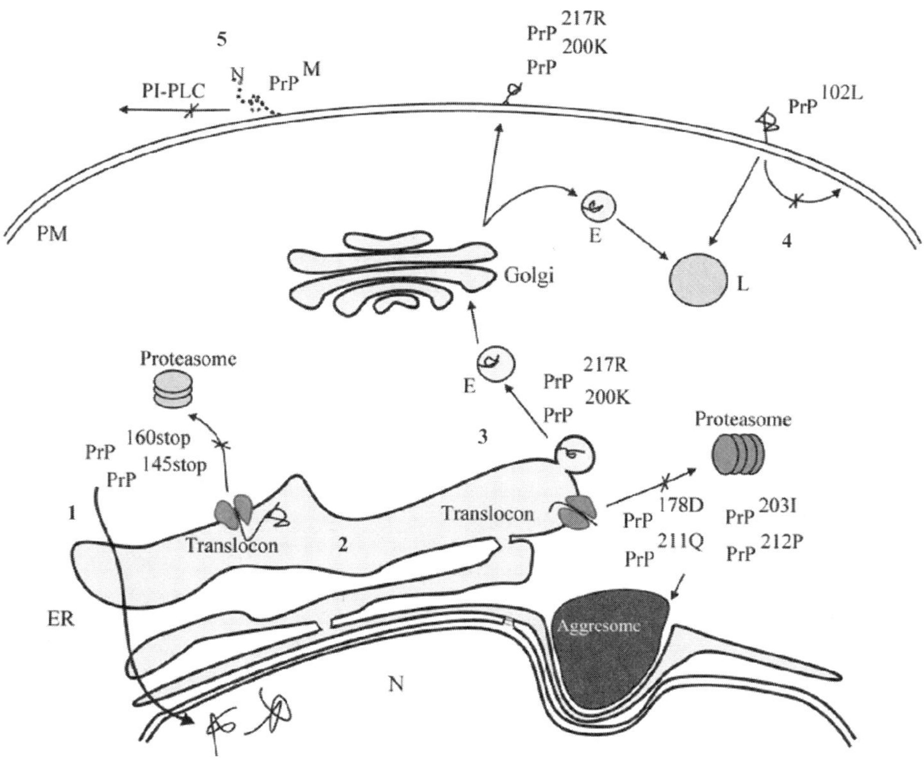

Figure 12.5. Upregulation of ^{Ctm}PrP by intracellular aggregates. Micro-aggregates of $PrP^{106-126}$ bind to the plasma membrane and initiate the aggregation of PrP^{C} (1). Aggregated proteins are endocytosed in large vesicular structures, and transported to lysosomes (2). Intracellular PrP aggregates induce the synthesis of ^{Ctm}PrP through *trans*-activating factors (3). Through as yet unknown mechanism, the N-terminal region of ^{Ctm}PrP is cleaved by cytosolic enzymes (4), and the C-terminal 20kDa fragment of ^{Ctm}PrP is transported to the cell surface along the secretory path (5). Truncated ^{Ctm}PrP is inserted in the outer leaflet of the plasma membrane through the GPI anchor (6). Aggregation of mutant PrP^{S231T} in the ER also results in upregulation of ^{Ctm}PrP (7). PM: plasma membrane; ER: endoplasmic reticulum; ^{Ctm}PrP: C-transmembrane PrP.

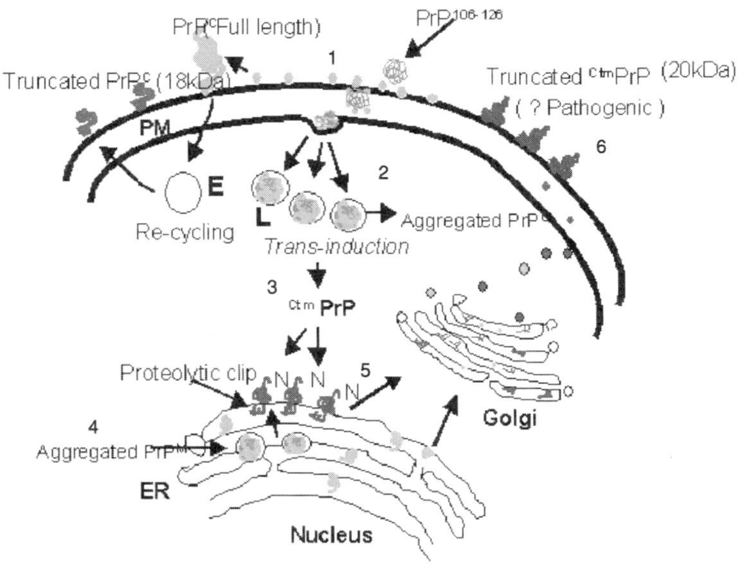

In addition, misfolded PrP has the potential to mediate the co-aggregation of additional, normal PrP molecules on the aggregated 'seed', thereby amplifying the toxic signal. One such example is the accumulation of mutant PrP^{S231T} in the ER, which mediates the co-aggregation of PrP^{C} and upregulation of ^{Ctm}PrP[52]. While accumulation of PrP^{Sc} would certainly be deleterious to cellular function, it is possible that cytotoxicity is instead mediated by ^{Ctm}PrP. A similar upregulation of ^{Ctm}PrP is observed due to intracellular accumulation of aggregated PrP106-126 fragment in neuroblastoma cells[40] (Figure 12.5). The mechanism of toxicity by ^{Ctm}PrP is unclear at present, and its elucidation poses a significant future challenge.

Several truncated PrP forms and fragments are generated during the processing of PrP^{C} and PrP^{M}. In addition to the 18 and 20kDa C-terminal fragments of PrP^{C} mentioned above, other fragments resulting from mis-processing of PrP^{M} have been reported[55]. Although PrP fragments have also been purified from prion disease affected human brains, the role of such fragments in neurotoxicity is unclear at present. In patients with GSS F198S (PrP^{198}) and Q217R (PrP^{217}), 7 and 11kDa peptides of mutant PrP spanning residues 81-150 and 58-150 have been recovered from amyloid plaques[56]. In GSS P102L (PrP^{102}), bands of 15–20 and 25–30kDa including epitopes in the PrP region 90-165 have been isolated[57]. In GSS Y145stop (PrP^{145}), a 7.5kDa fragment that immunoreacts with antibodies to the PrP region 90-147, and a C-terminal fragment of PrP from the normal allele is detected[17,58]. In several of these cases such as PrP^{198}, PrP^{217}, or PrP^{178}, PrP fragments derived only from the mutant allele are recovered from amyloid plaques, and are detergent-insoluble and PK-resistant. In CJD due to PrP^{200K} or insertional mutation, even the corresponding normal allele is detergent-insoluble (but PK-sensitive) (59-61). Similar observations have been made on transfected CHO cells expressing PrP^{200K} or neuroblastoma cells modeling GSS Q217R, although the PK-resistance in these case is much lower as compared to the brain samples[55,46,47]. In CJD associated with PrP^{210}, both the normal and mutant allele are converted to PrP^{Sc} (61). It appears that the conversion of normal allele by modified PrP^{M} is specific to the mutation. Whether this difference is due to inefficient dimerization of the converted PrP^{M} with the normal allele, or due to the generation of different truncated forms of PrP^{M} with a different potential for converting PrP^{C} is unclear at present. Thus, diverse processing pathways of PrP^{C} and PrP^{M} could initiate neurotoxicity secondarily by accumulating in abnormal cellular compartments, by mediating the upregulation of ${}^{Ctm}PrP$, by generating fragments that accumulate in the cytosol and the nucleus, and probably by other less defined pathways described below.

12.6. Alternative pathways of cell death in familial prion disorders

The ER quality control appears to play a paradoxical role in the pathogenesis of prion disorders; efficient function would abrogate disease by eliminating aberrant, misfolded forms, whereas sub-optimal function or mistargeting would accentuate the pathogenic process by sequestering these forms in abnormal cellular compartments where they may perturb cellular function by undefined, unconventional pathways. In long-lived, non-diving cells like the neurons, accumulation of even small amounts of PrP in an abnormal cellular compartment may allow its slow build up over time, with eventual toxicity.

In order to identify the determinants of transport within the PrP molecule, we generated a series of C-terminally truncated PrP fragments

Figure 12.6. Transport pathways of truncated PrP forms. A GFP tag has been inserted between residues 39 and 40 of PrP to study its transport and processing in living cells. To investigate the effect of glycosylation on PrP transport, the first 114, 180, 190 and 200 amino acids of PrP have been ligated to GFP (Clontech). The resulting chimeric PrP-GFP proteins contain no glycans, or one and two glycans respectively. None of the chimeras contain a membrane anchor. When expressed in neuroblastoma cells, $PrP^{114-GFP}$ is degraded by proteasomes, but accumulates in the nucleus in the presence of proteasomal inhibitors. $PrP^{180-GFP}$ is very stable, and accumulates in the nucleus regardless of the presence or absence of proteasomal inhibitors. $PrP^{190-GFP}$ accumulates in the ER, whereas $PrP^{200-GFP}$ is transported along the secretory path though a significant amount is retained in the ER.

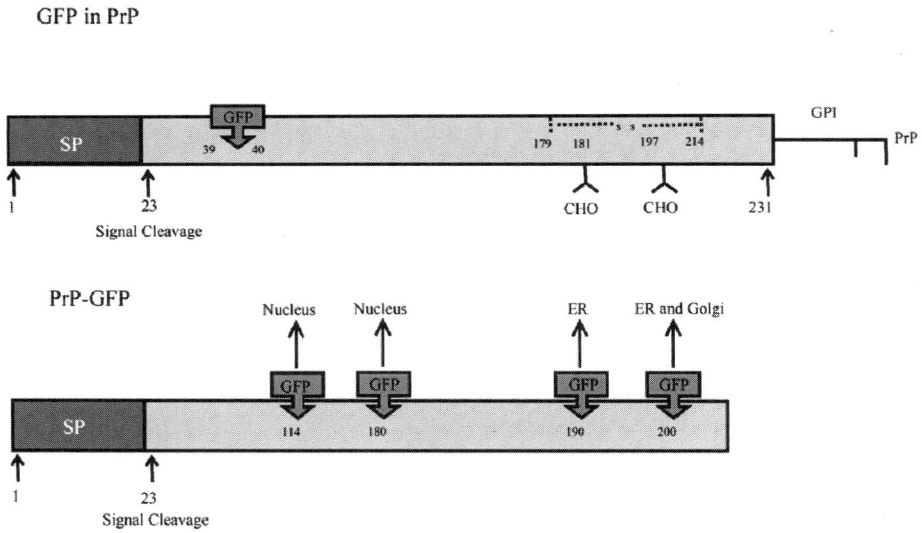

linked to the green fluorescent protein (GFP), and followed their transport in transfected neuroblastoma cells (Figure 12.6) (62). We noticed that N-terminal fragments lacking both N-glycans were all transported to the nucleus. Shorter fragments were less stable and required the presence of proteasomal inhibitors for significant accumulation, whereas $PrP^{180-GFP}$ was considerably more stable, and accumulated in significant amounts in the nucleus. N-terminal fragments with one glycan were retained in the nucleus, whereas the addition of a second glycan facilitated partial folding and transport out of the ER. Addition of the GPI anchor resulted in normal transport to the cell surface. Thus, in the absence of N-linked glycans and the GPI-anchor, the nuclear localization signals (NLS) become operative. The addition of glycans probably promotes interaction with ER chaperones, and retention of PrP in the ER till it achieves a mature, transport competent conformation (Figure 12.6).

The striking observation demonstrating the nuclear accumulation of $PrP^{145stop}$ and $PrP^{160stop}$ associated with GSS and CJD respectively demonstrates the versatile nature of PrP and the cryptic NLS signals that become active under certain conditions. Thus, the N-terminal fragments of PrP generated as a result of a variety of conditions may up regulate or down regulate a variety of genes, and perturb cellular function. In fact, PrP forms that are diverted to the cytosol for proteasomal degradation[21] would be an ideal substrate for nuclear translocation. The ^{Ctm}PrP is another PrP form that could potentially release an N-terminal fragment in the cytosol. In addition, cleavage of PrP within the octapeptide repeat region has been reported under physiological conditions[61] and on exposure to reactive oxygen species[62], releasing the N-terminal fragment of PrP. Several mutant PrP forms also accumulate in the cytosol, often as an aggregate, when proteasomal function is inhibited. These forms may be translocated to the nucleus as the aggregates disperse due to partial proteolysis. Thus, nuclear transport of PrP appears to be a common and deliberate diversion from the normal path, not an erroneous destination of a stray molecule. Why then is PrP not detected in the nucleus of neurons isolated from mutant PrP transgenic mice or mice infected with PrP^{Sc}? One possible explanation is that reasonable accumulation of PrP in the nucleus requires significant inhibition of both cytosolic and nuclear proteasomes. While such conditions may arise in non-dividing neurons of an aging brain, they are difficult to reproduce in a cell culture or a mouse model where the proteasome-mediated degradative pathways may differ from an aging human brain.

The processing and transport of transmembrane forms of PrP has been a contentious subject that has triggered considerable amount of interest, but little progress has been made to clarify its biogenesis in cells or its role in the neurodegenerative process. The main focus of recent studies has been on the C-transmembrane form (^{Ctm}PrP) that has been implicated in the pathogenesis of familial and infectious prion disorders[11,12]. Studies in cell models show convincingly that ^{Ctm}PrP is linked with a GPI anchor, and is degraded by the proteasomal pathway[38,39]. These observations are logical, considering that the N-terminal domain of this form faces the cytosol, and would be an easy target for cytosolic proteasomes. The same study contends that ^{Ctm}PrP contains an uncleaved N-terminal signal peptide in the cytosol, and is not transported from the ER. Contradictory studies argue for the transport of ^{Ctm}PrP beyond the cis-medial Golgi compartment based on the resistance of its glycans to endoglycosidase-H digestion[11,12]. Yet another study proposes proteolytic cleavage of ^{Ctm}PrP at the ER membrane, and transport of the C-terminal fragment to the cell surface[40,41]. The representation of ^{Ctm}PrP fragment on the cell surface has been reported to increase in response to intracellular aggregates of PrP and in association with the PrP^{102L} mutation, although by distinct mechanisms[40,41].

The above observations are provocative, and need further investigations to clarify the biogenesis and processing of the Ctm and Ntm PrP forms.

The persistent N-terminal signal peptide in $PrP^{145stop}$ predisposes the protein to aggregate more readily. Interestingly, a proportion of the full-length PrP^{C} in neuroblastoma cells also retains the N-terminal signal peptide (unpublished observations). Whether this omission affects the processing, transport, and turnover of this form in a significant way is unclear at present. Since co- or post-translational cleavage of the N-terminal signal peptide is an obligatory step in the correct folding and transport of membrane and secretory proteins, retention of the signal may result in atypical processing of the protein releasing N-terminal fragments with an increased potential to aggregate. Although most proteins with an uncleaved N-terminal signal peptide are retained in the ER, PrP1-140 is efficiently secreted in the medium[53]. Further investigations are needed to evaluate if a similar phenomenon occurs for full-length PrP, and the consequences of such a species on PrP function and cell viability.

12.7. Summary and future directions

Although one can argue that the concept of multiple potential pathways of neurotoxicity by the prion protein, at times promoted by the ER quality control, contests the simple and singular hypothesis of PrP^{Sc}-mediated cell death, multiplicity may allow us to better understand the apparently distinct but inter-related and complex pathways of PrP-mediated neurotoxicity. As further evidence becomes available, the above mechanisms may prove to be an exception, or may ultimately converge to a final common biochemical pathway, the identity of which poses a challenging and an important question. The identification of such a pathway will not only improve our understanding of the mechanisms of neurotoxicity in these disorders, but open the future for prion therapeutics, which at the present time is limited to indirect evaluation of drugs and chemical compounds on experimental models[65]. Thus, future studies on this mysterious and ever-challenging set of neurodegenerative disorders will further clarify the intricate and delicate interaction between prion protein metabolism, the cellular quality control, and neuronal cell death.

12.8. References

1. S. B. Prusiner, Molecular Biology and Genetics of Prion Diseases. In *Cold Spring Harbor Symposia on Quantitative Biology LXI*, Cold Spring Harbor Laboratory Press, (1996) pp. 473–493.

2. S. B. Prusiner, Prions. *Proc. Natl. Acad. Sci. U.S.A.* **95**, 13363–13383 (1998).

3. S. B. Prusiner, M. R. Scott, S. J. DeArmond, and F. E. Cohen, Prion protein biology. *Cell* **93**, 337–348 (1998).

4. A. L. Horwich, and J. S. Weissman, Deadly conformations-Protein misfolding in Prion disease. *Cell* **89**, 495–510 (1997).

5. S. B. Prusiner, Inherited prion diseases. *Proc Natl. Acad. Sci. U.S.A.* **91**, 4611–4614 (1994).

6. J. Collinge, Prion diseases of humans and animals: Their causes and molecular basis. *Annu. Rev. Neurosci.* **24**, 519–550 (2001).

7. R. Chiesa, and D. Harris, Prion diseases: What is the neurotoxic molecule? *Neurobiol.Dis.* **8**, 743–763 (2001).

8. D. Harris, Trafficking, turnover and membrane topology of PrP. *British Medical Bulletin* **66**, 71–85 (2003).

9. K. K. Hsiao, M. M. Scott, D. Foster, D. F. Groth, S. J. Groth, S. J. DeArmond, and S. B. Prusiner, Spontaneous neurodegeneration in transgenic mice with mutant prion protein. *Science* **250**, 1587–1590 (1990).

10. C. I. Lasmezas, J. P. Deslys, O. Robain, A. Jaegly, V. Beringue, J. M. Peyrin, J. G. Fournier, J. J. Hauw, J. Rossier, and D. Dormont, Transmission of the BSE agent to mice in the absence of detectable abnormal prion protein. *Science* **275**, 402–5 (1997).

11. R. S. Hegde, J. A. Mastrianni, M. R. Scott, K. A. DeFea, P. Tremblay, M. Torchia, S. J. DeArmond, S. B. Prusiner, and V. R. Lingappa, A transmembrane form of the prion protein in neurodegenerative disease. *Science* **279**, 827–834 (1998).

12. R. S. Hegde, P. Trembly, D. Groth, S. J. DeArmond, S. B. Prusiner, and V. R. Lingappa, Transmissible and genetic prion diseases share a common pathway of neurodegeneration. *Nature* **402**, 822–826 (1999).

13. J. Ma, and S. Lindquist, De novo generation of a PrP^{Sc}-like conformation in living cells. *Nature Cell Boil.* **1**, 358–361 (1999).

14. J. Ma, and S. Lindquist, Wild type PrP and a mutant associated with prion disease are subject to retrograde transport and proteasome degradation. *Proc. Natl. Acad. Sci. U.S.A.* **98**, 14955–60 (2001).

15. J. Ma, and S. Lindquist, Conversion of PrP to a self-perpetuating PrP^{Sc}-like conformation in the cytosol. *Science* **298**, 1785–1788 (2002).

16. J. Ma, R. Wollmann, and S. Lindquist, Neurotoxicity and neurodegeneration when PrP accumulates in the cytosol. *Science* **298**, 1781–1785 (2002).

17. T. Kitamoto, R. Iizuka, and J. Tateishi, An amber mutation of prion protein in Gerstmann-Sträussler-Scheinker syndrome with mutant PrP plaques. *Biochem. Biophys. Res. Comm.* **192**, 525–531 (1993).

18. B. Drisaldi, R. S. Stewart, C. Adles, L. R. Stewart, E. Quaglio, E. Biasini, L. Fioriti, R. Chiesa, and D. A. Harris, Mutant PrP is delayed in its exit from the endoplasmic reticulum, but neither wild-type nor mutant PrP undergoes retrotranslocation prior to proteasomal degradation. *J Biol Chem.* **278**, 21732–43 (2003).

19. C. Hammond, and A. Helenius, Quality control in the secretory pathway. *Curr. Opin. Cell Biol.* **7**, 523–529 (1995).

20. X. Roucou, Q. Guo, Y. Zhang, C. Goodyer, and A. LeBlanc, Cytosolic prion protein is not toxic and protects against Bax-mediated cell death in human primary neurons. *J. Biol. Chem.* **278**, 40877–40881 (2003).

21. Y. Yedidia, L. Horonchik, S. Tzaban, A. Yanai, and A. Taraboulos, Proteasomes and ubiquitin are involved in the turnover of the wild-type prion protein. *EMBO J.* **20**, 5383–5391 (2001).

22. R. S. Mishra, S. Bose, Y. Gu, R. Li, and N. Singh, Aggresome formation by mutant prion proteins: the unfolding role of proteasomes in familial prion disorders. *J. Alzheimers Dis.* **5**, 15–23. (2002).

23. G. Mallucci, A. Dickinson, J. Linehan, P. C. Klohn, S. Brandner, and J. Collinge, Depleting neuronal PrP in prion infection prevents disease and reverses spongiosis. *Science* **302**, 871–4 (2003).

24. M. Jeffrey, C. M. Goodsir, R. E. Race, and B. Chesebro, Scrapie-specific neuronal lesions are independent of neuronal PrP expression. *Ann Neurol.* **55**, 781–92 (2004).

25. A. J. Raeber, R. E. Race, S. Brandner, S. A. Priola, A. Sailer, R. A. Bessen, L. Mucke, J. Manson, A. Aguzzi, M. B. Oldstone, C. Weissmann, and B. Chesebro, Astrocyte-specific expression of hamster prion protein (PrP) renders PrP knockout mice susceptible to hamster scrapie. *EMBO J.* **16**, 6057–65 (1997).

26. M. Marella, and J. Chabry, Neurons and astrocytes respond to prion infection by inducing microglia recruitment. *J. Neurosci.* **24**, 620–7 (2004).

27. N. Stahl, M. A. Baldwin, R. Hecker, K. M. Pan, A. L. Burlingame, and S. B. Prusiner, Glycosylinositol phospholipid anchors of the scrapie and cellular prion proteins contain sialic acid. *Biochemistry* **31**, 5043–53 (1992).

28. B. Caughey, R. E. Race, D. Ernst, M. J. Buchmeier, and B. Chesebro, Prion protein biosynthesis in scrapie-infected and uninfected neuroblastoma cells. *J. Virol.* **63**, 175–181 (1989).

29. S. L. Shyng, M. T. Huber, and D. A. Harris, A prion protein cycles between the cell surface and an endocytic compartment in cultured neuroblastoma cells. *J. Biol. Chem.* **268**, 15922–8 (1993).

30. M. Muniz, and H. Riezman, Intracellular transport of GPI-anchored proteins. *EMBO J.* **19**, 10–15 (2000).

31. S. Udenfriend, and K. Kodukula, How glycosyl-phosphatidylinositol-anchored membrane proteins are made. *Ann. Rev. Biochem.* **64**, 563–591 (1995).

32. A. Taraboulos, M. Scott, A. Semenov, D. Avrahami, L. Laszlo, and S. B. Prusiner, Cholesterol depletion and modification of COOH-terminal targeting sequence of the prion protein inhibit formation of the scrapie isoform. *J. Cell Biol.* **129**, 121–32 (1995).

33. S. G. Chen, D. B. Teplow, P. Parchi, J. K. Teller, P. Gambetti, and L. Autilio-Gambetti. Truncated forms of the human prion protein in normal brain and in prion diseases. *J. Biol. Chem* **270**, 19173–19180 (1995).

34. S. L. Shyng, J. E. Heuser, and D. A. Harris, A glycolipid-anchored prion protein is endocytosed via clathrin-coated pits. *J. Cell. Biol.* **125**, 1239–50 (1994).

35. S. L. Shyng, K. L. Moulder, A. Lesko, and D. A. Harris. The N-terminal domain of a glycolipid-anchored prion protein is essential for its endocytosis via clathrin-coated pits. *J. Biol. Chem.* **270**, 14793–800 (1995).

36. P. C. Pauly, and D. A. Harris, Copper stimulates endocytosis of the prion protein. *J. Biol. Chem.* **273**, 33107–10 (1998).

37. S. L. Shyng, S. Lehmann, K. L. Moulder, and D. A. Harris. Sulfated glycans stimulate endocytosis of the cellular isoform of the prion protein, PrPC, in cultured cells. *J. Biol. Chem.* **270**, 30221–29 (1995).

38. R. S. Stewart, and D. A. Harris, Most pathogenic mutations do not alter the membrane topology of the prion protein. *J. Biol. Chem.* **276**, 2212–20 (2001).

39. R. S. Stewart, B. Drisaldi, and D. A. Harris, A transmembrane form of the prion protein contains an uncleaved signal peptide and is retained in the endoplasmic Reticulum. *Mol. Biol. Cell* **12**, 881–889 (2001).

40. Y. Gu, H. Fujioka, R. S. Mishra, R. Li, and N. Singh, Prion peptide 106-126 modulates the aggregation of cellular prion protein and induces the synthesis of potentially neurotoxic transmembrane PrP. *J. Biol. Chem.* **277**, 2275–2286 (2002).

41. R. S. Mishra, Y. Gu, S. Bose, S. Verghese, S. Kalepu, and N. Singh, Cell surface accumulation of a truncated transmembrane prion protein in Gerstmann-Straussler-Scheinker disease P102L. *J. Biol. Chem.* **277**, 24554–61 (2002).

42. L. Ellgaard, M. Molinari, A. Helenius, Setting the standards: quality control in the secretory pathway. *Science* **286**, 1882–8 (1999).

43. J. L. Brodsky, and A. A. McCracken, ER-associated and proteasome-mediated protein degradation: how two topologically restricted events came together. *TICB* **7**, 151–156 (1997).

44. S. Capellari, S. I. A. Zaidi, A. C. Long, E. E. Kwon, and R. B. Petersen, The Thr183A mutation, not the loss of the first glycosylation site, alters the physical properties of the prion protein. *J. Alzheimers Dis* **2**, 27–35 (2000).

45. S. Capellari, P. Parchi, C. M. Russo, J. Sanford, M. S. Sy, P. Gambetti, and R. B. Petersen, Effect of the E200K mutation on prion protein metabolism. *Am. J. Pathol* **157**, 613–22 (2000).

46. S. Lehmann, and D. A. Harris, Two mutant prion proteins expressed in cultured cells acquire biochemical properties reminiscent of the scrapie isoform. *Proc. Natl. Acad. Sci. USA* **93**, 5610–5614 (1996).

47. S. Lehmann, and D. A. Harris, Mutant and infectious prion proteins display common biochemical properties in cultured cells. *J. Biol. Chem.* **271**, 1633–1637 (1996).

48. M. D. Delahunty, F. J. Stafford, L. C. Yuan, D. Shaz, J. S. Bonifacino, Uncleaved Signals for Glycosylphosphatidylinositol Anchoring Cause Retention of Precursor Proteins in the Endoplasmic Reticulum. *J. Biol. Chem.* **268**, 12017–12027 (1993).

49. M. C. Field, P. Moran, W. Li, G. Keller, I. W. Caras, Retention and Degradation of Proteins Containing an Uncleaved Glycosylphosphatidylinositol Signal. *J. Biol. Chem.* **269**, 10830–10837 (1994).

50. D. A. Kocisko, J. H. Come, S. A. Priola, B. Chesebro, G. J. Raymond, P. T. Lansbury, and B. Caughey, Cell-free formation of protease-resistant prion protein. *Nature* **370**, 471–4 (1994).

51. T. Jin, Y. Gu, G. Zanusso, M. Sy, A. Kumar, M. Cohen, P. Gambetti, and N. Singh, The chaperone protein BiP binds to a mutant prion protein and mediates its degradation by the proteasome. *J Biol Chem.* **275**, 38699–704 (2000).

52. Gu, Y., Verghese, S., Mishra, R.S., Xu, X., Shi, Y., and Singh, N. (2003) Mutant prion protein (PrP) mediated aggregation of normal PrP in the endoplasmic reticulum: Implications for prion propagation and neurotoxicity. *J. Neurochem.* **84**, 10–22.

53. G. Zanusso, R. B. Petersen, T. Jin, Y. Jing, R. Kanoush, S. Ferrari, P. Gambetti, and N. Singh, Proteasomal degradation and N-terminal protease resistance of the codon 145 mutant prion protein. *J. Biol. Chem.* **274**, 23396–404 (1999).

54. H. Lorenz, O. Windl, and H. A. Kretzschmar, Cellular phenotyping of secretory and nuclear prion proteins associated with inherited prion diseases. *J. Biol. Chem.* **277**, 8508–16 (2002).

55. N. Singh, G. Zanusso, S. G. Chen, H. Fujioka, S. Richardson, P. Gambetti, and R. B. Petersen, Prion protein aggregation reverted by low temperature in transfected cells carrying a prion protein gene mutation. *J. Biol. Chem.* **272**, 28461–70 (1997).

56. F. Tagliavini, F. Prelli, J. Ghiso, O. Bugiani, D. Serban, S. B. Prusiner, M. R. Farlow, B. Ghetti, and B. Frangione, Amyloid protein of Gertsmann-Straussler-Schinker disease (Indiana Kindred) is an 11kd fragment of prion protein with an N-terminal glycine at codon 58. *EMBO. J.* **10**, 513–519 (1991).

57. P. Piccardo, B. Ghetti, D. W. Dickson, H. V. Vinters, G. Giaccone, O. Bugiani, F. Tagliavini, K. Young, S. R. Dlouhy, and C. Seiler, Gertsmann-Straussler-Scheinker disease (PRNP P102L): amyloid deposits are best recognised by antibodies directed to epitopes in PrP region 90-165. *J. Neuropathol. Exp. Neurol.* **54**, 790–801 (1995).

58. B. Ghetti, P. Piccardo, M. G. Spillantini, Y. Ichimiya, M. Porro, F. Perini, T. Kitamoto, J. Tateishi, C. Seiler, B. Frangione, O. Bugiani, G. Giaccone, F. Prelli, M. Goedert, S. R. Dlouhy, and F. Tagliavini, Vascular variant of prion protein cerebral amyloidosis with Tau-positive neurofibrillary tangles: The phenotype of the stop codon 145 mutation in PRNP. *Proc. Natl. Acad. Sci. USA.* **93**, 744–748 (1996).

59. R. Gabizon, G. Telling, Z. Meiner, M. Halimi, I. Kahana, and S. B. Prusiner, Insoluble wild type and protease-resistant mutant prion protein in brains of patients with inherited prion disease. *Nat. Med.* **2**, 59–64 (1996).

60. S. G. Chen, P. Parchi, P. Brown, S. Capellari, W. Zou, E. J. Cochran, C. L. Vnencak-Jones, J. Julien, C. Vital, J. Mikol, E. Lugaresi, L. Autilio-Gambetti, and P. Gambetti, Allelic origin of the abnormal prion protein isoform in familial prion diseases. *Nat. Med.* **3**, 1009–1015 (1997).

61. M. C. Silvestrini, F. Cardone, B. Maras, P. Pucci, D. Barra, M. Brunori, and M. Pocchiari, Identification of the prion protein allotypes which accumulate in the brain of sporadic and familial Creutzfeldt-Jakob disease patients. *Nat. Med.* **5**, 521–5 (1997).

62. Y. Gu, J. Hinnerwisch, R. Fredricks, S. Kalepu, R. S. Mishra, and N. Singh, Identification of cryptic nuclear localization signals in the prion protein. *Neurobiol. Dis.* **12**, 133–149 (2002).

63. A. Jimenez-Huete, P. M. Lievens, R. Vidal, P. Piccardo, B. Ghetti, F. Tagliavini, B. Frangione, F. Prelli, Endogenous proteolytic cleavage of normal and disease-associated isoforms of the human prion protein in neural and non-neural tissues. *Am. J Pathol.***153**, 1561–72 (1998).

64. H. E. McMahon, A. Mange, N. Nishida, C. Creminon, D. Casanova, and S. Lehmann, Cleavage of the amino-terminus of the prion protein by reactive oxygen species. *J. Biol. Chem.* **276**, 2286–2291 (2000).

65. Y. Gu, and N. Singh, Doxycycline and protein folding agents rescue the abnormal phenotype of CJD H187R in a cell model. *Brain Res Mol Brain Res.* **123**, 37–44. (2004).

Chapter 13

SIGNALING PATHWAYS CONTROLING PRION NEUROTOXICITY: ROLE OF ENDOPLASMIC RETICULUM STRESS-MEDIATED APOPTOSIS

Rodrigo Morales[1], Claudio Hetz[2] and Claudio Soto[1]

[1]*Department of Neurology, University of Texas Medical Branch, Galveston, TX USA*
[2]*Instituto de Ciencias Biomedicas, Universidad de Chile, Santiago, Chile*

13.1. Neurodegenerative diseases, protein misfolding and apoptosis

Cells can die by diverse mechanisms depending upon the stimulus triggering the death process. Among these mechanisms it is possible to include necrosis, apoptosis, autophagia, mitotic catastrophe, and others[1,2]. Apoptosis has been implicated in diseases affecting the nervous system such as neurodegenerative disorders and ischemia (review in 3). Programmed cell death[4] and its morphological manifestation "apoptosis"[5] is a conserved pathway that in its basic features appears to be operative in all metazoans. During embryonic developmental apoptosis is essential for successful organogenesis, and participates in the control of cellular populations (review in 6). Apoptosis also operates in adult organisms to maintain normal cellular homeostasis. This is an especially critical process in long-lived mammals, in which cells integrate multiple physiological as well as pathological signals.

Neurodegenerative diseases are some of the most debilitating disorders, affecting abstract thinking, skilled movements, emotional feelings, cognition, memory, and other abilities that distinguish human beings from other mammals. The analysis of the neuropathological characteristics of several neurodegenerative diseases has revealed common features underlying the mechanism of the disease initiation and progression. Compelling evidence accumulated in the last few years suggest that the misfolding, aggregation and cerebral deposition of proteins play a central role on the pathogenesis of neurodegenerative diseases[7]. As we will discuss later on, the protein structural changes are associated with neuronal damage and the appearance of the disease-associated clinical symptoms. The most common neurodegenerative disorder is Alzheimer's disease (AD) where the extracellular accumulation of misfolded amyloid β peptide is one of the principal features that lead to apoptotic neuronal loss[7]. Other pathologies, like Parkinson Disease (PD), Huntington disease (HD) and Amyotrophic Lateral Sclerosis (ALS) are related to the accumulation of other misfolded proteins such as α-synuclein in PD, huntingtin in HD and superoxide dismutase (SOD)-1 in ALS[7]. Transmissible Spongiform Encephalopathies (TSEs) also known as prion disorders are the rarest, but perhaps the most famous neurodegenerative diseases. In TSEs the misfolded protein is not only associated to the disease, but is also the major or even the only component of the infectious agent[8].

13.2. TSE pathogenesis and misfolding of the prion protein

TSEs are a group of clinically diverse, but mechanistically similar neurological diseases affecting humans and animals. The group includes Creutzfeldt-Jakob (CJD), fatal familial insomnia (FFI), Gerstmann-Straussler-Scheinker (GSS) and kuru in humans as well as bovine spongiform encephalopathy (BSE), scrapie and chronic wasting disease (CWD) in animals[9]. The distinguishing pathological features of TSEs are the spongiform degeneration of the brain, accompanied by extensive neuronal loss, astrogliosis, and cerebral accumulation of a misfolded and protease-resistant form of the normal prion protein (PrP^{C}), termed PrP^{SC} (ref 10). No sequence or post-translational differences have been detected between the normal host cell surface PrP^{C} and the misfolded PrP^{SC}(ref 10). Based on structural studies, it has been proposed that during the pathogenesis of TSEs PrP^{C} undergoes a conformational transition from α helical to β sheet structure, resulting in the formation of PrP^{SC} (ref 11). This phenomenon leads to an "autocatalytic process", which replicates the abnormal protein structure and propagates new pathogenic prions in the brain of affected individuals.

TSEs can be initiated by different causes, including hereditary, sporadic or infectious[9]. It has been described that at least 10% of the CJD cases and all cases of GSS and FFI have an inherited origin. In these cases mutations on the PrP gene stabilize the anomalous conformation of PrP, triggering the pathology[12]. Most of the CJD cases have been described as sporadic, where the cause that triggers the disease progression remains without a satisfactory explanation[9]. When brain homogenate of sick animals is injected into healthy animals the disease is transmitted in a dose-dependent manner[10]. Misfolded prion protein accumulates gradually in the brain at expense of the host PrP^C. In humans it has been described that treating patients with infected surgical supplies, cornea transplant or human growth hormone administration extracted from pineal gland of infected individuals, is possible to induce the pathology. An infectious origin for human prion diseases was also observed in the transmission of kuru by cannibalism in tribes from New Guinea and the recent transmission of BSE to human beings producing a new disease, named variant CJD (vCJD)[13,14].

A great deal of effort has been made to understand the remarkable biology of the prion replication process and the nature of the infectious agent[8]. The high β-sheet content of PrP^{SC} confers this protein distinct physicochemical properties from PrP^C, which are reflected in its insolubility in non-denaturating detergents, its partial resistance to proteolysis, and its ability to form fibrillar structures *in vitro*[10,12]. PrP^C is essential for the development of prion diseases since ablation of the gene which codifies PrP^C renders the mice resistant to prion infection[15]. Under certain conditions, the conversion of PrP^C into PrP^{SC} can be achieved in a cell-free replication assay, supporting the "protein-only hypothesis"[16,17].

Two alternative models have been proposed to explain the process of aggregation in TSE[18]. One of them is the "Nucleation-Polimerization" hypothesis, where PrP^{SC} is a multimeric particle, acting as a nucleus to induce and stabilize the misfolding of the monomeric protein by incorporating it into the oligomer. Another model termed "Template Assisted Conversion" postulates that PrP^{SC} serves as a template to direct the misfolding of a partially unfolded intermediate produced by interaction of PrPC with a yet unknown pathological chaperone named protein X.

13.3. Neuronal loss in TSE is mediated by apoptosis

A number of studies indicate that neuronal dysfunction in humans and animals affected with TSEs occurs through apoptosis[19–32]. The detection of cells with DNA degradation and the morphological characterization of the brain areas affected by prion infection have shown that neuronal apoptosis is observed mostly in terminally ill animals, and in those brain areas that show vacuolation[19]. Neuronal loss and apoptosis have also been described in experimental models for CJD in mice[21,33].

Interestingly, several groups have shown that in post-mortem samples of humans affected with FFI[22] and CJD[23–25], apoptotic cell death of neurons does not correlate well with the deposition of PrP (reviewed in 29). These findings suggest that the mechanism relating PrP^{SC} generation and neuronal loss is a complex phenomenon. It was proposed that the dissociation between neuronal damage and the amount of prion deposition only reflects variations in the selective neuronal vulnerability to PrP^{SC} toxicity[29].

Different strategies have been developed to understand the relationship between PrP misfolding and neuronal dysfunction. A transgenic mice model of familial prion diseases has been developed by expressing the PrP homologue of a nine-octapeptide insertional mutation described in human patients affected with prion diseases[34]. This insertional mutation is genetically linked with a disease characterized by dementia and ataxia, and by the presence of PrP-containing amyloid plaques in the cerebellum and basal ganglia[35–37]. These transgenic mice showed accumulation of protease-resistant PrP^{SC} and apoptotic cell death of the cerebellar granule cells, in addition to progressive ataxia[38]. On the other hand, transgenic mice expressing PrP fragments die spontaneously by ataxia, showing an accumulation of protease resistant PrP within neuronal dendrites and cell bodies, apparently causing apoptosis[39,40]. Finally, in some inherited cases of prion diseases, the predominant form of mutant PrP detectable in the brain is ^{CTM}PrP, a transmembrane form of the prion protein. Transgenic mice expressing this particular PrP mutation also developed neurodegeneration[41]. In summary, neuronal cell death is a common feature observed in different forms of the disease, which can be reproduced when mutant PrP genes are expressed in transgenic mice or in mice experimentally infected with scrapie prions.

The general idea supported by many studies is that the structural conversion affecting PrP^{C} is a fundamental step in the neurodegeneration process, and at least two major hypotheses have been proposed to explain this causative relation: PrP^{SC} formation might be associated with the gain of a neurotoxic activity of this protein, or alternatively, the conversion of PrP^{C} into PrP^{SC} could cause neurodegeneration through the loss of the normal biological function of PrP^{C} (Figure13.1). We have recently reviewed the literature supporting each of these two hypotheses (review in 42, 43).

13.4. Pathways controlling neuronal apoptosis

Apoptosis can be induced by the ligation of plasma membrane death receptors, which constitute the "extrinsic" pathway, or by the perturbation of intracellular homeostasis, known as the "intrinsic" pathway[44–46].

Figure 13.1. Putative cellular pathways for neurodegeneration in TSEs. Conformational changes of PrPC induced by prion infections, mutations or unknown factors, lead to the production and accumulation of the misfolded PrPSC protein. The pathological protein may interact with different neuronal cell-surface receptors and with microglia and astrocytes, triggering signal transduction cascades which result in cellular stress and neuronal dysfunction. In addition, the conformational transition could lead to the lost of a beneficial activity of the natively folded protein.

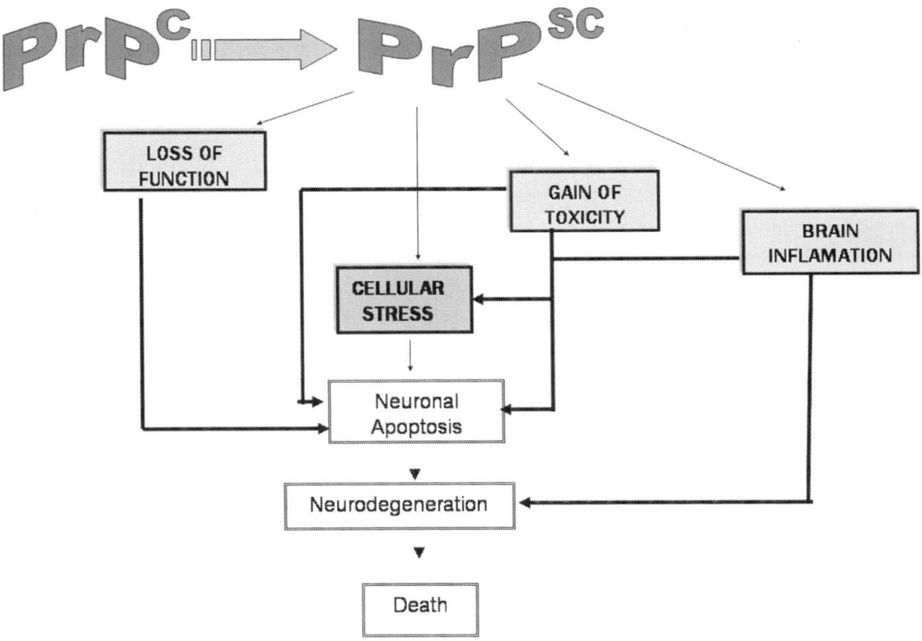

The viability of a cell strictly depends on the functional and structural integration of a number of subcellular organelles like the nucleus, the mitochondria, the lysosomes and the endoplasmic reticulum[47]. Each organelle can sense stressful cellular conditions and initiate cellular responses either to adapt or to activate specific cell death signalling pathways, if a critical threshold of damage has been reached (reviewed in 48). In spite of the absence of large organelle ultra-structural changes in dying cells, it is now well established that most organelles manifest subtle biochemical alterations, such as permeabilization of membranes and changes on the concentration of messenger molecules, including calcium. In general the process of cell death has two phases: A terminal stage that is mediated by executer common molecules in which the different apoptotic signals converge; and an activator phase mediated by initiator molecules that are up-stream of executor molecules, and are associated with particular cell death stimuli. These complex processes involve cross talk between many signaling pathways and include different molecular components that regulate the cell death response. In general, the apoptotic process can be subdivided in four different steps:

1. Initiation process: Under certain conditions, apoptosis can be triggered by cell death receptors binding to their corresponding ligands. This signal transduction event is known as the extrinsic pathway and is generally related with the triggering of Fas, TNF receptor or TRAIL[49]. Other extracellular signaling that disrupts ion-channel activity at the cell surface can play a central role in neuronal loss under pathological conditions. In addition, each organelle can sense and initiate cell death responses when the affected element can not restore the homeostasis under stress conditions. In general those events include mitochondrial stress, endoplasmic reticulum stress, DNA damage and others.
2. Amplification process: Independent on the initiator mechanism of cell death, common signaling pathways regulates the apoptotic process that leads to the execution of cell death. This involves phosphorylation steps, multiple proteolytic processes, calcium mobilization, activation of specific transcriptional factors and many others molecular events[48,50,51].
3. The cell death process which is an irreversible step and involves the self-degradation of macromolecules leading to DNA fragmentation, chromatin condensation and proteolytic degradation of many structural proteins and enzymes[1].
4. Finally a phagocytic event is involved in the clearance of the apoptotic bodies. Several cell types participates in this process, including dendritic cells, macrophages, and brain microglia, which recognize membrane receptors in the apoptotic cells and initiate their degradation.

The induction of apoptosis depends on the activation of cysteine proteases of the caspase family (review in 52). Caspases are produced as inactive zymogens, and after activation they cleave their substrates at aspartic acid residues contained within a tetrapeptide recognition motif. Currently, more than 14 caspases have been cloned and partially characterized in mammals, some of which are not involved in apoptosis but rather mediate inflammation and cytokine processing (such as caspase-1 and caspase-11)[53]. Activation of initiator caspases (such as pro-caspase-8, pro-caspase-9 and pro-caspase-12) leads to the proteolytic activation of downstream executor caspases (such as caspase-3), which cleave multiple substrates culminating with the morphological manifestation of apoptosis (review in 53).

The best studied apoptosis signaling pathways involve the activation of death receptors by their ligands and the mitochondrial stress leading to the release of apoptogenic factors from this organelle. In neuronal death, most of the studies have been focused on the mitochondrial pathway. The central event in the regulation of apoptosis by this pathway is the release of mitochondrial proteins, such as cytochrome *c*, which triggers the activation of caspases through the formation of the "apoptosome" complex[54]. This protein complex is formed when cytosolic cytochrome c binds to the adaptor protein apaf-1, recruiting the inactive form of caspase-9 and triggering its self-proteolytic activation[55]. Cytocrome c release depends upon the opening of a mitochondrial pore termed "permeability transition pore" or PTP[56]. The opening of the PTP is highly regulated by the Bcl-2 family proteins, representing a critical intracellular checkpoint upstream of the caspase cascade (review in 57). The Bcl-2 family is comprised of pro- and anti-apoptotic members. Anti-apoptotic Bcl-2 family members include Bcl-2 and Bcl-X_L. Pro-apoptotic Bcl-2 members include for example Bax and Bak. Apoptotic signals trigger the conformational activation of Bax and Bak, inducing their oligomerization in the mitochondria and resulting in the opening of the PTP. Anti-apoptotic Bcl-2 proteins inhibit the activation of Bax and Bak. In general, the mitochondrial apoptotic pathway has been shown to be activated in neurons by growth factors deprivation, oxidative stress, DNA damage or by changes in the expression levels of Bcl-2 family proteins[59].

Although apoptosis probably participates in the development of all cell lineages, aberrations in genes encoding pro- or anti-apoptotic proteins have been implicated in the initiation of a variety of human diseases, such as cancer through a process of "cell inmortalization", while accelerated cell death is evident in immunodeficiency (review in 58, 59). In this sense, gain- and loss-of-function mice models for genes encoding proteins of the core apoptotic pathways indicate that the violation of cellular death homeostasis is a primary pathogenic event that results in disease (review in 60).

The analysis of signaling pathways involved in neuronal apoptosis in neurodegenerative diseases associated with the misfolding and accumulation of protein aggregates in the brain, has provided data for a novel apoptosis pathway implicating endoplasmic reticulum (ER) stress and the unfolding protein response process.

13.5. Caspase-12 and endoplasmic reticulum stress

Cellular alterations leading to a general accumulation of unfolded proteins in the ER, or modifying the homeostasis of calcium in this subcellular compartment results in a condition denominated ER stress. Some of these triggering events include alterations on protein maturation, glucose depletion, expresion of mutant proteins, heat shock, and oxidative stress among others[61–65]. Experimentally the pharmacological targeting of the ER homeostasis with molecules such as brefeldin A (which block the trafficking between ER-Golgi), thapsigargin (an inhibitor of the ER calcium pump SERCA) and tunicamycin (an inhibitor of N-glycosylation) are able to induce ER stress and apoptosis.

It has been also described that under certain conditions, ER stress can be triggered by the accumulation of misfolded proteins outside of the reticular compartment[66]. The protein biosynthesis and degradation processes are tightly associated, determining that normal proteins synthesized in the ER and abnormally folded ones are recognized by the ER quality control[67,68]. During this process the abnormally folded proteins are subjected to ER-associated degradation (ERAD)-pathway, which includes the recognition of the protein by specific chaperones, deglycosylation, ubiquitination and translocation to the cytoplasm for degradation by the proteasome. The overload of the proteasome in the cytosol triggers a delay in the folding and degradation process inducing ER stress. Another known mechanism in which misfolded protein accumulation can involve reticular responses is the ER Overload Response (EOR), produced when the ER lumen is overloaded with proteins that are not transported to the Golgi apparatus, probably by a general saturation of the biosynthesis pathways[68].

The ER-stress response has mainly two phases: One anti-apoptotic phase mediated by a general decrease in protein synthesis and increases in the expression of different chaperones and folding enzymes of the glucose regulated family proteins (GRPs)[61,68]. This response is mediated by a signalling cascade known as the "unfolding protein response" or UPR. This process attenuates the toxicity of misfolded proteins in the ER by refolding the proteins or by promoting their degradation through the proteasome pathway[69]. The best characterized chaperon proteins involved in this process are the calcium-binding chaperones Grp78/Bip (involved in protein refolding processes) and Grp94 (involved in ER lumen calcium homeostasis). Experimentally, the overexpression of Grp78/Bip and Grp94 has been shown to protect cells against ischemia, and the pharmacological induction of ER stress. Conversely, the inhibition of the expression of these chaperones renders cells more susceptible to ER stress[65,70–76]. If the damage is too strong and the homeostasis cannot be restored, a second phase is initiated, which is mediated by several pro-apoptotic components triggering cellular death. This includes the activation of an ER-resident caspase, the induction of the pro-apoptotic transcriptional factor GADD153/CHOP, and the activation of several regulatory kinases (review in 47).

The induction of apoptosis by ER stress is dependent upon the activation of an ER-resident caspase, termed caspase-12[77]. Caspase-12 is ubiquitously expressed and is synthesized as an inactive pro-enzyme. Upon proteolytical processing, the active form of caspase-12 is generated consisting in a regulatory pro-domain and two catalytic (p20 and p10) subunits. The mechanism of caspases-12 activation is unclear but, unlike other caspases, caspase-12 is remarkably specific to insults that elicit ER stress[77]. Accordingly, caspase-12-null cells are resistant to apoptosis induced by ER stress, but not to other apoptotic stimuli related, for example, with normal cellular death during development or with the maintenance of the tissues homeostasis[77]. These findings may explain why caspase-12 knock out animals are viable.

It has been suggested that caspase-12 activation is linked to the ER stress pathway through the ER transmembrane kinase Ire1α and the adapter protein TRAF2.[68,78]. Ire1α is normally maintained in an inactive state through an association between its N-terminal lumenal domain and the chaperone Grp78/Bip. Under conditions of ER stress, Grp78/Bip dissociates to bind unfolded proteins and Ire1α undergoes homo-oligomerization, stimulating a trans-autophosphorylation within its serine/threonine kinase domains (review in 47). Upon activation, the cytosolic tail of Ire1α can recruit the adaptor protein TRAF-2[79]. TRAF-2 interacts with caspase-12 and induces its oligomerization and cleavage[78]. Moreover, Ire1α induces apoptosis when it is overexpressed, presumably due to caspase-12 activation[80]. Finally, overexpression of full-length caspase-12 induces its oligomerization and self-cleavage between the p20 and p10 subunits at position D318[81,82].

Alternatively, a second model for caspase-12 activation has been proposed. In mouse glial cells undergoing ER stress, caspase-12 was cleaved by calpain. *In vitro*, m-calpain cleaved caspase-12 at T132 and K158, which released the pro-domain from the catalytic subunits, increasing the enzymatic activity of this protease[83]. Thus, in this second model of activation, extensive intracellular calcium increases may trigger m-calpain activation with the subsequent proteolitic activation of caspase-12 at the ER membrane. Following ER stress, caspase-12, as a member of initiator caspasas group, may directly process downstream caspases in the cytosol or target, as yet unidentified substrates that influence the progression of apoptosis. Two groups have recently reported that caspase-12 directly cleaves caspase-9, leading to caspase-9-dependent activation of caspase-3[70,84]. This phenomenon does not require the expression of the adaptor protein apaf-1[70]. In addition, a direct activation of caspase-3 by caspase-12 through a protein complex formation has been described in other experimental systems[85,86,87].

13.6. ER stress involvement in TSEs neuronal apoptosis

In studies using cell lines which express a mutant form of PrP genetically linked with GSS (mutation PrPY145stop), retention of this protein in the ER and Golgi compartments was described[88]. The mutant PrPQ217R, which is also linked with GSS, was shown to be accumulated and aggregated in the ER[89,90]. This mutant form, as wild type PrP, is also subjected to ERAD, ubiquitinated and degraded by the proteasome system[90]. Moreover, it was described that after proteasome inhibition with different compounds, wild type misfolded PrP is accumulated in the cell leading to cytotoxicity[91–93]. This abnormal PrP^C molecule exhibits some of the biochemical properties of PrP^{SC}, such as insolubility in non-ionic detergents, increased aggregation and partial resistance to protease degradation[91,94,95]. Given these results, it was postulated that in sporadic forms of prion diseases, the conditions in which the proteasome ability to degrade PrP is compromised, like cellular stress and events associated with aging, the accumulation of unfolded PrP^C derived from the ER might promote neuronal degeneration.

Recently, it has been described that other PrP mutants involved in hereditary forms of the disease (14-octarepeats PrP and PrPD177N) are significantly delayed in their transit along the early part of the secretory pathway through the ER-Golgi[96]. This event opens the possibility that alteration of PrP maturation upon genetic mutations may be the triggering step in the toxicity of some abnormal PrP molecules. However, the mechanism involved in the neurotoxic effects of mutant PrP is not known. In summary, the overall of these data suggest that in sporadic and hereditary forms of TSEs, the ER is a key subcellular compartment where pathological PrP forms are generated and exert their lethal effects.

It has been shown that brain derived PrP^{SC} is cytotoxic *in vitro* in several experimental systems, however the toxicity mechanism was not investigated[97–100]. Recently, we have shown that treatment of mouse neuroblastoma cell cultures with nanoMolar concentrations of brain-derived PrP^{SC} purified from scrapie infected mice is able to induce ER stress, reflected in an increase expression of several ER chaperones and release of ER calcium. In addition, in dying cells, the induction of apoptosis by PrP^{SC} was associated with caspase-12 activation, confirming the participation of ER stress as a cytotoxic mechanism. Finally, the activation of the caspase-12 downstream target, caspase-3, was observed in these cells.

Experimentally, the targeting of anti-apoptotic protein Bcl-2 to the ER membrane was shown to decrease the susceptibility of neuroblastoma cells to PrP^{SC} toxicity. The protective effect was associated with the inhibition of caspase-12 activation. In addition, expression of a dominant negative form of caspase-12 decreases the induction of apoptosis by PrP^{SC}.

Surprisingly, in post-mortem human samples of patients affected with sCJD or vCJD a high increase in the expression levels of the ER chaperones Grp58, Grp78/BiP and Grp94 was observed. Similar observations were described after a proteomic analysis of CJD brain samples, showing that Grp58, was the protein that showed the highest induction under disease conditions. However, no alteration in the expression levels of other chaperones such as Hsp70 was observed. In addition, a similar pattern to caspase-12 activation was observed in the same human samples reinforcing the hypothesis that ER stress is a central signaling pathway mediating neurodegeneration and neuronal loss. In agreement with these observations, the analysis of scrapie-infected mice revealed that the ER stress pathway correlates with the disease progression and brain damage. After the analysis of different brain samples (distinct brain regions and different times during the disease progression), it was shown that PrP^{SC} levels directly correlated with the rate of Grp58 upregulation (Hetz et al, manuscript submitted). Moreover, only in the brain regions that showed extensive neuronal loss, active caspase-12 fragments were detected. We are currently studying the exact contribution of ER chaperones to the cell death process and how their upregulation is related to the prion replication process. We have found that the upregulation of Grp58 levels occurs early in the disease progression, during the presymptomatic phase of the disease. *In vitro*, Grp58 was shown to be neuroprotectine against ER stress conditions and PrP^{SC} in vitro (Hetz et al, manuscript submitted). This observation is in agreement with previous findings showing that the expression of the chaperones Grp78 and Grp94 are protective against cell death caused by disturbances of ER homeostasis[65,70–76,101,102]. In addition, two close homologues of Grp58, PDI and EndoPDI, are induced during ischemia *in vivo*, and have a protective activity against cell death[73,101,102].

Interestingly, conditions that affect the formation of disulfide bridges induce PrP to adopt some PrP^{SC}–like properties, such as proteinase K-resistance and insolubility in non-denaturating detergents[103]. In addition, *in vitro* amplification of PrP^{SC} was shown to be dependent on the presence of free sulfhydryl groups[104] and reshuffling of cystein bridges from intra-molecular to inter-molecular forms has been proposed to play a role on PrP conversion and on the stabilization of the misfolded protein aggregates[105]. These findings may suggest that the protective activity of Grp58 against PrP^{SC} neurotoxicity may be mediated by a direct interaction between the two proteins, resulting in reduction of PrP^{SC} misfolding. We are currently investigating this possibility.

13.7. Endoplasmic reticulum stress in neurodegenerative diseases

Several other neurodegenerative diseases associated to the misfolding and cerebral accumulation of a particular protein have been shown to be related to ER stress, which has been proposed to mediate the neuronal cell death process observed in these diseases. This is the case of Alzheimer disease (AD), Parkinson's disease (PD) and Huntington disease (HD)[8,68]. These neurological diseases are characterized by accumulation of misfolded protein aggregates in the brain[7], which suggests that the triggering of ER stress could be a general mechanism related with the cellular toxicity of abnormally folded proteins. The main observations associated with the occurrence of ER stress in neurological diseases are the followings:

Alzheimer's disease—Caspase-12 activation has been reported in experimental models of AD and linked with neurotoxicity of amyloid fibrils. For instance, caspase-12 knockout neurons are less sensitive to Alzheimer's amyloid β cytotoxicity[77], and caspase-4 partially mediates the toxicity of this peptide in human neurons[87]. The injection of amyloid β in the hippocampus of aged rabbits induced the expression of GADD153/CHOP and its translocation into the nucleus, with the concomitant decrease of Bcl-2 expression[106]. In the same experimental system, the intracerebral injection of amyloid β triggered the activation of caspase-3 and caspase-12[107]. Additionally, in AD transgenic mice models, an increased susceptibility to ER stress induction was detected[108]. The observation described in these animals included an increased activation of caspase-12 and abnormal ER-calcium signaling after stimulation of apoptosis[109]. Also, increased expression of GADD153/CHOP in pharmacological paradigms of ER stress[110], and altered expression of Grp78/Bip[111] have been reported.

Huntington's disease and spinocerebral ataxias—The expansion of CAG trinucleotide encoding polyglutamine is the underlying cause of at least nine inherited human neurodegenerative disorders, including HD and spinocerebral ataxias. The expression of expanded polyglutamine repeats has been shown to induce ER stress in several cell lines, which is reflected as the activation of the Ire1α/TRAF-2 pathway[112], the induction of Grp78/Bip and the activation of caspase-12[113,114]. On the other hand, the expression of mutant ataxin-3, a polyglutamine repeat-containing protein involved in spinocerebellar ataxia[115], induces apoptosis and caspase-12 activation[114].

Parkinson's disease—In models of PD, oxidative stress and mitochondrial dysfunction are believed to be central players in the mechanism of neuronal dysfunction (review in 116, or in 117). However, recent reports point out that the ER stress pathway is also involved in the disease process. Treatment of dopaminergic neurons with 6-hydroxydopamine, a parkinsonism-inducing neurotoxin, activates the UPR pathway reflected by the induction of several chaperones, such as Grp58, Grp78/Bip, GADD153/CHOP, as well as Ire1α activation[118]. Similar results have been described by another group, showing that toxic agents affecting the viability of dopaminergic neurons activate the ER-stress pathway, associated with a massive upregulation of chaperones (including calnexin, heat shock proteins and Grp chaperones), the expression of GADD153/CHOP, and the phosphorylation of Ire1α and other related ER kinases[119]. Hereditary parkinsonism has been genetically linked with mutations in the genes encoding the proteins parkin or α-synuclein[120]. ER stress-induced by mutant α-synuclein is decreased by the expression of wild type parkin.[114,117].

In addition, ER stress has been observed in other pathological conditions affecting the brain, such as ischemia, traumatic brain injury and retrovirus induced spongiform brain degeneration. The type of brain damage triggered by these retroviral infections closely resembles several pathological features of TSEs. At the molecular level, it has been described that the degeneration of the brain was associated with the induction of GADD153/CHOP, Grp58, Grp78/Bip, Grp94 and calreticulin[121]. Also, the activation of caspase-12 and caspase-3 was observed in this experimental system[122]. Finally, up-regulation of ER stress markers and activation of caspase-12 were described in animal models of neuronal loss by ischemia[123,124] and traumatic brain injury[125].

13.8. Concluding remarks

In the last few years it has become clear that ER-stress mediated apoptosis plays an important role in diverse diseases and in particular in neurodegenerative disorders associated with the misfolding and brain deposition of proteins. TSEs are a prototype of these diseases where the central role of the misfolded protein is widely accepted[7]. Data generated in our laboratory shows that an atypical form of ER stress features the pathogenesis of TSEs[126]. Figure 13.2 shows a schematic diagram of the cellular events occurring after treatment of neurons with PrP^{SC}. This signaling includes the release of calcium from the ER, the induction of the pro-apoptotic caspase-12 and the up-regulation of certain ER chaperones (Grp58, Grp78 and Grp94), but not others commonly detected under ER stress conditions (such as GAD153/CHOP, calreticulin, Hsp70). The initial molecular events associated with the apoptotic effect of PrP^{SC} remains unclear. We speculate that PrP^{SC} may interact with a

Figure 13.2. A working hypothesis for PrP^{SC} induced apoptosis in neuronal cells. Interaction of PrP^{SC} with an unknown receptor activates a signaling pathway which induces the release of calcium from the ER though the ryanodine receptors (RyR) and the IP3-receptors (IP3R). The alteration in calcium homeostasis promotes ER stress, which in turn triggers the accumulation of misfolded proteins in the ER, leading to a neuroprotective response associated with the induction of chaperones of the Grp family. Ultimately, ER stress leads to the activation of caspase-12, which in turn activates the executioner caspase-3, resulting in neuronal apoptosis. Experimentally, this process can be modulated *in vitro* by: expression of Bcl-2 in the ER membrane (Bcl-2ER), expressing a dominant negative form of caspase-12 (caspase-12DN), treating the cells with caspase-3 inhibitors (Ac-DEVD-fmk).

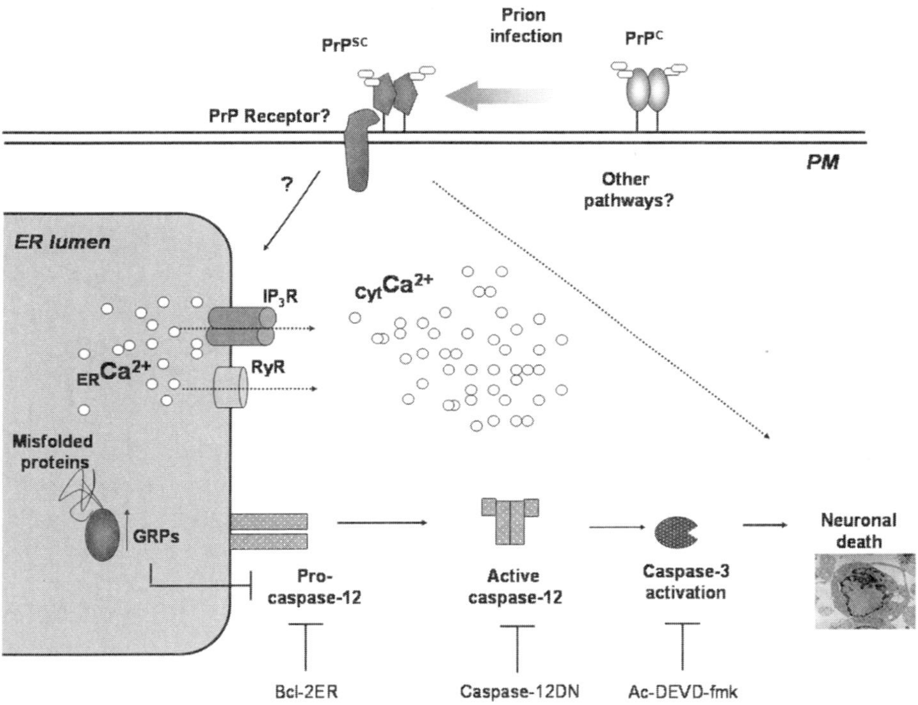

yet unknown receptor, triggering an abnormal release of calcium from the ER, activating the cell death program.

An alternative possibility that has been arising in recent years is that PrP^{SC} neurotoxic effect is mediated through PrP^{C} signalling. Mallucci and co-workers have found that the depletion of endogenous neuronal PrP^{C} in a post-natal knockout mice after prion infection, leads to a reversion of the early spongiform changes of the brain and prevents neuronal loss and progression to clinical disease[127]. This occurred despite the presence of extracellular PrP^{SC} deposition in the brain similar to the levels observed in terminally ill wild-type animals. In a similar study,

Brandner and colleagues showed that PrP-null brain tissue surrounding prion-infected *Prnp*$^{+/+}$ neurografts does not develop prion neuropathological changes[128]. In both experimental systems, PrP^{SC} seems unable to trigger neuronal death in the absence of PrP^{C}. These results suggest that PrP^{SC} requires the presence of PrP^{C} to be neurotoxic. This interpretation finds support in recent data from Solforosi et al., which shows that neurodegeneration could be directly triggered through a cross-linking of PrP^{C} by a monoclonal antibody[129]. The extrapolation of these findings is that PrP^{SC} could be the activator of a PrP^{C}–mediated signaling pathway[8]. This is in agreement with the observation that PrP knockout cells are resistant to the toxic activity of partially purified PrP^{SC} (review in 18). Hence, the possibility that normal PrP^{C} is the receptor for PrP^{SC} remains open for future research. An alternative model to explain PrP^{SC} cytotoxicity suggests that the misfolded protein might be transported to the ER and, by itself, induces ER stress. This mechanism has been described for bacterial toxins, like the cholera toxin, which binds to lipid rafts structures in the plasma membrane and is then internalised by endocytosis toward the ER[130]. PDI has been implicated in the regulation of the pathogenic effects of cholera toxin, since it recognizes the toxin in the ER and release it from the internalization-protein complex enabling its toxic effects[131]. This could be an interesting possibility to explore, since it has recently been shown that in neuroblastoma cells infected with scrapie, modification of intracellular trafficking between Golgi-ER or endosome-Golgi induces an accumulation of PrP^{SC} in the ER[132], opening the possibility that PrP^{SC} can reach the ER and be recognized by Grp58 as a stress factor.

The elucidation of the mechanism of neuronal apoptosis in TSE has clear implications for the development of TSE treatments directed to prevent neurodegeneration. Since caspases are central to both normal programmed cell death and injury-dependent apoptosis, inhibition of these proteases usually results in serious adverse effects. However, caspase-12 appears not to be essential for normal development or physiological cell death, but rather its activation seems confined to some specific pathological stress signals[50]. Indeed, caspase-12-deficient mice have no noticeable developmental or behavioral defects[77,133]. Therefore, inhibition of caspase-12 activation might provide a novel therapeutic strategy for TSEs and other neurodegenerative diseases initiated by protein misfolding. However, the participation of caspase-12 in human pathologies has been controversial, since the caspase-12 gene is not functional in humans[134]. Recent reports have shown that caspase-4 is the human caspase-12 homolog in terms of structure, function and subcellular localization[87]. Our data suggests that targeting caspase-4 or other components of the ER-stress mediated apoptosis pathway may lead to therapeutic benefits for TSEs. Similarly, pharmacological treatment that induces the expression of Grp58 may have neuroprotective effects. An example of this class of drugs is valproate, which is used to treat brain-related alterations such as bipolar disorders. This compound is known to induce the expression of the chaperone Grp78/Bip without activating ER stress or any detectable ER-associated damage (review in 135). Also, recent reports suggest that under certain conditions, known endogenous ER-inducible chaperones can be expressed in the absence of any ER stress signal[61,136], suggesting that an strategy designed to generate drugs that induce Grps expression may be possible. Finally, the close relationship between Grp58 up-regulation and PrP^{SC} accumulation, which is even detected during the pre-symptomatic stages of the disease, may be exploited to generate a specific and early diagnosis for prion disorders. A biochemical and non-invasive diagnosis of these diseases is a high priority to minimize further spreading of the disease[7]. Therefore, the identification of a new putative surrogate marker for prion disease is of potentially high importance.

13.9. References

1. G. Majno and I. Joris, Apoptosis, Oncosis, and Necrosis—An Overview of Cell-Death. *Am. J. Pathol.* **146**, 3–15 (1995).

2. M. Castedo, J. L. Perfettini, T. Roumier, K. Andreau, R. Medema, and G. Kroemer, Cell death by mitotic catastrophe: a molecular definition. *Oncogene* **23**, 2825–2837 (2004).

3. J. Yuan, M. Lipinski, and A. Degterev, Diversity in the mechanisms of neuronal cell death. *Neuron* **40**, 401–413 (2003).

4. R. A. Lockshin and C. M. Williams, Programmed cell death I—Cytology of degeneration in the intersegmental muscles of the Pernyi Silknoth. *J. Insect Physiol* **11**, 123–133 (1965).

5. J. F. Kerr, A. H. Wyllie, and A. R. Currie, Apoptosis: a basic biological phenomenon with wide-ranging implications in tissue kinetics. *Br. J. Cancer* **26**, 239–257 (1972).

6. D. L. Vaux and S. J. Korsmeyer, Cell death in development. *Cell* **96**, 245–254 (1999).

7. C. Soto, Unfolding the role of Protein Misfolding in Neurodegenerative Diseases. *Nature Rev. Neurosci.* **4**, 49–60 (2003).

8. J. Castilla, C. Hetz, and C. Soto, Molecular mechanisms of neurotoxicity of pathological prion protein. *Curr. Mol. Med.* **4**, 397–403 (2004).

9. J. Collinge, Prion diseases of humans and animals: their causes and molecular basis. *Annu. Rev. Neurosci.* **24**, 519–550 (2001).

10. S. B. Prusiner, Prions. *Proc. Natl. Acad. Sci. U.S.A.* **95**, 13363–13383 (1998).

11. K. M. Pan, M. Baldwin, J. Nguyen, M. Gasset, A. Serban, D. Groth, I. Mehlhorn, Z. Huang, R. J. Fletterick, F. E. Cohen and S. B. Prusiner, Conversion of alpha-helices into beta-sheets features in the formation of the scrapie prion proteins. *Proc. Natl. Acad. Sci. U.S.A.* **90**, 10962–10966 (1993).

12. F. E. Cohen and S. B Prusiner,. Pathologic conformations of prion proteins. *Annu. Rev. Biochem.* **6**7, 793–819 (1998).

13. N. Streichenberger, D. Jordan, I. Verejan, C, Souchier, F. Philippeau, E. Gros, C. Mottolese, K. Ostrowsky, A. Perret-Liaudet, J. L. Laplanche, M. Hermier, J. P. Deslys, G. Chazot and N. Kopp, The first case of new variant Creutzfeldt-Jakob disease in France: clinical data and neuropathological findings. *Acta Neuropathol. (Berl)* **99**, 704–708 (2000).

14. G. Chazot, E. Broussolle, C. Lapras, T. Blattler, A. Aguzzi and N. Kopp, New variant of Creutzfeldt-Jakob disease in a 26-year-old French man. *Lancet* **347**, 1181 (1996).

15. H. Büeler, A. Aguzzi, A. Sailer, R. A. Greiner, P. Autenried , M. Aguet and C. Weissmann, Mice devoid of PrP are resistant to scrapie. *Cell* **73**, 1339–1347 (1993).

16. D. A. Kocisko, J. H. Come, S. A. Priola, B. Chesebro, G. J. Raymond, P. T. Lansbury and B. Caughey, Cell-free formation of protease-resistant prion protein. *Nature* **370**, 471–474 (1994).

17. G. P. Saborio, B. Permanne and C. Soto, Sensitive detection of pathological prion protein by cyclic amplification of protein misfolding. *Nature* **411**, 810–813 (2001).

18. C. Hetz and C. Soto, Protein misfolding and disease: the case of prion disorders. *Cell Mol. Life Sci.* **60**, 133–143 (2003).

19. P. J. Lucassen, A. Williams, W. C. Chung and H. Fraser, Detection of apoptosis in murine scrapie. *Neurosci. Lett.* **198**, 185–188 (1995).

20. A. Giese, M. H. Groschup, B. Hess and H. A. Kretzschmar, Neuronal cell death in scrapie-infected mice is due to apoptosis. *Brain Pathol.* **5**, 213–221 (1995).

21. D. Jesionek-Kupnicka, R. Kordek, J. Buczynski and P. P. Liberski, Apoptosis in relation to neuronal loss in experimental Creutzfeldt-Jakob disease in mice. *Acta Neurobiol. Exp. (Wars.)* **61**, 13–19 (2001).

22. A. Dorandeu, L. Wingertsmann, F. Chretien, M. B. Delisle, C. Vital, P. Parchi, P. Montagna, E. Lugaresi, J. W. Ironside H. Budka, P. Gambetti and F. Gray, Neuronal apoptosis in fatal familial insomnia. *Brain Pathol.* **8**, 531–537 (1998).

23. I. Ferrer, Nuclear DNA fragmentation in Creutzfeldt-Jakob disease: does a mere positive in situ nuclear end-labeling indicate apoptosis? *Acta Neuropathol. (Berl)* **97**, 5–12 (1999).

24. F. Gray, H. Adle-Biassette, F. Chretien, T. Ereau, M. B. Delisle and C. Vital, [Neuronal apoptosis in human prion diseases]. *Bull. Acad. Natl. Med.* **183**, 305–320 (1999).

25. F. Gray, F. Chretien, H. Adle-Biassette, A. Dorandeu, T. Ereau, M. B. Delisle, N. Kopp, J. W. Ironside, C. Vital, Neuronal apoptosis in Creutzfeldt-Jakob disease. *J. Neuropathol. Exp. Neurol.* **58**, 321–328 (1999).

26. A. Williams, P. J. Lucassen, D. Ritchie and M. Bruce, PrP deposition, microglial activation, and neuronal apoptosis in murine scrapie. *Exp. Neurol.* **144**, 433–438 (1997).

27. D. W. Fairbairn, K. G. Carnahan, R. N. Thwaits, R. V. Grigsby, G. R. Holyoak and K. L.O'Neill, Detection of apoptosis induced DNA cleavage in scrapie-infected sheep brain. *FEMS Microbiol. Lett.* **115**, 341–346 (1994).

28. D. Theil, R. Fatzer, R. Meyer, M. Schobesberger, A. Zurbriggen and M. Vandevelde, Nuclear DNA fragmentation and immune reactivity in bovine spongiform encephalopathy. *J. Comp Pathol.* **121**, 357–367 (1999).

29. F. Chretien, A. Dorandeu, H. Adle-Biassette, T. Ereau, L. Wingertsmann, F. Brion, F. Gray, [A process of programmed cell death as a mechanisms of neuronal death in prion diseases]. *Clin. Exp. Pathol.* **47**, 181–191 (1999).

30. D. Jesionek-Kupnicka, J. Buczynski, R. Kordek, T. Sobow, I. Kloszewska, W. Papierz and P. P. Liberski, Programmed cell death (apoptosis) in Alzheimer's disease and Creutzfeldt-Jakob disease. *Folia Neuropathol.* **35**, 233–235 (1997).

31. T. Ookohchi, H. Ito, T. Serikawa and K. Sato, Detection of apoptosis in the brain of the zitter rat with genetic spongiform encephalopathy. *Biochem. Mol. Biol. Int.* **41**, 279–284 (1997).

32. S. Siso, B. Puig, R. Varea, E. Vidal, C. Acin, M. Prinz, F. Montrasio, J. Badiola, A. Aguzzi, M . Pumarola and I. Ferrer. Abnormal synaptic protein expression and cell death in murine scrapie. *Acta Neuropathol. (Berl)* **103**, 615–626 (2002).

33. D. Jesionek-Kupnicka, J. Buczynski, R. Kordek and P. P. Liberski, Neuronal loss and apoptosis in experimental Creutzfeldt-Jakob disease in mice. *Folia Neuropathol.* **37**, 283–286 (1999).

34. R. Chiesa, P. Piccardo, B. Ghetti and D. A. Harris, Neurological illness in transgenic mice expressing a prion protein with an insertional mutation. *Neuron* **21**, 1339–1351 (1998).

35. S. Lehmann and D. A. Harris, A mutant prion protein displays an aberrant membrane association when expressed in cultured cells. *J. Biol. Chem.* **270**, 24589–24597 (1995).

36. S. Lehmann and D. A. Harris, Two mutant prion proteins expressed in cultured cells acquire biochemical properties reminiscent of the scrapie isoform. *Proc. Natl. Acad. Sci. U.S.A.* **93**, 5610–5614 (1996).

37. N. Daude, S. Lehmann and D. A. Harris, Identification of intermediate steps in the conversion of a mutant prion protein to a scrapie-like form in cultured cells. *J. Biol. Chem.* **272**, 11604–11612 (1997).

38. R. Chiesa, B. Drisaldi, E. Quaglio, A. Migheli, P. Piccardo, B. Ghetti and D. A. Harris, Accumulation of protease-resistant prion protein (PrP) and apoptosis of cerebellar granule cells in transgenic mice expressing a PrP insertional mutation. *Proc. Natl. Acad. Sci .U.S.A.* **97**, 5574–5579 (2000).

39. S. Supattapone, P. Bosque, T. Muramoto, H. Wille, C. Aagaard, D. Peretz, H. O. Nguyen, C. Heinrich, M. Torchia, J. Safar, F. E. Cohen, S. J. DeArmond, S. B. Prusiner and M. Scott, Prion protein of 106 residues creates an artificial transmission barrier for prion replication in transgenic mice. *Cell*, **96**, 869–878 (1999).

40. S. Supattapone, E. Bouzamondo, H. L. Ball, H. Wille, H. O. Nguyen, F. E. Cohen, S. J. DeArmond, S. B. Prusiner and M. Scott, A protease-resistant 61-residue prion peptide causes neurodegeneration in transgenic mice. *Mol. Cell Biol.* **21**, 2608–2616 (2001).

41. R. S. Hegde, J. A. Mastrianni, M. R. Scott, K. D. Defea, P. Tremblay, M. Torchia, S. J. DeArmond, S. B. Prusiner and V. R. Lingappa, A transmembrane form of the prion protein in neurodegenerative disease. *Science* **279**, 827–834 (1998).

42. C. Hetz, K. Maundrell and C. Soto, Is loss of function of the prion protein the cause of prion disorders? *Trends Mol. Med.* **9**, 237–243 (2003).

43. C. Hetz, S. Benavent, S. Bieler and C. Soto, (2003). Prion Protein Misfolding and Brain Degeneration. *Curr. Med. Chem.***3**, 137–147.

44. A. Eastman, Survival Factors, Intracellular Signal-Transduction, and the Activation of Endonucleases in Apoptosis. *Seminars Cancer Biol.* **6**, 45–52 (1995).

45. A. M. Davies, Regulation of neuronal survival and death by extracellular signals during development. *EMBO J.* **22**, 2537–2545 (2003).

46. A. Ashkenazi and V. M. Dixit, Death receptors: Signaling and modulation. *Science* **281**, 1305–1308 (1998).

47. K. F. Ferri and G. Kroemer, Organelle-specific initiation of cell death pathways. *Nature Cell Biology* **3**, E255–E263 (2001).

48. T. Dragovich, C. M. Rudin and C. B. Thompson, Signal transduction pathways that regulate cell survival and cell death., *Oncogene*. 1998 Dec 24;**17**(25): 3207–13.

49. L. E. French and J. Tschopp, Protein-based therapeutic approaches targeting death receptors. *Cell Death and Differentiation* **10**, 117–123 (2003).

50. H. Mehmet, Caspases find a new place to hide. *Nature* **403**, 29–30 (2000).

51. M. S. Forman, V. M. Lee and J. Q. Trojanowski, 'Unfolding' pathways in neurodegenerative disease. *Trends Neurosci.* **26**, 407–410 (2003).

52. A. Degterev, M. Boyce and J. Y. Yuan, A decade of caspases. *Oncogene* **22**, 8543–8567 (2003).

53. S. J. Kang, S. Wang, H. Hara, E. P. Peterson, S. Namura, S. Amin-Hanjani, Z. Huang, A. Srinivasan, K. J. Tomaselli, N. A. Thornberry, M. A. Moskowitz and J. Yuan, Dual role of caspase-11 in mediating activation of caspase-1 and caspase-3 under pathological conditions. *J. Cell Biol.* **149**, 613–622 (2000).

54. Y. Tsujimoto, Role of Bcl-2 family proteins in apoptosis: apoptosomes or mitochondria?, *Genes Cells*. **3**, 697–707 (1998).

55. K. Kuida, Caspase-9. *Int. J. Biochem. Cell Biol.* **32**, 121–124 (2000).

56. J. G. Pastorino, M. Tafani, R. J. Rothman, A. Marcinkeviciute, J. B. Hoek, J. L. Farber and A. Marcineviciute, Functional consequences of the sustained or transient activation by Bax of the mitochondrial permeability transition pore. *J. Biol. Chem.* **274**, 31734–31739 (1999).

57. S. J. Korsmeyer, (1999). Programmed cell death and the regulation of homeostasis. *Harvey Lect.* **95**, 21–41.

58. N. N. Danial, and S. J. Korsmeyer, Cell death: Critical control points. *Cell* **116**, 205–219 (2004).

59. J. T. Opferman and S. J. Korsmeyer, Apoptosis in the development and maintenance of the immune system. *Nature Immunol.* **4**, 410–415 (2003).

60. A. M. Ranger, B. A. Malynn, and S. J. Korsmeyer, Mouse models of cell death. *Nature Gen.* **28**, 113–118 (2001).

61. J. N. Gass, K. E. Gunn, R. Sriburi and J. W. Brewer, Stressed-out B cells? Plasma-cell differentiation and the unfolded protein response. *Trends Immunol.* **25**, 17–24 (2004).

62. R. J. Kaufman, Orchestrating the unfolded protein response in health and disease. *J. Clin. Invest.* **110**, 1389–1398 (2002).

63. D. T. Rutkowski and R. J. Kaufman, A trip to the ER: coping with stress. *Trends Cell Biol.* **14**, 20–28 (2004).

64. A. J. L. Macario, Heat-Shock Proteins and Molecular Chaperones—Implications for Pathogenesis, Diagnostics, and Therapeutics. *Int. J. Clin. Lab. Res.* **25**, 59–70 (1995).

65. H. Liu, R. C. Bowes 3rd, B. van de Water, C. Sillence, J. F. Nagelkerke and J. L. Stevens, Endoplasmic reticulum chaperones GRP78 and calreticulin prevent

oxidative stress, Ca2+ disturbances, and cell death in renal epithelial cells. *J. Biol. Chem.* **272**, 21751–21759 (1997).

66. K. J. Travers, C. K. Patil, L. Wodicka, D. J. Lockhart, J. S. Weissman and P. Walter, Functional and genomic analyses reveal an essential coordination between the unfolded protein response and ER-associated degradation. *Cell* **101**, 249–258 (2000).

67. M. Molinari and A. Helenius, Chaperone selection during glycoprotein translocation into the endoplasmic reticulum. *Science* **288**, 331–333 (2000).

68. W. Paschen, Endoplasmic reticulum: a primary target in various acute disorders and degenerative diseases of the brain. *Cell Calcium* **34**, 365–383 (2003).

69. D. T. Rutkowski and R. J. Kaufman, A trip to the ER: coping with stress. *Trends Cell Biol.* **14**, 20–28 (2004).

70. R. V. Rao, S. Castro-Obregon, H. Frankowski, M. Schuler, V. Stoka, G. del Rio, D. E. Bredesen and H. M. Ellerby, Coupling endoplasmic reticulum stress to the cell death program: role of the ER chaperone GRP78. *FEBS Lett.* **514**, 122–128 (2002).

71. R. K. Reddy, C. Mao, P. Baumeister, R. C. Austin, R. J. Kaufman, A. S. Lee, Endoplasmic reticulum chaperone protein GRP78 protects cells from apoptosis induced by topoisomerase inhibitors—Role of ATP binding site in suppression of caspase-7 activation. *J. Biol. Chem.* **278**, 20915–20924 (2003).

72. R. K. Reddy, J. Lu and A. S. Lee, The endoplasmic reticulum chaperone glycoprotein GRP94 with Ca(2+)-binding and antiapoptotic properties is a novel proteolytic target of calpain during etoposide-induced apoptosis. *J. Biol. Chem.* **274**, 28476–28483 (1999).

73. D. C. Sullivan, L. Huminiecki, J. W. Moore, J. J. Boyle, R. Poulsom, D. Creamer, J. Barker and R. Bicknell, EndoPDI, a novel protein-disulfide isomerase-like protein that is preferentially expressed in endothelial cells acts as a stress survival factor. *J. Biol. Chem.* **278**, 47079–47088 (2003).

74. Y. Bando, T. Katayama, K. Kasai, M. Taniguchi, M. Tamatani and M. Tohyama, GRP94 (94 kDa glucose-regulated protein) suppresses ischemic neuronal cell death against ischemia/reperfusion injury. *Eur. J. Neurosci.* **18**, 829–840 (2003).

75. H. Miyake, I. Hara, S. Arakawa and S. Kamidono, Stress protein GRP78 prevents apoptosis induced by calcium ionophore, Ionomycin, but not by glycosylation inhibitor, tunicamycin, in human prostate cancer cells. *J. Cell Biochem* **77**, 396–408 (2000).

76. Z. Yu, H. Luo, W. Fu and M. P. Mattson, The endoplasmic reticulum stress-responsive protein GRP78 protects neurons against excitotoxicity and apoptosis: suppression of oxidative stress and stabilization of calcium homeostasis. *Exp. Neurol.* **155**, 302–314 (1999).

77. T. Nakagawa, H. Zhu, N. Morishima, E. Li, J. Xu, B. A.Yankner and J. Yuan, Caspase-12 mediates endoplasmic-reticulum-specific apoptosis and cytotoxicity by amyloid-beta. *Nature* **403**, 98–103 (2000).

78. T. Yoneda, K. Imaizumi, K. Oono, D. Yui, F. Gomi, T. Katayama and M. Tohyama, Activation of caspase-12, an endoplastic reticulum (ER) resident caspase, through tumor necrosis factor receptor-associated factor 2-dependent mechanism in response to the ER stress. *J. Biol. Chem.* **276**, 13935–13940 (2001).

79. F. Urano, X. Wang, A. Bertolotti, Y. Zhang, P. Chung, H. P. Harding and D. Ron, Coupling of stress in the ER to activation of JNK protein kinases by transmembrane protein kinase IRE1. *Science* **287**, 664–666 (2000).

80. X. Z. Wang, H. P. Harding, Y. Zhang, E. M. Jolicoeur, M. Kuroda and D. Ron, (1998). Cloning of mammalian Ire1 reveals diversity in the ER stress responses. *EMBO J.* **17**, 5708–5717.

81. R. V. Rao, A. Peel, A. Logvinova, G. del Rio, E. Hermel, T. Yokota, P. C. Goldsmith, L. M. Ellerby, H. M. Ellerby and D. E. Bredesen, Coupling endoplasmic reticulum stress to the cell death program. Mechanism of caspase activation. *J. Biol. Chem.* **276**, 33869–33874 (2001).

82. E. Fujita, Y. Kouroku, A. Jimbo, A. Isoai, K. Maruyama and T. Momoi, Caspase-12 processing and fragment translocation into nuclei of tunicamycin-treated cells. *Cell Death Diff.* **9**, 1108–1114 (2002).

83. T. Nakagawa and J. Yuan, Cross-talk between two cysteine protease families. Activation of caspase-12 by calpain in apoptosis. *J Cell Biol.* **150**, 887–894 (2000).

84. N. Morishima, K. Nakanishi, H. Takenouchi, T. Shibata and Y. Yasuhiko, An endoplasmic reticulum stress-specific caspase cascade in apoptosis—Cytochrome c-independent activation of caspase-9 by caspase-12. *J. Biol. Chem.* **277**, 34287–34294 (2002).

85. M. Kilic, R. Schafer, J. Hoppe and U. Kagerhuber, Formation of noncanonical high molecular weight caspase-3 and -6 complexes and activation of caspase-12 during serum starvation induced apoptosis in AKR-2B mouse fibroblasts. *Cell Death. Differ.* **9**, 125–137 (2002).

86. J. Hoppe, M. Kilic, V. Hoppe, A. Sachinidis, and U. Kagerhuber, Formation of caspase-3 complexes and fragmentation of caspase-12 during anisomycin-induced apoptosis in AKR-2B cells without aggregation of Apaf-1. *Eur. J. Cell Biol.* **81**, 567–576 (2002).

87. J. Hitomi, T. Katayama, Y. Eguchi, T. Kudo, M. Taniguchi, Y. Koyama, T. Manabe S.Yamagishi, Y. Bando, K. Imaizumi, Y. Tsujimoto and M.Tohyama, Involvement of caspase-4 in endoplasmic reticulum stress-induced apoptosis and A beta-induced cell death. *J. Cell Biol.* **165**, 347–356 (2004).

88. G. Zanusso, R. B. Petersen, T. Jin, Y. Jing, R. Kanoush, S. Ferrari, P. Gambetti, and N. Singh, Proteasomal degradation and N-terminal protease resistance of the codon 145 mutant prion protein. *J. Biol. Chem.* **274**, 23396–404 (1999).

89. N. Singh, G. Zanusso, S. G. Chen, H. Fujioka, S. Richardson, P. Gambetti, and R. B. Petersen, Prion protein aggregation reverted by low temperature in transfected cells carrying a prion protein gene mutation. *J. Biol. Chem.* **272**, 28461–70 (1997).

90. T. Jin, Y. Gu, G. Zanusso, M. Sy, A. Kumar, M. Cohen, P. Gambetti, and N. Singh, The chaperone protein BiP binds to a mutant prion protein and mediates its degradation by the proteasome. *J Biol Chem.* **275**, 38699–704 (2000).

91. J. Ma, R. Wollmann and S. Lindquist, Neurotoxicity and neurodegeneration when PrP accumulates in the cytosol. *Science* **298**, 1781–1785 (2002).

92. J. Ma and S. Lindquist, Conversion of PrP to a self-perpetuating PrPSc-like conformation in the cytosol. *Science* **298**, 1785–1788 (2002).

93. D. E. Dimcheff J. L. Portis, and B. Caughey, Prion proteins meet protein quality control. *Trends Cell Biol.* **13**, 337–340 (2003).

94. Y. Yedidia, L. Horonchik, S. Tzaban, A. Yanai and A. Taraboulos, Proteasomes and ubiquitin are involved in the turnover of the wild-type prion protein. *EMBO J.* **20**, 5383–5391 (2001).

95. J. Ma and S. Lindquist, Wild-type PrP and a mutant associated with prion disease are subject to retrograde transport and proteasome degradation. *Proc. Natl. Acad. Sci. U.S.A.* **98**, 14955–14960 (2001).

96. B. Drisaldi, R. S. Stewart, C. Adles, L. R. Stewart, E. Quaglio, E. Biasini, L. Fioriti, R. Chiesa and D. A. Harris, Mutant PrP is delayed in its exit from the endoplasmic reticulum, but neither wild-type nor mutant PrP undergoes retrotranslocation prior to proteasomal degradation. *J. Biol. Chem.* **278**, 21732–21743 (2003).

97. C. Bate, S. Reid and A. Williams, Killing of prion-damaged neurones by microglia. *Neuroreport* **12**, 2589–2594 (2001).

98. K. Post, D. R. Brown, M. Groschup, H. A. Kretzschmar and D. Riesner, Neurotoxicity but not infectivity of prion proteins can be induced reversibly *in vitro*. *Arch. Virol. Suppl* **16** 265–273 (2000).

99. C. Bate, M. Salmona, L. Diomede and A. Williams, Squalestatin cures prion-infected neurons and protects against prion neurotoxicity. *J. Biol. Chem.* **279**, 14983–14990 (2004).

100. W. E. Müller, H. Ushijima, H. C. Schroder, J. M. Forrest, W. F. Schatton, P. G. Rytik and M. Heffner-Lauc, Cytoprotective effect of NMDA receptor antagonists on prion protein (PrionSc)-induced toxicity in rat cortical cell cultures. *Eur. J. Pharmacol.* **246**, 261–267 (1993).

101. S. Tanaka, T. Uehara and Y. Nomura, Up-regulation of protein-disulfide isomerase in response to hypoxia/brain ischemia and its protective effect against apoptotic cell death. *J. Biol. Chem.* **275**, 10388–10393 (2000).

102. H. S. Ko, T. Uehara and Y. Nomura, Role of ubiquilin associated with protein-disulfide isomerase in the endoplasmic reticulum in stress-induced apoptotic cell death. *J. Biol. Chem.* **277**, 35386–35392 (2002).

103. J. Ma and S. Lindquist, De novo generation of a PrPSc-like conformation in living cells. *Nat. Cell Biol.* **1**, 358–361 (1999).

104. R. Lucassen, K. Nishina and S. Supattapone, In vitro amplification of protease-resistant prion protein requires free sulfhydryl groups. *Biochemistry* **42**, 4127–4135 (2003).

105. Tompa, P., Tusnady, G. E., Friedrich, P., and Simon, I. The role of dimerization in prion replication. *Biophys. J.* **82**, 1711–1718 (2002).

106. O. Ghribi, M. M. Herman, D. A. DeWitt, M. S. Forbes and J. Savory, Abeta(1–42) and aluminum induce stress in the endoplasmic reticulum in rabbit hippocampus, involving nuclear translocation of gadd 153 and NF-kappaB. *Brain Res. Mol. Brain Res.* **96**, 30–38 (2001).

107. O. Ghribi, M. M. Herman, and J. Savory, Lithium inhibits Abeta-induced stress in endoplasmic reticulum of rabbit hippocampus but does not prevent oxidative damage and tau phosphorylation. *J. Neurosci. Res.* **71**, 853–862 (2003).

108. F. Terro, C. Czech, F. Esclaire, W. Elyaman, C. Yardin, M. C. Baclet, N. Touchet, G. Tremp, L. Pradier and J. Hugon, Neurons overexpressing mutant presenilin-1 are more sensitive to apoptosis induced by endoplasmic reticulum-Golgi stress. *J.Neurosci.Res.* **69**, 530–539 (2002).

109. S. L. Chan, C. Culmsee, N. Haughey, W. Klapper and M. P. Mattson, Presenilin-1 mutations sensitize neurons to DNA damage-induced death by a mechanism involving perturbed calcium homeostasis and activation of calpains and caspase-12. *Neurobiol. Dis.* **11**, 2–19 (2002).

110. O. Milhavet, J. L. Martindale, S. Camandola, S. L. Chan, D. S. Gary, A. Cheng, N. J. Holbrook, M. P. Mattson, Involvement of Gadd153 in the pathogenic action of presenilin-1 mutations. *J. Neurochem.* **83**, 673–681 (2002).

111. Y. Yasuda, T. Kudo, T. Katayama, K. Imaizumi, M. Yatera, M. Okochi, H. Yamamori, N. Matsumoto, T. Kida, A. Fukumori, M. Okumura, M. Tohyama and M.Takeda, FAD-linked presenilin-1 mutants impede translation regulation under ER stress. *Biochem. Biophys. Res. Commun.* **296**, 313–318 (2002).

112. H. Nishitoh, A. Matsuzawa, K. Tobiume, K. Saegusa, I. Takeda K, . Inoue, S. Hori, A. Kakizuka and H. Ichijo, ASK1 is essential for endoplasmic reticulum stress-induced neuronal cell death triggered by expanded polyglutamine repeats. *Genes and Development* **16**, 1345–1355 (2002).

113. Y. Kouroku, E. Fujita, A. Jimbo, T. Kikuchi, T. Yamagata, M. Y. Momoi, E. Kominami, K. Kuida, K. Sakamaki, S. Yonehara and T. Momoi, Polyglutamine aggregates stimulate ER stress signals and caspase-12 activation. *Hum. Mol. Genet.* **11**, 1505–1515 (2002).

114. Tsai, Y. C., Fishman, P. S., Thakor, N. V., and Oyler, G. A. Parkin facilitates the elimination of expanded polyglutamine proteins and leads to preservation of proteasome function. *J. Biol. Chem.* **278**, 22044–22055 (2003).

115. H. L. Paulson, M. K. Perez, Y. Trottier, J. Q. Trojanowski, S. H. Subramony, S. S. Das, P. Vig, J. L. Mandel, K. H. Fischbeck and R. N. Pittman, Intranuclear inclusions of expanded polyglutamine protein in spinocerebellar ataxia type 3. *Neuron* **19**, 333–344 (1997).

116. T. M. Dawson and V. L. Dawson, Molecular pathways of neurodegeneration in Parkinson's disease. *Science* **302**, 819–822 (2003).

117. W. Dauer and S. Przedborski, Parkinson's disease: Mechanisms and models. *Neuron* **39**, 889–909 (2003).

118. W. A. Holtz and K. L. O'Malley, Parkinsonian mimetics induce aspects of unfolded protein response in death of dopaminergic neurons. *J. Biol. Chem.* **278**, 19367–19377 (2003).

119. E. J. Ryu, H. P. Harding, J. M. Angelastro, O. V. Vitolo, D. Ron and L. A. Greene, (2002). Endoplasmic reticulum stress and the unfolded protein response in cellular models of Parkinson's disease. *J. Neurosci.* **22**, 10690–10698.

120. R. E. Burke, Recent advances in research on Parkinson disease: Synuclein and parkin. *Neurologist* **10**, 75–81 (2004).

121. D. E. Dimcheff, S. Askovic, A. H. Baker, C. Johnson-Fowler and J. L. Portis, Endoplasmic reticulum stress is a determinant of retrovirus-induced spongiform neurodegeneration. *J.Virol.* **77**, 12617–12629 (2003).

122. H. T. Kim, K. Waters, G. Stoica, W. Qiang, N. Liu, V. L. Scofield and P. K. Wong, (2004). Activation of endoplasmic reticulum stress signaling pathway is associated with neuronal degeneration in MoMuLV-ts1-induced spongiform encephalomyelopathy. *Lab Invest.*

123. M. Shibata, H. Hattori, T. Sasaki, J. Gotoh J. Hamada and Y. Fukuuchi, Activation of caspase-12 by endoplasmic reticulum stress induced by transient middle cerebral artery occlusion in mice. *Neuroscience* **118**, 491–499 (2003).

124. G. Mouw, J. L. Zechel, J. Gamboa, W. D. Lust, W. R. Selman and R. A. Ratcheson, Activation of caspase-12, an endoplasmic reticulum resident caspase, after permanent focal ischemia in rat. *Neuroreport* **14**, 183–186 (2003).

125. S. F. Larner, R. L. Hayes, D. M. McKinsey, B. R. Pike, and K. K. W.Wang, Increased expression and processing of caspase-12 after traumatic brain injury in rats. *J.Neurochem.* **88**, 78–90 (2004).

126. C. Hetz, M. Russelakis-Carneiro, K. Maundrell, J. Castilla, and C. Soto, Caspase-12 and endoplasmic reticulum stress mediate neurotoxicity of pathological prion protein. *EMBO J.* **22**, 5435–5445 (2003).

127. G. Mallucci, A. Dickinson, J. Linehan, P. C. Klohn, S. Brandner and J. Collinge, Depleting neuronal PrP in prion infection prevents disease and reverses spongiosis. *Science.* **302**, 871–874 (2003).

128. S. Brandner, S. Isenmann , A. Raeber, M. Fischer, A. Sailer, Y. Kobayashi, S. Marino, C. Weissmann and A. Aguzzi, Normal host prion protein necessary for scrapie-induced neurotoxicity. *Nature*, **379**, 339–343, (1996).

129. L. Solforosi, J. R. Criado, D. B. McGavern, S. Wirz, M. Sanchez-Alavez, S. Sugama, L. A. DeGiorgio, B. T. Volpe, E. Wiseman, G. Abalos, E. Masliah, D. Gilden, M. B. Oldstone, B. Conti and R. A. Williamson, Cross-linking cellular prion protein triggers neuronal apoptosis *in vivo*. *Science* **303**, 1514–1516 (2004).

130. K. Sandvig and B. van Deurs, Transport of protein toxins into cells: pathways used by ricin, cholera toxin and Shiga toxin. *FEBS Lett.* **529**, 49–53 (2002).

131. B. Tsai and T. A. Rapoport, Unfolded cholera toxin is transferred to the ER membrane and released from protein disulfide isomerase upon oxidation by Ero1. *J Cell Biol.* **159**, 207–216 (2002).

132. F. Beranger, A. Mange, B. Goud and S. Lehmann, Stimulation of PrP(C) retrograde transport toward the endoplasmic reticulum increases accumulation of PrP(Sc) in prion-infected cells. *J.Biol.Chem.* **277**, 38972–38977 (2002).

133. T. Nakagawa and J. Yuan, Cross-talk between two cysteine protease families. Activation of caspase-12 by calpain in apoptosis. *J. Cell Biol.* **150**, 887–894 (2000).

134. H. Fischer, U. Koenig, L. Eckhart and E. Tschachler, Human caspase 12 has acquired deleterious mutations. *Biochem. Biophys. Res. Commun.* **293**, 722–726 (2002).

135. C. D. Bown, J. F. Wang, B. Chen and L. T. Young, Regulation of ER stress proteins by valproate: therapeutic implications. *Bipolar Disorders* **4**, 145–151 (2002).

136. M. Flores-Diaz, J. C. Higuita, I. Florin, T. Okada, P. Pollesello, T. Bergman, M. Thelestam, K. Mori and A. Alape-Giron, A cellular UDP-glucose deficiency causes overexpression of glucose/oxygen-regulated proteins independent of the endoplasmic reticulum stress elements. *J. Biol. Chem.* **279**, 21724–21731 (2004).

Chapter 14

CELL CULTURE MODELS TO UNRAVEL PRION PROTEIN FUNCTION AND ABERRANCIES IN TSE

Katarina Bedecs

Department of Biochemistry and Biophysics, Stockholm University, S-10691 Stockholm, Sweden

14.1. Introduction

Already in the very beginning of prion research, tissue cultures that could support and propagate the scrapie agent were sought for. The earliest attempts were explants from brains of infected mice, and their growth and morphological characteristics were compared to those from uninfected mice[1]. Using the explant technique, several investigators reported increased cell growth in cultures established from scrapie-sick brain compared to cultures from normal mice[1–4]. These are odd findings in the light of the massive neuronal cell death known to occur in scrapie-infected brains. However the cell types responsible for the increased cell growth in the scrapie-explants most probably were not neuronal.

The first successful cell culture established in this way, in which the scrapie agent was serially and continuously passaged beyond the initial explant, was in the SMB culture (scrapie mouse brain)[5], which is still used today[6,7].

In this chapter, I will go through the generation and use of chronically prion-infected cell lines as cell culture models of prion diseases. These cell lines have been crucial for the current understanding of the cell biology of both the normal (PrP^{C}) and the pathogenic isoform (PrP^{Sc}) of the prion protein. They have also been useful in the development of anti-prion drugs, prospectively used for therapy of prion diseases, and offer an alternative approach for transmission/infectivity assays normally performed by mouse bioassay. Cell culture models have also been used to study prion-induced cytopathological changes, which could explain the typical spongiform neurodegeneration in prion diseases.

In summary, I will go through:

1) The generation of scrapie-infected cell lines
2) Detection of PrP^{Sc} in cell lines
3) Cell biology and putative functions of PrP^{C}
   a) Cellular localization of PrP^{C}
   b) Putative PrP^{C} binding partners
   c) PrP^{C} knock-out mice
   d) Involvement of PrP^{C} in copper metabolism and oxidative stress
   e) The role of PrP^{C} in cell survival and differentiation
   f) Signal transduction by PrP^{C}
4) Aberrancies in cell biology and biochemistry in prion-infected cells

14.2. Generation of scrapie-infected cells

As described above, the initial attempts to create scrapie-propagating cell cultures were performed by cultivating explants from scrapie-infected brains[1–4]. However, only the SMB cells remain infectious after years of in vitro passages[5–7]. The apparent disadvantage of generating scrapie-infected cell cultures by this technique, is the lack of uninfected controls. Therefore, alternative methods were developed. One such method was to intravenously inoculate splenotropic tumour cell lines derived from e. g. macrophage, B- or T-lymphocytes and erythroid lineages, into scrapie-infected mice (which after a few weeks after infection produce high titers of scrapie in the spleen). The presumably scrapie-infected tumors were subsequently explanted and their in vitro passaging was resumed. Sadly, none of the cultures established in this way were infected[8]. After these primary attempts, cells from several sources and a variety of experimental approaches have been used to establish scrapie-infected cultures. The most straightforward approach i.e. the exposure of cell monolayers or cell suspensions to homogenates of scrapie-infected brains or partly or highly purified preparations, proved to be a successful one, to many if not all cell types[8–11]. Many of the currently used chronically prion-infected cell cultures have been generated this way (Table 14.1).

Of all the available cell cultures today, I have focused on the N2a and GT1 cells because PrP^{Sc} accumulates mainly in neurons and these are the targets for prion-induced neurodegeneration. However, non-neuronal cells, eg. the recently developed Schwann cell cultures, may be useful to study the peripheral steps of prion invasion[22,23].

Table 14.1. Cell lines supporting prion-infection

Cell type	Species	Tissue or cell of origin	Prion isolate	Comments	References
Neuronal cell types					
N2a	Mouse	Neuroblastoma	Chandler		8,11,12
C-1300	Mouse	Neuroblastoma	Chandler		8,12
N1E-115	Mouse	Neuroblastoma	Chandler, C506		8,13,14
N2a #58	Mouse	Neuroblastoma	Chandler, FU	N2a 6x over-expressing PrP^C	15,16
SHSY-5Y	Human	Neuroblastoma	CJD		17
GT1-1	Mouse	Hypothalamic neuronal cell subclone 1	Chandler,RML, FU	Large T immortalized cells	16,18
GT1-7	Mouse	Hypothalamic neuronal cellsubclone 7	Chandler, 139A, 22L, FU, SY	Large T immortalized cells	15,16,18,19
PC12	Rat	Pheochromocytoma	139A/ME7	Neuronal differentiation in presence of NGF	20,21
Non-neuronal cell types					
MSC-80	Mouse	Schwann cell	Chandler		22
MovS6/S2	TgMouse	Schwann cell-like, DRG	PG127	Ovine PrP in Tg Mouse	23
HaB	Hamster	Non-neuronal hamster brain cell	Sc237 rods	Spontaneously immortalized cells	24
Glial cell	Rat	Glial cells from rat tri-geminal ganglion	Chandler	Ethylnitrosourea-induced tumour	25
SMB	Mouse	Mesodermal cells from scrapie-infected mouse brain (Chandler)		Still propagating scrapie	5,26
SMB-PS	Mouse		Chandler, 22F	Pentosan sulfate-cured SMB	7
L-fibroblast	Mouse	Subclone of L929 fibroblast cell	Chandler		27
L23	Mouse	Subclone of L929 fibroblast cell	Compton ME7		28
L929	Mouse	Fibroblast cell	RML, 22L, ME7		29
NIH/3T3	Mouse	Fibroblast cell	22L		29
NS1	Mouse	Spleen cell from scrapie-sick mouse fused w. NS1 cell	Chandler		10
Rov	Rabbit	Kidney epithelial cell	PG127, LA404	RK-13 expressing ovine PrP^C	30,31

* = FU = mouse-adapted Fukuoka-1 (familial GSS), SY = mouse-adapted sporadic CJD

Looking at this table, a striking fact is that most of the cell lines susceptible for persistent prion-infection are of non-neuronal origin. Among the murine neuronal cell lines capable of continuously replicating prions are the N2a, N1E-115 and C-1300 neuroblastoma cell lines. Interestingly, all these cell lines have a common origin, derived from a spontaneous tumor arising in A/J mice, a *Prnpa/a* mouse strain[32], but with different passage histories. This particular cell lineage, therefore appears to be especially susceptible to scrapie infection when infected with mouse–adapted prion isolates. This is clearly apparent in view of the large number of neuronal or neural cell lines which could not be infected[8–11,13,33]. Except for the C-1300-derived clones, only one other murine neuronal cell line is easily infected. That is the GT1 cell line, originating from hypothalamic neurons and immortalized by genetically targeted tumorigenesis in transgenic mice[34]. The GT1 cells are highly differentiated gonadotropin-releasing hormone neurons, and are in contrast to the C-1300 cells, susceptible to prions other than the Chandler/RML isolate[18]. The GT1 cells have shown to replicate both the scrapie-derived 139A and 22L prions, as well as fGSS- and sCJD- derived FU and SY prions, respectively, whereas only transfected N2a cells (N2a#58) overexpressing PrP^{C} are possible to infect with isolates other than Chandler/RML[15,16].

The GT1 cells offer several important advantages over the use of N2a cells, especially for studying cytopathological effects provoked by prion infection. To begin with they express approximately eight times higher levels of endogenous PrP^{C}[15] and are much more susceptible to prion-infection than N2a cells (in our lab every single attempt to RML-infect GT1 cells have succeeded), providing a simple means to produce several independently prion-infected ScGT1 cell lines.

In contrast, in typical N2a cultures exposed to prions, $<2\%$ of the cells become infected and only low levels of prions are produced. Also, these cultures frequently loose infectivity within 10–15 passages[11,12,35]. In order to obtain persistently prion-infected cultures that produce sufficient quantities of PrP^{Sc}, the prion-infected N2a cultures (ScN2a) must be subcloned[11,36]. By this method, ScN2a lines have been obtained, in which 80–90% of the cells are infected. These ScN2a subclones have been very useful for studying the cell biology of prion replication i.e. to determine the subcellular location of PrP^{Sc} as well as the metabolism and kinetics of PrP^{Sc} formation[24,37–40]. ScN2a subclones are also suitable in the search for inhibitors of PrP^{Sc} formation and in the development of antiprion therapeutic agents[41–45]. However, for studies aiming at elucidating potential neurotoxic changes induced by prion infection, these subclones are less suitable, as these studies may suffer from the fact that ScN2a cells are subcloned from a population of cells, to which they are then compared. The observed differences between ScN2a and the uninfected N2a population, might just represent cloning artifacts or be due to a selection artifact during the prion infection. An elegant way to avoid this possibility was to derive highly susceptible N2a sublines (by subcloning the N2a population prior to prion infection) from which prion-infected cultures were generated, without further subcloning[35]. These infected sublines can be compared to the corresponding uninfected ones, without interference from potential cloning artifacts.

The GT1 cells, however, do not need to be subcloned after prion-infection, as a much higher proportion of the cells become infected[15,18], and therefore represent an excellent culture model, avoiding clonal differences. In addition, GT1 cells are the only CNS-derived neuronal cells susceptible to prion-infection today.

14.3. Detection of PrP^{Sc} in infected cells—the definition of prion infection in cell cultures

Though not fully understood, the infectious agent is largely, if not exclusively an abnormal form of the host's own PrP^{C} (reviewed in[46]). The pathogen-associated form, PrP^{Sc} differs from the normal form in its biochemical and biophysical properties, including protease-resistance, solubility in non-denaturing detergents and resistance to phosphatidylinositol specific phospholipase C[47,48].

The most straightforward approach to detect PrP^{Sc} would be to use a PrP^{Sc}-specific antibody. Despite years of intense effort, only two PrP^{Sc}-specific antibodies have been described—a 1997 report that has neither been confirmed nor extended in the literature[49], and a recent report by the group of Cashman[50].

Instead, infectivity has traditionally been linked to the presence of a proteinase K (PK)-resistant core of PrP^{Sc}, PrP27-30[47], which can be detected by Western blot analysis, using C-terminally directed anti-PrP antibodies. However, in some experimental setups, protease-resistant PrP could not be found in samples that contained prion infectivity[51–55]. Conversely, not all PK-resistant PrP species are associated with prion infectivity[56,57]. Thus it is important to recognize that the presence of PK-resistant PrP species is a mere surrogate marker for prion-infection. However, in most of the cell cultures described in Table 14.1, a correlation between PK-resistant PrP and prion infectivity was confirmed, by inoculating mice with cell culture extracts (after sufficient passages to rule out remaining infectivity from the intitial inoculum). Several techniques with different sensitivities to detect PrP^{Sc} in prion-infected cultures, have been developed (Figure 14.1).

With the generation of more specific PrP-antibodies, the standard PK-digestion-Western blot assay[58] is progressively replaced by quicker and more sensitive techniques. Whereas Western blot could detect PrP^{Sc}

Figure 14.1. Different methods to detect PrP^{Sc} in prion-infected cells. (A) Western blot. Lysates of GT1 or ScGT1 are treated with proteinase K (PK) and analyzed by SDS–PAGE and Western blotting with anti-PrP antibody. PrP27-30, the resistant core of PrP^{Sc} is detected only in infected cells. (B) Cell blot. GT1 and ScGT1 cells grown on cover-slip are blotted directly on to a nitrocellulose membrane and PrP^{Sc} is detected, after PK-digestion and denaturation with guanidine isothiocyanate of the membrane, by immunoblotting with anti-PrP antibody. (C) Dot blot. Cell lysates of GT1 and ScGT1 cells are filtered through a nitrocellulose membrane with a dot blot device and PrP^{Sc} is detected as described for B.

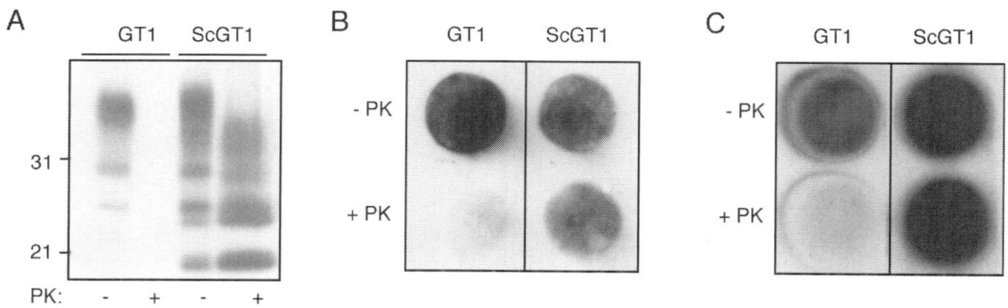

when 10% of the cells in a 6-cm diameter dish were infected, a sensitive cell blot technique could detect PrP^{Sc} when only 1% of the cells in a single well of a 24-well plate were infected, corresponding to a 150-fold increase in sensitivity[35]. Another method is the filter-retention assay (slot–blot), which takes into account two diagnostic criteria in combination; the protease resistance and the presence of detergent-insoluble aggregates of PrP^{Sc}[59]. This assay, in addition to classical ELISA[60,61], are well suited assays for high throughput screenings of therapeutic compounds. With none of these methods can the percentage of cells producing PrP^{Sc} be determined. However, a post-embedded method has been developed, which enables the detection of a single infected cell[30].

14.4. Cell biology and putative functions of PrP^c

14.4.1. Cellular localization of PrP^c

The development of cell cultures has been fundamental for the understanding of the cell biology of PrP^c, including its intracellular trafficking, localization and interaction with other proteins/molecules. Several techniques, including immunofluorescence, metabolic labeling, cell surface biotinylation, Western blot, immunoprecipitation, and cross-linking techniques have been used to determine the trafficking and localization of PrP^c. PrP^c is a glycosylphosphatidylinositol-anchored glycoprotein, generally found at the cell membrane associated with cholesterol and glycolipid-rich microdomains, called lipid rafts or DRM (detergent-resistant microdomains)[39,40,62]. Once on the cell surface, PrP^c is endocytosed, cleaved in its central domain and then recycled to the surface or degraded[63,64]. Despite many important discoveries, the cellular biology of PrP^C remains unclear. A comprehensive review on the endocytic pathway of PrP^C, its degradation or the importance of the cleavage for its function is found in chapter 15.

14.4.2. Putative PrP^c binding partners

One way to elucidate the cellular function of PrP^c has been the search of putative interacting ligands. The two-hybrid system of yeast, which can be used to demonstrate protein-protein interaction, have identified the anti-apoptotic Bcl-2[65,66], the chaperone Hsp60[67], the 37-kDa laminin receptor precursor (LRP)[68], as well as the adaptor protein Grb2, Synapsin1 and an unknown protein Pint1 as possible binding partners[69], when screening HeLa cell or mouse brain cDNA libraries. Screening of a mouse brain cDNA expression library with a soluble PrP-probe revealed an interaction with NF-E2 related factor 2 (Nrf2) transcription factor and amyloid precursor-like protein 1(Aplp1)[70]. As PrP^C is predominantly expressed as a GPI-anchored plasma membrane protein, an interaction with many of these proteins may present a logistic problem. In addition, the biological significance of many of these interactions remains unclear and confirmation in functional assays has yet to be established.

However, two proteins singled out from these studies as putative binding partner; the 37-kDa LRP and its mature 67 kDa form, termed the high affinity laminin receptor LR[68,71,72]. The LRP/LR was suggested to act as a putative PrP receptor as they were shown, by immunofluorescence studies, to colocalize with PrP^C on the surface of both N2a and baby hamster kidney (BHK) cells. Cell-binding assays with exogenously applied PrP on these cell cultures, revealed an LRP/LR-dependent binding and internalization of PrP^C, which was inhibited in the presence of an anti-LRP antibody[71]. Furthermore, the LRP/LR was also shown to be required for PrP^{Sc} propagation in ScN2a and ScGT1 cells, as I expression of an antisense LRP RNA-expression plasmid, II transfection with small interfering RNAs specific for the LRP mRNA or III incubation with an anti-LRP/LR antibody inhibited the accumulation of PrP^{Sc} in these cells. These findings suggest that the LRP/LR may not only act as a cell surface receptor for PrP^C, possibly involved in the endocytic pathway of PrP^c, but might also represent the portal of entry for PrP^{Sc} in prion infection[73]. Another report also suggests PrP^c to be a ligand to a yet unknown heterophilic receptor mediating neuronal recognition, as shown by increased neurite outgrowth from primary neurons on a PrP^c-coated substratum, compared to non-PrP coated dishes[74].

Alternative methods to find interacting partners are co-immunoprecipitation or chemical crosslinking, to demonstrate either a direct physical interaction or at least co-localization. PrP^C could be immunoprecipitated with antibodies directed against stress inducible protein-1 (STI-1) from PrP^C transfected HEK 293T cells[75] and Grb2, Synapsin1 and Pint1 were coimmunoprecipitated from BHK cells, showing that PrP^C can specifically interact with these proteins not only in a yeast model but also in mammalian cells[69]. Co-immunoprecipitation from hamster brain extacts, in presence of detergents that either preserve or completely dissociate lipid rafts, showed that PrP^C interacts strongly with neuronal nitric oxide synthase (nNOS) and dystroglycan, a transmembrane protein that is the core of the dystrophin-glycoprotein complex[76]. Binding of PrP^C to neural cell adhesion molecules (N-CAMs) was demonstrated in N2a and ScN2a cells by use of mild formaldehyde crosslinking, followed by SDS-PAGE and HPLC/mass spectrometry. This interaction was further shown to occur in raft domains[77].

Many studies indicate that heparan sulfate (HS) proteoglycans play a role in the life cycle of PrP^C and possibly also in the formation of PrP^{Sc} (reviewed in[78]). First, HS interacts with PrP^C[79–81], an interaction possibly involving the N-terminal octarepeat region of PrP^C and which is weakened in the presence of Cu(II) ions[80], although disputed by others[82,83]. Second, a variety of sulfated glycans, such as heparan sulfate mimetics[79,84,85], pentosan polysulfate and dextran sulfate[86,87], reduce the formation of PrP^{Sc} in infected ScN2a and ScGT1 cultures, and in some cases prolong the incubation time of experimental prion diseases[88,89]. HS was also shown to accumulate in cerebral prion amyloid plaques[90] and was associated with the more diffuse PrP^{Sc} deposits that appear in early stages of prion diseases[91]. Functionally, HS and sulfated cell-surface glycans seem to play a role in PrP^C internalization, as treatment of N2a cells with sulfated glycans stimulate endocytosis of PrP^C[82,92]. This may account for the antiprion effect of sulfated glycans. Alternatively, these findings suggest that exogenously applied sulfated glycans inhibit PrP^{Sc} formation by competing with the binding of PrP^C and/or PrP^{Sc} to a putative cellular HS proteoglycan. The involvement of a cellular HS proteoglycan was recently demonstrated by Taraboulos's group, showing that heparinase treatment[93] or long term incubation of ScN2a with chlorate (which is a global sulfation inhibitor), reduced PrP^{Sc} levels[79,93].

14.4.3. PrP^c knock-out mice

Another common route to determine protein function is to ablate the gene of interest and examine homozygous null mice for novel phenotypes. However, there is always a degree of uncertainty in this approach since compensatory expression of other genes may occur during embryogenesis. Several lines of mice devoid of PrP^C have been generated by homologous recombination in embryonic stem cells, using either of two strategies. One in which the disruptive modification is restricted to the open reading frame (ORF) generated the Npu[94] and Zrch1[95] lines. Mice homozygous for the inactivated gene (Prnp-/-) develop normally, show no striking pathology, and are resistant to prion infection.

Behavioural studies of the Npu and Zrch1 mice, revealed no significant differences compared to wild-type[94,96], except for alterations in circadian activity[97,98] and synaptic behaviour in brain slices[99], although synaptic changes have not been confirmed by others[100,101]. However, multiple biochemical changes were found in cerebellar cultures from these mice, such as increased levels of nuclear factor kappa B (NFκB), increased Mn superoxide dismutase (SOD) levels, decreased level of Cu/Zn SOD activity, decreased p53 and altered melatonin levels[102].

The other strategy involves deletion of not only the reading frame, but also its flanking regions. The three lines generated this way; Ngsk[103], Rcm0[104] and Zrch2[105] also develop normally, but exhibit severe ataxia and Purkinje cell loss later in life[106–108]. In these mice, Purkinje cell loss and ataxia were shown to be due to ectopic expression of an unknown protein, doppel, which is not expressed in the Npu and Zrch1 mice[106,109]. Doppel (German for double) or Dpl (downstream of the Prnp locus) was expressed because of an intergenic splicing event that placed the Dpl gene under the control of the Prnp promotor. Thus, ectopic expression of Dpl in the absence of PrP^C, rather than absence of PrP^C itself causes Purkinje cell loss.

To circumvent the problems involved in the interpretation of PrP^C-null mice, two lines of conditional knock-out mice have been generated, with an intact PrP^C expression during embryogenesis which can be knocked down in the adult mouse. Unfortunately, these mice did not reveal the function of PrP^C as no obvious effects upon neuronal viability, neuropathological abnormalities or neurological status were observed[110,111].

14.4.4. Involvement of PrP^c in copper metabolism and oxidative stress

PrP^c is a metal ion binding protein, binding copper ions with high affinity, as well as nickel, zinc and manganese cations, but with lower affinities. Full-length recombinant PrP^c can potentially bind up to five copper ions, four in the highly conserved N-terminal octarepeat region (PHGGGWGQ)[112–114], with a fifth Cu-binding site around residues His-96 and His-111[114–116]. Several studies have shown that the octarepeat segment selectively binds Cu over other divalent metal ion species[112,117,118].

Copper binding affinity has been a controversial issue. Earlier studies reported that PrP^C binds Cu with low μM to nM affinity, in a pH–dependent and cooperative manner via the octarepeats[112,118–121]. However, additional high-affinity binding sites have been characterized around His 96 and His 111 in the unstructured region of PrP^C, with *Kd* for Cu in the nM[122] and fM[115] range.

Several lines of evidence suggest a functional role of PrP^C in cellular copper metabolism and maintenance of the proper oxidative balance, possibly through a regulation of intracellular copper transport. A key finding was that copper treatment of N2a cells, stimulates the internalization of PrP^C from the cell surface[123]. This effect is temperature-dependent and rapidly reversible, and is also seen with Zn^{2+}, but not with the other transition metals. The copper-induced endocytosis is dependent on the presence of the N-terminal octarepeats. One publication suggests that specifically His68 and His76 in the central two repeats are necessary for this Cu-induced endocytosis, because mutations of these residues abolished the copper-induced endocytosis of PrP^C expressed in human neuroblastoma SH-SY5Y cells[124]. However, some other recent work has suggested that this endocytosis only requires a single histidine in the octameric repeat regions (D.R. Brown, personal communication). The importance of an intact PrP^C N-terminal for copper-induced endocytosis, was confirmed in murine septal, SN56 cells, expressing different GFP-PrP^C-constructs[125]. These findings thus suggest that the transport of copper from the extra- to intracellular compartment is performed through the internalization of PrP^c. Interestingly, Schätzl's group recently showed that the same N-terminal region is important for the basal copper-independent endocytosis of PrP^C in N2a cells[126].

Findings supporting a role of PrP^C in copper transportation and in cellular antioxidant defenses is that PrP^C-deficient mice exhibit a 50% lower copper concentration in synaptosomal fractions[120], together with reduced Cu/Zn SOD (SOD1) and glutathione reductase activities[120,127,128], although these findings have been debated[129,130]. The reduced SOD1 activity in PrP-deficient mice could be attributed to decreased Cu-delivery by PrP^C to SOD1 in these mice, with a subseqently reduced SOD1 activity. In line with this assumption, is the finding that re-introducing Prnp in an immortalized Prnp-deficient neuronal cell line results in an upregulated SOD activity[131]. Then again, a recent study of SOD activity in genetically defined crosses of mice lacking the

Sod1 gene with mice lacking PrP^C, hemizygous or homozygous Tg20 transgenic mice overexpressing PrP^C failed to detect any correlation between Prnp gene dosage on SOD activity in synaptosome-enriched brain fractions[132].

To add some complexity, it has been reported that recombinant and native PrP^C *per se* possesses SOD-like activity, when refolded in the presence of copper chloride[133]. Comparison of the SOD activity in brain lysates from wild-type mice before and after PrP^C depletion with immobilized anti-PrP^C antibodies, showed a reduced level of total SOD activity after removal of PrP^C[134]. Thus, an alternative and/or additional explanation for the decreased SOD activity in Prnp-/- mice, could be the absence of PrP's inherent SOD-like activity.

14.4.5. The role of PrP^c in cell survival and differentiation

Several lines of evidence suggest that PrP^C may play an important role in neuronal survival and/or differentiation, possibly acting like a cell-surface receptor itself. A putative role of PrP^C in neuronal differentiation was demonstrated by laminin-induced neuritogenesis of primary neurons from wild-type but not PrP-null (Zrch1) mice[135]. This finding was confirmed in the PC12 cell model, showing that antibodies against PrP^C inhibit cell adhesion to laminin-coated dishes and laser-induced ablation of PrP^C inhibits laminin-induced differentiation as well as promotes retraction of pre-formed neurites[136]. Additional results, supporting a role of PrP^C in cell-cell interaction and differentiation show that the PrP^C expression in a rat neuroblastoma cell B104 is tightly regulated both as a function of cell density and during neuronal differentiation[137].

A neuroprotective function of PrP^C has been suggested as PrP-null (Ngsk) neuronal cell lines are more vulnerable to serum deprivation than their PrP^C-expressing counterparts[138], and PrP^C could protect human primary neurons against Bax-mediated apoptosis, with the same neuroprotective potency as Bcl-2[139]. Interesting, in the light of the finding that PrP^C can bind Bcl-2[65,66]. Moreover, antibody-mediated ligation of PrP^C or binding of a PrP^C-binding peptide protects retinal neuroblastic layer cells from anisomycin-induced apoptosis from wild-type, but not PrP^C-null mice, in a cAMP/PKA dependent manner[140]. In addition, PrP^C was shown to be dramatically upregulated in tumour necrosis factor α (TNFα)-resistant human breast carcinoma MCF7 sub-clones and overexpression of PrP^C in nomally TNFα-sensitive MCF7 cells protects them against TNFα-induced apoptosis[141]. This anti-apoptotic function could be due to a PrP^C-mediated decrease in the expression of several pro-apoptotic proteins, as PrP^C expression in a PrP^C-null neuronal cell line decreased the levels of p53, Bax, and caspase-3 and increased Bcl-2 levels[142]. These data are contradicted by others, who on the contrary suggest a pro-apoptotic function of PrP^c. These authors reported an increase in the expression and activity of p53 and caspase-3 when expressing PrP^C in TSM1 and HEK239 cells[143,144] or in primary cultures from PrP^C null-mice[144].

These apparently paradoxical findings may be explained by the use of neuronal cells derived from PrP^c null mice generated by the two different strategies (described in the PrP^C knock-out mice section). Thus, Kim et. al used the Ngsk mouse[142], which besides ablated PrP^C expression also express the doppel protein, whereas Paitel et al. derived their cells from the Zrch1 mouse[144], which does not express doppel. However, the expression of doppel in these cells or the influence of doppel on the measured parameters, was not addressed in these papers. The above shows that the results depend on which type of knockout mouse provided the PrP-null cell line. Therefore, the interpretation of these results would gain from reconsideration.

14.4.6. Signal transduction by PrP^c

Because PrP^c may act as a cell surface receptor for a still hypothetical extra-cellular ligand, various studies aiming at unraveling its signalling potential has been performed. As the ligand is still unknown, one way to activate or stimulate PrP^c, was to cross-link PrP^C by use of specific anti-PrP^c antibodies, directed against PrP(142-160). Using this technique it was demonstrated that PrP^c can function in a signal transduction cascade upstream of the protein tyrosine kinase Fyn. A neuroectodermal progenitor cell line (1C11) was used and a PrP^c-dependent Fyn activation was observed when the 1C11 cells had been differentiated to their serotonergic or noradrenergic progenies. Because PrP^C and Fyn are bound to opposing faces of the membrane, a transmembrane factor that could function as a link between PrP^c and Fyn was searched for. Caveolin-1 was presented as a candidate based on its coimmunoprecipitative behavior and the finding that cellular bombardment with caveolin-1-directed antibodies abolished PrP^C dependent Fyn activation[145]. Interesting to note is that cross-linking of PrP^C *in vivo,* with antibodies directed against PrP(95-105), was found to trigger rapid and extensive apoptosis in hippocampal and cerebellar neurons, but not an antibody directed against PrP(133-157)[146], the same epitope recognized by the anti-PrP antibody shown to induce Fyn activation in 1C11 cells[145].

In a follow-up study, the same authors recently showed that following ligation of PrP^C, NADPH oxidase, a major cellular generator of reactive oxygene species (ROS) and extracellular regulated kinase 1/2 (ERK1/2) members of the mitogen-activated protein kinase (MAPK) family, are targets of the caveolin-dependent Fyn activation, but only in the fully differentiated serotonergic or adrenergic 1C11[147]. In the 1C11 progenitor, hypothalamic GT1-7 and BW5147 lymphoid cells, antibody-mediated ligation of PrP^C also induced activation of NADPH oxidase and ERK1/2. However the involvement of Fyn kinase in NADPH oxidase and ERK1/2 activation was not adressed in these cells[147]. These findings are attractive because of the potential role of PrP^C in neuronal survival and/or differentiation, as both ROS and ERK1/2 are important mediators in these cellular processes. They may also clarify the potential link between PrP^C and the cellular redox state. Several other studies have also proposed a link between PrP^C and Fyn or Src, the prototype member of the Src-family kinases. In enterocytes PrP^C and Src were shown to colocalize in raft domains in cell-cell junctional complexes and a direct physical interaction with Src was also demonstrated by coimmunoprecipitaiton of PrP^C[148]. In the human T-cell line CEM, PrP^C colocalizes with and coimmunoprecipitates Fyn, and was also shown to interact with the tyrosine kinase ZAP70 after T-cell activation, suggesting that PrP^C is a component of the signaling complex involved in T-cell activation[149]. An interesting finding in this context is that STI571 (Gleevec®, a potent inhibitor of the BCR-Abl tyrosine kinase), potently inhibits PrP^{Sc} replication in ScN2a, ScGT1 and SMB cells, via lysosomal degradation of existing PrP^{Sc}[150].

As some neuronal cell lines, such as the GT1-7[147,151] and N2a[39,152,153], do not express caveolin, the existence of an alternative signaling route from PrP^C to Fyn is suggested. One such candidate could be N-CAM, which was shown to associate with PrP^C[77]. This scenario is supported by data showing a direct interaction of N-CAM with Fyn and the selective inhibition of N-CAM-dependent neurite outgrowth in neurons from Fyn–null mice[154,155].

Alternatively, no transmembrane intermediate may be necessary at all, suggested by previous studies showing that antibody-ligation of the glycosphingolipids in rafts[156,157], as well as that of GPI-anchored proteins, induces a transient increase in the tyrosine phosphorylation of several substrates. This model thus propose that glycosphingolipid-crosslinking may induce a redistribution and clustering of signaling components in rafts, including Src-family kinases, on the opposite cytoplasmic leaflet resulting in the activation of these kinases (reviewed in[158]).

Consistent with a role of PrP^C in signal transduction is the finding that PrP^C interacts with Grb2, as shown by coimmunoprecipitation from

PrP^C-and Grb2-transfected HEK 293T cells[69]. This interaction might be the result of their colocalization in rafts, as both these proteins, together with other signaling intermediates in the MAPK pathway, such as Shc, Ras, phosphatidylinositol 3-kinase (PI3K) and ERK1/2 are concentrated to rafts[159,160]. Taken together, all these findings suggest that PrP^C has a significant role in signal transduction via lipid rafts, and the next question is how signal transduction may be affected by PrP^{Sc} accumulation in general and in rafts in particular.

14.5. Abberancies in cell biology and biochemistry in prion-infected cells—a long list of biochemical abnormalities

One of the major objectives in the development of prion-infected cell cultures was to look for morphological and cytopathological manifestations of the prion infection. However, one of the striking features was the lack of any obvious signs of cell death. In fact several studies described an increased cell growth in cultures established from scrapie-sick brain compared to cultures from uninfected mice[1–4]. Increased cell growth and transformation was also reported when N1E-115 was inoculated with brain homogenate from C506-infected mice[13]. Unfortunately, it is not clear if the changes described were necessarily due only to the scrapie agent rather than to other factors present in the brain homogenate. Although no gross morphological differences have been observed, alterations in the expression and/or function of several proteins in scrapie-infected cells have been described, possibly corresponding to the pre-clinical and subacute manifestations in prion diseases. NGF-differentiated scrapie-infected PC12 cells display a slightly modified phenotype with a decreased activity of choline acetyltransferase and acetylcholinesterase[21].

In ScN2a and ScHab cells, bradykinin- and PDGF-induced calcium responses were significantly reduced, compared to uninfected cells[161,162]. Although the number of ^{125}I-bradykinin binding sites was increased 4-fold in ScN2a, their binding affinity was reduced 10-fold, probably explaining the decreased Ca^{2+} responses[163]. These changes were further ascribed to a significant reduction of plasma membrane fluidity[163]. The authors hypothesize that the conversion of PrP^C to PrP^{Sc} and the subsequent shunting to secondary lysosomes, may alter protein and lipid trafficking pathways sufficiently to change the composition and properties of the plasma membrane, resulting in abnormal receptor-mediated functions.

We have speculated that prion-infection may alter the expression, processing and/or function of neurotrophic receptors and thereby contribute to neurodegeneration in prion diseases by weakening the trophic support in prion-infected neurons. Indeed, our results show that scrapie infection induces a 2- and 4-fold increase in insulin receptor (IR)[164] and insulin-like growth factor-1 receptor (IGF-1R)[14] protein levels, respectively, in two independently scrapie-infected cells lines; the ScN2a and ScN1E-115. However, despite the increased IR/IGF-1R expression in ScN2a, receptor binding studies revealed an important decrease in ^{125}I-insulin and ^{125}I-IGF-1 binding sites compared to the amount of immunoreactive receptors. In the case of the IGF-1R, the absence of increased number of binding sites was due to a 7-fold decrease in IGF-1R binding affinity in ScN2a compared to N2a cells[14]. However, binding studies revealed no change in IR binding affinity, rather indicating a complete functional loss of a sub-population of IR[164]. In addition, ScN2a showed no significant difference in cell proliferation in the presence of insulin or IGF-1 as the only mitogen, despite the increased receptor expression, probably explained by the decrease in IR/IGF-1R binding sites.

Further studies revealed that the apparent loss of insulin binding sites was due to an increased formation of IR/IGF-1R hybrid receptors, with high affinity to IGF-1 but a 10-20 fold lower affinity to insulin than the homotypic IR[165]. Moreover, the IR α- and β-subunits are aberrantly processed with apparent molecular weights of 128 and 85 kD in ScN2a, as compared to 136 and 95 kD in uninfected N2a cells[164,165], and the reason for this was shown to be due to altered glycosylation, as shown by combined enzymatic or chemical deglycosylation of the IR[164,165]. In addition to these differences in IR properties, the basal ERK2 activity was significantly elevated and the insulin stimulated-ERK2 phosphorylation was subsequently decreased in ScN2a. This is interesting in the light of the finding that antibody-mediated ligation of PrP^C induces activation of NADPH oxidase and ERK1/2[147]. Thus, it can be speculated that the propagation and presence of PrP^{Sc} in raft-domains may crosslink and oligomerize colocalized PrP^C, resulting in an uncontrolled NADPH oxidase and ERK1/2 activation.

The observed changes in IR/IGF-1R expression and function in ScN2a and ScN1E-115 cells, suggest that although these receptors are expressed, their folding, trafficking or processing is disturbed by scrapie infection, resulting in decreased function and/or decreased levels of functional receptors, which may contribute to neuronal cell death in prion diseases. Altered expression, processing, localization, and function have previously been described for several proteins in scrapie-infected cells and brains[161,163,166–169].

Autoradiographic studies of hippocampus from scrapie-infected mice revealed a marked decrease in neuropeptide Y2-receptor binding, although corresponding Y2 mRNA levels were essentially unchanged[166], before the appearance of behavioral symptoms. Abnormal intracellular localization linked to functional impairment of heat shock proteins[167,168] nNOS[169] has also been demonstrated in scrapie-infected brain and in ScN2a.

In addition, ScN2a cells do not respond to lipopolysaccharide (LPS)-stimulation with increased nitric oxide (NO) production, in contrast to N2a cells, which upon LPS-stimulation respond with a robust and time-dependent increase in inducible NOS (iNOS) expression[170].

LPS-signalling is initiated by binding of LPS to the LPS binding protein (LBP), present in serum (Figure 14.2) . This complex is then transferred to the major LPS receptor, CD14, a GPI-anchored protein localized to rafts[171]. LPS binding to CD14 further induces clustering and interaction with the transmembrane Toll-like receptor 4 (Tlr4) in rafts[172,173], which activates associated Ser/Thr kinase complexes, finally activating the transcription factor nuclear factor κB (NFκB)[174]. Activated NFκB then translocates to the nucleus and induces the expression of several pro-inflammatory genes eg. iNOS and cyclooxygenase 2 (COX-2).

In ScN2a, the absence of LPS-induced NO production was not due to abolished enzymatic activity of iNOS, but due to a complete inhibition of the LPS-induced iNOS gene expression as measured by reverse transcription-polymerase chain reaction (RT-PCR) and western blot[170]. In a first attempt to explain the loss of LPS-receptor signalling in ScN2a, we studied the expression of CD14 in ScN2a, since CD14 acts as the principal LPS-recognizing component in the LPS receptor complex. RT-PCR and amplification of genomic DNA, revealed a complete lack of CD14 mRNA expression in ScN2a, although the gene encoding CD14 is still present, excluding a chromosomal difference (Figure 14.3).

Although CD14 is not a signalling molecule, it is necessary for LPS to induce an efficient response[171,175]. In ScN2a the absence of CD14 expression, prevents further interaction with the transmembrane and signalling Tlr4 and thereby disrupt the LPS-induced signal transduction, most probably explaining the absence of LPS-stimulated iNOS expression in ScN2a cells. Interestingly, downregulation of CD14 transcription has been reported in human alveolar macrophages as a defense against cytomegalovirus infection[176]. Thus the absence of CD14 mRNA in ScN2a may represent an adaptive response to protect against scrapie-infection.

The only prion-infected cell line exhibiting morphological signs of neurodegeneration and vacuolation is the ScGT1 cell line[18]. In scrapie-infected GT1 cells, stably transfected with trkA (the high affinity nerve

Figure 14.2. LPS receptor signalling. Toll-like receptor-4 activates several intracellular pathways including NFκB activation, resulting in expression of e.g. iNOS and COX-2. LBP, LPS binding protein; MyD88, myeloid differentiation factor 88; IRAK, IL-1 receptor-associated kinase; TRAF6, TNF-α receptor-associated factor 6; ECSIT, evolutionary conserved signaling intermediate in Toll pathways; NIK, NIK = NFκB-inducing kinase; IKK = IκB kinase, NFκB = nuclear factor κB, IκB = inhibitor of NFκB

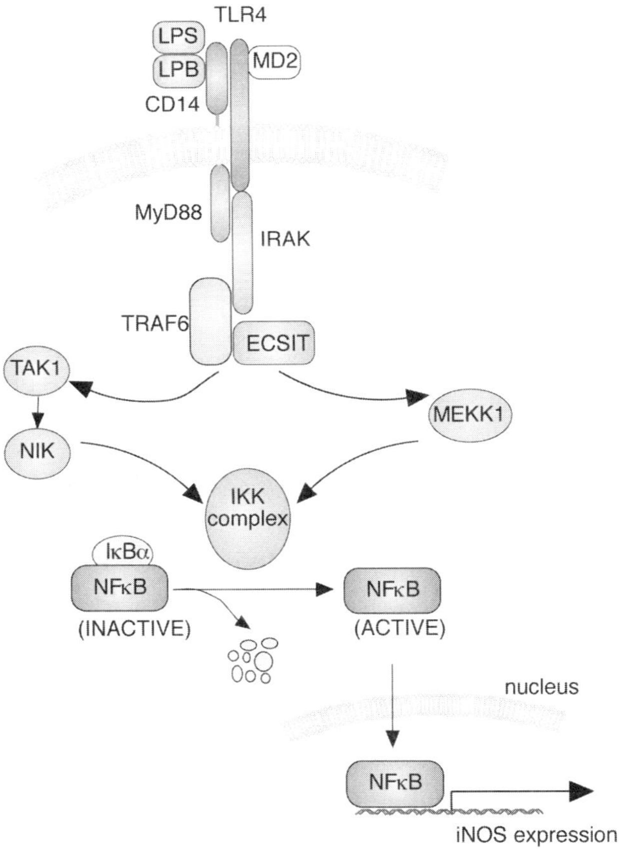

Figure 14.3. Genomic DNA was isolated or cDNA was prepared from each cell type and amplified by PCR with CD14 or GAPDH specific primers at 58°C for 28 cycles. BV-2 is a microglia cell line responding to LPS by iNOS expression and was used as a positive control for CD14 mRNA expression. Three independent experiments were performed. A representative image is shown.

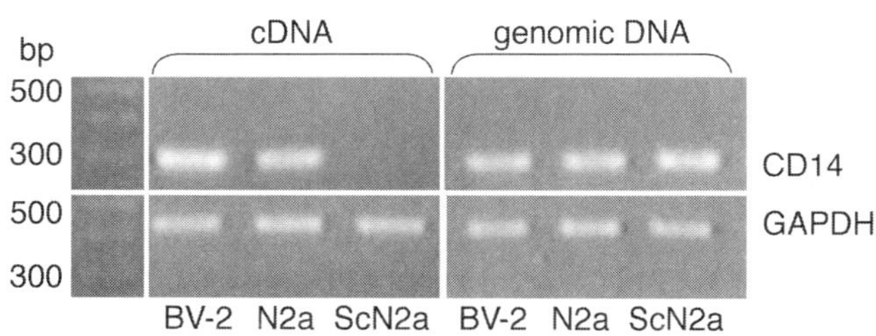

growth factor (NGF) receptor)—ScGT1-TrkA, NGF-treatment increased the viability and reduced the morphological signs of apoptosis. ScGT1 cells also display a higher sensitivity to induced oxidative stress than GT1 cells[19], together with an increased lipid peroxidation and reduction in the activities of the glutathione-dependent and SOD antioxidant systems. These findings are indeed very important because ScGT1 cells reproduce some of the major pathological changes found in prion-infected brains and therefore represent a good system for studying the molecular mechanisms underlying prion-induced neurodegeneration. Using the same cell line, it was also shown that scrapie-infection inhibits the voltage-dependent N-type calcium channel, a dysfunction which was exacerbated slowly over time after scrapie-infection in GT1-7 cells[177].

14.6. Concluding remarks

Although the establishment of scrapie-infected cell lines or cell lines expressing mutant PrP^{C}s linked to hereditary prion diseases have been crucial for the understanding of the biogenesis and metabolism of PrP^{Sc}, the weak point in using scrapie-infected cell cultures to study molecular mechanisms underlying prion-induced neurodegeneration is the lack of apparent neurotoxicity in these cultures.

The discrepancy between *in vivo* and *in vitro* PrP^{Sc} neurotoxicity might be due to the transformed phenotypes of the available cell culture models today, masking the PrP^{Sc} neurotoxicity which may manifest itself only in finally differentiated cells, resembling more the post-mitotic character of the adult central nervous system. This may explain why ScGT1 cells, which has a more differentiated neuronal phenotype than ScN2a and ScN1E-115 cells, do display signs of neurodegeneration whereas ScN2a and ScN1E-115 cells do not.

Alternatively, the formation and/or accumulation of PrP^{Sc} is not neurotoxic *per se* but "neuro-stressant" and requires an interaction with microglia/astrocytes to be neurotoxic. Although the cell types involved in scrapie pathology have not been completely identified the emerging picture indicates a glial-derived inflammatory component in prion pathogenesis[178]. Only circumstantial evidence suggest that PrP^{Sc} is neurotoxic *per se* because its accumulation in scrapie infected brains correlates to areas of vacuolation and astrogliosis. Studies addressing the spatial and temporal relationships between PrP^{Sc} accumulation, glial activation, neuronal vacuolation and apoptosis, indicate that microglia and astroglia activation precedes that of spongiform degeneration[179,180]. Altogether this indicates an indirect glial neurotoxic effect, possibly mediated by micro- and astroglial cytokines and reactive oxygen species in scrapie pathogenesis—an inflammatory "vicious cycle" also suggested as causative/contributory in other chronic

Figure 14.4. Involvement of an inflammatory loop in prion-induced neurodegeneration

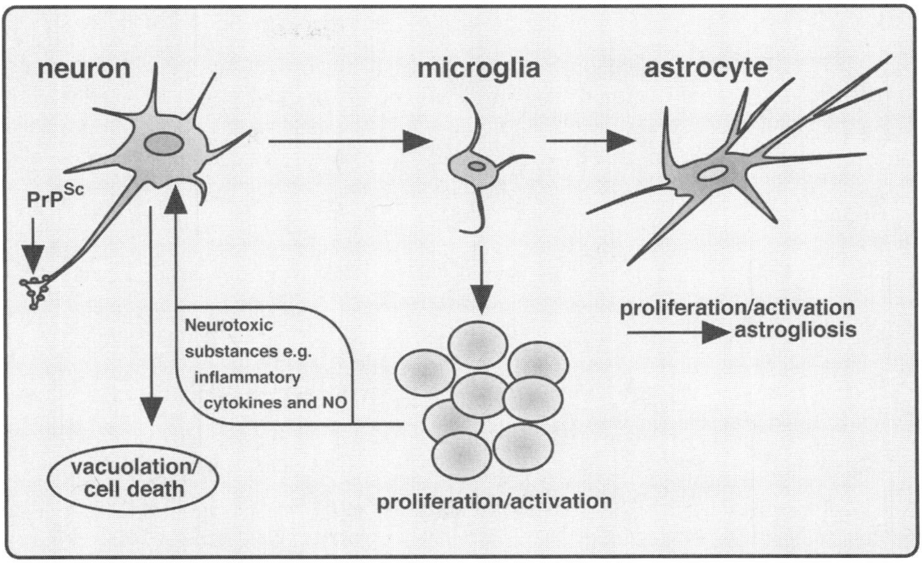

neurodegenerative diseases such as Alzheimer's and amyotrophic lateral sclerosis (ALS) (Figure 14.4).

I hypothesize that the formation and accumulation of PrP^{Sc} in combination with an inflammatory glial response has a synergistic deteriorating effect on neurons—together adding up to a fatal threshold level of cellular stress resulting in neuronal apoptosis. Therefore a truer cell-culture model of prion-induced neurodegeneration might require the creation of a "mini-brain" in the culture-dish, constituting besides neuronal cells also glial cells.

14.7. References

1. E. J. Field and G. D. Windsor, Cultural Characters of Scrapie Mouse Brain. *Res Vet Sci* **35**, 130–132 (1965).

2. D. A. Haig and I. H. Pattison, *In vitro* growth of pieces of brain from scrapie-affected mice. *J Pathol Bacteriol* **93**,724–727 (1967).

3. E. A. Caspary and T. M. Bell, Growth potential of scrapie mouse brain *in vitro*. *Nature* **229**, 269–270 (1971).

4. G. M. Buening and D. P. Gustafson, Growth characteristics of scrapie agent-infected mouse brain cell cultures. *Am J Vet Res* **32**, 953–958 (1971).

5. M. C. Clarke and D. A. Haig, Evidence for the multiplication of scrapie agent in cell culture. *Nature* **225**, 100–101. (1970).

6. C. Bate, M. Salmona, L. Diomede and A. Williams, Squalestatin cures prion-infected neurons and protects against prion neurotoxicity. *J Biol Chem* **279**, 14983–14990 (2004).

7. C. R. Birkett, R. M. Hennion, D. A. Bembridge, M. C. Clarke, A. Chree, M. E. Bruce and C. J. Bostock, Scrapie strains maintain biological phenotypes on propagation in a cell line in culture. *Embo J* **20**, 3351–3358. (2001).

8. R. Race, The scrapie agent in vitro. *Curr Top Microbiol Immunol* **172**, 181–193. (1991).

9. B. Chesebro, K. Wehrly, B. Caughey, J. Nishio, D. Ernst and R. Race, Foreign PrP expression and scrapie infection in tissue culture cell lines. *Dev Biol Stand* **80**, 131–140. (1993).

10. C. J. Elleman, Attempts to establish the scrapie agent in cell lines. *Vet Res Commun* **8**, 309–316. (1984).

11. D. A. Butler, M. R. Scott, J. M. Bockman, D. R. Borchelt, A. Taraboulos, K. K. Hsiao, D. T. Kingsbury and S. B. Prusiner, Scrapie-infected murine neuroblastoma cells produce protease-resistant prion proteins. *J Virol* **62**, 1558–1564. (1988).

12. R. E. Race, L. H. Fadness and B. Chesebro, Characterization of scrapie infection in mouse neuroblastoma cells. *J Gen Virol* **68**, 1391–1399. (1987).

13. P. Markovits, C. Dautheville, D. Dormont, L. Dianoux and R. Latarjet, In vitro propagation of the scrapie agent. I. Transformation of mouse glia and neuroblastoma cells after infection with the mouse-adapted scrapie strain c-506. *Acta Neuropathol (Berl)* **60**, 75–80. (1983).

14. P. Östlund, H. Lindegren, C. Pettersson and K. Bedecs, Up-regulation of functionally impaired insulin-like growth factor-1 receptor in scrapie-infected neuroblastoma cells. *J Biol Chem* **276**, 36110–36115. (2001).

15. N. Nishida, D. A. Harris, D. Vilette, H. Laude, Y. Frobert, J. Grassi, D. Casanova, O. Milhavet and S. Lehmann, Successful transmission of three mouse-adapted scrapie strains to murine neuroblastoma cell lines overexpressing wild-type mouse prion protein. *J Virol* **74**, 320–325. (2000).

16. A. Arjona, L. Simarro, F. Islinger, N. Nishida and L. Manuelidis, Two Creutzfeldt-Jakob disease agents reproduce prion protein-independent identities in cell cultures. *Proc Natl Acad Sci USA* **101**, 8768–8773 (2004).

17. A. Ladogana, Q. Liu, Y. G. Xi and M. Pocchiari, Proteinase-resistant protein in human neuroblastoma cells infected with brain material from Creutzfeldt-Jakob patient. *Lancet* **345**, 594–595. (1995).

18. H. M. Schätzl, L. Laszlo, D. M. Holtzman, J. Tatzelt, S. J. DeArmond, R. I. Weiner, W. C. Mobley and S. B. Prusiner, A hypothalamic neuronal cell line persistently infected with scrapie prions exhibits apoptosis. *J Virol* **71**, 8821–8831. (1997).

19. O. Milhavet, H. E. McMahon, W. Rachidi, N. Nishida, S. Katamine, A. Mange, M. Arlotto, D. Casanova, J. Riondel, A. Favier and S. Lehmann, Prion infection impairs the cellular response to oxidative stress. *Proc Natl Acad Sci USA* **97**, 13937–13942. (2000).

20. R. Rubenstein, R. I. Carp and S. M. Callahan, *In vitro* replication of scrapie agent in a neuronal model: infection of PC12 cells. *J Gen Virol* **65**, 2191–2198. (1984).

21. R. Rubenstein, H. Deng, R. E. Race, W. Ju, C. L. Scalici, M. C. Papini, R. J. Kascsak and R. I. Carp, Demonstration of scrapie strain diversity in infected PC12 cells. *J Gen Virol* **73**, 3027–3031. (1992).

22. J. Follet, C. Lemaire-Vieille, F. Blanquet-Grossard, V. Podevin-Dimster, S. Lehmann, J. P. Chauvin, J. P. Decavel, R. Varea, J. Grassi, M. Fontes and J. Y. Cesbron, PrP expression and replication by Schwann cells: implications in prion spreading. *J Virol* **76**, 2434–2439. (2002).

23. F. Archer, C. Bachelin, O. Andreoletti, N. Besnard, G. Perrot, C. Langevin, A. Le Dur, D. Vilette, A. Baron-Van Evercooren, J. L. Vilotte and H. Laude, Cultured peripheral neuroglial cells are highly permissive to sheep prion infection. *J Virol* **78**, 482–490 (2004).

24. A. Taraboulos, D. Serban and S. B. Prusiner, Scrapie prion proteins accumulate in the cytoplasm of persistently infected cultured cells. *J Cell Biol* **110**, 2117–2132. (1990).

25. V. M. Roikhel, G. I. Fokina, V. M. Lisak, L. I. Kondakova, M. B. Korolev and V. V. Pogodina, Persistence of the scrapie agent in glial cells from rat Gasserian ganglion. *Acta Virol* **31**, 36–42. (1987).

26. D. A. Haig and M. C. Clarke, Multiplication of the scrapie agent. *Nature* **234**, 106–107. (1971).

27. M. C. Clarke and G. C. Millson, Infection of a cell line of mouse L fibroblasts with scrapie agent. *Nature* **261**, 144–145. (1976).

28. N. Cherednichenko Yu, G. R. Mikhailova, J. Rajcani and V. M. Zhdanov, *In vitro* studies with the scrapie agent. *Acta Virol* **29**, 285–293. (1985).

29. I. Vorberg, A. Raines, B. Story and S. A. Priola, Susceptibility of common fibroblast cell lines to transmissible spongiform encephalopathy agents. *J Infect Dis* **189**, 431–439 (2004).

30. D. Vilette, O. Andreoletti, F. Archer, M. F. Madelaine, J. L. Vilotte, S. Lehmann and H. Laude, Ex vivo propagation of infectious sheep scrapie agent in heterologous epithelial cells expressing ovine prion protein. *Proc Natl Acad Sci USA* **98**, 4055–4059. (2001).

31. E. Sabuncu, S. Petit, A. Le Dur, T. Lan Lai, J. L. Vilotte, H. Laude and D. Vilette, PrP Polymorphisms Tightly Control Sheep Prion Replication in Cultured Cells. *J Virol* **77**, 2696–2700. (2003).

32. R. J. Klebe and F. H. Ruddle, Neuroblastoma: Cell culture analysis of a differentiating stem cell system. *J Cell Biol* **43**, (1969).

33. R. T. Yanagihara, D. M. Asher, C. J. Gibbs, Jr. and D. C. Gajdusek, Attempts to establish cell cultures infected with the viruses of subacute spongiform encephalopathies. *Proc Soc Exp Biol Med* **165**, 298–305 (1980).

34. P. L. Mellon, J. J. Windle, P. C. Goldsmith, C. A. Padula, J. L. Roberts and R. I. Weiner, Immortalization of hypothalamic GnRH neurons by genetically targeted tumorigenesis. *Neuron* **5**, 1–10 (1990).

35. P. J. Bosque and S. B. Prusiner, Cultured cell sublines highly susceptible to prion infection. *J Virol* **74**, 4377–4386. (2000).

36. R. E. Race, B. Caughey, K. Graham, D. Ernst and B. Chesebro, Analyses of frequency of infection, specific infectivity, and prion protein biosynthesis in scrapie-infected neuroblastoma cell clones. *J Virol* **62**, 2845–2849. (1988).

37. D. R. Borchelt, M. Scott, A. Taraboulos, N. Stahl and S. B. Prusiner, Scrapie and cellular prion proteins differ in their kinetics of synthesis and topology in cultured cells. *J Cell Biol* **110**, 743–752. (1990).

38. B. Caughey and G. J. Raymond, The scrapie-associated form of PrP is made from a cell surface precursor that is both protease- and phospholipase-sensitive. *J Biol Chem* **266**, 18217–18223. (1991).

39. A. Gorodinsky and D. A. Harris, Glycolipid-anchored proteins in neuroblastoma cells form detergent-resistant complexes without caveolin. *J Cell Biol* **129**, 619–627. (1995).

40. A. Taraboulos, M. Scott, A. Semenov, D. Avrahami, L. Laszlo, S. B. Prusiner and D. Avraham, Cholesterol depletion and modification of COOH-terminal targeting sequence of the prion protein inhibit formation of the scrapie isoform. *J Cell Biol* **129**, 121–132. (1995).

41. B. Caughey and R. E. Race, Potent inhibition of scrapie-associated PrP accumulation by congo red. *J Neurochem* **59**, 768–771. (1992).

42. K. F. Winklhofer and J. Tatzelt, Cationic lipopolyamines induce degradation of PrPSc in scrapie-infected mouse neuroblastoma cells. *Biol Chem* **381**, 463–469. (2000).

43. A. Mange, N. Nishida, O. Milhavet, H. E. McMahon, D. Casanova and S. Lehmann, Amphotericin B inhibits the generation of the scrapie isoform of the prion protein in infected cultures. *J Virol* **74**, 3135–3140. (2000).

44. C. Korth, B. C. May, F. E. Cohen and S. B. Prusiner, Acridine and phenothiazine derivatives as pharmacotherapeutics for prion disease. *Proc Natl Acad Sci USA* **98**, 9836–9841. (2001).

45. S. Supattapone, K. Nishina and J. R. Rees, Pharmacological approaches to prion research. *Biochem Pharmacol* **63**, 1383–1388. (2002).

46. S. B. Prusiner, Prions. *Proc Natl Acad Sci USA* **95**, 13363–13383. (1998).

47. D. C. Bolton, M. P. McKinley and S. B. Prusiner, Molecular characteristics of the major scrapie prion protein. *Biochemistry* **23**, 5898–5906. (1984).

48. B. Caughey, K. Neary, R. Buller, D. Ernst, L. L. Perry, B. Chesebro and R. E. Race, Normal and scrapie-associated forms of prion protein differ in their sensitivities to phospholipase and proteases in intact neuroblastoma cells. *J Virol* **64**, 1093–1101. (1990).

49. C. Korth, P. Streit and B. Oesch, Monoclonal antibodies specific for the native, disease-associated isoform of the prion protein. *Methods Enzymol* **309**, 106–122. (1999).

50. E. Paramithiotis, M. Pinard, T. Lawton, S. LaBoissiere, V. L. Leathers, W. Q. Zou, L. A. Estey, J. Lamontagne, M. T. Lehto, L. H. Kondejewski, G. P. Francoeur, M. Papadopoulos, A. Haghighat, S. J. Spatz, M. Head, R. Will, J. Ironside, K. O'Rourke, Q. Tonelli, H. C. Ledebur, A. Chakrabartty and N. R. Cashman, A prion protein epitope selective for the pathologically misfolded conformation. *Nat Med* **9**, 893–899 (2003).

51. L. Manuelidis, T. Sklaviadis and E. E. Manuelidis, Evidence suggesting that PrP is not the infectious agent in Creutzfeldt-Jakob disease. *EMBO J* **6**, 341–347. (1987).

52. Y. G. Xi, L. Ingrosso, A. Ladogana, C. Masullo and M. Pocchiari, Amphotericin B treatment dissociates *in vivo* replication of the scrapie agent from PrP accumulation. *Nature* **356**, 598–601. (1992).

53. K. K. Hsiao, D. Groth, M. Scott, S. L. Yang, H. Serban, D. Rapp, D. Foster, M. Torchia, S. J. Dearmond and S. B. Prusiner, Serial transmission in rodents of neurodegeneration from transgenic mice expressing mutant prion protein. *Proc Natl Acad Sci USA* **91**, 9126–9130. (1994).

54. C. I. Lasmezas, J. P. Deslys, O. Robain, A. Jaegly, V. Beringue, J. M. Peyrin, J. G. Fournier, J. J. Hauw, J. Rossier and D. Dormont, Transmission of the BSE agent to mice in the absence of detectable abnormal prion protein. *Science* **275**, 402–405. (1997).

55. S. Tzaban, G. Friedlander, O. Schonberger, L. Horonchik, Y. Yedidia, G. Shaked, R. Gabizon and A. Taraboulos, Protease-sensitive scrapie prion protein in aggregates of heterogeneous sizes. *Biochemistry* **41**, 12868–12875. (2002).

56. A. F. Hill, M. Antoniou and J. Collinge, Protease-resistant prion protein produced *in vitro* lacks detectable infectivity. *J Gen Virol* **80**, 11–14. (1999).

57. G. M. Shaked, G. Fridlander, Z. Meiner, A. Taraboulos and R. Gabizon, Protease-resistant and detergent-insoluble prion protein is not necessarily associated with prion infectivity. *J Biol Chem* **274**, 17981–17986. (1999).

58. R. K. Meyer, M. P. McKinley, K. A. Bowman, M. B. Braunfeld, R. A. Barry and S. B. Prusiner, Separation and properties of cellular and scrapie prion proteins. *Proc Natl Acad Sci USA* **83**, 2310–2314. (1986).

59. K. F. Winklhofer, F. U. Hartl and J. Tatzelt, A sensitive filter retention assay for the detection of PrP(Sc) and the screening of anti-prion compounds. *FEBS Lett* **503**, 41–45. (2001).

60. D. Serban, A. Taraboulos, S. J. DeArmond and S. B. Prusiner, Rapid detection of Creutzfeldt-Jakob disease and scrapie prion proteins. *Neurology* **40**, 110–117. (1990).

61. D. Peretz, R. A. Williamson, Y. Matsunaga, H. Serban, C. Pinilla, R. B. Bastidas, R. Rozenshteyn, T. L. James, R. A. Houghten, F. E. Cohen, S. B. Prusiner and D. R. Burton, A conformational transition at the N terminus of the prion protein features in formation of the scrapie isoform. *J Mol Biol* **273**, 614–622. (1997).

62. A. Taraboulos, M. Scott, A. Semenov, D. Avrahami and S. B. Prusiner, Biosynthesis of the prion proteins in scrapie-infected cells in culture. *Braz J Med Biol Res* **27**, 303–307 (1994).

63. B. Caughey, G. J. Raymond, D. Ernst and R. E. Race, N-terminal truncation of the scrapie-associated form of PrP by lysosomal protease(s): implications regarding the site of conversion of PrP to the protease-resistant state. *J Virol* **65**, 6597–6603. (1991).

64. S. L. Shyng, M. T. Huber and D. A. Harris, A prion protein cycles between the cell surface and an endocytic compartment in cultured neuroblastoma cells. *J Biol Chem* **268**, 15922–15928. (1993).

65. C. Kurschner and J. I. Morgan, The cellular prion protein (PrP) selectively binds to Bcl-2 in the yeast two-hybrid system. *Brain Res Mol Brain Res* **30**, 165–168 (1995).

66. C. Kurschner and J. I. Morgan, Analysis of interaction sites in homo- and heteromeric complexes containing Bcl-2 family members and the cellular prion protein. *Brain Res Mol Brain Res* **37**, 249–258 (1996).

67. F. Edenhofer, R. Rieger, M. Famulok, W. Wendler, S. Weiss and E. L. Winnacker, Prion protein PrPc interacts with molecular chaperones of the Hsp60 family. *J Virol* **70**, 4724–4728 (1996).

68. R. Rieger, F. Edenhofer, C. I. Lasmezas and S. Weiss, The human 37-kDa laminin receptor precursor interacts with the prion protein in eukaryotic cells. *Nat Med* **3**, 1383–1388 (1997).

69. C. Spielhaupter and H. M. Schätzl, PrP^C directly interacts with proteins involved in signaling pathways. *J Biol Chem* **276**, 44604–44612 (2001).

70. F. Yehiely, P. Bamborough, M. Da Costa, B. J. Perry, G. Thinakaran, F. E. Cohen, G. A. Carlson and S. B. Prusiner, Identification of candidate proteins binding to prion protein. *Neurobiol Dis* **3**, 339–355 (1997).

71. S. Gauczynski, J. M. Peyrin, S. Haik, C. Leucht, C. Hundt, R. Rieger, S. Krasemann, J. P. Deslys, D. Dormont, C. I. Lasmezas and S. Weiss, The 37-kDa/67-kDa laminin

receptor acts as the cell-surface receptor for the cellular prion protein. *EMBO J* **20**, 5863–5875 (2001).

72. C. Hundt, J. M. Peyrin, S. Haik, S. Gauczynski, C. Leucht, R. Rieger, M. L. Riley, J. P. Deslys, D. Dormont, C. I. Lasmezas and S. Weiss, Identification of interaction domains of the prion protein with its 37-kDa/67-kDa laminin receptor. *EMBO J* **20**, 5876–5886 (2001).

73. C. Leucht, S. Simoneau, C. Rey, K. Vana, R. Rieger, C. I. Lasmezas and S. Weiss, The 37 kDa/67 kDa laminin receptor is required for PrP(Sc) propagation in scrapie-infected neuronal cells. *EMBO Rep* **4**, 290–295 (2003).

74. S. Chen, A. Mange, L. Dong, S. Lehmann and M. Schachner, Prion protein as trans-interacting partner for neurons is involved in neurite outgrowth and neuronal survival. *Mol Cell Neurosci* **22**, 227–233 (2003).

75. S. M. Zanata, M. H. Lopes, A. F. Mercadante, G. N. Hajj, L. B. Chiarini, R. Nomizo, A. R. Freitas, A. L. Cabral, K. S. Lee, M. A. Juliano, E. de Oliveira, S. G. Jachieri, A. Burlingame, L. Huang, R. Linden, R. R. Brentani and V. R. Martins, Stress-inducible protein 1 is a cell surface ligand for cellular prion that triggers neuroprotection. *EMBO J* **21**, 3307–3316 (2002).

76. G. I. Keshet, O. Bar-Peled, D. Yaffe, U. Nudel and R. Gabizon, The cellular prion protein colocalizes with the dystroglycan complex in the brain. *J Neurochem* **75**, 1889–1897 (2000).

77. G. Schmitt-Ulms, G. Legname, M. A. Baldwin, H. L. Ball, N. Bradon, P. J. Bosque, K. L. Crossin, G. M. Edelman, S. J. DeArmond, F. E. Cohen and S. B. Prusiner, Binding of neural cell adhesion molecules (N-CAMs) to the cellular prion protein. *J Mol Biol* **314**, 1209–1225 (2001).

78. J. Diaz-Nido, F. Wandosell and J. Avila, Glycosaminoglycans and beta-amyloid, prion and tau peptides in neurodegenerative diseases. *Peptides* **23**, 1323–1332 (2002).

79. R. Gabizon, Z. Meiner, M. Halimi and S. A. Ben-Sasson, Heparin-like molecules bind differentially to prion-proteins and change their intracellular metabolic fate. *J Cell Physiol* **157**, 319–325 (1993).

80. R. G. Warner, C. Hundt, S. Weiss and J. E. Turnbull, Identification of the heparan sulfate binding sites in the cellular prion protein. *J Biol Chem* **277**, 18421–18430 (2002).

81. B. Caughey, K. Brown, G. J. Raymond, G. E. Katzenstein and W. Thresher, Binding of the protease-sensitive form of PrP (prion protein) to sulfated glycosaminoglycan and congo red [corrected]. *J Virol* **68**, 2135–2141 (1994).

82. T. Pan, B. S. Wong, T. Liu, R. Li, R. B. Petersen and M. S. Sy, Cell-surface prion protein interacts with glycosaminoglycans. *Biochem J* **368**, 81–90 (2002).

83. R. Gonzalez-Iglesias, M. A. Pajares, C. Ocal, J. C. Espinosa, B. Oesch and M. Gasset, Prion protein interaction with glycosaminoglycan occurs with the formation of oligomeric complexes stabilized by Cu(II) bridges. *J Mol Biol* **319**, 527–540 (2002).

84. O. Schonberger, L. Horonchik, R. Gabizon, D. Papy-Garcia, D. Barritault and A. Taraboulos, Novel heparan mimetics potently inhibit the scrapie prion protein and its endocytosis. *Biochem Biophys Res Commun* **312**, 473–479 (2003).

85. K. T. Adjou, S. Simoneau, N. Sales, F. Lamoury, D. Dormont, D. Papy-Garcia, D. Barritault, J. P. Deslys and C. I. Lasmezas, A novel generation of heparan sulfate mimetics for the treatment of prion diseases. *J Gen Virol* **84**, 2595–2603 (2003).

86. B. Caughey and G. J. Raymond, Sulfated polyanion inhibition of scrapie-associated PrP accumulation in cultured cells. *J Virol* **67**, 643–650 (1993).

87. D. B. Brimacombe, A. D. Bennett, F. S. Wusteman, A. C. Gill, J. C. Dann and C. J. Bostock, Characterization and polyanion-binding properties of purified recombinant prion protein. *Biochem J* **342 Pt 3**, 605–613 (1999).

88. A. Ladogana, P. Casaccia, L. Ingrosso, M. Cibati, M. Salvatore, Y. G. Xi, C. Masullo and M. Pocchiari, Sulphate polyanions prolong the incubation period of scrapie-infected hamsters. *J Gen Virol* **73**, 661–665 (1992).

89. B. Ehlers and H. Diringer, Dextran sulphate 500 delays and prevents mouse scrapie by impairment of agent replication in spleen. *J Gen Virol* **65**, 1325–1330 (1984).

90. A. D. Snow, R. Kisilevsky, J. Willmer, S. B. Prusiner and S. J. DeArmond, Sulfated glycosaminoglycans in amyloid plaques of prion diseases. *Acta Neuropathol (Berl)* **77**, 337–342 (1989).

91. P. A. McBride, M. I. Wilson, P. Eikelenboom, A. Tunstall and M. E. Bruce, Heparan sulfate proteoglycan is associated with amyloid plaques and neuroanatomically targeted PrP pathology throughout the incubation period of scrapie-infected mice. *Exp Neurol* **149**, 447–454 (1998).

92. S. L. Shyng, S. Lehmann, K. L. Moulder and D. A. Harris, Sulfated glycans stimulate endocytosis of the cellular isoform of the prion protein, PrPC, in cultured cells. *J Biol Chem* **270**, 30221–30229 (1995).

93. O. Ben-Zaken, S. Tzaban, Y. Tal, L. Horonchik, J. D. Esko, I. Vlodavsky and A. Taraboulos, Cellular heparan sulfate participates in the metabolism of prions. *J Biol Chem* **278**, 40041–40049 (2003).

94. H. Bueler, M. Fischer, Y. Lang, H. Bluethmann, H. P. Lipp, S. J. DeArmond, S. B. Prusiner, M. Aguet and C. Weissmann, Normal development and behaviour of mice lacking the neuronal cell-surface PrP protein. *Nature* **356**, 577–582 (1992).

95. J. C. Manson, A. R. Clarke, M. L. Hooper, L. Aitchison, I. McConnell and J. Hope, 129/Ola mice carrying a null mutation in PrP that abolishes mRNA production are developmentally normal. *Mol Neurobiol* **8**, 121–127 (1994).

96. H. P. Lipp, M. Stagliar-Bozicevic, M. Fischer and D. P. Wolfer, A 2-year longitudinal study of swimming navigation in mice devoid of the prion protein: no evidence for neurological anomalies or spatial learning impairments. *Behav Brain Res* **95**, 47–54 (1998).

97. I. Tobler, S. E. Gaus, T. Deboer, P. Achermann, M. Fischer, T. Rulicke, M. Moser, B. Oesch, P. A. McBride and J. C. Manson, Altered circadian activity rhythms and sleep in mice devoid of prion protein. *Nature* **380**, 639–642 (1996).

98. I. Tobler, T. Deboer and M. Fischer, Sleep and sleep regulation in normal and prion protein-deficient mice. *J Neurosci* **17**, 1869–1879 (1997).

99. J. Collinge, M. A. Whittington, K. C. Sidle, C. J. Smith, M. S. Palmer, A. R. Clarke and J. G. Jefferys, Prion protein is necessary for normal synaptic function. *Nature* **370**, 295–297 (1994).

100. P. M. Lledo, P. Tremblay, S. J. DeArmond, S. B. Prusiner and R. A. Nicoll, Mice deficient for prion protein exhibit normal neuronal excitability and synaptic transmission in the hippocampus. *Proc Natl Acad Sci USA* **93**, 2403–2407 (1996).

101. J. W. Herms, H. A. Kretzchmar, S. Titz and B. U. Keller, Patch-clamp analysis of synaptic transmission to cerebellar purkinje cells of prion protein knockout mice. *Eur J Neurosci* **7**, 2508–2512 (1995).

102. D. R. Brown, R. S. Nicholas and L. Canevari, Lack of prion protein expression results in a neuronal phenotype sensitive to stress. *J Neurosci Res* **67**, 211–224 (2002).

103. S. Sakaguchi, S. Katamine, N. Nishida, R. Moriuchi, K. Shigematsu, T. Sugimoto, A. Nakatani, Y. Kataoka, T. Houtani, S. Shirabe, H. Okada, S. Hasegawa, T. Miyamoto and T. Noda, Loss of cerebellar Purkinje cells in aged mice homozygous for a disrupted PrP gene. *Nature* **380**, 528–531 (1996).

104. R. Moore, Gene targeting studies at the mouse prion protein locus. *PhD thesis, University of Edinburgh, Edinburgh, Scotland* (1997).

105. C. Weissmann and A. Aguzzi, Perspectives: neurobiology. PrP's double causes trouble. *Science* **286**, 914–915 (1999).

106. R. C. Moore, I. Y. Lee, G. L. Silverman, P. M. Harrison, R. Strome, C. Heinrich, A. Karunaratne, S. H. Pasternak, M. A. Chishti, Y. Liang, P. Mastrangelo, K. Wang, A. F. Smit, S. Katamine, G. A. Carlson, F. E. Cohen, S. B. Prusiner, D. W. Melton, P. Tremblay, L. E. Hood and D. Westaway, Ataxia in prion protein (PrP)-deficient mice is associated with upregulation of the novel PrP-like protein doppel. *J Mol Biol* **292**, 797–817 (1999).

107. G. L. Silverman, K. Qin, R. C. Moore, Y. Yang, P. Mastrangelo, P. Tremblay, S. B. Prusiner, F. E. Cohen and D. Westaway, Doppel is an N-glycosylated, glycosylphosphatidylinositol-anchored protein. Expression in testis and ectopic production in the brains of Prnp(0/0) mice predisposed to Purkinje cell loss. *J Biol Chem* **275**, 26834–26841 (2000).

108. D. Rossi, A. Cozzio, E. Flechsig, M. A. Klein, T. Rulicke, A. Aguzzi and C. Weissmann, Onset of ataxia and Purkinje cell loss in PrP null mice inversely correlated with Dpl level in brain. *EMBO J* **20**, 694–702 (2001).

109. A. Li, S. Sakaguchi, R. Atarashi, B. C. Roy, R. Nakaoke, K. Arima, N. Okimura, J. Kopacek and K. Shigematsu, Identification of a novel gene encoding a PrP-like protein expressed as chimeric transcripts fused to PrP exon 1/2 in ataxic mouse line with a disrupted PrP gene. *Cell Mol Neurobiol* **20**, 553–567 (2000).

110. P. Tremblay, Z. Meiner, M. Galou, C. Heinrich, C. Petromilli, T. Lisse, J. Cayetano, M. Torchia, W. Mobley, H. Bujard, S. J. DeArmond and S. B. Prusiner, Doxycycline control of prion protein transgene expression modulates prion disease in mice. *Proc Natl Acad Sci USA* **95**, 12580–12585 (1998).

111. G. R. Mallucci, S. Ratte, E. A. Asante, J. Linehan, I. Gowland, J. G. Jefferys and J. Collinge, Post-natal knockout of prion protein alters hippocampal CA1 properties, but does not result in neurodegeneration. *EMBO J* **21**, 202–210 (2002).

112. J. H. Viles, F. E. Cohen, S. B. Prusiner, D. B. Goodin, P. E. Wright and H. J. Dyson, Copper binding to the prion protein: structural implications of four identical cooperative binding sites. *Proc Natl Acad Sci USA* **96**, 2042–2047 (1999).

113. E. Aronoff-Spencer, C. S. Burns, N. I. Avdievich, G. J. Gerfen, J. Peisach, W. E. Antholine, H. L. Ball, F. E. Cohen, S. B. Prusiner and G. L. Millhauser, Identification of the Cu^{2+} binding sites in the N-terminal domain of the prion protein by EPR and CD spectroscopy. *Biochemistry* **39**, 13760–13771 (2000).

114. C. S. Burns, E. Aronoff-Spencer, G. Legname, S. B. Prusiner, W. E. Antholine, G. J. Gerfen, J. Peisach and G. L. Millhauser, Copper coordination in the full-length, recombinant prion protein. *Biochemistry* **42**, 6794–6803 (2003).

115. G. S. Jackson, I. Murray, L. L. Hosszu, N. Gibbs, J. P. Waltho, A. R. Clarke and J. Collinge, Location and properties of metal-binding sites on the human prion protein. *Proc Natl Acad Sci USA* **98**, 8531–8535 (2001).

116. K. Qin, Y. Yang, P. Mastrangelo and D. Westaway, Mapping Cu(II) binding sites in prion proteins by diethyl pyrocarbonate modification and matrix-assisted laser desorption ionization-time of flight (MALDI-TOF) mass spectrometric footprinting. *J Biol Chem* **277**, 1981–1990 (2002).

117. R. P. Bonomo, G. Imperllizzeri, G. Pappalardo, E. Rizzarelli and G. Tabbi, Copper(II) binding modes in the prion octapeptide PHGGGWGQ: a spectroscopic and voltammetric study. *Chemistry* **6**, 4195–4202 (2000).

118. J. Stöckel, J. Safar, A. C. Wallace, F. E. Cohen and S. B. Prusiner, Prion protein selectively binds copper(II) ions. *Biochemistry* **37**, 7185–7193 (1998).

119. M. P. Hornshaw, J. R. McDermott, J. M. Candy and J. H. Lakey, Copper binding to the N-terminal tandem repeat region of mammalian and avian prion protein: structural

studies using synthetic peptides. *Biochem Biophys Res Commun* **214**, 993–999 (1995).

120. D. R. Brown, K. Qin, J. W. Herms, A. Madlung, J. Manson, R. Strome, P. E. Fraser, T. Kruck, A. von Bohlen, W. Schulz-Schaeffer, A. Giese, D. Westaway and H. Kretzschmar, The cellular prion protein binds copper *in vivo*. *Nature* **390**, 684–687. (1997).

121. M. L. Kramer, H. D. Kratzin, B. Schmidt, A. Romer, O. Windl, S. Liemann, S. Hornemann and H. Kretzschmar, Prion protein binds copper within the physiological concentration range. *J Biol Chem* **276**, 16711–16719 (2001).

122. C. E. Jones, S. R. Abdelraheim, D. R. Brown and J. H. Viles, Preferential copper2+ coordination by His96 and His111 induces beta -sheet formation in the unstructured amyloidogenic region of the prion protein. *J Biol Chem* (2004).

123. P. C. Pauly and D. A. Harris, Copper stimulates endocytosis of the prion protein. *J Biol Chem* **273**, 33107–33110. (1998).

124. W. S. Perera and N. M. Hooper, Ablation of the metal ion-induced endocytosis of the prion protein by disease-associated mutation of the octarepeat region. *Curr Biol* **11**, 519–523 (2001).

125. K. S. Lee, A. C. Magalhaes, S. M. Zanata, R. R. Brentani, V. R. Martins and M. A. Prado, Internalization of mammalian fluorescent cellular prion protein and N-terminal deletion mutants in living cells. *J Neurochem* **79**, 79–87 (2001).

126. M. Nunziante, S. Gilch and H. M. Schätzl, Essential Role of the Prion Protein N Terminus in Subcellular Trafficking and Half-life of Cellular Prion Protein. *J Biol Chem* **278**, 3726–3734 (2003).

127. F. Klamt, F. Dal-Pizzol, M. J. Conte da Frota, R. Walz, M. E. Andrades, E. G. da Silva, R. R. Brentani, I. Izquierdo and J. C. Fonseca Moreira, Imbalance of antioxidant defense in mice lacking cellular prion protein. *Free Radic Biol Med* **30**, 1137–1144 (2001).

128. D. R. Brown, W. J. Schulz-Schaeffer, B. Schmidt and H. A. Kretzschmar, Prion protein-deficient cells show altered response to oxidative stress due to decreased SOD-1 activity. *Exp Neurol* **146**, 104–112 (1997).

129. D. J. Waggoner, B. Drisaldi, T. B. Bartnikas, R. L. Casareno, J. R. Prohaska, J. D. Gitlin and D. A. Harris, Brain copper content and cuproenzyme activity do not vary with prion protein expression level. *J Biol Chem* **275**, 7455–7458 (2000).

130. D. R. Brown, Prion protein expression modulates neuronal copper content. *J Neurochem* **87**, 377–385 (2003).

131. A. Sakudo, D. C. Lee, K. Saeki, Y. Nakamura, K. Inoue, Y. Matsumoto, S. Itohara and T. Onodera, Impairment of superoxide dismutase activation by N-terminally

truncated prion protein (PrP) in PrP-deficient neuronal cell line. *Biochem Biophys Res Commun* **308**, 660–667 (2003).

132. G. Hutter, F. L. Heppner and A. Aguzzi, No superoxide dismutase activity of cellular prion protein *in vivo*. *Biol Chem* **384**, 1279–1285 (2003).

133. D. R. Brown, B. S. Wong, F. Hafiz, C. Clive, S. J. Haswell and I. M. Jones, Normal prion protein has an activity like that of superoxide dismutase. *Biochem J* **344 Pt 1**, 1–5 (1999).

134. B. S. Wong, T. Pan, T. Liu, R. Li, P. Gambetti and M. S. Sy, Differential contribution of superoxide dismutase activity by prion protein *in vivo*. *Biochem Biophys Res Commun* **273**, 136–139 (2000).

135. E. Graner, A. F. Mercadante, S. M. Zanata, O. V. Forlenza, A. L. Cabral, S. S. Veiga, M. A. Juliano, R. Roesler, R. Walz, A. Minetti, I. Izquierdo, V. R. Martins and R. R. Brentani, Cellular prion protein binds laminin and mediates neuritogenesis. *Brain Res Mol Brain Res* **76**, 85–92 (2000).

136. E. Graner, A. F. Mercadante, S. M. Zanata, V. R. Martins, D. G. Jay and R. R. Brentani, Laminin-induced PC-12 cell differentiation is inhibited following laser inactivation of cellular prion protein. *FEBS Lett* **482**, 257–260 (2000).

137. C. Monnet, V. Marthiens, H. Enslen, Y. Frobert, A. Sobel and R. M. Mege, Heterogeneity and regulation of cellular prion protein glycoforms in neuronal cell lines. *Eur J Neurosci* **18**, 542–548 (2003).

138. C. Kuwahara, A. M. Takeuchi, T. Nishimura, K. Haraguchi, A. Kubosaki, Y. Matsumoto, K. Saeki, T. Yokoyama, S. Itohara and T. Onodera, Prions prevent neuronal cell-line death. *Nature* **400**, 225–226 (1999).

139. Y. Bounhar, Y. Zhang, C. G. Goodyer and A. LeBlanc, Prion protein protects human neurons against Bax-mediated apoptosis. *J Biol Chem* **276**, 39145–39149 (2001).

140. L. B. Chiarini, A. R. Freitas, S. M. Zanata, R. R. Brentani, V. R. Martins and R. Linden, Cellular prion protein transduces neuroprotective signals. *EMBO J* **21**, 3317–3326 (2002).

141. M. Diarra-Mehrpour, S. Arrabal, A. Jalil, X. Pinson, C. Gaudin, G. Pietu, A. Pitaval, H. Ripoche, M. Eloit, D. Dormont and S. Chouaib, Prion protein prevents human breast carcinoma cell line from tumor necrosis factor alpha-induced cell death. *Cancer Res* **64**, 719–727 (2004).

142. B. H. Kim, H. G. Lee, J. K. Choi, J. I. Kim, E. K. Choi, R. I. Carp and Y. S. Kim, The cellular prion protein (PrPC) prevents apoptotic neuronal cell death and mitochondrial dysfunction induced by serum deprivation. *Brain Res Mol Brain Res* **124**, 40–50 (2004).

143. E. Paitel, R. Fahraeus and F. Checler, Cellular prion protein sensitizes neurons to apoptotic stimuli through Mdm2-regulated and p53-dependent caspase 3-like activation. *J Biol Chem* **278**, 10061–10066 (2003).

144. E. Paitel, C. Sunyach, C. Alves da Costa, J. C. Bourdon, B. Vincent and F. Checler, Primary cultured neurons devoid of cellular prion display lower responsiveness to staurosporine through the control of p53 at both transcriptional and post-transcriptional levels. *J Biol Chem* **279**, 612–618 (2004).

145. S. Mouillet-Richard, M. Ermonval, C. Chebassier, J. L. Laplanche, S. Lehmann, J. M. Launay and O. Kellermann, Signal transduction through prion protein. *Science* **289**, 1925–1928. (2000).

146. L. Solforosi, J. R. Criado, D. B. McGavern, S. Wirz, M. Sanchez-Alavez, S. Sugama, L. A. DeGiorgio, B. T. Volpe, E. Wiseman, G. Abalos, E. Masliah, D. Gilden, M. B. Oldstone, B. Conti and R. A. Williamson, Cross-linking cellular prion protein triggers neuronal apoptosis *in vivo*. *Science* **303**, 1514–1516 (2004).

147. B. Schneider, V. Mutel, M. Pietri, M. Ermonval, S. Mouillet-Richard and O. Kellermann, NADPH oxidase and extracellular regulated kinases 1/2 are targets of prion protein signaling in neuronal and nonneuronal cells. *Proc Natl Acad Sci USA* **100**, 13326–13331 (2003).

148. E. Morel, S. Fouquet, D. Chateau, L. Yvernault, Y. Frobert, M. Pincon-Raymond, J. Chambaz, T. Pillot and M. Rousset, The cellular prion protein PrPc is expressed in human enterocytes in cell-cell junctional domains. *J Biol Chem* **279**, 1499–1505 (2004).

149. V. Mattei, T. Garofalo, R. Misasi, A. Circella, V. Manganelli, G. Lucania, A. Pavan and M. Sorice, Prion protein is a component of the multimolecular signaling complex involved in T cell activation. *FEBS Lett* **560**, 14–18 (2004).

150. A. Ertmer, S. Gilch, S. W. Yun, E. Flechsig, B. Klebl, M. Stein-Gerlach, M. A. Klein and H. M. Schätzl, The tyrosine kinase inhibitor STI571 induces cellular clearance of PrPSc in prion-infected cells. *J Biol Chem* (2004).

151. S. Prioni, N. Loberto, A. Prinetti, V. Chigorno, F. Guzzi, R. Maggi, M. Parenti and S. Sonnino, Sphingolipid metabolism and caveolin expression in gonadotropin-releasing hormone-expressing GN11 and gonadotropin-releasing hormone-secreting GT1-7 neuronal cells. *Neurochem Res* **27**, 831–840 (2002).

152. S. L. Shyng, J. E. Heuser and D. A. Harris, A glycolipid-anchored prion protein is endocytosed via clathrin-coated pits. *J Cell Biol* **125**, 1239–1250 (1994).

153. P. Oh and J. E. Schnitzer, Segregation of heterotrimeric G proteins in cell surface microdomains. G(q) binds caveolin to concentrate in caveolae, whereas G(i) and G(s) target lipid rafts by default. *Mol Biol Cell* **12**, 685–698 (2001).

154. H. E. Beggs, P. Soriano and P. F. Maness, NCAM-dependent neurite outgrowth is inhibited in neurons from Fyn-minus mice. *J Cell Biol* **127**, 825–833 (1994).

155. H. E. Beggs, S. C. Baragona, J. J. Hemperly and P. F. Maness, NCAM140 interacts with the focal adhesion kinase p125(fak) and the SRC-related tyrosine kinase p59(fyn). *J Biol Chem* **272**, 8310–8319 (1997).

156. K. Kasahara, Y. Watanabe, T. Yamamoto and Y. Sanai, Association of Src family tyrosine kinase Lyn with ganglioside GD3 in rat brain. Possible regulation of Lyn by glycosphingolipid in caveolae-like domains. *J Biol Chem* **272**, 29947–29953 (1997).

157. V. Horejsi, M. Cebecauer, J. Cerny, T. Brdicka, P. Angelisova and K. Drbal, Signal transduction in leucocytes via GPI-anchored proteins: an experimental artefact or an aspect of immunoreceptor function? *Immunol Lett* **63**, 63–73 (1998).

158. K. Kasahara and Y. Sanai, Possible roles of glycosphingolipids in lipid rafts. *Biophys Chem* **82**, 121–127 (1999).

159. R. M. Smith, S. Harada, J. A. Smith, S. Zhang and L. Jarett, Insulin-induced protein tyrosine phosphorylation cascade and signalling molecules are localized in a caveolin-enriched cell membrane domain. *Cell Signal* **10**, 355–362 (1998).

160. C. B. Wu, S. Butz, Y. S. Ying and R. G. W. Anderson, Tyrosine kinase receptors concentrated in caveolae-like domains from neuronal plasma membrane. *J Biol Chem* **272**, 3554–3559 (1997).

161. S. J. DeArmond, K. Kristensson and R. P. Bowler, PrPSc causes nerve cell death and stimulates astrocyte proliferation: a paradox. *Prog Brain Res* **94**, 437–446 (1992).

162. K. Kristensson, B. Feuerstein, A. Taraboulos, W. C. Hyun, S. B. Prusiner and S. J. DeArmond, Scrapie prions alter receptor-mediated calcium responses in cultured cells. *Neurology* **43**, 2335–2341 (1993).

163. K. Wong, Y. Qiu, W. Hyun, R. Nixon, J. VanCleff, J. Sanchez-Salazar, S. B. Prusiner and S. J. DeArmond, Decreased receptor-mediated calcium response in prion-infected cells correlates with decreased membrane fluidity and IP3 release. *Neurology* **47**, 741–750 (1996).

164. P. Östlund, H. Lindegren, C. Pettersson and K. Bedecs, Altered insulin receptor processing and function in scrapie-infected neuroblastoma cell lines. *Brain Res Mol Brain Res* **97**, 161–170. (2001).

165. D. Nielsen, H. Gyllberg, P. Östlund, T. Bergman and K. Bedecs, Increased levels of insulin and insulin-like growth factor-1 hybrid receptors and decreased glycosylation of the insulin receptor alpha- and beta-subunits in scrapie-infected neuroblastoma N2a cells. *Biochem J* **380**, 571–579 (2004).

166. M. Diez, J. Koistinaho, S. J. Dearmond, D. Groth, S. B. Prusiner and T. Hokfelt, Marked decrease of neuropeptide Y Y2 receptor binding sites in the hippocampus in murine prion disease. *Proc Natl Acad Sci USA* **94**, 13267–13272 (1997).

167. J. Tatzelt, J. Zuo, R. Voellmy, M. Scott, U. Hartl, S. B. Prusiner and W. J. Welch, Scrapie prions selectively modify the stress response in neuroblastoma cells. *Proc Natl Acad Sci USA* **92**, 2944–2948 (1995).

168. J. Tatzelt, R. Voellmy and W. J. Welch, Abnormalities in stress proteins in prion diseases. *Cell Mol Neurobiol* **18**, 721–729 (1998).

169. H. Ovadia, H. Rosenmann, E. Shezen, M. Halimi, I. Ofran and R. Gabizon, Effect of scrapie infection on the activity of neuronal nitric-oxide synthase in brain and neuroblastoma cells. *J Biol Chem* **271**, 16856–16861 (1996).

170. H. Lindegren, P. Östlund, H. Gyllberg and K. Bedecs, Loss of lipopolysaccharide-induced nitric oxide production and inducible nitric oxide synthase expression in scrapie-infected N2a cells. *J Neurosci Res* **71**, 291–299. (2003).

171. S. Akashi, H. Ogata, F. Kirikae, T. Kirikae, K. Kawasaki, M. Nishijima, R. Shimazu, Y. Nagai, K. Fukudome, M. Kimoto and K. Miyake, Regulatory roles for CD14 and phosphatidylinositol in the signaling via toll-like receptor 4-MD-2. *Biochem Biophys Res Commun* **268**, 172–177 (2000).

172. A. Pfeiffer, A. Bottcher, E. Orso, M. Kapinsky, P. Nagy, A. Bodnar, I. Spreitzer, G. Liebisch, W. Drobnik, K. Gempel, M. Horn, S. Holmer, T. Hartung, G. Multhoff, G. Schutz, H. Schindler, A. J. Ulmer, H. Heine, F. Stelter, C. Schutt, G. Rothe, J. Szollosi, S. Damjanovich and G. Schmitz, Lipopolysaccharide and ceramide docking to CD14 provokes ligand-specific receptor clustering in rafts. *Eur J Immunol* **31**, 3153–3164. (2001).

173. P. Y. Wang, R. L. Kitchens and R. S. Munford, Bacterial lipopolysaccharide binds to CD14 in low-density domains of the monocyte-macrophage plasma membrane. *J Inflamm* **47**, 126–137. (1995).

174. K. V. Anderson, Toll signaling pathways in the innate immune response. *Curr Opin Immunol* **12**, 13–19 (2000).

175. F. Stelter, Structure/function relationships of CD14. *Chem Immunol* **74**, 25–41. (2000).

176. H. A. Hopkins, M. M. Monick and G. W. Hunninghake, Cytomegalovirus inhibits CD14 expression on human alveolar macrophages. *J Infect Dis* **174**, 69–74 (1996).

177. M. K. Sandberg, P. Wallen, M. A. Wikstrom and K. Kristensson, Scrapie-infected GT1-1 cells show impaired function of voltage-gated N-type calcium channels (Ca(v) 2.2) which is ameliorated by quinacrine treatment. *Neurobiol Dis* **15**, 143–151 (2004).

178. W. K. Ju, K. J. Park, E. K. Choi, J. Kim, R. I. Carp, H. M. Wisniewski and Y. S. Kim, Expression of inducible nitric oxide synthase in the brains of scrapie-infected mice. *J Neurovirol* **4**, 445–450 (1998).

179. A. Williams, P. J. Lucassen, D. Ritchie and M. Bruce, PrP deposition, microglial activation, and neuronal apoptosis in murine scrapie. *Exp Neurol* **144**, 433–438 (1997).

180. A. Williams, A. M. Van Dam, D. Ritchie, P. Eikelenboom and H. Fraser, Immunocytochemical appearance of cytokines, prostaglandin E2 and lipocortin-1 in the CNS during the incubation period of murine scrapie correlates with progressive PrP accumulations. *Brain Res* **754**, 171–180 (1997).

Chapter 15

INSIGHTS INTO THE CELLULAR TRAFFICKING OF PRION PROTEINS

Max Nunziante, Sabine Gilch and Hermann M. Schätzl

Institute of Virology Prion Research Group, Technical University of Munich, 80802 Munich, Germany

The detailed understanding of the trafficking and the subcellular localization of the cellular prion protein (PrP^c) will contribute to a better understanding of the prion conversion process and might provide new insights into the possible functions of PrP^c. It has been known for many years that a preceding cell surface localization of PrP^c appears to be required for the conversion into the pathological isoform PrP^{Sc} in prion-infected cells[1,2]. Recently, new and often unexpected findings in the cell biology of PrP^c, the pathogenesis of PrP^{Sc}, and in cellular quality control mechanisms have been accumulating. Many of these findings were obtained by analysing the trafficking of cellular and pathological prion proteins in cell culture models, either in the presence or absence of infectious prions.

15.1. The prion protein in the secretory pathway

Upon import into the endoplasmic reticulum (ER), PrP^c is transported to the plasma membrane within 1 hour[1,2]. The transport into the ER is mediated by a N-terminal 22 amino acid (aa) signal peptide, which is co-translationally cleaved off. Furthermore, the C-terminal globular domain of PrP^c also seems to be necessary for crossing the ER membrane[3]. If this part of PrP was removed, the protein chain was not imported into the ER. Consequently, two disease-associated stop codon mutants (W145Stop and Q160Stop) remained almost entirely in the cytosol without having the signal peptide cleaved off. During the passage through the secretory pathway, two N-linked carbohydrate chains are attached to Asp 181 and Asp 197 residues (numbering related to[4]), resulting in a heavy and very complex glycosylation, and a disulphide bond is formed between the cysteine residues at position 177 and 213. Presumably 10 % of the nascent PrP^c is subjected to the endoplasmic reticulum-associated degradation (ERAD) pathway[5]. For this ubiquitous pathway, misfolded proteins are recognized by an ER-based quality control mechanism and subsequently retrograde-translocated into the cytosol for degradation by the proteasome. In addition, several disease-associated mutant forms of the prion protein have been shown to be subjected to proteasomal degradation[6–8].

For anchoring at the outer leaflet of the plasma membrane, a complex glycosyl-phosphatidyl-inositol-(GPI) anchor is co-translationally added and the C-terminal signal peptide (i. e. aa 232-254) is removed[9]. Several years ago, transmembrane forms of the prion protein were discovered, in part linked to some forms of GSS[10]. Recently, the importance of the N-terminal part of the prion protein (aa 23-90) for the transport through the secretory pathway to the plasma membrane has been highlighted[11]. It was shown that progressive deletions within the N-terminus lead to a delayed appearance of the protein at the plasma membrane. As there was no retardation in the glycosylation kinetics, it could be assumed that the delay occurred in post-ER compartments.

15.2. The membrane sorting of PrP

An important sorting station for proteins travelling through the secretory pathway on their way to the plasma membrane is the Golgi compartment. Here, the synthesis of so-called rafts is initiated. Rafts are membrane microdomains characterized by a high content of cholesterol and sphingolipids, which can be biochemically characterized by their insolubility in cold detergent, therefore also known as detergent-resistant microdomains (DRM)[12]. In membranes, they form an ordered phase where GPI-anchored proteins and proteins involved in signal transduction are preferentially localized[13]. Rafts might also offer the spatial proximity for protein interactions. For PrP^c, the proposed interaction with the still unidentified factor or protein X, assumed to be important for prion conversion, might take place in these compartments[14].

Several groups have shown that PrP, like many other GPI-anchored proteins, partitions in DRMs[15–17]. It could be demonstrated that in neuroblastoma cells and their prion-infected counterparts not only PrP^c, but also PrP^{Sc} is contained within DRMs. These results favour the hypothesis that rafts are important for prion conversion. Previously, it was shown that the inhibition of raft synthesis by cholesterol depletion, resulting in re-distribution of PrP^c at the plasma membrane, compromises prion conversion in cell culture[18]. Further evidence was achieved when the distribution of PrP^c and PrP^{Sc} regarding the co-fractionation in DRMs was investigated in cell culture and in brain homogenates of uninfected or prion-infected hamsters[16]. Again, PrP^c and PrP^{Sc} of cultured cells as well as of brain material were contained in the same gradient fractions if the lysis was performed at 4°C. Interestingly, the distribution was different in some infected cell lines and in infected hamster brain tissue if the lysates were incubated at 37°C and then subjected to the standard flotation assay. Under these conditions, PrP^{Sc} and PrP^c segregated to peaks of different density, providing an elegant tool to separate PrP^c and PrP^{Sc} for the investigation of interactors of the PrP isoforms. It was suggested that this might be due to the distribution of PrP^c and PrP^{Sc} in rafts with a different lipid composition. These diverse membrane attributes might correspond to different subcellular structures, as PrP^c is mainly found at the cell surface whereas PrP^{Sc} localises more intracellularly and accumulates finally in secondary lysosomes[19].

In order to investigate whether the intracellular sorting of PrP^c is dependent on rafts, polarized Fischer rat thyroid (FRT) and Madin Darby canine kidney (MDCK) cells were stably transfected with PrP[17]. These cells are well characterized for the intracellular trafficking and sorting of GPI-anchored proteins. PrP^c turned out to be the first example of a GPI-anchored protein localized basolaterally in both cell lines. The protein was associated with rafts, and this association was dependent on cholesterol, as it was prevented by the inhibition of cholesterol synthesis. On the other hand, the polarized sorting was not influenced by cholesterol depletion, indicating that DRMs are not necessary for the sorting and transport of PrP to the plasma membrane.

The association with rafts can be facilitated by the interaction of membrane lipids with acyl chains of a GPI-anchor[20], transmembrane domains[21], or by the binding of a protein to a protein resident in rafts. PrP represents an example of a protein containing, in addition to the GPI-anchor, an exoplasmic raft determinant within the N-terminus[22], a flexible and highly conserved PrP domain[4]. When the signal peptide of PrP necessary for GPI-anchoring was replaced with the transmembrane and cytosolic domain e. g. of the amyloid precursor protein (APP), the N-terminal PrP was sufficient to direct this transmembrane form of PrP to rafts. APP, which is usually found in clathrin-coated pits, segregated with raft marker proteins in sucrose gradients upon fusion to the PrP N-terminus. Unfortunately, in this study N-terminal deletion was only performed with the transmembrane mutant of PrP instead of the physiological GPI-anchored form. Therefore, the question remains open whether the GPI-anchor or the PrP N-terminus contain the dominant raft determinant. In addition, these data are contrary to previous reports in which the GPI-anchor was replaced by the transmembrane domains of several proteins, resulting in a non-raft membrane topology of PrP[18,23].

15.3. The mysteries of internalization

Although the localization of PrP^c in lipid rafts at a particular time during its life cycle seems to be fairly corroborated, the pathway for the internalization of the protein remains still controversial. Overall, the internalization of GPI-anchored proteins is poorly understood, as cytoplasmic domains, which contain sorting motifs mediating the interaction with adaptor proteins to facilitate proper sorting, are missing. For the prion protein either clathrin-dependent internalization or internalization via rafts have been discussed. The chicken homologue of the mammalian prion protein (chPrP) has been described to be internalized via clathrin-coated pits and to cycle between the plasma membrane and an intracellular compartment of the endocytic pathway[24,25]. This could be facilitated by the interaction of PrP^c at the plasma membrane with the extracellular part of a hypothetical transmembrane protein containing internalization and sorting signals in its cytoplasmic tail. Again, the importance of the PrP N-terminus was emphasized in mediating this proposed binding and the subsequent endocytosis not only of chPrP, but also of murine PrP[25]. Of note, this mechanism has already been described for the GPI-anchored urokinase receptor binding to the transmembrane low-density lipoprotein receptor-related protein for internalization[26]. Interestingly, the cellular uptake of extracellularly added recombinant prion protein appears to be mediated by the transmembrane laminin receptor[27]. Increased endocytosis of PrP^c was seen due to binding of copper to the N-terminus[28–31]. In cell culture experiments, PrP^c was rapidly endocytosed upon exposure of neuronal cells to physiological amounts of Cu^{2+} or Zn^{2+}, whereas deletion of four octarepeats abolished endocytosis[31]. Moreover, N-terminal deletion mutants previously shown to cause ataxia and neuronal degeneration[32] failed to internalise in response to Cu^{2+}.

The internalization of proteins requires the generation of several types of endocytic vesicles. The fission of these organells from the plasma membrane is controlled by large GTPases of the dynamin family[33]. To characterize intermediate steps in the internalization of PrP, PrP^c fused to green fluorescent protein (GFP) was expressed in SN56 cells[34]. Distinct steps of the endocytic pathway were interrupted by the overexpression of transdominant-negative mutants of dynamin I and Rab5, and the localization of GFP-PrP^c was compared to a GPI-anchored GFP (GFP-GPI), which served as a control for the clathrin-independent internalization mechanism of GPI-anchored proteins. In contrast to GFP-GPI, the internalization of GFP-PrP^c was dependent on dynamin I and it was found in Rab5-positive early endosomes. Proteins of the Rab-family are small GTPases which regulate vesicular transport in endocytosis and exocytosis[35]. Different Rab proteins are associated with various domains of endosomes. Rab5 is linked to clathrin-coated vesicles and early endosomes. Proteins co-localizing with Rab5-positive endosomes probably have been internalized via the classical endosomal pathway rather than a clathrin-independent mechanism. This leads to the conclusion that GPI-anchored proteins can be internalized by different mechanisms, which is in agreement with the observation that the internalization of Thy-I, a major GPI-anchored protein expressed on neuronal cells, is distinct from that of PrP^c [36]. Thy-I is also an example for the finding that various GPI-anchored proteins can occupy diverse types of rafts composed of different lipids, to be distinguished by their solubility in non-ionic detergent[37]. Fusion of Thy-I to the PrP N-terminus results in the ability of Thy-I to enter clathrin-coated pits, similar to PrP^c[36]. This finding is in sharp contrast to the results of Walmsley et al.[22], who claimed that the PrP N-terminus mediates the targeting of the protein to rafts. Interestingly, PrP^c seems to be localized in rafts and might leave them to enter clathrin-coated pits for internalization[36].

Mutation of the highly conserved cluster of basic amino acids between position 23 and 28 abrogated endocytosis[36], in line with an earlier report about the analysis of N-terminal deletion mutants which were deceleratedly internalized[11]. These findings argue against endocytosis of PrP stimulated by binding of Cu^{2+} to the octarepeat region, even more as the effect of N-terminal deletion could be competed by the insertion of the N-terminus of *X. laevis* prion protein[11], which does not harbour an octarepeat element at all[38]. In summary, there are clear hints that the N-terminal part of PrP acts as a sorting and internalization signal for PrP^c.

In contrast to the repeated reports about PrP^c being internalized via clathrin-coated pits, but supporting earlier findings[15,16,29], is a recent electron microscopic study showing that PrP^c is enriched in caveolae and caveolae-containing endocytic vesicles (caveosomes) in CHO cells[39]. Caveolae, which are flask-shaped plasma membrane invaginations, are involved in cellular processes like signal transduction[40]. The regulated endocytosis of caveolae has been demonstrated to require dynamin[41]. This could also explain the inhibited endocytosis of GFP-PrP^c upon overexpression of a transdominant-negative dynamin-I mutant[36], therefore this observation does not provide definite evidence for internalization by clathrin-coated pits. However, it should be taken into consideration that for the electron microscopy study a non-neuronal cell line has been used, which could explain the observed differences. Thus, neuronal cell models might provide a more physiological tool to study the trafficking of PrP. Nevertheless, in a murine neuronal cell differentiation model, a functional connection of PrP^c with caveolin has been observed. Upon antibody-mediated cross-linking of PrP^c, a caveolin-1 dependent association with Fyn kinase has been shown, arguing for a role of the prion protein in signal transduction[42].

The pathway of PrP^c internalization remains still a matter of debate. The association of PrP with lipid rafts and the role of the N-terminus acting as a determinant for the trafficking seem to be convincing, and there is growing evidence that PrP is internalized via the classical endosomal pathway rather than in a clathrin-independent manner.

15.4. The finaldestiny—degradation of PrP

The degradation of PrP^c involves several steps. The half-life time of the protein in cultured cells is around 3–5 hours[2,11,43]. The proteolysis can be inhibited by ammonium chloride (NH_4Cl) treatment, a substance which raises the pH within acidic vesicles and thereby inactivates proteases. This indicates that the proteolysis occurs in such organelles[18].

The degradation of PrP is initiated during its cycling between the plasma membrane and an intracellular compartment by the removal of the amino terminal moiety of the protein[18,44–47]. This cleavage gives rise to an N-terminal fragment of about 11.5 kDa, denominated N1, and a C-terminal half with a molecular weight of approximately 17 kDa, referred to as C1[45] or PrP-II[18]. Both the amino- and the carboxy-terminal fragments are re-cycled to the plasma membrane, and whereas the short N-terminal fragment is released into the culture medium, the C-terminal part accumulates at the plasma membrane[44]. The compartment where the cleavage proceeds remains ambiguous. ChPrP expressed in the murine neuroblastoma cell line N2a was cleaved in a NH_4Cl-sensitive manner, suggesting acidic vesicles[44]. In contrast, the processing of mouse PrP investigated in the same cell line was insensitive to NH_4Cl treatment, but was inhibited by perturbation of the cholesterol synthesis. The C-terminal part was, like full-length PrP^c, found to be associated with DRMs[18]. The N-terminal truncation appears to be a physiological degradation step. C-terminal cleavage products with a molecular weight consistent with that found in cell culture have also been detectable in uninfected and infected human brain tissue[48,49]. Interestingly, amino terminal sequencing of C1 revealed that the cleavage occurs between aa 110/111 and 112[48,45]. Thereby, the normal processing of PrP^c is initiated by the cleavage within the amino acid stretch between residues 106 and 126, for which several studies indicated neurotoxic properties[50–52].

Additionally, this degradation excludes PrP^c from the conversion into PrP^{Sc}. In prion-infected cells, and also in brain tissue, the protease resistant core of PrP^{Sc} is N-terminally truncated in an acidic compartment around residue 90, generating a C-terminal fragment of approximately 19 kDa[53,48].

The processing of PrP at aa 110/111 is reglulated by protein kinase C (PKC), and is up-regulated by effectors of PKC[45]. Further investigations employing inhibitors of distinct proteases revealed a contribution of the metalloprotease ADAM10 (a disintegrin and metalloprotease) and TACE (tumor necrosis factor α-converting enzyme) to the cleavage[46]. Of note, the proteolysis of PrP^c in brain material is also blocked by inhibitors of metalloproteases[49]. Strikingly, the cleavage by a metalloprotease and the subsequent disruption of the neurotoxic PrP peptide finds a parallel in the proteolysis of APP. This comprises the cleavage of APP within the $A\beta$-peptide, the production of which is central to the pathology of Alzheimers disease[54]. Upon physiological processing within the $A\beta$-peptide by proteases of the disintegrin family (α-secretase) the neurotoxic potential is compromised[55]. In a recent report, another N-terminal cleavage site of PrP has been identified, producing N- and C-terminal fragments of about 7 and 20 kDa, respectively[47]. The small fragment was, similar to N1, found in the cell culture medium and in cell lysates. This might have functional implications, and it has been shown previously with recombinant PrP that this cleavage, occuring in the octarepeat region, is dependent on Cu^{2+} and oxidative stress. This processing might turn PrP^c into a functionally active form. The rapid subsequent cleavage between residues 110/111 and 112 would then inactivate the biological activity. However, the final proof for this hypothesis remains to be established.

In summary, the degradation of PrP demonstrates an interesting similarity with the processing of APP, another precursor protein linked to a neurodegenerative disease. Interestingly, enhancement of α-secretase cleavage in $A\beta$ overproducing mice led to a 50% inhibition of $A\beta$ production in the brain[56], suggesting in analogy that increasing the N1/C1 cleavage might reduce the generation of PrP^{Sc}.

15.5. The investigation of pathogenic PrP mutants in cell culture studies

As the numerous studies performed in cell culture and in vivo models have helped shedding more light on the metabolism of the prion protein, it has become clear that the trafficking of this protein is far from being a simple default process. Several specific sorting signals seem to be required for correct transport from the Golgi to the cell surface and for endocytosis. Understanding how the cellular events underlying neuronal disfunction and cell death might relate to an abnormal processing and trafficking of this protein represents an important goal in prion research. The establishment of appropriate cell culture models for the analysis of mutant PrPs associated with familial prion diseases has significantly facilitated this investigation. These diseases account for 10% of the cases of CJD and almost all cases of GSS and FFI and are linked to dominantly inherited mutations in the PrP gene. Point mutations linked to either CJD, GSS, or FFI occur mainly in the C-terminal half of the PrP molecule, while insertions associated with a variable phenotype that can include features of CJD or GSS comprise 2 to 9 additional copies of the octapeptide repeats. These phenomena are presumed to favor spontaneous conversion of the protein to the PrP^{Sc} state without necessarily interfering with an exogenous infectious agent. Insertion of nine octapeptide repeats in addition to the five normally present at the N-terminal half of the human protein sequence is associated with a variant phenotype characterized by dementia, ataxia, and PrP-containing amyloid plaques. A related neurodegenerative disorder has been characterized in mice[57] and the corresponding mouse-PrP mutation (PG14) has been widely studied in cell culture. Several other mutated PrPs based on the mouse sequence have been investigated in this system. The mouse PrP-D177N is homologous to the human mutation D178N-Met129 associated with FFI (Val129 in correlation with the same mutated allele relates to familial CJD), whereas E199K (E200K in humans) is linked to CJD. When expressed in different cell lines, these mutant PrPs acquire several biochemical features reminiscent of PrP^{Sc}, including insolubility in non-ionic detergents and mild resistance to digestion by proteinase K[58–60]. These biochemical features probably reflect changes in PrP conformation and/or aggregation, which, possibly depending on the cell line in which they are expressed, result in an abnormal association with the plasma membrane. This phenomenon is reflected by reduced accessibility by the bacterial phospholipase PIPLC, which might give evidence of the earliest step along the pathway leading to conversion from PrP^{c} to PrP^{Sc}. Particularly evident is the correlation between the number of octapeptides and the increased insolubility/aggregation of the mutant PrP[60], a characteristic reminiscent of the polyglutamin repeats expansion in Huntington's disease[61].

The subsequent investigation of the precise subcellular localization of mutant PrPs by immunofluorescence and electron microscopy techniques has made clear that misfolding and subsequent aggregation characterizing several forms of familial TSEs often lead to perturbation of the intracellular trafficking of the prion protein. Altered

Figure 15.1. Subcellular trafficking of the prion protein. Newly synthesised PrP^c (blue circles) is transported along the secretory pathway through the endoplasmic reticulum (ER) and the Golgi. At the cell surface, it is localised in cholesterol rich domains (rafts or caveolae) by its GPI-anchor. PrP^c is endocytosed and recycles to the cell surface or reaches lysosomes for final degradation. Conversion into PrP^{Sc} (red squares) occurs close to the cell surface in rafts or in compartments of the endocytic pathway. PrP^{Sc} is not efficiently degraded in lysosomes, and accumulates. Misfolded PrP (blue squares) can be retro-translocated into the cytosol and degraded by the proteasome or accumulate and induce neurotoxicity. Treatment of cells with Cyclosporin A (CsA) can induce aggregation of PrP molecules as aggresomes (blue triangles). ERGIC: ER-Golgi intermediate compartment; TGN: trans-Golgi network

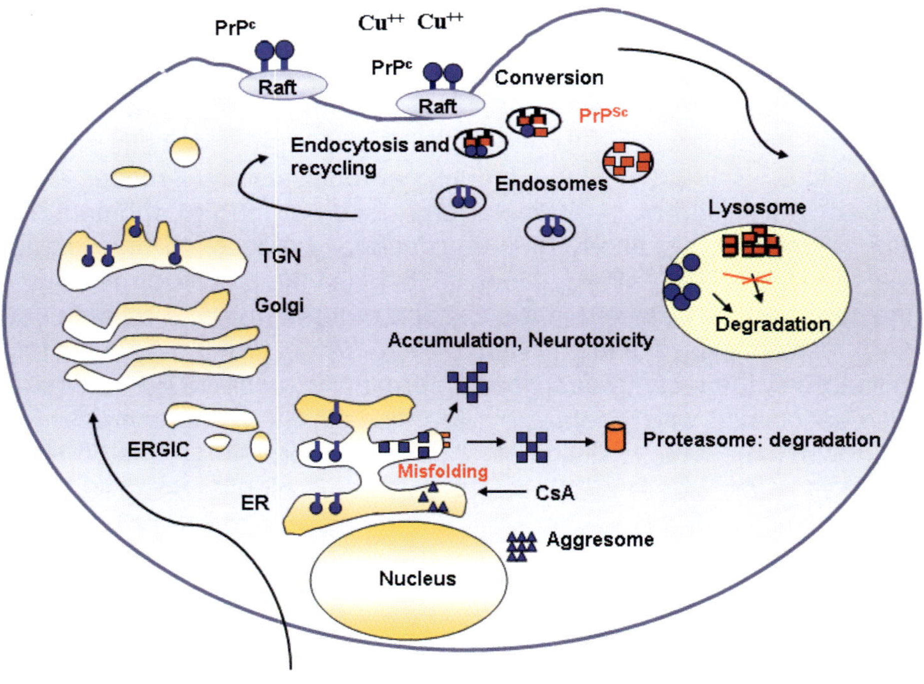

subcellular distribution appears to be a common phenomenon for a large number of disease- associated PrP mutants. Several of these molecules share reduced cell surface expression and accumulate in the Golgi or the ER[8]. Biochemical analysis of mutated PrP E200K and PrP D178N showed impaired transport to the cell surface mainly of the unglycosylated isoform[62] and enhanced localisation in the ER[63]. With regard to their altered biochemical features, to a significant shorter metabolic half-life measured for several of the mutant PrPs compared with wtPrP, the common intracellular distribution accounts for a possible involvement of the ubiquitous ER-based quality control mechanisms in the metabolism of these proteins. This pathway ensures that only newly synthezised proteins that have undergone correct co- and post-translational processing and folding are transported to their target organelles and cellular compartments. Incorrectly assembled proteins are normally retained in the ER and subjected to the ER-associated degradation (ERAD) pathway. This includes retrograde-translocation through the ER membrane into the cytosol by the hetero-trimer Sec61 complex (translocon)[64]. The polypeptide chain is then deglycosylated by a cytosolic N-glycanase and often covalently bound to the lysine residues of the conserved protein ubiquitin. This step ensures its degradation by the 26S multi-protein proteasome complex[65]. Assumed to be of little relevance in the past, the role of the ER-related degradation pathway in the metabolism of the prion protein, and its implication in prion diseases have become topics of research and debate in recent years[66].

A common pattern involving ER-retention and proteasome degradation has been described for PrP Q217R and Y145stop, both mutants linked to GSS. Expressed in a human neuroblastoma cell model, a significant proportion of these molecules does not lead to spontaneous PrP^{Sc} formation, is poorly post-translationally processed and is subjected to the ERAD pathway. Upon inhibition of the proteasomal activity, they are partially retained in the ER and other membrane bound compartments in an insoluble and weakly PK-resistant form[6,7]. A similar fate is shared by the recently described C-terminally truncated PrPQ160stop also associated with inheritable prion diseases[67]. When expressed in neuroblastoma cells, the homologous murine protein Q159stop is located at a low extent in the cytoplasm and is highly concentrated in the nucleus. With a very short half-life comparable to PrP145stop, this mutant is also sorted out of the secretory pathway at distinct sites of the ER memrane and rapidly degraded by the proteasomal system in the cytosol. Nuclear localization was assessed regardless of proteasomal inhibition, arguing for a direct influx of this mutated prion protein. This represents an interesting finding if one takes into account that the N-terminal region of human and ovine PrP has been described to harbor nuclear localization signals[68] and nucleic acid binding properties[69]. Both mutants, Y145stop and Q159stop, partially retain their N-terminal signal peptide[6,67]. A recent report identified a new regulatory element for ER-import within the C-terminus of the prion protein[3]. In this study, C-terminally deleted PrPs like Y145stop and Q159stop localized in the cytosol due to compromised ER-translocation.

Since almost all familial cases of prion disorders are heterozygous, and patients express both the normal and the mutated form of PrP, the exact role of wild-type (wt) PrP^{c} in the pathogenic process is not completely understood. Analysis of plaques isolated from brains yielded varying results. Only the mutant protein was recovered in the

PK-resistant fraction in some cases, while an intermediate detergent-insoluble but PK-sensitive PrP was found in others. In other patients, wtPrP^{c} was also converted into PrP^{Sc}. Recent studies with GFP-tagged prion proteins in human neuroblastoma cells have now assessed that seeds of aggregated PrP-mutants can indeed cause normal PrP^{c} to coalesce[70]. These results provide new insight into the molecular mechanisms of prion aggregation and support a nucleation hypothesis of PrP^{Sc} formation.

The two transmembrane topological variants of the prion protein designated (Ntm)PrP and (Ctm)PrP contain a conserved hydrophobic sequence that can span the lipid bilayer in either direction[10]. Mutations in the membrane-spanning segment of PrP can lead to increased generation of these transmembrane forms. Several pieces of evidence have linked (Ctm)PrP to neurodegeneration in transgenic mice and to some heritable prion diseases. This isoform has been hypothesized to represent a key intermediate in the pathway of prion-induced neurodegeneration[10,71]. In cultured cells, a mutated form of PrP synthesized exclusively with a (Ctm)PrP topology contains an uncleaved N-terminal signal peptide, is retained in the ER and is degraded by the proteasome[72]. Taking into account the degradation pathway of this protein, the fact that a good portion of the molecules is exposed to the cytoplasm and the toxicity of cytosolic PrP, these results provide new clues as to possible mechanisms of neurodegeneration. These findings probably link prion diseases to other human neurodegenerative syndromes such as Huntington's and Parkinson diseases as well as to conditions affecting peripheral organs[73].

A recent series of studies have explored the role of the proteasome and ubiquitin also in the metabolism of wtPrP^{c}[66,5]. Here, a small percentage of nascent wtPrP accumulates in the cytosol in the presence of different proteasome inhibitors and comprise detergent-soluble and -insoluble molecules. Insoluble aggregates are composed of unglycosylated PrP partially resistant to proteinase K treatment, a feature usually associated with PrP^{Sc}, and include ubiquitylated PrP-species. This PrP population seems to have undergone N- and C-terminal proteolytic cleavage occurring during normal processing in the ER. Such events account for delivery to the cytoplasm by retrograde transport from the ER. These molecules also seem to be able to sustain and promote misfolding of newly synthesised PrP^{c}. Retrograde translocation of misfolded PrP, which could interact with other PrP molecules and promote conversion into a PrP^{Sc}-like conformation, should therefore be implicated in the initial steps of conversion in rare spontaneous PrP^{Sc} scenarios and in generating toxic PrP species. The described results find additional evidence in cell culture models and *in vivo* experiments performed in transgenic mice[74]. In these studies, the accumulation of even small amounts of cytosolic prion protein results in extreme neurotoxicity, with a pathology very similar to that seen in animals expressing mutant forms of PrP associated with human prion conditions[75,32]. Since no PrP^{Sc} was detected in these mice, the described findings are in favour of an intrinsic toxicity of PrP^{c}, possibly by activating cell death signaling pathways and by inducing neuronal damage. Efficient ER-quality control mechanisms normally shunt and degrade unfolded or misfolded PrP. Aging and biological traumas might compromise the natural capacity of this system leading to accumulation in the cytosol. Pathogenic mutations in the PrP coding sequence also lead to an increase in cytosolic misfolded PrP.

A different explanation for the localization of unglycosylated PrP in the cytosol is that proteasomal inhibition and increase in mRNA amount upon over-expression of PrP in transfected cells lead to inefficient translocation into the ER of a small fraction of newly synthesized PrP-population. These molecules are rapidly degraded by the proteasome[76]. According to this model, treatment of cells with proteasomal inhibitors has no effect on the maturation or turnover of either wild-type or mutated PrP molecules. Cytosolic PrP molecules harbor an intact N-terminal signal peptide and lack a GPI-anchor and N-linked glycans, all features indicating that the protein has not undergone processing in the ER. This model argues against a pathogenic role of cytosolic PrP and supports the idea that known forms of inherited prion disorders may rather be related to the toxic effects of misfolded PrP mutants accumulating in the lumen of the ER. Whether the severe neurodegeneration caused by cytosolic PrP is related to some forms of prion disease is therefore still a matter of debate.

The importance of correct folding and of the cellular quality control in the metabolism of proteins in general and of the prion protein in particular is once more stated in studies done on cyclophilins, a group of peptidylprolyl isomerases (PPIases) expressed in most cellular compartments[77]. Hampering the activity of cyclophilins with the fungal immunosuppressant Cyclosporin A (CsA) in different cell lines leads to accumulation of a PrP population with prion-like properties. These aggregated PrP molecules form aggresomes, perinuclear microtubule-dependent inclusion bodies located at the centrosome and characterizing several neurodegenerative disorders related to toxic proteins[78], and also described in the pathogenesis of known familial forms of prion disease[79]. ER-cyclophilins might therefore intervene directly on the folding and processing of wtPrP by ensuring the correct conformation of non-native isomers spontaneously forming. Inhibition of this PPIase activity of the cyclophilins with CsA therefore leads to accumulation of non-native peptides, which are re-translocated into the cytosol. According to this model, the natural weakening of cyclophilin activity by aging could contribute to formation of the "prion seed" required for initiation of familial or sporadic prion diseases.

15.6. Anti-prion compounds and cell culture

The search for strategies aimed at the inhibition of prion propagation involves experimental transmission of TSEs to laboratory animals. Several classes of drugs effectively delay the appearance of clinical symptoms and prolong survival time of laboratory animals experimentally infected with scrapie. Understanding the molecular and cellular mechanisms of action of these compounds and how they affect PrP^c/ PrP^{Sc} trafficking and localization nevertheless remains an important goal for the achievement of more efficient therapeutic approaches. Cell culture models represent an important and versatile system in which the activity of structurally heterogeneous compounds on the subcellular pathways leading to conversion of PrP^c into PrP^{Sc} can be analysed in detail.

Anti-prion agents are principally represented by polysulfonated polyanionic agents as Congo red, pentosan polysulfate and Suramin, polyene antibiotics like amphotericin B and MS-8209, and the antracycline iododoxorubicin[80–84]. Treatment with the amyloid-binding dye Congo red modifies the biochemical properties of PrP^c, which becomes insoluble and partially PK-resistant[85]. The rate of endocytosis is also enhanced and the molecules targeted to late endosomes and lysosomes[86]. This redistribution of PrP^c localization from the plasma membrane (or other early endocytic compartments) where conversion is meant to occur, makes PrP no longer available as a substrate for prion conversion. An additional explanation for the therapeutic properties of Congo red lies in its ability to bind and overstabilize the structure of PrP^{Sc} and, possibly, PrP^c[87], restraining the direct contact between molecules necessary for prion conversion. This action can also prevent the binding of PrP^{Sc} to endogenous glycosaminoglycans necessary for amyloid PrP^{Sc} plaques formation in natural TSEs and in scrapie infected mice[88].

Others among the polyanionic sulfonated glycans shown to prolong the life span of animals inoculated with scrapie are pentosan polysulfate and dextran sulfate. These chemotherapeutic agents, originally used in DNA and RNA virus therapy, in some cases completely prevent the development of symptoms when administered prophylactically to mice and hamsters[80,89,90]. There is evidence that these polyanions prevent early uptake and replication of the scrapie agent (rather than destabilization of the pre-existing PrP^{Sc}) in the lymphoreticular system or nerve endings leading to lack of PrP^{Sc} in brain tissue. At a subcellular level, density of sulphatation and molecular size are factors influencing their anti-prion activity[81]. As seen for Congo red, these inhibitors may competitively block interactions between endogenous glycosaminoglycans and PrP essential for accumulation in a protease-resistant form. Also these compounds cause a redistribution of the prion protein to intracellular compartments probably unfavorable for prion conversion, but have no effect on synthesis or degradation of the molecule[86]. Of note, cross-linking of cell surface proteins with antibodies and other ligands rapidly stimulates their internalization[91]. The described compounds might therefore also induce oligomerization of PrP^{c}.

A conformational change leading to aggregation of PrP^{c} is the mechanism accounting for inhibition of conversion to PrP^{Sc} by another polysulfonated compound, the naphtylurea drug Suramin[43]. Used in the past to treat trypanosomiasis and as an antiviral drug, when applied in mouse bioassays at the time of peripheral prion inoculation, this agent significantly prolonged the onset of disease. In prion infected neuronal cells, Suramin shows a phenotype similar to that described for other sulfonated compounds, with induction of insoluble, PK-sensitive PrP-aggregates at the cell surface[92]. A more intriguing aspect in the action of Suramin, opening interesting perspectives in the trafficking of proteins, is the finding that aggregates of misfolded PrP are found in intracellular compartments along the secretory pathway[43]. These aggregates trigger post-ER quality control mechanisms and are diverted directly to late endosomes and lysosomes for final degradation. Localization at the plasma membrane and in the compartments of conversion is thereby bypassed. Further studies assessed that PrP molecules devoid of the pre-octapeptide domain (residues 23-50), although becoming insoluble upon treatment with Suramin, still have access to the cell surface[93]. This segment of the prion protein therefore plays an important role in controlling the transport of misfolded PrP-molecules. Misfolding/aggregation of the prion protein and its recognition by cellular quality control mechanisms therefore represent distinct aspects in the metabolism and targeting of the protein.

Polyene antibiotics like Amphotericin B (AmB) and MS-8209 have been extensively studied in in vivo models and, in contrast to most anti-prion compounds tested, prolong incubation even long periods of time after experimental transmission in laboratory animals has occurred[94]. Although the molecular mechanisms are still not completely clear, cell culture studies on wild type[95] and mutated PrP[96] hint at the ability of AmB to modify the properties of sphingolipid and cholesterol rich domains at the surface of cells and disorganise endosomal trafficking to account for the inhibition of PrP^{Sc} accumulation. This drug, however, was able to reduce but not completely stop conversion for mouse PrP^{c} in infected cells[95]. In sucrose gradients and flotation experiments, AmB formes stable complexes with cholesterol which alter the structure of DRM (a property responsible for its antifungal activity) and the metabolic equilibrium inside the cell. A direct interaction of the drug with PrP^c seems unlikely. More probable is an alteration of the lipid and protein composition of DRM which might have impact on the trafficking of PrP^c and inhibit interactions of PrP^c with a putative protein X, a receptor, or with PrP^{Sc}. It was recently shown that AmB can selectively bind to β-sheet rich protein fibrils and oligomers, a property which might also account for its anti-prion effect by preventing or slowing nucleation and propagation[97].

Another potent polyene antibiotic inhibiting PrP^{Sc} formation in chronically infected neuroblastoma cells is filipin[29]. By binding to cholesterol, filipin alters the caveolar structure and function. As a consequence, and in contrast to other agents of the same class, filipin limits endocytosis of PrP^c and reduces the amount of membrane bound PrP^c by massively releasing the protein into the medium. This process decreases the amount of PrP available at conversion sites and therefore reinforces the role of lipid rafts in the metabolism of the prion protein.

For a variety of other compounds, their ability to interfere in prion propagation has been mainly tested in *in vivo* and cell-free studies. The exact cellular mechanism of action on the trafficking of PrP^c/PrP^{Sc} needs to be characterized. Of note, in contrast to most substances described above, some of them, like branched polyamines[98], tetracycline[99] and tetrapyrroles[100] seem to act primarily on PrP^{Sc}, binding and reverting PK-resistance of PrP^{Sc} aggregates and fibrils, thereby enabling cells to degrade them.

Despite the large number of experimental approaches, of which only few have been summarized above, the way to therapy and prophylaxis in prion diseases still encounters massive obstacles. As for other neurodegenerative disorders, effective therapy needs delivery to the central nervous system (CNS) and therefore crossing of the blood-brain barrier, an ability shared by a minority of compounds. Their application remains, so far, restricted to prophylactic and post-exposure scenarios, especially in cases with peripheral infection and accumulation of prions before invasion of the CNS. In light of the long incubation time of these diseases, extra-CNS therapy/prophylaxis can therefore be conceivable in these situations.

15.7. Manipulating the trafficking of the prion protein in cell culture

Knowledge of transport, localization and compartment(s) of conversion for the prion protein have often been acquired by the analysis of the metabolism of artificially designed mutated PrPs. N-terminal deletion mutants have helped shedding light on the mechanisms and the sequences necessary for PrP internalization. On the other hand, PrP-mutants in which the two short β-strands (residues 127-130 and 160-163, respectively, in mouse) or the first α-helix (residues 143-153) had been deleted were used to characterize the role of distinct elements of the PrP secondary structure in the conformational transitions occurring during prion conversion[101]. Removal of the second β-strand or of the first α-helix significantly alter processing and cellular localization of PrP^c, which is mainly retained intracellularly in pre/medial Golgi compartments. Deletion of the first β-strand has no effect on the processing and trafficking of the PrP^c. All these mutants, however, inhibit formation of endogenous PK-resistant PrP in mouse neuroblastoma cells and greatly decrease the ability to form PrP^{Sc} in cell-free conversion assay. With regard to the reduced cell surface localization of some of these mutants, it is interesting to speculate about the subcellular compartments of PrP^c/PrP^{Sc} interference/conversion, whether these processes occur in the same compartments, whether the PrP^c surface localization is mandatory for conversion or about the interference of mutant PrPs with other auxiliary factors.

C-terminal deletions and shift in the plasma membrane localization of PrP^c are the central theme of studies assessing the importance of the GPI-lipid anchor and the correct localization of PrP^c for conversion into PrP^{Sc}[102,18,23]. Already identified as an important cellular sorting motif, GPI also directs distribution of PrP^c to rafts or caveolae. Replacement of the GPI-addition signal with the transmembrane and cytoplasmic regions of different molecules targets PrP to clathrin-coated pits (as assessed in Triton X-100 solubility assays), an event which prevents conversion of this molecule into the PK-resistant isoform. These results favor a model according to which formation of PrP^{Sc} is restricted to rafts/caveolae and requires involvement of auxiliary factors.

Mutations introduced into the PrP coding sequence have also been used to study the role of glycosylation in the determination of the strain specific properties of prions in cell cultures. Oligosaccharide residues attached to glycoproteins are known to promote correct folding and maturation of nascent polypeptide chains and to interact with the cellular quality control machinery[103]. PrP glycosylation has also been shown to have an impact on the conformational transition of PrP^c into PrP^{Sc}, to influence the selective neuronal targeting of PrP^{Sc}, and to contribute to the phenomenon of strain diversity[104,105] The mere inhibition of glycosylation with tunycamycin in scrapie-infected cells results in protease-resistant unglycosylated PrP^{Sc}[106–108]. PrP molecules in which both consensus sites are mutated can also sustain PrP^{Sc} conversion[107]. N-linked glycans therefore seem to be dispensable for the conversion process. Glycan chains nevertheless seem to modulate the efficiency with which

PrP^{Sc} is formed, as low molecular weight isoforms and a physiological intermediate glycoform of PrP^{c} seem to be preferred substrates for conversion[109,110]. Cell culture studies have dealt with several mutations leading to lack of glycosylation in the mouse PrP-consensus sites (N180XT, N196XT)[108,111]. PrP-molecules mutated at T182 (mutation homologous to the T183A found in cases of familial forms of TSEs) alone or at both T182 and T196 fail to reach the cell surface and localize close to the mid-Golgi stack, while those mutated at Thr196 or synthesized in the presence of tunicamycin can be detected at the plasma membrane. All mutants acquire biochemical features reminiscent of PrP^{Sc}[108]. A different set of glycosylation mutants (N180Q, N196Q) also show plasma membrane distribution[111]. N-linked glycans *per se*, although stabilizing protein structure are therefore not strictly required for normal biosynthetic trafficking of PrP.

Modulation of the normal trafficking of PrP offers several possibilities to understand routes and compartments underlying prion conversion. An elegant example is provided by overexpression in mouse neuroblastoma cells of mutant GTPases of the Rab family which play a role in the accurate targeting and docking of transport vesicles with their acceptor membranes[112]. When a dominant-negative mutant of the Rab4-GDP, normally implicated in regulation of membrane recycling from early endosomes, or a constitutively active Rab6a-GDP, which stimulates retrograde transport of Golgi- resident proteins to the ER, are overexpressed in scrapie-infected N2a cells, an increase in the amount of PrP^{Sc} can be detected. Using immunofluorescence and subcellular fractionation techniques a redistribution of PrP^{c} to intracellular compartments including the ER can be thereby assessed. A still not completely cleared Rab6a-controlled retrograde pathway of PrP from the Golgi to the ER could therefore play a role in the physiological trafficking of PrP^{c} and, possibly, of PrP^{Sc}.

15.8. Outlook

In the meantime a huge variety of cell culture models exists which are capable of propagating infectious prions. This will facilitate the analysis of the cellular uptake of prions, their mode of cellular propagation, how infectivity is transferred from one cell to another, and how cells are finally able to clear prion infectivity. Up to now it is not known what makes a cell susceptible to prion infection and what is needed to transfer an acute infection into a persistent one. Huge progress was made towards the potential use of cell culture models for therapeutic and diagnostic purposes. It remains to be established whether cell culture models can be used for screening anti-prion compounds in high trough-put systems or for replacing cumbersome diagnostic bioassays in rodents. Despite these open questions and although a lot has to be done in the future, cell culture models proved to be excellent and indispensable model systems for studying the biogenesis and pathogenesis of prion proteins and prions.

15.9. References

1. B. Caughey, R. E. Race, D. Ernst, M. J. Buchmeier, and B. Chesebro, Prion protein biosynthesis in scrapie-infected and uninfected neuroblastoma cells. *J. Virol.* **63**, 175–181 (1989)

2. D. R. Borchelt, M. Scott, A. Taraboulos, N. Stahl, and S. B. Prusiner, Scrapie and cellular prion proteins differ in their kinetics of synthesis and topology in cultured cells. *J. Cell Biol.* **110**, 743–752 (1990)

3. J. Heske, U. Heller, K. F. Winklhofer, and J. Tatzelt, The C-terminal globular domain of the prion protein is necessary and sufficient for import into the endoplasmic reticulum. *J. Biol. Chem.* **279**, 5435–5443 (2004)

4. F. Wopfner, G. Weidenhofer, R. Schneider, A. von Brunn, S. Gilch, T. F. Schwarz, T. Werner, and H. M. Schatzl, Analysis of 27 mammalian and 9 avian PrPs reveals high conservation of flexible regions of the prion protein. *J. Mol. Biol.* **289**, 1163–1178 (1999)

5. Y. Yedidia, L. Horonchik, S. Tzaban, A. Yanai, and A. Taraboulos, Proteasomes and ubiquitin are involved in the turnover of the wild-type prion protein. *EMBO J.* **20**, 5383–5391 (2001)

6. G. Zanusso, R. B. Petersen, T. Jin, Y. Jing, R. Kanoush, S. Ferrari, P. Gambetti, and N. Singh, Proteasomal degradation and N-terminal protease resistance of the codon 145 mutant prion protein. *J. Biol. Chem.* **274**, 23396–23404 (1999)

7. T. Jin, Y. Gu, G. Zanusso, M. Sy, A. Kumar, M. Cohen, P. Gambetti, and N. Singh, The chaperone protein BiP binds to a mutant prion protein and mediates its degradation by the proteasome. *J. Biol. Chem.* **275**, 38699–38704 (2000)

8. L. Ivanova, S. Barmada, T. Kummer, and D. A. Harris, Mutant prion proteins are partially retained in the endoplasmic reticulum. *J. Biol. Chem.* **276**, 42409–42421 (2001)

9. N. Stahl, D. R. Borchelt, K. Hsiao, and S. B. Prusiner, Scrapie prion protein contains a phosphatidylinositol glycolipid. *Cell* **51**, 229–240 (1987)

10. R. S. Hegde, J. A. Mastrianni, M. R. Scott, K. A. DeFea, P. Tremblay, M. Torchia, S. J. DeArmond, S. B. Prusiner, and V. R. Lingappa, A transmembrane form of the prion protein in neurodegenerative disease. *Science* **279**, 827–834 (1998)

11. M. Nunziante, S. Gilch, and H. M. Schatzl, Essential role of the prion protein N terminus in subcellular trafficking and half-life of cellular prion protein. *J. Biol. Chem.* **278**, 3726–3734 (2003)

12. K. Simons and E. Ikonen, Functional rafts in cell membranes. *Nature* **387**, 569–572 (1997)

13. D. A. Brown and E. London, Functions of lipid rafts in biological membranes. *Annu. Rev. Cell Dev. Biol.* **14**, 111–136 (1998)

14. K. Kaneko, L. Zulianello, M. Scott, C. M. Cooper, A. C. Wallace, T. L. James, F. E. Cohen, and S. B. Prusiner, Evidence for protein X binding to a discontinuous epitope on the cellular prion protein during scrapie prion propagation. *Proc. Natl. Acad. Sci. U. S. A.* **94**, 10069–10074 (1997)

15. M. Vey, S. Pilkuhn, H. Wille, R. Nixon, S. J. DeArmond, E. J. Smart, R. G. Anderson, A. Taraboulos, and S. B. Prusiner, Subcellular colocalization of the cellular and scrapie prion proteins in caveolae-like membranous domains. *Proc. Natl. Acad. Sci. U. S. A.* **93**, 14945–14949 (1996)

16. N. Naslavsky, R. Stein, A. Yanai, G. Friedlander, and A. Taraboulos, Characterization of detergent-insoluble complexes containing the cellular prion protein and its scrapie isoform. *J. Biol. Chem.* **272**, 6324–6331 (1997)

17. D. Sarnataro, S. Paladino, V. Campana, J. Grassi, L. Nitsch, and C. Zurzolo, PrPC is sorted to the basolateral membrane of epithelial cells independently of its association with rafts. *Traffic.* **3**, 810–821 (2002)

18. A. Taraboulos, M. Scott, A. Semenov, D. Avrahami, L. Laszlo, S. B. Prusiner, and D. Avraham, Cholesterol depletion and modification of COOH-terminal targeting sequence of the prion protein inhibit formation of the scrapie isoform. *J. Cell Biol.* **129**, 121–132 (1995)

19. A. Taraboulos, D. Serban, and S. B. Prusiner, Scrapie prion proteins accumulate in the cytoplasm of persistently infected cultured cells. *J. Cell Biol.* **110**, 2117–2132 (1990)

20. K. Simons and D. Toomre, Lipid rafts and signal transduction. *Nat. Rev. Mol. Cell Biol.* **1**, 31–39 (2000)

21. P. Scheiffele, M. G. Roth, and K. Simons, Interaction of influenza virus haemagglutinin with sphingolipid-cholesterol membrane domains via its transmembrane domain. *EMBO J.* **16**, 5501–5508 (1997)

22. A. R. Walmsley, F. Zeng, and N. M. Hooper, The N-terminal region of the prion protein ectodomain contains a lipid raft targeting determinant. *J. Biol. Chem.* **278**, 37241–37248 (2003)

23. K. Kaneko, M. Vey, M. Scott, S. Pilkuhn, F. E. Cohen, and S. B. Prusiner, COOH-terminal sequence of the cellular prion protein directs subcellular trafficking and

controls conversion into the scrapie isoform. *Proc. Natl. Acad. Sci. U. S. A.* **94**, 2333–2338 (1997)

24. S. L. Shyng, J. E. Heuser, and D. A. Harris, A glycolipid-anchored prion protein is endocytosed via clathrin-coated pits. *J. Cell Biol.* **125**, 1239–1250 (1994)

25. S. L. Shyng, K. L. Moulder, A. Lesko, and D. A. Harris, The N-terminal domain of a glycolipid-anchored prion protein is essential for its endocytosis via clathrin-coated pits. *J. Biol. Chem.* **270**, 14793–14800 (1995)

26. M. Conese, A. Nykjaer, C. M. Petersen, O. Cremona, R. Pardi, P. A. Andreasen, J. Gliemann, E. I. Christensen, and F. Blasi, Alpha-(2)-Macroglobulin Receptor Ldl-Receptor-Related-Protein (Lrp)-Dependent Internalization of the Urokinase Receptor. *J. Cell Biol.* **131**, 1609–1622 (1995)

27. S. Gauczynski, J. M. Peyrin, S. Haik, C. Leucht, C. Hundt, R. Rieger, S. Krasemann, J. P. Deslys, D. Dormont, C. I. Lasmezas, and S. Weiss, The 37-kDa/67-kDa laminin receptor acts as the cell-surface receptor for the cellular prion protein. *EMBO J.* **20**, 5863–5875 (2001)

28. P. C. Pauly and D. A. Harris, Copper stimulates endocytosis of the prion protein. *J. Biol. Chem.* **273**, 33107–33110 (1998)

29. M. Marella, S. Lehmann, J. Grassi, and J. Chabry, Filipin prevents pathological prion protein accumulation by reducing endocytosis and inducing cellular PrP release. *J. Biol. Chem.* **277**, 25457–25464 (2002)

30. K. S. Lee, A. C. Magalhaes, S. M. Zanata, R. R. Brentani, V. R. Martins, and M. A. Prado, Internalization of mammalian fluorescent cellular prion protein and N-terminal deletion mutants in living cells. *J. Neurochem.* **79**, 79–87 (2001)

31. W. Sumudhu, S. Perera, and N. M. Hooper, Ablation of metal ion-induced endocytosis of the prion protein by disease-associated mutation of the octarepeat region. *Curr. Biol.* **11**, 519–523 (2001)

32. D. Shmerling, I. Hegyi, M. Fischer, T. Blattler, S. Brandner, J. Gotz, T. Rulicke, E. Flechsig, A. Cozzio, C. von Mering, C. Hangartner, A. Aguzzi, and C. Weissmann, Expression of amino-terminally truncated PrP in the mouse leading to ataxia and specific cerebellar lesions. *Cell* **93**, 203–214 (1998)

33. H. Damke, D. D. Binns, H. Ueda, S. L. Schmid, and T. Baba, Dynamin GTPase domain mutants block endocytic vesicle formation at morphologically distinct stages. *Mol. Biol. Cell* **12**, 2578–2589 (2001)

34. A. C. Magalhaes, J. A. Silva, K. S. Lee, V. R. Martins, V. F. Prado, S. S. Ferguson, M. V. Gomez, R. R. Brentani, and M. A. Prado, Endocytic intermediates involved with the intracellular trafficking of a fluorescent cellular prion protein. *J. Biol. Chem.* **277**, 33311–33318 (2002)

35. M. Zerial and H. McBride, Rab proteins as membrane organizers. *Nat. Rev. Mol. Cell Biol.* **2**, 107–117 (2001)

36. C. Sunyach, A. Jen, J. Deng, K. T. Fitzgerald, Y. Frobert, J. Grassi, M. W. McCaffrey, and R. Morris, The mechanism of internalization of glycosylphosphatidylinositol-anchored prion protein. *EMBO J.* **22**, 3591–3601 (2003)

37. N. Madore, K. L. Smith, C. H. Graham, A. Jen, K. Brady, S. Hall, and R. Morris, Functionally different GPI proteins are organized in different domains on the neuronal surface. *EMBO J.* **18**, 6917–6926 (1999)

38. B. Strumbo, S. Ronchi, L. C. Bolis, and T. Simonic, Molecular cloning of the cDNA coding for Xenopus laevis prion protein. *FEBS Lett.* **508**, 170–174 (2001)

39. P. J. Peters, A. Mironov, Jr., D. Peretz, E. van Donselaar, E. Leclerc, S. Erpel, S. J. DeArmond, D. R. Burton, R. A. Williamson, M. Vey, and S. B. Prusiner, Trafficking of prion proteins through a caveolae-mediated endosomal pathway. *J. Cell Biol.* **162**, 703–717 (2003)

40. R. G. Parton, Caveolae—from ultrastructure to molecular mechanisms. *Nat. Rev. Mol. Cell Biol.* **4**, 162–167 (2003)

41. P. Oh, D. P. McIntosh, and J. E. Schnitzer, Dynamin at the neck of caveolae mediates their budding to form transport vesicles by GTP-driven fission from the plasma membrane of endothelium. *J. Cell Biol.* **141**, 101–114 (1998)

42. S. Mouillet-Richard, M. Ermonval, C. Chebassier, J. L. Laplanche, S. Lehmann, J. M. Launay, and O. Kellermann, Signal transduction through prion protein. *Science* **289**, 1925–1928 (2000)

43. S. Gilch, K. F. Winklhofer, M. H. Groschup, M. Nunziante, R. Lucassen, C. Spielhaupter, W. Muranyi, D. Riesner, J. Tatzelt, and H. M. Schatzl, Intracellular re-routing of prion protein prevents propagation of PrP(Sc) and delays onset of prion disease. *EMBO J.* **20**, 3957–3966 (2001)

44. S. L. Shyng, M. T. Huber, and D. A. Harris, A prion protein cycles between the cell surface and an endocytic compartment in cultured neuroblastoma cells. *J. Biol. Chem.* **268**, 15922–15928 (1993)

45. B. Vincent, E. Paitel, Y. Frobert, S. Lehmann, J. Grassi, and F. Checler, Phorbol ester-regulated cleavage of normal prion protein in HEK293 human cells and murine neurons. *J. Biol. Chem.* **275**, 35612–35616 (2000)

46. B. Vincent, E. Paitel, P. Saftig, Y. Frobert, D. Hartmann, B. De Strooper, J. Grassi, E. Lopez-Perez, and F. Checler, The disintegrins ADAM10 and TACE contribute to the constitutive and phorbol ester-regulated normal cleavage of the cellular prion protein. *J. Biol. Chem.* **276**, 37743–37746 (2001)

47. A. Mange, F. Beranger, K. Poec'h, T. Onodera, Y. Frobert, and S. Lehmann, Alpha- and beta- cleavages of the amino-terminus of the cellular prion protein. *Biology of the Cell* **96**, 125–132 (2004)

48. S. G. Chen, D. B. Teplow, P. Parchi, J. K. Teller, P. Gambetti, and L. Autilio-Gambetti, Truncated forms of the human prion protein in normal brain and in prion diseases. *J. Biol. Chem.* **270**, 19173–19180 (1995)

49. A. Jimenez-Huete, P. M. Lievens, R. Vidal, P. Piccardo, B. Ghetti, F. Tagliavini, B. Frangione, and F. Prelli, Endogenous proteolytic cleavage of normal and disease-associated isoforms of the human prion protein in neural and non-neural tissues. *Am. J. Pathol.* **153**, 1561–1572 (1998)

50. G. Forloni, N. Angeretti, R. Chiesa, E. Monzani, M. Salmona, O. Bugiani, and F. Tagliavini, Neurotoxicity of a prion protein fragment. *Nature* **362**, 543–546 (1993)

51. D. R. Brown, Prion protein peptide neurotoxicity can be mediated by astrocytes. *J. Neurochem.* **73**, 1105–1113 (1999)

52. M. F. Jobling, L. R. Stewart, A. R. White, C. McLean, A. Friedhuber, F. Maher, K. Beyreuther, C. L. Masters, C. J. Barrow, S. J. Collins, and R. Cappai, The hydrophobic core sequence modulates the neurotoxic and secondary structure properties of the prion peptide 106–126. *J. Neurochem.* **73**, 1557–1565 (1999)

53. A. Taraboulos, A. J. Raeber, D. R. Borchelt, D. Serban, and S. B. Prusiner, Synthesis and trafficking of prion proteins in cultured cells. *Mol. Biol. Cell* **3**, 851–863 (1992)

54. D. J. Selkoe, Normal and Abnormal Biology of the Beta-Amyloid Precursor Protein. *Ann. Rev. Neurosci.* **17**, 489–517 (1994)

55. F. Checler and B. Vincent, Alzheimer's and prion diseases: distinct pathologies, common proteolytic denominators. *Trends Neurosci.* **25**, 616–620 (2002)

56. M. J. Savage, S. P. Trusko, D. S. Howland, L. R. Pinsker, S. Mistretta, A. G. Reaume, B. D. Greenberg, R. Siman, and R. W. Scott, Turnover of amyloid beta-protein in mouse brain and acute reduction of its level by phorbol ester. *J. Neurosci.* **18**, 1743–1752 (1998)

57. R. Chiesa, B. Drisaldi, E. Quaglio, A. Migheli, P. Piccardo, B. Ghetti, and D. A. Harris, Accumulation of protease-resistant prion protein (PrP) and apoptosis of cerebellar granule cells in transgenic mice expressing a PrP insertional mutation. *Proc. Natl. Acad. Sci. U. S. A.* **97**, 5574–5579 (2000)

58. S. Lehmann and D. A. Harris, Mutant and infectious prion proteins display common biochemical properties in cultured cells. *J. Biol. Chem.* **271**, 1633–1637 (1996)

59. R. Narwa and D. A. Harris, Prion proteins carrying pathogenic mutations are resistant to phospholipase cleavage of their glycolipid anchors. *Biochemistry* **38**, 8770–8777 (1999)

60. S. A. Priola and B. Chesebro, Abnormal properties of prion protein with insertional mutations in different cell types. *J. Biol. Chem.* **273**, 11980–11985 (1998)

61. M. E. MacDonald, C. M. Ambrose, M. P. Duyao, R. H. Myers, C. Lin, L. Srinidhi, G. Barnes, S. A. Taylor, M. James, N. Groot, H. Macfarlane, B. Jenkins, M. A. Anderson, N. S. Wexler, J. F. Gusella, G. P. Bates, S. Baxendale, H. Hummerich, S. Kirby, M. North, S. Youngman, R. Mott, G. Zehetner, Z. Sedlacek, A. Poustka, A. M. Frischauf, H. Lehrach, A. J. Buckler, D. Church, L. Doucettestamm, M. C. Odonovan, L. Ribaramirez, M. Shah, V. P. Stanton, S. A. Strobel, K. M. Draths, J. L.

Wales, P. Dervan, D. E. Housman, M. Altherr, R. Shiang, L. Thompson, T. Fielder, J. J. Wasmuth, D. Tagle, J. Valdes, L. Elmer, M. Allard, L. Castilla, M. Swaroop, K. Blanchard, F. S. Collins, R. Snell, T. Holloway, K. Gillespie, N. Datson, D. Shaw, and P. S. Harper, A novel gene containing a trinucleotide repeat that is expanded and unstable on huntingtons-disease chromosomes. *Cell* **72**, 971–983 (1993)

62. R. B. Petersen, P. Parchi, S. L. Richardson, C. B. Urig, and P. Gambetti, Effect of the D178N mutation and the codon 129 polymorphism on the metabolism of the prion protein. *J. Biol. Chem.* **271**, 12661–12668 (1996)

63. A. Negro, C. Ballarin, A. Bertoli, M. L. Massimino, and M. C. Sorgato, The metabolism and imaging in live cells of the bovine prion protein in its native form or carrying single amino acid substitutions. *Mol. Cell Neurosci.* **17**, 521–538 (2001)

64. M. Pilon, R. Schekman, and K. Romisch, Sec61p mediates export of a misfolded secretory protein from the endoplasmic reticulum to the cytosol for degradation. *EMBO J.* **16**, 4540–4548 (1997)

65. J. L. Brodsky and A. A. McCracken, ER protein quality control and proteasome-mediated protein degradation. *Sem. Cell Dev. Biol.* **10**, 507–513 (1999)

66. J. Ma and S. Lindquist, Wild-type PrP and a mutant associated with prion disease are subject to retrograde transport and proteasome degradation. *Proc. Natl. Acad. Sci. U. S. A.* **98**, 14955–14960 (2001)

67. H. Lorenz, O. Windl, and H. A. Kretzschmar, Cellular phenotyping of secretory and nuclear prion proteins associated with inherited prion diseases. *J. Biol. Chem.* **277**, 8508–8516 (2002)

68. Y. Gu, J. Hinnerwisch, R. Fredricks, S. Kalepu, R. S. Mishra, and N. Singh, Identification of cryptic nuclear localization signals in the prion protein. *Neurobiol. Dis.* **12**, 133–149 (2003)

69. C. Gabus, E. Derrington, P. Leblanc, J. Chnaiderman, D. Dormont, W. Swietnicki, M. Morillas, W. K. Surewicz, D. Marc, P. Nandi, and J. L. Darlix, The prion protein has RNA binding and chaperoning properties characteristic of nucleocapsid protein NCP7 of HIV-1. *J. Biol. Chem.* **276**, 19301–19309 (2001)

70. Y. Gu, S. Verghese, R. S. Mishra, X. Xu, Y. Shi, and N. Singh, Mutant prion protein-mediated aggregation of normal prion protein in the endoplasmic reticulum: implications for prion propagation and neurotoxicity. *J. Neurochem.* **84**, 10–22 (2003)

71. R. S. Hegde, P. Tremblay, D. Groth, S. J. DeArmond, S. B. Prusiner, and V. R. Lingappa, Transmissible and genetic prion diseases share a common pathway of neurodegeneration. *Nature* **402**, 822–826 (1999)

72. R. S. Stewart, B. Drisaldi, and D. A. Harris, A transmembrane form of the prion protein contains an uncleaved signal peptide and is retained in the endoplasmic reticulum. *Mol. Biol. Cell* **12**, 881–889 (2001)

73. M. Y. Sherman and A. L. Goldberg, Cellular defenses against unfolded proteins: A cell biologist thinks about neurodegenerative diseases. *Neuron* **29**, 15–32 (2001)

74. J. Ma, R. Wollmann, and S. Lindquist, Neurotoxicity and neurodegeneration when PrP accumulates in the cytosol. *Science* **298**, 1781–1785 (2002)

75. R. Chiesa, P. Piccardo, B. Ghetti, and D. A. Harris, Neurological illness in transgenic mice expressing a prion protein with an insertional mutation. *Neuron* **21**, 1339–1351 (1998)

76. B. Drisaldi, R. S. Stewart, C. Adles, L. R. Stewart, E. Quaglio, E. Biasini, L. Fioriti, R. Chiesa, and D. A. Harris, Mutant PrP is delayed in its exit from the endoplasmic reticulum, but neither wild-type nor mutant PrP undergoes retrotranslocation prior to proteasomal degradation. *J. Biol. Chem.* **278**, 21732–21743 (2003)

77. E. Cohen and A. Taraboulos, Scrapie-like prion protein accumulates in aggresomes of cyclosporin A-treated cells. *EMBO J.* **22**, 404–417 (2003)

78. J. P. Taylor, F. Tanaka, J. Robitschek, C. M. Sandoval, A. Taye, S. Markovic-Plese, and K. H. Fischbeck, Aggresomes protect cells by enhancing the degradation of toxic polyglutamine-containing protein. *Hum. Mol. Genet.* **12**, 749–757 (2003)

79. R. S. Mishra, S. Bose, Y. Gu, R. Li, and N. Singh, Aggresome formation by mutant prion proteins: the unfolding role of proteasomes in familial prion disorders. *J. Alzheimers. Dis.* **5**, 15–23 (2003)

80. C. F. Farquhar and A. G. Dickinson, Prolongation of scrapie incubation period by an injection of dextran sulphate 500 within the month before or after infection. *J. Gen. Virol.* **67**, 463–473 (1986)

81. B. Caughey and R. E. Race, Potent inhibition of scrapie-associated PrP accumulation by congo red. *J. Neurochem.* **59**, 768–771 (1992)

82. S. A. Priola, B. Caughey, G. J. Raymond, and B. Chesebro, Prion protein and the scrapie agent: *in vitro* studies in infected neuroblastoma cells. *Infect. Agents Dis.* **3**, 54–58 (1994)

83. F. Tagliavini, R. A. McArthur, B. Canciani, G. Giaccone, M. Porro, M. Bugiani, P. M. Lievens, O. Bugiani, E. Peri, P. Dall'Ara, M. Rocchi, G. Poli, G. Forloni, T. Bandiera, M. Varasi, A. Suarato, P. Cassutti, M. A. Cervini, J. Lansen, M. Salmona, and C. Post, Effectiveness of anthracycline against experimental prion disease in Syrian hamsters. *Science* **276**, 1119–1122 (1997)

84. R. Demaimay, R. Race, and B. Chesebro, Effectiveness of polyene antibiotics in treatment of transmissible spongiform encephalopathy in transgenic mice expressing Syrian hamster PrP only in neurons. *J. Virol.* **73**, 3511–3513 (1999)

85. O. Milhavet, A. Mange, D. Casanova, and S. Lehmann, Effect of Congo red on wild-type and mutated prion proteins in cultured cells. *J. Neurochem.* **74**, 222–230 (2000)

86. S. L. Shyng, S. Lehmann, K. L. Moulder, and D. A. Harris, Sulfated glycans stimulate endocytosis of the cellular isoform of the prion protein, PrPC, in cultured cells. *J. Biol. Chem.* **270**, 30221–30229 (1995)

87. S. Caspi, M. Halimi, A. Yanai, S. B. Sasson, A. Taraboulos, and R. Gabizon, The anti-prion activity of Congo red. Putative mechanism. *J. Biol. Chem.* **273**, 3484–3489 (1998)

88. P. A. McBride, M. I. Wilson, P. Eikelenboom, A. Tunstall, and M. E. Bruce, Heparan sulfate proteoglycan is associated with amyloid plaques and neuroanatomically targeted PrP pathology throughout the incubation period of scrapie-infected mice. *Exp. Neurol.* **149**, 447–454 (1998)

89. A. Ladogana, P. Casaccia, L. Ingrosso, M. Cibati, M. Salvatore, Y. G. Xi, C. Masullo, and M. Pocchiari, Sulphate polyanions prolong the incubation period of scrapie-infected hamsters. *J. Gen. Virol.* **73 (Pt 3)**, 661–665 (1992)

90. V. Beringue, K. T. Adjou, F. Lamoury, T. Maignien, J. P. Deslys, R. Race, and D. Dormont, Opposite effects of dextran sulfate 500, the polyene antibiotic MS-8209, and Congo red on accumulation of the protease-resistant isoform of PrP in the spleens of mice inoculated intraperitoneally with the scrapie agent. *J. Virol.* **74**, 5432–5440 (2000)

91. E. W. Marsh, P. L. Leopold, N. L. Jones, and F. R. Maxfield, Oligomerized transferrin receptors are selectively retained by a lumenal sorting signal in a long-lived endocytic recycling compartment. *J. Cell Biol.* **129**, 1509–1522 (1995)

92. S. Kiachopoulos, J. Heske, J. Tatzelt, and K. F. Winklhofer, Misfolding of the prion protein at the plasma membrane induces endocytosis, intracellular retention and degradation. *Traffic* **5**, 426–436 (2004)

93. S. Gilch, M. Nunziante, A. Ertmer, F. Wopfner, L. Laszlo, and H. M. Schatzl, Recognition of lumenal prion protein aggregates by post-ER quality control mechanisms is mediated by the preoctarepeat region of PrP. *Traffic* **5**, 300–313 (2004)

94. R. Demaimay, K. T. Adjou, V. Beringue, S. Demart, C. I. Lasmezas, J. P. Deslys, M. Seman, and D. Dormont, Late treatment with polyene antibiotics can prolong the survival time of scrapie-infected animals. *J. Virol.* **71**, 9685–9689 (1997)

95. A. Mange, N. Nishida, O. Milhavet, H. E. McMahon, D. Casanova, and S. Lehmann, Amphotericin B inhibits the generation of the scrapie isoform of the prion protein in infected cultures. *J. Virol.* **74**, 3135–3140 (2000)

96. A. Mange, O. Milhavet, H. E. McMahon, D. Casanova, and S. Lehmann, Effect of amphotericin B on wild-type and mutated prion proteins in cultured cells: putative mechanism of action in transmissible spongiform encephalopathies. *J. Neurochem.* **74**, 754–762 (2000)

97. S. C. Hartsel and T. R. Weiland, Amphotericin B binds to amyloid fibrils and delays their formation: a therapeutic mechanism? *Biochemistry* **42**, 6228–6233 (2003)

98. S. Supattapone, H. O. Nguyen, F. E. Cohen, S. B. Prusiner, and M. R. Scott, Elimination of prions by branched polyamines and implications for therapeutics. *Proc. Natl. Acad. Sci. U. S. A.* **96**, 14529–14534 (1999)

99. F. Tagliavini, G. Forloni, L. Colombo, G. Rossi, L. Girola, B. Canciani, N. Angeretti, L. Giampaolo, E. Peressini, T. Awan, L. De Gioia, E. Ragg, O. Bugiani, and M. Salmona, Tetracycline affects abnormal properties of synthetic PrP peptides and PrP(Sc) *in vitro*. *J. Mol. Biol.* **300**, 1309–1322 (2000)

100. S. A. Priola, A. Raines, and W. S. Caughey, Porphyrin and phthalocyanine antiscrapie compounds. *Science* **287**, 1503–1506 (2000)

101. I. Vorberg, K. Chan, and S. A. Priola, Deletion of beta-strand and alpha-helix secondary structure in normal prion protein inhibits formation of its protease-resistant isoform. *J. Virol.* **75**, 10024–10032 (2001)

102. M. Rogers, F. Yehiely, M. Scott, and S. B. Prusiner, Conversion of truncated and elongated prion proteins into the scrapie isoform in cultured cells. *Proc. Natl. Acad. Sci. U. S. A.* **90**, 3182–3186 (1993)

103. A. Helenius, How N-linked oligosaccharides affect glycoprotein folding in the endoplasmic reticulum. *Mol. Biol. Cell* **5**, 253–265 (1994)

104. J. Collinge, K. C. Sidle, J. Meads, J. Ironside, and A. F. Hill, Molecular analysis of prion strain variation and the aetiology of 'new variant' CJD. *Nature* **383**, 685–690 (1996)

105. S. J. DeArmond, Y. Qiu, H. Sanchez, P. R. Spilman, A. Ninchak-Casey, D. Alonso, and V. Daggett, PrPc glycoform heterogeneity as a function of brain region: implications for selective targeting of neurons by prion strains. *J. Neuropathol. Exp. Neurol.* **58**, 1000–1009 (1999)

106. M. Rogers, A. Taraboulos, M. Scott, D. Groth, and S. B. Prusiner, Intracellular accumulation of the cellular prion protein after mutagenesis of its Asn-linked glycosylation sites. *Glycobiology* **1**, 101–109 (1990)

107. A. Taraboulos, M. Rogers, D. R. Borchelt, M. P. McKinley, M. Scott, D. Serban, and S. B. Prusiner, Acquisition of protease resistance by prion proteins in scrapie-infected cells does not require asparagine-linked glycosylation. *Proc. Natl. Acad. Sci. U. S. A.* **87**, 8262–8266 (1990)

108. S. Lehmann and D. A. Harris, Blockade of glycosylation promotes acquisition of scrapie-like properties by the prion protein in cultured cells. *J. Biol. Chem.* **272**, 21479–21487 (1997)

109. B. Caughey, G. J. Raymond, D. Ernst, and R. E. Race, N-terminal truncation of the scrapie-associated form of PrP by lysosomal protease(s): implications regarding the site of conversion of PrP to the protease-resistant state. *J. Virol.* **65**, 6597–6603 (1991)

110. K. F. Winklhofer, U. Heller, A. Reintjes, and J. Tatzelt, Inhibition of complex glycosylation increases the formation of PrPsc. *Traffic.* **4**, 313–322 (2003)

111. C. Korth, K. Kaneko, and S. B. Prusiner, Expression of unglycosylated mutated prion protein facilitates PrP(Sc) formation in neuroblastoma cells infected with different prion strains. *J. Gen. Virol.* **81**, 2555–2563 (2000)

112. F. Beranger, A. Mange, B. Goud, and S. Lehmann, Stimulation of PrP(C) retrograde transport toward the endoplasmic reticulum increases accumulation of PrP(Sc) in prion-infected cells. *J. Biol. Chem.* **277**, 38972–38977 (2002)

Chapter 16

THE MOLECULAR BASIS OF PRION PROTEIN-MEDIATED NEURONAL DAMAGE

Ramanujan S. Hegde and Neena S. Rane

Cell Biology and Metabolism Branch, NICHD, National Institutes of Health, Bethesda, MD 20892, USA

16.1. Introduction

The most seductive questions in prion biology have always centered around the unusual nature of disease transmissibility[1,2]. What is the transmissible agent? What are the routes of transmission from one organism to another? How does the transmissible agent spread within an organism? What is the mechanism of replication? What is the nature of the barrier to transmission between species? Although the answers to these and related questions are far from clear, intense investigation over the past several decades has led to a working framework for understanding how the prion protein (PrP), in the absence of nucleic acids, could mediate the transmission and spread of neurodegenerative disease[3–6].

By contrast, another series of questions has received far less attention within the prion field. This is in part because, rather than being specific and unique to prion diseases, this second set of questions is shared among a wide range of neurodegenerative diseases associated with abnormal protein accumulation. These include Alzheimer's, Parkinson's, and Huntington's diseases, among many others. The vital questions of interest in all of these areas concerns the molecular pathways that underlie the selective and devastating neuronal damage occurring during the pathogenesis of each of these diseases. Hence, one desires to know what the inciting event(s) or protein species are that lead to neurodegeneration. What are the causal relationships between protein aggregation and cellular toxicity? What pathways of cellular homeostasis are disturbed and how does this lead to cell death? Why does pathology occur in only some but not other cell types?

It is this latter, poorly studied set of questions in prion biology that define the scope of this chapter. Unlike a traditional review article however, we will refrain from cataloging the bewildering array of toxicities, functions, and pathologies that have been ascribed to PrP over the years[7,8]. Such observations have often been conflicting, contentious, and limited in scope. Thus, an attempt to reconcile these claims, either with each other or to the pathogenesis of prion diseases is premature at the present time. We shall instead concentrate our efforts on delineating a systematic and logical *approach* for the study of PrP-mediated neurodegeneration. In developing these ideas, we have taken many cues from analogous studies of the other neurodegenerative diseases. Unencumbered by the issues of transmission, scientists in these other areas have delved more broadly and deeply into the mechanisms contributing to neuronal dysfunction during disease pathogenesis[9–13]. Conceptual lessons from these studies, although not specifically considered in this chapter, significantly shape the concepts articulated herein.

In any discussion of prion diseases, the issues of transmissibility (the first set of questions above) and neurodegeneration (the second set of questions) are inextricably linked. However, this linkage in thinking has significantly clouded meaningful investigations into the mechanism of neurodegeneration. In the first part of the chapter, we examine the basis for historically considering transmission and neurodegeneration as coupled aspects of disease pathogenesis. We then present the rationale, based on available studies, for asserting that the issues of neurodegeneration are not obligatorily coupled to transmission. And finally, we argue that for a productive investigation into PrP-mediated neurodegeneration, it should first be studied, both conceptually and experimentally, in isolation from the complexities of transmissibility issues.

In the second part of this chapter, we apply the ideas developed in the first half to formulating one way in which PrP-mediated neurodegeneration can be studied in the absence of PrP-templated transmissibility. We then present a framework for identifying and validating the molecular basis of neuronal damage during prion disease pathogenesis. In so doing, we will argue for the importance of a quantitative understanding of PrP cell biology, including its biosynthesis, trafficking, and metabolism. The best available information and candidates for PrP-mediated neuropathology will be discussed and the key steps for future studies aimed at their validation will be outlined. While the topic of discussion will be focused exclusively on diseases involving PrP, we hope that the general cell biological approach prescribed herein will provide a framework within which to evaluate the pathogenesis of related neurodegenerative diseases that face remarkably similar issues once the complicating variables of prion transmissibility are stripped away.

16.2. Definitions

The nomenclature in the prion field is both heterogeneous and inconsistent. To avoid confusion, it is prudent to begin with clear definitions of the terminology used in this chapter. Transmissible spongiform encephalopathies (or TSEs) and prion diseases are synonymous: both refer to the spectrum of slowly progressing, *transmissible* neurodegenerative diseases with characteristic clinical and pathological sequalae[2–6]. The term prion, derived from the words **pro**teinaceous and **in**fectious, is *by definition* the transmissible agent in these diseases[2]. Importantly however, a specific molecular description of the prion (in terms of a precise composition) remains elusive. While most in the field would agree that a prion is free of large nucleic acids[2–6], the potential presence or roles for small ligands, micro-RNAs, and other components have not been clearly defined. Thus, a prion is an agent, composed mostly if not exclusively of protein, responsible for the transmission of prion diseases (or TSEs) from one organism to another.

The prion protein (PrP), to be clearly distinguished from a prion, is the name for the most abundant and consistently identified protein in the purest available preparations of prions (and hence, the name originally given to this protein). This single protein has been described in numerous forms and locations. Thus, "PrP" is often appended with a modifier to more specifically refer to one particular species among several that include: PrP^{C}, PrP^{Sc}, ${}^{Ctm}PrP$, ${}^{Ntm}PrP$, ${}^{sec}PrP$, cyPrP, PrP-res, and PrP-sen. In this chapter, the general term PrP is used to refer to this protein when no single form is being specified (e.g., as in PrP-mediated neurodegeneration). PrP^{Sc} is used to specify the form that progressively accumulates during the course of prion disease and is therefore also the most abundant species in prions. This form is widely[2–6], but not universally[14–17] thought to be the transmissible agent in prions. PrP^{C} refers to the normal, endogenously expressed, cellular PrP: a GPI-anchored protein found on the surface of many cell types, most abundantly neurons. The terms PrP-res and PrP-sen, which define species of PrP on the basis of the biochemical property of protease resistance, will not be used in this chapter due to the ambiguity of what conditions distinguish 'sensitivity' versus 'resistance' to proteases. The remaining forms of PrP (${}^{sec}PrP$, ${}^{Ctm}PrP$, ${}^{Ntm}PrP$, and cyPrP) make distinctions based on the cellular locale and topological orientation of PrP relative to a membrane. Each of these is defined in further detail in the subsequent sections where they are discussed.

And finally, a word about the term pathogenesis. Since this term broadly encompasses all of the events that go awry during the course of a disease, it is often insufficiently specific in describing a particular facet of disease. In the case of the prion diseases, there are at least two distinct phases that we wish to distinguish: i) *transmission*—acquisition, replication, and accumulation of prions, and ii) *neurodegeneration*—the processes that result in the observed neuronal damage, pathology and clinical symptoms. Since the term pathogenesis includes both of these processes, we shall refrain from its use unless referring to the entire disease process. Otherwise, we shall use the more specific terms related to the aspect of disease pathogenesis under discussion.

16.3. Transmission versus neurodegeneration during prion disease pathogenesis

Prion diseases are by definition transmissible: they are either acquired by transmission and/or can be subsequently transmitted to another individual. This necessarily means that the disease involves the acquisition and replication of prions. It is entirely reasonable therefore to presume that the neurodegenerative disease that ensues must obligatorily be caused by the replication and accumulation of prions. Hence, the nature of the prion and the mechanisms underlying its replication would appear to be central questions in not only the transmissibility of prion diseases, but also in the neurodegeneration that results. For these reasons, an implicit assumption has generally been that an understanding of prion replication would also reveal the molecular basis of prion-mediated neurodegeneration. Unfortunately, this supposition has thus far not proven to be the case despite significant advances in defining both the nature of the prion and the mechanism of its propagation.

After identifying a protein, PrP^{Sc}, as the major component of prions[2], a crucial advance was the subsequent discovery that this protein is encoded by a normal host gene[18,19]. The host encoded form of PrP, termed PrP^{C}, was found to be identical in sequence but distinct in conformation than PrP^{Sc} (ref. 20). These findings had two major consequences for the understanding of prion diseases. First, it immediately suggested a mechanistic model of prion propagation in which PrP^{Sc} would mediate the conformational conversion of PrP^{C} into additional copies of PrP^{Sc} (the so-called 'protein-only' or 'prion' hypothesis)[2]. In the two decades since this model was proposed, it has gained tremendous support and is the generally accepted paradigm of prion propagation and transmission[3–6].

Figure 16.1. Two conceptually different ways (models A and B) of reconciling the relationship between transmissible and inherited forms of neurodegeneration mediated by PrP.

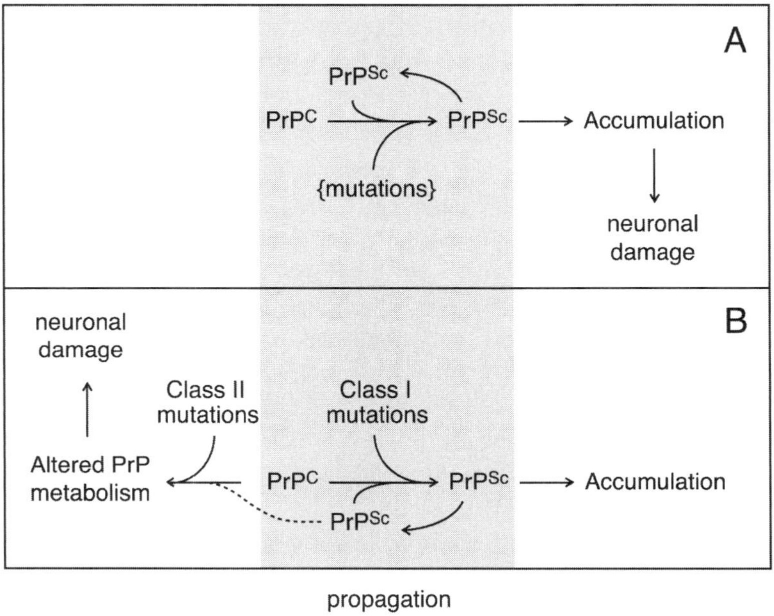

The second consequence of cloning the PrP gene was the resolution of a previously long-standing conundrum in prion diseases: how is it that a disease could have both familial and transmissible forms? The answer lie in the finding that the familial forms of these neurodegenerative diseases were caused by mutations in the gene encoding PrP (ref. 21–24). Thus, PrP can apparently cause disease in two ways. In transmissible forms of the disease, PrP^{Sc} can induce the misfolding of host PrP^{C} into additional copies of PrP^{Sc}. Alternatively, mutations in host PrP can lead directly to disease.

In hindsight, these observations can be used to reconcile the transmissible and inherited forms of disease and unify them under a single paradigm in one of *two* (non-mutually exclusive) ways (Figure 16.1). The first possibility (model A) is that mutations in PrP destabilize its folding to predispose the spontaneous generation of the PrP^{Sc} form. In this model, the spontaneously generated PrP^{Sc} could then facilitate its further propagation from PrP^{C}, accumulate, and cause disease by the same mechanisms involved in transmissible prion diseases. The second less obvious, but equally plausible possibility (model B) is that neuronal damage results from a derangement in some aspect of normal PrP^{C} metabolism. In this view such altered cellular metabolism of PrP could be caused either *directly* by certain PrP mutations, or *indirectly* as a secondary consequence of PrP^{Sc} generation or accumulation. Thus, the most proximal cause of neurodegeneration is focused on PrP^{C} in model B, but PrP^{Sc} in model A.

Both models share several important features. First, the conversion of PrP^{C} to PrP^{Sc} (indicated in gray shading) is the central event in the propagation and accumulation of PrP^{Sc}, and therefore essential for *transmission* of disease. And second, both models posit that the transmissible and familial forms of the disease converge on a single proximal cause of neurodegeneration. Such a shared final pathway of neuronal damage would explain the similarities in clinical course and pathological findings common to the various neurodegenerative diseases mediated by PrP. Yet, the two views make dramatically different predictions about the key events leading to neurodegeneration. The first model proposes that the accumulation of PrP^{Sc} not only generates more transmissible agent, but is the direct cause of neurotoxicity. By contrast, the second model proposes that some feature of PrP^{C} metabolism causes neurodegeneration when misregulated (either directly by mutation, or indirectly by the accumulation of PrP^{Sc}).

Historically, only the first of these two models has been articulated or seriously considered. The reason for this prejudice is because at the time of the identification of inherited PrP mutations, PrP^{C} to PrP^{Sc} conversion was already well-established as the central event in both transmission and neurodegeneration. It was therefore logical and simpler to propose that PrP^{Sc} accumulation acted *directly* to cause neuronal damage (model A) rather than *indirectly* via some other aspect of PrP metabolism (model B). In fact, a desire to experimentally discriminate between these two models was largely obscured by the strong bias in favor of the first model. This was further complicated by the rarity of the human genetic forms of these diseases, making their molecular analysis difficult. Thus, the two views have never been systematically or directly tested for their validity. Instead, an *ad hoc* series of experiments, often designed for other purposes, must be evaluated to help discriminate between the different views of PrP-mediated neurodegeneration.

One of the key predictions of model A is that human neurodegenerative diseases caused by mutations in PrP should accumulate PrP^{Sc}, generate transmissible agent (i.e., prions), and should therefore be transmissible. Testing definitively for PrP^{Sc} or prion accumulation in diseased human tissue is complicated by several confounding variables. While protease resistance and relative insolubility provide valuable surrogate markers of PrP^{Sc}, neither of these properties is unique to PrP^{Sc} (that is, there are several ways to make PrP insoluble or protease resistant without necessarily generating disease-associated PrP^{Sc}; see for example, ref. 25). Conversely, the lack of these properties would not necessarily rule out the presence of PrP^{Sc}, especially since digestion and solubilization conditions are operationally defined and variable from one laboratory to another. Nonetheless, such analyses have been done on at least some of the familial cases of PrP-mediated neurodegenerative disease with mixed results. Some mutations (e.g., D178N, E200K, or V210I; ref. 26 and 27) result in the abundant accumulation in brain of protease-resistant PrP whose digestion properties and fragments are characteristic of PrP^{Sc}. Other mutants (e.g., P102L or F198S) appear to accumulate PrP in a state distinguishable from normal PrP^{C}, but do not have the same biochemical properties of PrP^{Sc}. For example, they may be only partially resistant to proteases, or yield digestion fragments of different sizes[28–35]. Yet other mutants (e.g., A117V) show no obvious biochemical evidence of PrP^{Sc} accumulation[33,34].

Meanwhile, the most definitive test for prions, a bioassay, was being performed in parallel on comparable samples[36–39]. However, due to a potential species barrier between the source (human) and host (usually rodents, or in some instances non-human primates), the interpretation of transmission results are often complicated. In these experiments, disease transmissibility appears to correlate more or less with the biochemical studies: samples showing clear evidence of PrP^{Sc} (e.g., E200K) transmit disease with high frequency, while others (e.g., P102L or A117V) transmit disease to few or no recipients. While these observations further strengthen the case for PrP^{Sc} as the transmissible agent in prions, it was also the first indication that neurodegeneration could potentially be caused by PrP mutants that did not obligatorily generate either PrP^{Sc} or prions. It was therefore at least feasible that the development of PrP-mediated neurodegenerative disease could be uncoupled from the replication and accumulation of prions (i.e., suggested by model B).

Despite these results, the experimental difficulties and the inability to do conclusive studies without the confounding variables of species barriers left a definitive conclusion out of reach. It remained entirely possible that prions did form and accumulate in all of these PrP-mediated diseases, but were simply not readily detectable in some instances by the admittedly complicated assays being employed. Thus, the simplest view remained that *both* transmission and neurodegeneration depend absolutely on the replication and accumulation of prions, a process presumed to be synonymous with the conversion of PrP^{C} to PrP^{Sc} (i.e., model A). Indeed, an alternative view was essentially still not considered. What then was imagined to be the cause of neurodegeneration? Two obvious (non-mutually exclusive) possibilities were usually cited. Either neurodegeneration was a result of the depletion of PrP^{C} as a consequence of its conversion to PrP^{Sc}, or the accumulation in brain of the insoluble, aggregation-prone PrP^{Sc} form was proposed to be inherently harmful. However, both of these possibilities soon proved difficult to demonstrate.

The first surprising finding was the observation that PrP knockout mice are phenotypically normal[40]. As predicted, these mice are resistant to prion infection and propogation[41,42], consistent with a requirement for endogenous PrP^{C} in the replication of PrP^{Sc}. However, the fact that the absence of PrP^{C} did not directly cause neurodegeneration suggested that the depletion of PrP^{C} during prion replication may not be the cause of neurodegeneration. At the time, it remained possible that the knockout mice, having lacked PrP from the single-cell stage, did not mimic the acute PrP^{C} depletion that might occur during prion disease. However, recent studies in which no adverse consequences were observed upon post-natal disruption of the PrP gene strongly argue against the depletion of PrP^{C}, either acutely or chronically, as the cause of neurodegeneration[43,44].

This then left the proposed toxicity of PrP^{Sc} accumulation as the most obvious candidate in causing neurodegeneration. Unfortunately, the simplest variant of this hypothesis quickly became untenable as well. In a wonderfully elegant experiment, the brain tissue of PrP-expressing mice was grafted into the brain of PrP-knockout mice[45]. Upon inoculation of these grafted mice with prions, the PrP-expressing tissue replicated the prions and generated large amounts of PrP^{Sc} and transmissible agent. Despite its deposition throughout the brain, only the tissue actively expressing PrP succumbed to neurodegeneration; the PrP-knockout tissue remained completely unaffected even after prolonged exposure to PrP^{Sc} and transmissible prions. This conclusion has been confirmed more recently by a completely independent approach in which PrP was selectively depleted in neurons after the initiation of prion infection[44]. Even though the infection continued to generate prions and PrP^{Sc} (presumably via non-neuronal cells such as astrocytes), the neurons remained free from further damage. In fact, the degeneration present at the time of PrP depletion may even have been reversed at later points. Thus, it appears that the accumulation of PrP^{Sc} is not inherently toxic to neurons *per se*.

From these various observations, it has become increasingly clear that the cause of neurodegeneration in prion diseases cannot easily be explained by the most apparent events that accompany the replication of prions: the acute depletion of PrP^{C} or the accumulation of PrP^{Sc}. Instead, these observations suggest remarkably that the accumulation of PrP^{Sc} and of prions is neither necessary (e.g., in the case of some familial PrP mutants) nor sufficient (e.g., in the context of neurons not actively expressing PrP^{C}) for neurodegeneration. Yet, the evidence that conversion of PrP^{C} to PrP^{Sc} is the central event in prion replication and disease transmission is now overwhelming[3–6]. How then can these two conclusions be reconciled?

The simplest way would be to posit that the events that are of paramount importance to transmission (PrP^{C} to PrP^{Sc} conversion) are not necessarily the same ones that are critical for neurodegeneration. Clearly however, both facets of the disease involve host-encoded PrP, but apparently in different ways. For prion replication and transmission, host PrP is absolutely required as a source of substrate for PrP^{Sc} propagation. For neurodegeneration (in both transmissible and non-transmissible forms of disease), ongoing PrP expression in the cells that eventually succumb to disease is absolutely required. *Thus, the primary insight that we currently have into the basis of neurodegeneration in these diseases is that some aspect of active PrP expression or metabolism is required for its selective toxicity to neurons.* For these reasons, we argue that while historically, there have been good reasons to consider transmissibility and neurodegeneration as coupled events in prion disease pathogenesis, the fact that they can be uncoupled experimentally and naturally merits their consideration as separate and distinct phases of the disease. But should neurodegenerative processes be studied separately from transmissibility and prion replication, and if so, how?

16.4. The case for uncoupling neurodegeneration from transmission

There are several reasons, both conceptual and technical, to study the neurodegenerative processes of PrP-mediated disease independently of transmission. A particularly pragmatic reason relates to the biochemical properties of PrP^{Sc} that make the cell biological and biochemical study of other PrP isoforms difficult. First, the half-life of PrP^{Sc} in cultured cells or brain tissue is substantially longer than other PrP species[46,47], resulting in its much higher levels at steady state. Second, PrP^{Sc} is both highly aggregated and heterogeneous[48,49]. Together, with its high abundance during transmissible disease pathogenesis, these properties make biochemical fractionation of the various PrP isoforms exceedingly difficult. Thus, nearly all fractions of any separation method contain amounts of PrP^{Sc} that are comparable to or exceed other PrP isoforms. Third, sensitive reagents (e.g., antibodies) highly specific to the PrP^{Sc} form remain elusive despite extensive efforts and some claims of success[50–52]. Thus, its definitive and high resolution detection in individual fractions, within a cell (e.g., by immunofluorescence), or in tissue remain difficult. Fourth, PrP^{Sc} is generally highly resistant to protease digestion relative to the other forms of PrP. This, combined with the lack of specific antibodies for PrP^{Sc} make it difficult to remove selectively in instances where the other PrP isoforms need to be analyzed. And finally, PrP^{Sc} continues to elude a clear molecular or structural description. Many different 'strains' have been identified[53] that differ in poorly defined ways with respect to both transmissible and biochemical properties.

It is therefore clear that the presence and accumulation of PrP^{Sc} during PrP-mediated neurodegeneration makes the selective analysis of non-PrP^{Sc} forms of PrP difficult. This has been a principal reason that any role in neuronal damage for PrP isoforms other than PrP^{Sc} has been difficult to evaluate. Conversely, such non-PrP^{Sc} forms are relatively easily removed (for example, with protease digestion) to selectively reveal the more abundant PrP^{Sc}. While this has facilitated PrP^{Sc} analysis during transmissible disease progression, it has also obscured other PrP forms that may contribute to or cause neurodegeneration. Thus, an evaluation of any role for non-PrP^{Sc} forms in the development of neuronal damage would be facilitated greatly by systems in which PrP-mediated neurodegeneration is recapitulated in the absence of PrP^{Sc} accumulation.

A second reason to uncouple the neurodegenerative from transmissible phases of disease relates to the multi-factorial and complex parameters that influence transmission of prion diseases[3–6]. These factors include the 'strain' of prion involved, the passage history (i.e., in what species did it pass through), the primary sequences of the host versus exogenous PrP, yet undefined factor(s) needed for prion replication (e.g., a hypothetical protein X, among other factors), and incompletely defined modifiers of prion susceptibility and incubation time[54–58]. These parameters each influence the time course of the disease, the pathological features that are observed, and the cell-type specificity of involvement. Ideally, it is desirable to simplify these variables by analyzing disease in a model where a defined inciting event (such as a point mutation) leads as directly as possible to the pathway of neurodegeneration without influencing too many other events that would obscure the relevant pathogenic steps.

The third reason to study the later neurodegenerative steps in the absence of transmissible agent is the simple practicality of biosafety and containment. The study of prions in either cell culture or mouse models requires specific biosafety considerations that involve a substantial investment of resources. Equipment and space are generally dedicated to prion work, making their use for other studies impractical. This makes it difficult for investigators in other fields of study to initiate prion-related studies. However, the issues of PrP-mediated neurodegeneration, if recapitulated in the absence of transmissible agent, can be studied as any other cell-biological or pathological process. Since these downstream events are likely to involve aspects of basic cell biology, signal transduction, apoptosis, etc., their analysis would be markedly facilitated by the involvement of experts in these different fields. Thus, models of PrP-mediated neurodegeneration in isolation from the issues of transmission would reduce barriers to a multi-disciplinary approach to these problems. As argued in the previous section, there are now compelling reasons to believe that in fact, PrP-mediated neurodegeneration is not obligatorily linked to either PrP^{Sc} or the formation and accumulation of transmissible prions. Therefore, it is not only feasible, but desirable to experimentally uncouple these two phases of the disease to facilitate the mechanistic dissection of the neurodegenerative process.

16.5. Genetic PrP-mediated neurodegeneration as a model system

What is the most productive way to study the neurodegenerative phase of PrP-mediated disease in the absence of PrP^{Sc} or transmissible agent? We believe the answer to this question lies in a careful consideration of model B (Figure 16.1). In this model, prion disease pathogenesis is depicted in two phases: the replication and accumulation of prions, followed by the neurodegeneration induced by this process. Both phases are experimentally known to require ongoing PrP synthesis, but appear to involve different aspects of its metabolism. In the first phase (shaded in gray), PrP^{C} is needed as the substrate for the template-mediated conversion into PrP^{Sc}, an event thought to be essential for the generation of prions. In the second phase, the role of PrP expression is unknown, but is an absolute prerequisite for neuronal cell death. Thus, some feature of PrP metabolism, after its active *de novo* synthesis, is altered in a way that leads to neurodegeneration. The neurodegenerative events are initiated in one of two ways: either as an indirect consequence of PrP^{Sc} formation and accumulation during transmissible prion disease (indicated by the dotted line in model B), or due to inherited mutations in the PrP gene in genetic disease.

In the case of the genetic diseases, the mutation can be envisioned to act in one of two ways. In the first way (designated Class I; see Figure 16.1), the mutation may influence PrP^{C} folding in a manner that facilitates its spontaneous conversion to PrP^{Sc}. Once this spontaneous event occurs, PrP^{Sc} would mediate its templated self-propagation to not only generate more PrP^{Sc}, but to initiate the heretofore unknown events leading the neurodegeneration. Thus, these forms of genetic disease act by first generating PrP^{Sc} (and hence, are predicted to be transmissible) which then leads to neurodegeneration by the same mechanisms utilized in transmissible prion diseases. The second way PrP mutations could cause disease is to alter PrP metabolism in a way that recapitulates its ability to cause neuronal damage. In these instances (designated Class II), the effect is directly on PrP^{C} metabolism and therefore need not involve PrP^{Sc} formation. Thus, such inherited diseases would neither accumulate PrP^{Sc} nor be transmissible.

The phenotype of Class I PrP mutants, if recapitulated in model systems, would be ideal for studying PrP^{Sc} formation and replication. Here, a defined change in primary sequence facilitates not only spontaneous conversion to PrP^{Sc}, but its subsequent self-propagation in a way that reconstitutes disease pathogenesis. By contrast, Class II mutations should allow the steps of PrP-mediated neurodegeneration to be reconstituted in a model system *without the involvement of PrP* ^{Sc}. Clearly, the most insight into neurodegeneration with the least confounding variables would be to study such genetic lesions that bypass prion replication and directly modulate the aspects of PrP metabolism that initiate the molecular pathways leading to neuronal dysfunction and death. Thus, an important step lies in distinguishing between the two ways that inherited PrP mutations lead to disease, and identifying for study those that are likely to work by the second mechanism.

Simplifying the study of a complex, multi-factorial disease process by first focusing on genetic examples that may recapitulate key facets of pathogenesis is an approach with numerous important precedents. This is analogous to progress in other complex diseases such as Alzheimer's disease or breast cancer. In these instances, the analysis of rare genetic variants were instrumental in illuminating particular molecular players and mechanistic steps to guide a better understanding of the more commonly occurring forms of disease that were otherwise too heterogeneous to allow systematic analysis. Even though clinical and pathological features of the genetic variants often differ in significant ways from the non-heritable forms of the disease, the initial faith that they would all share at least some common mechanistic steps at the molecular level was eventually validated upon further study.

At this stage in our understanding of prion disease pathogenesis, a similar faith is needed in the commonality of the underlying mechanistic steps involved in the different disease variants. Clearly, there are many differences among the various genetic, sporadic, and transmissible forms of PrP-mediated diseases. Yet, they all share the involvement of PrP, certain pathologic features, their late onset followed by rapid disease progression, and the selective involvement of the central

Figure 16.2. Different biochemical properties of PrP among various disease-associated mutations. Brain tissues from the indicated human diseases were analyzed before ('-') or after protease digestion using mild ('M') or harsh ('H') conditions (as defined in ref. 33). Tissue from non-familial CJD was also analyzed in parallel. All samples were digested with PNGase to remove glycans prior to analysis by immunoblotting.

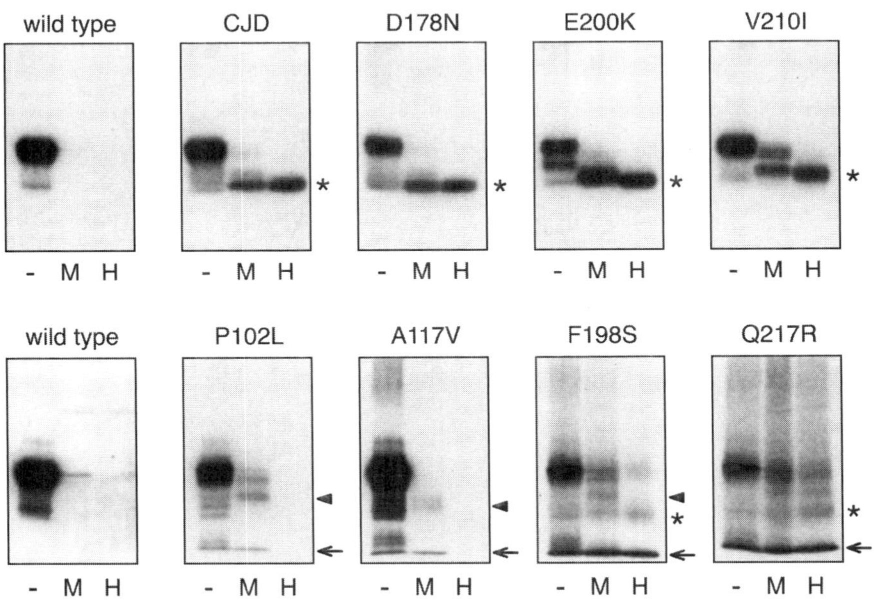

nervous system despite widespread PrP expression. For these reasons, a faith in at least some shared common features is probably not misguided, and more than offset by the potential for simplification by studying select inherited examples of PrP-mediated neurodegenerative disease. Even if the pathogenic events are not shared among the genetic and transmissible forms of disease, at the very least, insight into the pathogenic events of a model protein folding disease will be illuminated by studying PrP mutants that directly cause neurodegeneration.

There are roughly 30 choices among the genetic lesions in PrP that cause neurodegeneration[59,60]. The vast majority of these very rare mutations have not been studied in any significant detail. In at least a few instances however, sufficient analysis has been performed to evaluate whether they might be Class I or II mutants of PrP. Since one primary distinction between Class I and Class II mutants is whether they accumulate PrP^{Sc}, a biochemical analysis is perhaps the simplest way to initially categorize the mutants. As one such example (see Figure 16.2, which essentially recapitulates previous observations[26–35]), the susceptibility of PrP to different protease digestion conditions is evaluated. Here, a very high proportion of total PrP from transmissible CJD is characteristically resistant to 'harsh' protease digestion in a manner that generates *only* a resistant C-terminal domain (indicated by an asterisk in Figure 16.2). This behavior indicates that the majority of PrP is in the PrP^{Sc} form at the time of death from illness. Exactly this behavior, *both quantitatively and qualitatively*, is observed for several of the human PrP mutants (e.g., E200K, D178N, and V210I), suggesting that like transmissible disease, these familial forms also accumulate large amounts of PrP^{Sc}. Indeed, these are also the heritable diseases that have been easily transmitted to animal hosts with high efficiency[36–39]. Thus, they appear to be Class I mutants that may work by favoring the spontaneous generation of PrP^{Sc}.

By rather striking contrast, several other heritable PrP mutants show distinctly different behaviors. These mutants do not result in the accumulation of PrP forms whose C-terminal domain is highly resistant to 'harsh' protease digestion. Instead, some (such as P102L or A117V) appear to contain PrP forms that are only 'mildly' resistant to protease digestion (arrowheads in Figure 16.2). Others contain and accumulate smaller metabolic fragments of PrP (arrows, Figure 16.2) that seem to resist protease digestion and are not observed at comparable levels in normal brain tissue. These observations suggest that these mutants may cause a change in some aspect of PrP metabolism, but do not generate much if any PrP^{Sc}. Consistent with this interpretation, brain homogenates from such samples do not appear to transmit disease to experimental animals[36–39], suggesting that they may represent Class II mutations that directly lead to neurodegeneration in the absence of either transmissible agent or PrP^{Sc}. Similar observations of altered metabolism without apparent PrP^{Sc} or prion accumulation has also been made with other more gross mutations in PrP. These additional putative Class II mutants include premature stop codons and octapeptide (or octarepeat) insertions into the N-terminal domain of PrP (ref. 61–65). Hence, as a group, Class II mutants would appear to be the best candidates for examples of disease in which PrP metabolism is directly and specifically affected to cause neurodegeneration, without the complicating feature of prion replication and accumulation. Among them, point mutants may represent the best choices for a model system since their effects are presumably more selective than large insertions or premature truncations.

16.6. The importance of PrP cell biology

How can one use the genetic lesions that are hypothesized to selectively cause neurodegeneration to understand the mechanistic basis of the disease process? There are two qualitatively different, non-mutually exclusive directions. The first approach is to reconstitute the key events of cellular dysfunction that accompany neurodegeneration in a model system amenable to experimental manipulation. The second strategy involves comparative quantitative analyses of the cell biological behavior and metabolism of PrP and its disease-associated variants to generate testable hypotheses for the mechanisms involved in initiating neuronal damage. The rationale, utility, and current progress toward both of these strategies are discussed in turn below, with our argument for why the second, cell biological approach is particularly important at the present stage of progress in this field.

The first approach, that of selectively reconstituting PrP-mediated neuronal damage in a model system, would allow the evaluation and dissection of the steps leading from a defined lesion in PrP to eventual cell death. Ideally, the system of choice would be the simplest and most manipulable model, such as yeast or perhaps cultured cells that allow the combined use of genetic, molecular, and biochemical tools to easily modulate individual gene products, cellular pathways, and environmental parameters. For such a system to be useful, it must recapitulate at least some facets of the native disease process, such as the selective effects of known disease-associated mutations. At this point, no such model has been successfully developed or validated. This contrasts sharply with the process of PrP^{Sc} propagation, which has been both reproduced in several cell culture systems[66–69] and validated by the demonstration of infectivity in animal bioassays[69]. Such systems have proved quite valuable over the past 15 years in helping to uncover features important to PrP^{Sc} generation and propagation.

The difficulties of recapitulating the key features of neuronal dysfunction in simplified systems are many. These include the apparently extreme cell-type specificity of this process *in vivo*. Only neurons appear to be obviously affected during disease[70–71] despite widespread expression of PrP in multiple cell types both in and out of the nervous system[72–73]. Furthermore, specific and different subsets of neurons are affected in the different disease variants with no clear explanation[3–8,74,75]. These observations indicate that very precise cellular conditions that may not be easily recapitulated in other model systems (such as yeast or cultured cell lines) are necessary to manifest the downstream consequences associated with PrP mutations or PrP^{Sc} accumulation. In addition, the cellular context may play a currently unappreciated key role. For example, the *in vivo* situation of multiple interacting cell types and defined external cues such as hormones, growth factors, or extracellular matrix may substantially influence PrP-mediated neurotoxicity in ways that are difficult to reproduce in culture.

A second problem is that *in vivo*, the disease is temporally confined both in terms of the slow progression and defined age of onset. The basis of these observations is not known; why is it that despite expressing a mutant PrP gene at high levels for between 40–50 years, the disease is only manifest late in life? Thus, faithfully reproducing neurodegenerative events in a simplified system stripped of the *in vivo* context and under vastly different time scales, while potentially very useful and appealing, may be daunting. Tricks of PrP overexpression and the use of stressful conditions to tax the cells may be necessary to facilitate PrP-mediated toxicity in such model systems; however, such manipulations may make distinguishing effects of the normal from mutant proteins especially difficult. Furthermore, without specific intermediate markers of PrP dysfunction, one would need to reconstitute the entire downstream set of events that lead to detectable cellular damage. This may encompass too many steps to easily accomplish in simplified cellular systems. These considerations help to define the obstacles that one must consider in the establishment of model cellular systems, and perhaps explain why such systems have been very slow to develop in both PrP-mediated and other neurodegenerative diseases.

Many of these obstacles could potentially be overcome by the use of whole organisms that contain multiple differentiated cell types in which the likelihood of recapitulating the desired neurodegenerative pathways is increased. Again, one would seek to observe effects that are selective to both neurons and disease-associated PrP mutants. Although a homologue of PrP is not found in the genomes of either *Drosophila* or *C. elegans*, these are both attractive candidates for such an approach due to their genetic manipulability and the existence of tools for large scale loss-of-function screens using RNAi methodology. At present, little effort has been expended towards these goals, largely due to the focus on the transmissible features of the disease that have long been reconstituted in a cell culture system. Indeed, in the case of polyglutamine expansion-mediated protein aggregation and neurodegeneration (where transmissibility is not an issue), useful models have been developed and exploited in both *C. elegans* and *Drosophila*[76–78]. Similar models for the neurodegenerative phase of PrP-mediated disease should greatly facilitate both the testing of hypotheses related to the pathogenic events (see below) and the elucidation of the cellular pathways involved. Until such simpler model systems are developed and validated, one must either work within the confines of the modestly manipulable, slow time frames characteristic of transgenic mice, or take parallel alternative strategies to obtaining mechanistic insights into PrP-mediated neurodegeneration.

The principal parallel strategy that, in our opinion, offers the highest likelihood of success is founded on a thorough and quantitative understanding of PrP cell biology. In short, the logic is that in order to understand the causative basis of a disease-associated PrP mutation, the metabolism of the mutant PrP needs to be compared to and distinguished from wild type PrP in simplified biochemical and cell culture systems. In this manner, one can identify potential differences in the behavior of PrP mutants that may account for their biological consequences *in vivo*. It is important to note that in these experiments, the mutant PrP is not anticipated to necessarily induce the eventual consequences of cell damage in the model system. As discussed above, the downstream pathways are not likely to be easily recapitulated. Rather, this approach is intended to identify differences between wild-type and mutant PrPs that would represent *potential* initiating events for subsequent neurodegenerative sequelae. Once specific points of difference are identified in the pathways of PrP metabolism, hypotheses can be formulated regarding the role of such events in the neurodegenerative process. The readily manipulable biochemical and cell biological systems employed to initially identify the differences in PrP metabolism should also facilitate the development of tools to exaggerate or minimize the key step(s) in question. Finally, such tools would then be used in suitable model systems (such as transgenic mice) to either validate or negate hypotheses that propose key roles in inciting neurodegeneration. Thus, in this strategy, basic aspects of PrP cell biology are studied in easily manipulated and rapidly analyzed systems to generate hypotheses that are subsequently tested in the more laborious and slow *in vivo* setting only after specific tools and mechanistic insights are available.

A key step in this experimental strategy is to determine in molecular detail the cell biological properties and metabolism of PrP to facilitate its comparison to the mutants. This is not necessarily to learn the normal function of PrP, since loss of its still unknown functions[43] are not thought to be the key event leading to neurodegeneration. Rather, it is to facilitate the identification of an apparently dominant, gain of function feature imparted by the mutant that is likely to cause neurodegeneration. Put another way, how can one possibly figure out what goes 'wrong' without a clear description of what 'right' looks like? Thus, in the absence of a quantitative description of the steps in PrP biosynthesis and metabolism, one cannot reasonably hope to detect anything but the most dramatic effects caused by mutant PrP variants. However, dramatic consequences of PrP mutations are not particularly likely in light of the fact that the disease manifests over such a prolonged time frame. This principle is analogous to the effects of mutations in Alzheimer's precursor protein (APP) that cause neurodegeneration. The mutations do not have overtly obvious effects on APP metabolism; rather, most of them subtly influence specific processing events involved in the generation of a particular peptide fragment (a-beta) which, over many decades, has adverse consequences[9]. Such effects would have been very difficult to detect without well-defined and quantitative assays for the normal events in APP metabolism. Similar and analogous parallels can be drawn with many other slowly-developing diseases in which small biochemical effects are sufficient to cause disease over appropriate time scales in the correct *in vivo* context. It is therefore imperative that *in vitro* analyses should have the capability to quantitatively detect subtle differences in a variety of parameters related to PrP cell biology.

What then are the facets of PrP cell biology that should be the focus of our attention? Given that the normal function of PrP is neither known nor is believed to play a role in neurodegeneration, it is most reasonable to dissect the steps in PrP biosynthesis, maturation, trafficking, and degradation pathways. These metabolic events seem particularly relevant in light of the observation that even in cases where PrP^{Sc} accumulation is not occurring, various other forms of PrP are often deposited as plaques or other aggregates at the later stages of the disease[70,71]. While the causative role of such deposits remains uncertain, they do indicate that some facet of its normal metabolism has gone awry at some point during pathogenesis. Thus, the operant questions regarding PrP that should be asked and addressed include: what are the key steps during its biosynthesis? What machinery is required for its proper entry into the endoplasmic reticulum (ER)? What factors in the ER are involved in its modifications, folding, and maturation? How efficient are these various steps during its biogenesis? What happens to the population of PrP molecules that fail in their maturation? What are PrP's different destinations in the cell? How is it trafficked to these sites such as the cell surface, and what additional maturation steps occur en route? What are the pathways and time frame for its normal turnover? What machinery is involved, and what metabolic products are generated?

Thus, each facet of the life of PrP from its point of synthesis to its recycling into degradative products should be analyzed quantitatively. By analyzing these same events for various disease-associated mutations, specific steps that may be deranged, even very slightly, can be identified. Once these potential differences are found, they can then be studied to investigate the mechanistic basis of the effect and subsequently modified to either enhance or decrease the process in question. These tools can then be used *in vivo* to test specific hypotheses regarding the pathogenesis of prion diseases. While the complete sequence of investigation is far from complete in any aspect of PrP metabolism, some initial studies have identified potential candidates for being involved in PrP-mediated neurodegeneration. Below are described the historical development and current state of investigation into these aspects of PrP cell biology, their relationship to neurodegeneration, and some comments on what is now required to further our understanding.

16.7. ^{Ctm}PrP and the development of neurodegeneration

When the gene encoding PrP was first cloned, an obvious (and deceptively simple) question was to determine its normal biosynthetic pathway and cellular locale. Since PrP^{C} (and PrP^{Sc}) were known to be glycosylated, PrP was presumed to be trafficked through the secretory pathway. Indeed, sequence analysis of the full length PrP open reading frame suggested an N-terminal signal for targeting to the ER and two potential sites for N-linked glycosylation in the C-terminal domain. In addition, the sequence revealed a hydrophobic domain of ~20 residues and a downstream amphipathic region. These elements were incorporated together into a model of PrP as a double-spanning transmembrane protein in which the N- and C-termini were in the lumen[79]. Such a model was supported by the initial analysis of PrP topology upon its *in vitro* synthesis using wheat germ extracts and ER microsomes derived from canine pancreas[80].

The view of PrP as a transmembrane protein was very short lived. It was quickly realized that when synthesized in a *mammalian* translation system (rabbit reticulocyte lysate) with pancreatic ER microsomes, PrP was fully translocated across the membrane (similar to a secretory protein, and hence the operational designation of this topologic form as 'secretory'-PrP or ^{sec}PrP)[81]. Furthermore, in cells, the protein was found to be fully exposed on the extracellular surface, where it was discovered to be tethered to the plasma membrane by a C-terminal glycolipid anchor[82]. Thus, the original topology predictions and results from wheat germ translation systems were largely ignored, presumed to be an artifact of using a plant-based system to analyze a mammalian protein. Given the long-standing and widely held belief that each protein has a *single* 'correct' final configuration, it was concluded that normal cellular PrP is a GPI-anchored cell surface glycoprotein. All other observed forms were thought to represent either mistakes or artifacts (and hence, irrelevant to normal PrP function). This is the view that generally persists today.

Curiously however, the transmembrane form of PrP, while exaggerated in the wheat germ system (>80% of total PrP, depending on translation conditions), is nonetheless also observed (at an albeit lower level of ~5–10%) in the reticulocyte lysate system[33]. This topological heterogeneity had not been observed for any of numerous model secretory or membrane proteins that had been examined in the *in vitro* translocation systems: not only were these other proteins made faithfully in their predicted topology, but the same outcomes were obtained in the wheat germ, reticulocyte, and cell culture systems. Furthermore, the central hydrophobic domain of PrP that allows it to potentially span the membrane was subsequently found to be extremely well conserved across species[83]. These ~20 residues are absolutely invariant in all species including those as divergent as avians and reptiles (whose overall conservation is ~40% identity)[84,85]. This highly conserved, albeit unusual feature that is required for a proportion of PrP to be made as a membrane-spanning protein[86] suggested an alternative explanation for the transmembrane form. Perhaps the capability to make transmembrane PrP (at least under some conditions) may be both normal and important for some aspect of PrP biology. Unfortunately, the lack of a clear functional role for PrP made this hypothesis difficult to explore. Furthermore, a related idea that transmembrane PrP could somehow play a role in disease generally was not considered because at that time, the much more dramatic observation of PrP^{Sc} accumulation suggested a more obvious culprit.

However, several concurrent studies began to suggest that while PrP^{Sc} formation was clearly associated with disease transmission, its accumulation was not inherently toxic to neurons. The first hint was the observation that when mice heterozygous for the PrP gene ($PrP^{+/-}$) were inoculated with prions, PrP^{Sc} accumulation followed a course very similar to that observed in wild type mice (i.e., $PrP^{+/+}$). Yet, the progression to clinical disease resulting from neurodegeneration was markedly delayed[87]. This discordance between PrP^{Sc} and neuronal damage was particularly dramatic in brain grafting experiments[45] where PrP knockout neurons appeared impervious to any adverse consequences of PrP^{Sc} deposition. In parallel, the identification of the PrP gene made possible the discovery of a wide range of inherited PrP mutations that led to familial forms of PrP-mediated disease[59]. Biochemical analyses of tissue from such familial cases (e.g., as in Figure 16.2) suggested that while some of them had accumulated PrP^{Sc}, others were surprisingly devoid[26–35]. Such biochemical results were, over the course of several years, corroborated by extensive transmission studies[34–39]. Thus, by the mid-1990s, it was reasonable to consider the possibility that PrP-mediated neurodegeneration could be caused by means other than through PrP^{Sc} accumulation.

It was in the context of these studies that the question of the proximal causes of neurodegeneration was brought into slightly better focus. A reconsideration of a possible role for the originally observed transmembrane form of PrP was stimulated by the discovery that at least one disease-associated mutation (A117V) and a disease-influencing polymorphism (at codon 129) were in the highly conserved domain of PrP predicted to form a potential transmembrane segment. How then could one examine if and how this transmembrane form might play a role in disease? First, in preliminary experiments, it was observed that in fact, the A117V mutation influenced PrP biogenesis: a very subtle, but reproducibly detectable increase in transmembrane PrP (from ~5–10% to ~10–15%) was observed in translocation assays carried out in reticulocyte lysates. Motivated by this *in vitro* observation, it was then hypothesized that perhaps PrP-mediated neurodegeneration could be caused by transmembrane PrP. Indeed, it had been observed for some time that while PrP^{C} is largely released from the cell surface by cleavage of its GPI anchor, PrP^{Sc} remained cell-associated even after GPI anchor cleavage[88]. While this had many interpretations, one possibility was that PrP^{Sc} is membrane anchored by another mechanism, perhaps via a transmembrane topology. Unfortunately, the biochemical obstacles to analyzing PrP^{Sc} precluded a direct examination of this hypothesis.

Therefore, a different tact was taken: mutations that favor or disfavor the generation of the transmembrane form to different extents would be expressed in transgenic mice lacking their endogenous PrP to observe the consequences, if any, for neurodegeneration. Using the *in vitro* translocation assay, mutations (in or around the central hydrophobic domain of PrP) were assayed for their effect on topology. In the course of these studies, it was realized that rather remarkably, there was not one, but two transmembrane forms that were being generated[33]. One spanned the membrane with the N-terminus in the ER lumen and C-terminus in the cytosol, while the other was in exactly the reverse orientation. These were dubbed ^{Ntm}PrP and ^{Ctm}PrP, respectively (see Figure 16.3). Most of the mutations that increased the transmembrane forms in these original studies seemed to preferentially increase ^{Ctm}PrP (ref. 33, 35), and hence for somewhat arbitrary reasons, this become the form of interest while the ^{Ntm}PrP form has thus far been poorly studied.

When expressed in transgenic mice on a PrP-null background, PrP mutants that favor transmembrane forms of PrP (and in particular ^{Ctm}PrP) caused the development of neurodegenerative disease[33,35]. Upon the compilation of several such mutants (that ranged from ~10% to ~50% ^{Ctm}PrP generation, as compared to ~5% for wild-type), it became clear that there was a dose-response effect: the more heavily ^{Ctm}PrP is favored or the more highly expressed the ^{Ctm}PrP-favoring transgene, the earlier the development of neurodegeneration. Conversely, two different mutations that reduce or abolish the ability to generate either of the transmembrane forms did not lead to neurodegeneration[33] (although for reasons that are not yet clear, transgenic lines expressing these

Figure 16.3. Depiction of the different topologic forms of PrP: ^{sec}PrP, ^{Ctm}PrP, ^{Ntm}PrP, and cyPrP. At present, there remains some uncertainty regarding whether cyPrP and ^{Ctm}PrP contain uncleaved signal sequences[33,89–91]. Although an uncleaved signal can be detected on at least some ^{Ctm}PrP and cyPrP chains when they are generated by overexpression in cultured cells[89,91], it is less clear whether this is also the case under normal circumstances *in vivo*. Of these forms, ^{sec}PrP and ^{Ctm}PrP contain a C-terminal GPI anchor and two N-linked glycans[33,89] that are not found on either cyPrP or ^{Ntm}PrP.

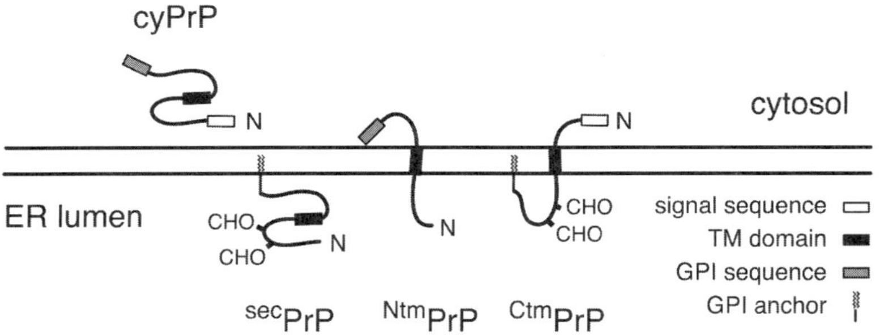

mutants proved more difficult to stably maintain). Thus, with the accumulated data from four different ^{Ctm}PrP-favoring mutations and two transmembrane-disfavoring mutations, each in multiple lines of transgenic mice at different expression levels, a very strong positive correlation can be made between the ability to generate ^{Ctm}PrP and neurodegenerative diseaes[33,35].

Biochemical analyses of brains from these transgenic mice revealed the presence of ^{Ctm}PrP (ranging from ~5–30% of total PrP) in those lines that both expressed ^{Ctm}PrP-favoring mutants and were prone to neurodegeneration[33]. Again, a dose-response relationship was observed between ^{Ctm}PrP in brain and severity of the neurodegenerative phenotype[33,35], validating in many ways the *in vitro* translocation systems in which these mutants (as well as the transmembrane forms themselves) were first identified. In a particularly satisfying experiment, one of the PrP constructs whose expression in mice led to neurodegeneration was the human disease-associated A117V mutant[35]. Even more remarkably, human tissue from such a patient contained detectable amounts of ^{Ctm}PrP, but not PrP^{Sc} (ref. 33). Indeed, this was one of the genetic diseases that was particularly puzzling because it was caused by a derangement in PrP, and yet was neither transmissible[34] nor contained PrP^{Sc}. Consistent with these original observations in humans, transgenic mice with ^{Ctm}PrP-mediated neurodegeneration lack PrP^{Sc} (ref. 33, 35). In extensive attempts at transmission involving hundreds of recipients, ^{Ctm}PrP-mediated neurodegeneration was shown to be non-transmissible[35].

These results established that inappropriate generation of ^{Ctm}PrP, even at only slightly elevated levels beyond wild type PrP, could result in neurodegeneration without the obligate generation of either PrP^{Sc} or infectious prions. Such a mechanism of PrP-mediated neurodegeneration was likely involved in at least two of the naturally occurring PrP mutants leading to heritable disease (P105L, in addition to A117V, leads to increased ^{Ctm}PrP generation)[31]. With the finding of at least one mechanism by which an alteration of PrP cell biology (in this case, its initial biogenesis at the ER) can cause disease, a more complicated question was raised: what role if any does ^{Ctm}PrP play in the pathogenesis of transmissible diseases? Addressing this issue, which presently remains unresolved, will be difficult and requires considerably more mechanistic knowledge about the biology of ^{Ctm}PrP. However, some potential insight into this issue can be obtained from the experiments performed thus far.

To begin examining the potential role for ^{Ctm}PrP in the neurodegeneration caused upon accumulation of PrP^{Sc} during transmissible disease, the various lines of ^{Ctm}PrP-altering transgenic mice were utilized in a different way. In these experiments, the susceptibility of each of the transgenic mouse lines to prion inoculation was assessed[35]. The logic was that if ^{Ctm}PrP was involved in the neurodegeneration caused by PrP^{Sc}, then modulating the propensity for ^{Ctm}PrP to be generated (by either favoring or disfavoring it with mutations in the transmembrane domain) should influence the progression of disease. By contrast, if PrP^{Sc} accumulation caused neuronal damage by mechanism(s) not involving ^{Ctm}PrP, then slight alterations in the ability to generate this form should have no effect on disease. Put another way, the experiment aimed to test whether in transmissible prion disease, the neurodegenerative phenotype correlated better with PrP^{Sc} accumulation or ^{Ctm}PrP-generating potential.

Despite numerous caveats and potential confounding variables, the experiment yielded a surprisingly clear result: increased propensity of PrP to be made in the ^{Ctm}PrP form sensitized mice to developing neurodegeneration during transmissible prion disease[35]. A particularly good illustration of this effect can be seen when mice expressing the A117V mutation (which very slightly favors ^{Ctm}PrP) are compared to the ΔSTE mutation (which decreases, although does not completely eliminate the generation of transmembrane forms of PrP). Here, the two lines of mice express the transgene at equal levels in a PrP-null background. Upon inoculation with prions, the A117V mice develop neurodegeneration in less than 60 days, at a time when only a relatively small amount of PrP^{Sc} has accumulated. By marked contrast, the ΔSTE mice do not develop signs of neurodegeneration for up to 350 days after inoculation. By this point, PrP^{Sc} has accumulated to levels more than 5–10 times that observed in the A117V mice at the time they became ill. Mice expressing wild type PrP at comparable levels get sick at an intermediate time of ~100 days[54]. Hence, it appeared that the ability of PrP^{Sc} accumulation to incite neurodegeneration is influenced by mutations that alter the ability of host PrP to be made in the ^{Ctm}PrP form[35]: the more easily ^{Ctm}PrP can be generated, the more potent the effect of PrP^{Sc}, while the inability to generate ^{Ctm}PrP seems to confer some degree of protection from accumulated PrP^{Sc}.

One expectation from such a model relating ^{Ctm}PrP to PrP^{Sc} accumulation is that during the course of transmissible disease, the amount of total ^{Ctm}PrP should rise (since its elevated levels are what is postulated to be the cause of neurodegeneration). Examining this idea directly poses a substantial technical hurdle because during the course of disease, the levels of PrP^{Sc} also rise dramatically. Since PrP^{Sc} at later points of disease progression is very abundant, highly heterogeneous in its fractionation properties, relatively insoluble, and protease resistant, the possibility of detecting small changes in ^{Ctm}PrP seem slim. Indeed, one would anticipate that only a small increase in ^{Ctm}PrP would be necessary to cause neurodegeneration given that some of the heritable disease mutations elevate ^{Ctm}PrP only slightly. In an attempt to circumvent such technical hurdles, double-transgenic mice expressing PrP from two different species were employed.

In this experiment, mice expressing both mouse PrP and hamster PrP were inoculated with mouse prions[35]. Given the species barrier to transmission[54], it was expected that only mouse PrP^{Sc} would be generated and accumulate. The hamster PrP, for which specific antibodies exist, would serve as a 'reporter' for measuring ^{Ctm}PrP. The question being asked was whether the accumulation of PrP^{Sc} (of mouse origin) leads to some change in host PrP metabolism that results in increased ^{Ctm}PrP (measured by examining the hamster PrP 'reporter'). Although indirect in its approach, a slight (~2–3 fold) increase in ^{Ctm}PrP was detected[35]. Indeed, based on the heritable mutations in both mice and humans, a mere 2-fold increase in ^{Ctm}PrP generation is potentially significant since it is clearly sufficient to cause neurodegeneration[33,35]. Thus, a working hypothesis, albeit based on indirect and complex experiments, is that the increased generation of ^{Ctm}PrP represents a step that is common to several types of PrP-mediated neurodegenerative diseases including the transmissible variety. One of the most important features of these studies and this hypothesis is that it provides a toehold into at least one direct cause of neurodegeneration, makes specific predictions about its generality, and is readily testable (as discussed further in section 9).

We therefore feel that at the present time, the generation of ^{Ctm}PrP is perhaps the most specific and well-defined event in PrP cell biology that has been directly linked to causing neurodegeneration. Indeed it remains the only proposed model that identifies a very specific neurotoxic molecule (^{Ctm}PrP), delineates the site (the ER) and mechanistic steps that can lead to its increased generation, demonstrates its presence *in vivo*, and tightly correlates its elevated presence in both experimental and naturally occurring PrP-mediated disease. Many of the other potentially toxic molecules (such as various fragments of PrP[28–30,92,93], or incompletely defined misfolded forms[94]) and events (such as PrP crosslinking at the cell surface[95], or increased PrP retention in the ER[91,96]) that have been proposed as a cause of neurodegeneration have yet to meet all (or in some cases, any) of these same criteria. Until this is achieved, sufficiently specific hypotheses and precise experimental tools cannot be generated to yet merit a serious consideration of their proposed roles in prion disease pathogenesis.

16.8. cyPrP and the development of neurodegeneration

Recently, a cytosolic form of PrP (cyPrP) has been discovered and suggested to play a role in PrP-mediated neurodegeneration. In this example, a specifically defined species of PrP has been identified, demonstrated to at least be *capable* of causing neurodegeneration, and mechanism(s) for its generation *in vivo* have been proposed based on well-established cellular pathways[90,97]. Thus, while a role for cyPrP in prion disease pathogenesis is even less established and more contentious than for ^{Ctm}PrP, sufficiently specific and testable hypotheses have been formulated to merit its careful consideration (reviewed in ref. 98).

The idea that PrP, which is normally co-translationally targeted to and translocated across the ER membrane, can reside in the cytosol has its genesis in *Saccharomyces cerevisiae*. It is ironic (and perhaps meaningful) that just as transmembrane PrP was initially discovered as a likely 'artifact' of expression in a heterologous wheat germ system, cyPrP was also first noticed when expression of mammalian PrP was attempted in the yeast system. In yeast cells, PrP appears to be very inefficiently translocated into the ER, even when a native signal sequence from the yeast Kar2 protein is used[99]. While this is perhaps not very surprising, what led to further investigation was the finding that the non-glycosylated, non-disulfide bonded, cytosolic PrP was prone to aggregation, insoluble, and partially resistant to protease digestion[90,99]. Although proteins in the wrong cellular compartment of a non-native organism are often misfolded, the superficial resemblance between PrP aggregates in the yeast cytosol and PrP^{Sc} in mammalian prion disease provided the basis for a provocative hypothesis[90,99]: perhaps even in mammalian cells, PrP in the cytosol could be the origin for the initial generation of PrP^{Sc}.

This hypothesis then raised the important questions of whether in mammalian cells, PrP (or disease-associated mutants) can ever reside in the cytosol, and if so, what relevance this would have for either PrP^{Sc} formation or disease pathogenesis. The issue of whether PrP can potentially reside in the cytosol was initially addressed indirectly by demonstrating that in cultured cells overexpressing PrP, a small proportion of it was degraded by a pathway that could be inhibited by proteasome inhibitors[90,100,101]. Thus, upon treatment of cells with such inhibitors, an unglycosylated, presumably cytosolic form of PrP accumulated. This form was found to be aggregation prone, insoluble in mild detergents, and at least partially resistant to protease digestion. Hence, under the appropriate conditions (overexpression and chronic proteasome inhibition), mammalian PrP could reside in the cytosol of mammalian cells.

Based largely on co-migration in SDS-PAGE with recombinant PrP lacking a signal or GPI anchoring sequence, it was thought that cyPrP had been subjected to processing by ER-lumenal signal peptidase and GPI-anchoring machinery[92]. Thus, cytosolic PrP was suggested to have originated from ER localized PrP that had failed to be properly folded[92]. In this view, the well-established (albeit incompletely understood) ER-associated degradation pathway[102] was being utilized by PrP molecules that had failed to meet the cellular quality control systems[103] in the ER lumen. It would then be retrotranslocated from the ER to the cytosol, deglycosylated by cytosolic N-glycanase, and degraded by the proteasome pathway. Hence, inhibition of the proteasome would cause accumulation of the species to be degraded, thereby explaining the appearance of cytosolic, unglycosylated PrP under these conditions.

The most compelling aspect of these studies, especially as it relates to disease, was the observation that a disease-associated PrP mutation (D178N) was found to a higher extent in the cytosol than wild type PrP under both normal and proteasome-inhibited conditions[90]. The supposition, which remains largely untested at present, was that this mutation was less likely to fold properly in the ER and therefore result in a higher proportion of it utilizing the ER-associated degradation pathway. This idea was especially attractive because it could potentially apply to many if not all disease-associated PrP mutants to provide a common mechanism for their adverse effects.

In order to provide support to these ideas, it was important to first provide proof of principle for two important predictions of the hypotheses linking cyPrP to prion diseases. First, if cyPrP is in fact involved in the de novo generation of PrP^{Sc} and/or transmissible prions, better evidence was needed in addition to rather non-specific biochemical features such as aggregation and protease resistance. And second, a role for cyPrP in neurodegeneration cannot even be considered without at least demonstrating that it *can* cause selective damage to neurons. The first charge was approached by attempting to determine if the most central feature of PrP^{Sc}, its self-propagation using host-encoded PrP^{C}, could also be demonstrated for cyPrP. In these experiments[101], an initial 'seed' of cyPrP was initiated by transient treatment of PrP-expressing cells with proteasome inhibitor. Then, the inhibitor was removed to determine whether this seed of cyPrP could grow (i.e., propagate itself) by recruitment of additional PrP molecules. Exactly such a phenomenon was observed, and used to support the proposition that at least one source of de novo PrP^{Sc} formation could be the cytosol[101]. Additional support is provided by the finding that the D178N mutant has both increased residence in the cytosol of cultured cells[90] and spontaneous PrP^{Sc} generation in human patients[26,37].

At present however, it remains to be seen whether the cyPrP aggregates are in fact transmissible and propagated when introduced into animals. Furthermore, the controls for complete removal of the proteasome inhibitor in the 'seeding' experiments were not particularly compelling since an unrelated protein did not resume degradation (although it did not accumulate like PrP)[101]. In addition, alternative interpretations are possible in which the apparent propagation is due to continued inhibition of the proteasome (at least partially) by the cyPrP aggregates in a manner observed for other non-transmissible proteins[104]. And finally, artifacts of overexpression in heterologous systems have also been suggested as an explanation[91]. Thus, while one interpretation of the data involves the spontaneous conversion of cyPrP to PrP^{Sc} in the cytosol, this remains unproven. However, the idea is readily testable by the appropriate infectivity assays using both cell culture derived cyPrP aggregates as well as material from brain tissue expressing cyPrP (see below).

The second prediction, that cyPrP is toxic to neurons, was demonstrated directly when PrP was forced to be expressed in the cytosol by removal of its N-terminal signal sequence (and C-terminal GPI anchoring sequence). In both cultured cells of neuronal origin and certain subsets of neurons in transgenic mice, forced expression of cyPrP resulted in neurodegeneration[97]. Thus in at least some (but clearly not all[97,105,106]) neurons under certain conditions, cyPrP *can* be detrimental. These results now provide sufficient key elements to propose a testable framework for neurodegeneration in prion diseases involving cyPrP[97,99,101].

In this model, a small proportion of PrP is always transiently trafficked through the cytosol prior to its rapid degradation by cytosolic proteasomes. This transient cytosolic population would be increased with either mutations (as in the case of heritable disease) or alterations the perturb the proteasome degradation pathway. Such perturbations would be postulated to result from PrP^{Sc} accumulation, old age, or both. If cyPrP accumulates above a certain threshold, it would not only cause cell death, but aggregate into a form that has self-perpetuating capability. This self-perpetuating aggregate, once released from the cell, would recruit PrP from the surface of other cells to generate the observed glycosylated PrP^{Sc} that is seen in prion diseases. While numerous questions remain unanswered in this rudimentary framework, it, like the ideas centered around ^{Ctm}PrP, draw heavily upon basic cell biological pathways to make specific and testable predictions regarding their respective roles in disease pathogenesis. Indeed, it is also quite plausible that the two sets of ideas share a common mechanistic feature given that in both models, a key feature involves exposure of at least a portion of PrP to the cytosolic environment.

16.9. Testing the roles of ^{Ctm}PrP and cyPrP in prion disease

In section 16.6, ('The importance of PrP cell biology'), we argued that a quantitative and mechanistic understanding of PrP cell biology may be the most productive way to identify facets of PrP metabolism that have potential importance for prion disease pathogenesis. Through such studies, we felt that insights and tools would be generated to allow selective manipulation of these steps in PrP metabolism *in vivo* to test the consequences for neurodegeneration. Where along this prescribed path do the studies of ^{Ctm}PrP and cyPrP stand, and what are the best future directions?

At this point, the relatively easy part has been accomplished. Rather drastic *exaggeration* of some facet of PrP metabolism with a fairly blunt manipulation (such as deleting the signal sequence in the case of cyPrP) has been used to cause disease in a whole organism[33,35,97]. By striking contrast, the far more difficult task will be to selectively *reduce* the propensity for this same event *in vivo* to test its possible role during prion infection and pathogenesis. Accomplishing this goal will either requires tremendous luck, or a significant degree of mechanistic insight into the molecular pathways involved in the respective aspects of normal PrP cell biology. For example, reducing the propensity for PrP to ever be in the cytosol would need sufficient insight into the pathways by which it is routed there normally so that one could selectively modulate this event. Such modulation is required to rigorously test whether the ability of PrP to be in the cytosol is important for the cell death that occurs during PrP^{Sc} accumulation. Such experiments would also help examine the (non mutually exclusive) proposal that PrP in the cytosol is actually protective during prion disease[105,106].

It should therefore be clear that in order to productively move forward in the ^{Ctm}PrP and cyPrP fields, two directions merit a high priority at the present time. First, the pathways for the generation, trafficking, and degradation of these molecules needs to be understood in mechanistic detail. Second, and of slightly lower initial priority, the interactions between these molecules and specific cellular pathways needs to be defined. During the course of these studies (particularly the first aim), valuable tools will emerge with which to selectively modulate the synthesis, metabolism, or function of ^{Ctm}PrP and cyPrP. Such tools can then be applied to precisely probe the complex problem of neurodegeneration during prion disease pathogenesis. Little progress has been made towards the second aim; the proteins that interact with either ^{Ctm}PrP or cyPrP, the pathways they influence, or the way in which they cause cell death all remain totally unknown. Fortunately, mechanistic studies of the biogenesis of ^{Ctm}PrP (and to a lesser extent, cyPrP) have begun to yield some insights which should facilitate their modulation *in vivo*.

After recognizing that PrP can be made in at least three distinct topological forms[33] (four if one includes cyPrP), a framework was needed to understand how a single polypeptide could acquire multiple outcomes during its synthesis at the ER. A key realization was that the topologic forms (see Figure 16.3) differ in *two* important ways. The first is the location of the N-terminus: either in the cytosol (as for cyPrP and ^{Ctm}PrP) or in the ER lumen (^{sec}PrP and ^{Ntm}PrP). The second is whether the central hydrophobic domain becomes membrane integrated (^{Ntm}PrP and ^{Ctm}PrP) or not (^{sec}PrP and cyPrP). Each of the four combinations of these two 'decisions' describes uniquely each of the four topologic outcomes (schematically depicted in Figure 16.4A). For example, ^{sec}PrP results from the decision to have the N-terminus translocated into the ER lumen, and a decision to not integrate the potential transmembrane domain into the membrane. In this model, heterogeneity at one or both of these decisions would lead to the generation of multiple topologic forms of PrP (see Figure 16.4B). Thus, to understand the basis of PrP biogenesis, it is crucial to decipher the mechanism by which these two decisions are made and regulated by the cell.

To begin addressing this issue, a mutational analysis was carried out to determine which domains of PrP are involved in localization of the N-terminus and membrane integration[107]. These experiments revealed that localization of the N-terminus (cytosol versus ER lumen) is influenced largely by the N-terminal signal sequence. Mutations in the signal

Figure 16.4. Schematic depiction of PrP topogenesis at the ER. Two decisions are combined to determine the final topologic outcome for PrP. The two decisions and four potential outcomes are depicted in chart format in Panel A and sequential order in Panel B. The first decision involves the N-terminal signal sequence and determines whether the N-terminal domain of PrP will be in the cytosol or ER lumen. The second decision involves the potential TMD, and determines whether the protein will be integrated into the membrane or remain soluble.

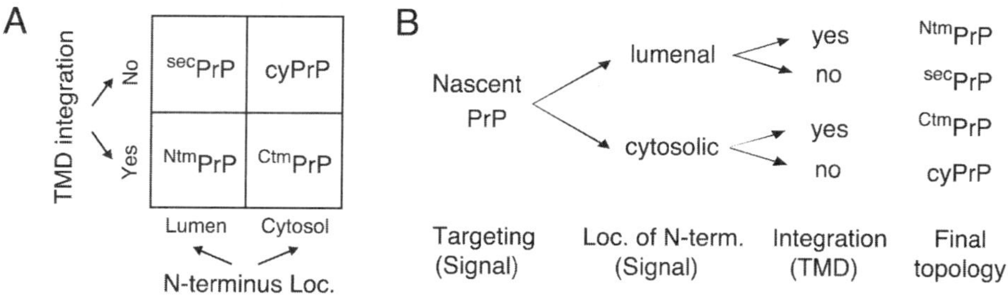

sequence could be identified which increase cytosolic localization (and hence, increase ^{Ctm}PrP and cyPrP relative to ^{Ntm}PrP and ^{sec}PrP) or increase lumenal localization (and hence, decrease ^{Ctm}PrP and cyPrP). By contrast, the second decision regarding membrane integration is influenced largely by the highly conserved potential transmembrane domain (TMD). In this case, changes which (even slightly) increase hydrophobicity of the TMD result in increased generation of the membrane integrated forms (^{Ctm}PrP and ^{Ntm}PrP) relative to the non-integrated forms (cyPrP and ^{sec}PrP). Thus, the signal sequence and TMD act together to allow the potential generation of four distinct topologic forms of PrP.

Experiments analyzing serial intermediates of increasing nascent chain length during PrP synthesis have revealed the sequential order of events involved in the determination of PrP topology[108] (Figure 16.5). These experiments demonstrated that the first step is targeting of ribosome-associated PrP nascent chains to the ER. This step requires a functional signal sequence, occurs by the time ~50–70 amino acids are synthesized, and presumably involves the well-characterized signal recognition particle (SRP) and SRP-receptor pathway[109]. Targeting to the ER appears to be essential for the generation of ^{Ctm}PrP, ^{Ntm}PrP, and ^{sec}PrP; in the absence of a functional signal sequence, PrP is made exclusively in the cytosol[108]. After targeting, but before the TMD is synthesized and emerges from the ribosome, there is a brief window of time during which a particularly critical step takes place. During this step, the signal sequence mediates the insertion of nascent PrP into the ER translocation channel. This facilitates the subsequent translocation of the N-terminus into the lumen, a prerequisite for the generation

Figure 16.5. A mechanistic depiction of the key steps during PrP biogenesis at the ER.

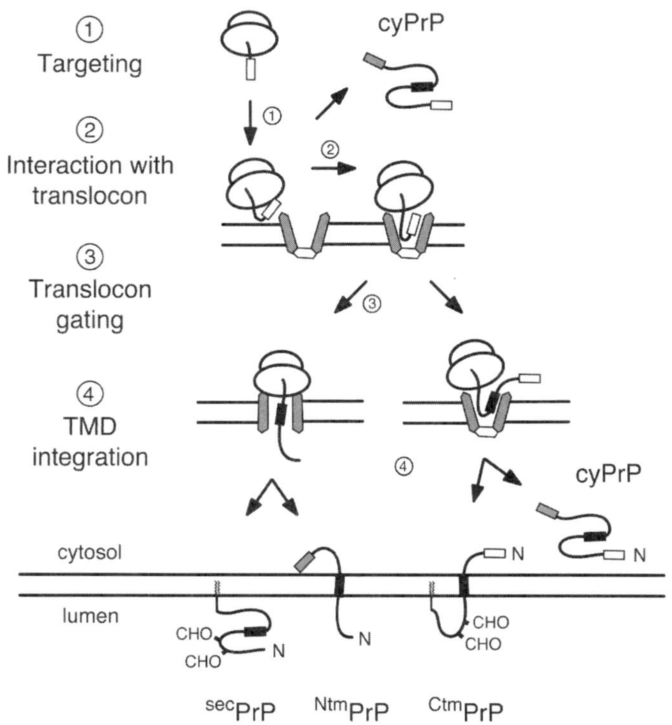

of ^{sec}PrP and ^{Ntm}PrP. For nascent polypeptides that fail to accomplish this step in a timely manner, the N-terminus remains on the cytosolic side of the membrane, although the ribosome-nascent chain complex remains in close proximity to the translocon while the remainder of PrP is synthesized.

As this key step is occurring, the TMD is synthesized and emerges from the ribosome. If the N-terminus has already been committed to the ER lumen, determinants in the TMD (primarily hydrophobicity) influence the propensity of the chain to become membrane integrated (to generate ^{Ntm}PrP) or fully translocated (to become ^{sec}PrP). If, when the TMD emerges, the N-terminus has not been committed to the ER lumen, the TMD then has an opportunity to interact with the translocon and become inserted into the membrane. Chains that insert in the membrane become ^{Ctm}PrP, while chains that do not can become cyPrP (if the N-terminus is not translocated by the time synthesis is completed). Mutational analysis suggests that one key determinant of this TMD-mediated integration step that generates ^{Ctm}PrP is hydrophobicity[107,108,110]. This appears to explain the mechanism of increased ^{Ctm}PrP generation for at least some disease-associated PrP mutants (e.g., A117V, P105L, and most recently, G131V)[31,34,111] that increase hydrophobicity of the TMD.

Taken together, the results summarized in Figure 16.5 reveal several important points. First, the key decisions that influence the outcome of PrP biogenesis (with respect to topology) are made during the synthesis of PrP (i.e., cotranslationally). Second, each step is influenced substantially by interactions between the translocon and elements in PrP (the signal sequence and TMD). Third, these interactions appear to occur with only moderate fidelity, a feature that is critical to the generation of topologic heterogeneity. And fourth, the strength of these interactions can be changed by mutations in the signal or TMD to influence the outcome of PrP topogenesis in predictable ways. These insights not only provide a framework for understanding PrP topogenesis, but facilitate the focusing of subsequent studies on the most important mechanistic steps of potential relevance to disease pathogenesis.

In the case of ^{Ctm}PrP and cyPrP, the critical step is now revealed to be the signal sequence-mediated translocation of the N-terminus into the ER lumen. The degree of inefficiency at this step determines the percent of nascent PrP chains that have the opportunity to be made as ^{Ctm}PrP and/or cyPrP. Detailed analysis of this step has demonstrated that it is surprisingly complex and involves several factors (Figure 16.6). First, it is clear that signal sequences from different proteins carry out

Figure 16.6. A key branch point in the biogenesis of the different topologic forms of PrP. Nascent PrP polypeptides at ER translocons can go down two pathways. The first involves an interaction between the PrP signal and components of the translocon to mediate translocation of the N-terminus into the ER lumen. This pathway is facilitated by the Sec61 complex, the TRAP complex, TRAM, and ER lumenal chaperones such as PDI. Following this pathway is a prerequisite for the generation of ^{sec}PrP or ^{Ntm}PrP, both of which have their N-terminus in the ER lumen. If this pathway is not followed, the N-terminus is not successfully translocated into the lumen, and remains in the cytosol. The can lead to the generation of ^{Ctm}PrP or cyPrP. This is the default pathway taken by PrP when the minimal translocon composed only of the Sec61 complex is available. Thus, a combination of features encoded in the nascent chain (e.g., the signal sequence) and accessory components of the translocon (such as TRAP and PDI) determine the amount of potentially cytotoxic forms of PrP (e.g., cyPrP and ^{Ctm}PrP) that are generated.

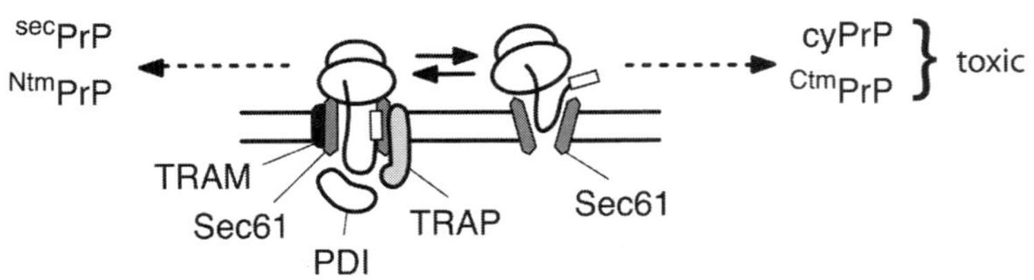

this step with markedly different efficiencies[112,113], with the PrP signal being roughly 'average' in this respect. Second, this step involves interactions between the signal sequence and the central component of the translocation channel, the Sec61 complex[114]. Third, the signal-translocon interaction appears to be influenced by at least two proteins termed TRAM[115,116] and the TRAP complex[117]. Fourth, not all signal sequences require TRAM and TRAP for efficient function; while most (including PrP) require at least one of these two complexes, a very small proportion of signal sequences can function well without either[114–117]. And finally, the availability of ER lumenal chaperones appears to influence translocation[118,119], particularly of PrP (our unpublished observations). Although it is not know which chaperones are most important, crosslinking studies indicate an interaction between the N-terminus of PrP and protein disulfide isomerase (PDI) at early steps during PrP translocation[108].

Thus, *avoiding* the generation of ^{Ctm}PrP and cyPrP requires the collective action of numerous determinants that include the signal sequence, Sec61 complex, TRAP complex, ER lumenal chaperones, and potentially yet unidentified factors. Conversely, when PrP is synthesized using proteoliposomes containing only the absolute minimal translocation machinery (composed of the SRP-receptor and Sec61 complex), essentially all of the polypeptides are made as either ^{Ctm}PrP and cyPrP (ref. 120). One of the few signal sequences capable of utilizing this minimal translocon efficiently comes from the protein prolactin[114–117]. All of this assembled information on the key steps of ^{Ctm}PrP and cyPrP synthesis now provides useful tools that can be used to modulate generation of these forms *in vivo*. For example, one can envision modulating the activity or expression levels of key factors such as the TRAP complex or PDI to influence PrP topogenesis.

Even simpler, at least initially, would be to modify the PrP signal sequence to alter its activity. Indeed, simply replacing the PrP signal with the prolactin signal substantially increases the efficiency of N-terminal translocation, and consequently, decreases generation of ^{Ctm}PrP (and as demonstrated in more recent unpublished experiments, cyPrP)[112,113]. Remarkably, this manipulation is so effective that, when assayed *in vitro*, it can totally reverse the increased ^{Ctm}PrP caused by disease-causing mutation in the TMD[108]. Such a manipulation is very valuable because it now allows the testing of hypotheses relating the ability to generate ^{Ctm}PrP (or cyPrP) to their proposed roles in heritable and transmissible disease. Importantly, the mature domain of PrP is not changed by such changes; only the relative amounts of its topologic forms. Several important ideas, discussed in earlier sections of this chapter, can and should be tested.

First, can the neurodegenerative phenotype of an otherwise disease-causing mutation (such as A117V) be pre-emptively avoided by increasing the efficiency of signal sequence-mediated translocation of the N-terminus? That is, is the neurodegeneration associated with the A117V mutation due to increased ^{Ctm}PrP generation, as is currently hypothesized, or to some other effect of this mutation? Second, does reducing generation of cyPrP during its translocation completely preclude generation of cyPrP *in vivo*? Here, the question is whether the majority of cyPrP is generated due to inefficient translocation, as has been suggested, or due to inefficient maturation in the ER followed by retrotranslocation? By eliminating one potential source (due to translocation), the contribution of the other potential source (due to retrotranslocation) can be isolated. Such an experiment will also allow the testing of a related hypothesis: do disease-associated mutations (such as D178N) result in increased retrotranslocation from the ER due to less efficient maturation? And finally, does reducing the propensity to generate ^{Ctm}PrP or cyPrP reduce susceptibility to neurodegeneration upon accumulation of prions and PrP^{Sc}? This is a key prediction of both the ^{Ctm}PrP and cyPrP frameworks, and can now be tested.

In general, testing each of these ideas will require relatively long term experiments involving the generation of suitable transgenic mice. However, a relatively high degree of confidence in the productiveness of such an approach is warranted by the *in vitro* analysis suggesting that the manipulations are selective and make specific predictions. In the case of cyPrP, where cell culture assays for its generation and accumulation (in the presence of proteasome inhibitors) have been developed, some of these hypotheses can also be examined in culture. Here, the results are striking. Simply increasing the efficiency of N-terminal translocation nearly completely eliminates cyPrP generation, even under conditions of prolonged proteasome inhibition (our unpublished results). This suggests that in cultured cells under normal conditions, little if any cyPrP is generated by the ER misfolding and retrotranslocation pathway. The consequence of avoiding cyPrP generation is increased resistance to PrP aggregate formation during proteasome inhibition and an accompanying resistance to cell death (our unpublished results). This illustrates the utility and power of a quantitative cell biological approach to understanding otherwise subtle, but potentially important aspects of PrP metabolism. It will now be of great interest to learn the results of currently ongoing transgenic mice studies in which the generation of ^{Ctm}PrP and cyPrP have been modulated. Will such manipulations influence the neurodegenerative phase of transmissible prion disease, and if so, how?

As more mechanistic insight is gained into each of the many other steps of PrP biosynthesis and metabolism, yet additional hypotheses and tools will be generated. These insights should be useful not only for the understanding of PrP biology and the diseases with which it is associated, but also for uncovering novel cell biological principles. In analogous fashion, another idea with roots in PrP biology and disease, that of information transfer mediated by protein elements, is now known to be far more generally applicable in other organisms and biological systems[121,122]. In fact, although not emphasized in this chapter, the studies on PrP biogenesis have helped identify functions for novel factors in protein translocation (e.g., the TRAP complex)[117], revealed the complexity and heterogeneity of signal sequences[112,113], and identified the protein translocon as a potential site for cellular regulation[123]. In the broader sense, the apparently unusual features of PrP (such as its ability to be made in multiple forms) may be a far more general but unappreciated area of cell biology.

16.10. References

1. D. C. Gajdusek, Unconventional viruses and the origin and disappearance of kuru. *Science* **197**, 943–960 (1977).

2. S. B. Prusiner, Novel proteinaceous infectious particles cause scrapie. *Science* **216**, 134–144 (1982).

3. S. B. Prusiner, Prions. *Proc. Natl. Acad. Sci.* **95**, 13363–13383 (1998).

4. A. Aguzzi and M. Polymenidou, Mammalian prion biology: one century of evolving concepts. *Cell* **116**, 313–327 (2004).

5. C. Weissmann, Molecular genetics of transmissible spongiform encephalopathies. *J. Biol. Chem.* **274**, 3–6 (1999).

6. J. Collinge, Prion diseases of human and animals: their causes and molecular basis. *Annu. Rev. Neurosci.* **24**, 519–550. (2001).

7. A. Giese and H. A. Kretzschmar, Prion-induced neuronal damage—the mechanisms of neuronal destruction in the subacute spongiform encephalopathies. *Curr. Top. Microbiol. Immunol.* **253**, 203–217 (2001).

8. R. Chiesa and D. A. Harris, Prion diseases: what is the neurotoxic molecule? *Neurobiol. Dis.* **8**, 103–112 (2001).

9. D. J. Selkoe, Alzheimer's disease: genes, proteins, and therapy. *Physiol. Rev.* **81**, 741–766 (2001).

10. B. I. Giasson and V. M. Lee, Parkin and the molecular pathways of Parkinson's disease. *Neuron* **31**, 885–888 (2001).

11. E. Bossy-Wetzel, R. Schwarzenbacher, and S. A. Lipton, Molecular pathways to neurodegeneration. *Nat. Med.* **10 Suppl**, S2–S9 (2004).

12. C. A. Ross, Polyglutamine pathogenesis: emergence of unifying mechanisms for Huntington's disease and related disorders. *Neuron* **35**, 819–822 (2002).

13. H. Y. Zoghbi and H. T. Orr, Glutamine repeats and neurodegeneration. *Annu. Rev. Neurosci.* **23**, 217–247 (2000).

14. B. Chesebro, Introduction to the transmissible spongiform encephalopathies or prion diseases. *Br. Med. Bull.* **66**, 1–20 (2003).

15. L. Manuelidis, Transmissible encephalopathies: speculations and realities. *Viral Immunol.* **16**, 123–139 (2003).

16. C. Soto and J. Castilla, The controversial protein-only hypothesis of prion propagation. *Nat. Med.* **10 Suppl**, S63–S67 (2004).

17. L. Manuelidis and Z. Y. Lu, Virus-like interference in the latency and prevention of Creutzfeldt-Jakob disease. *Proc. Natl. Acad. Sci.* **100**, 5360–5365 (2003).

18. B. Oesch, D. Westaway, M. Walchi, M. P. McKinley, S. B. Kent, R. Abersold, R. A. Barry, P. Tempst, D. B. Teplow, L. E. Hood, S. B. Prusiner, and C. Weissmann, A cellular gene encodes scrapie PrP 27–30 protein. *Cell* **40**, 735–746 (1985).

19. K. Basler, B. Oesch, M. Scott, D. Westaway, M. Walchi, D. F. Groth, M. P. McKinley, S. B. Prusiner, and C. Weissmann, Scrapie and cellular PrP isoforms are encoded by the same chromosomal gene. *Cell* **46**, 417–428 (1986).

20. N. Stahl, M. A. Baldwin, D. B. Teplow, L. Hood, B. W. Gibson, A. L. Burlingame, and S. B. Prusiner, Structural studies of the scrapie prion protein using mass spectrometry and amino acid sequencing. *Biochemistry* **32**, 1991–2002 (1993).

21. K. Hsiao, H. F. Baker, T. J. Crow, M. Poulter, F. Owen, J. D. Terwilliger, D. Westaway, J. Ott, and S. B. Prusiner, Linkage of a prion protein missense variant to Gerstmann-Straussler syndrome, *Nature* **338**, 342–345 (1989).

22. F. Owen, M. Poulter, R. Lofthouse, J. Collinge, T. J. Crow, D. Risby, H. F. Baker, R. M. Ridley, K. Haiao, and S. B. Prusiner, Insertion in prion protein gene in familial Creutzfeldt-Jakob disease. *Lancet* **1**, 51–52 (1989).

23. D. Goldgaber, L. G. Goldfarb, P. Brown, D. M. Asher, W. T. Brown, S. Lin, J. W. Teener, S. M. Feinstone, R. Rubenstein, and R. J. Kascsak, Mutations in familial Creutzfeldt-Jakob disease and Gerstmann-Straussler-Scheinker's syndrome. *Exp. Neurol.* **106**, 204–206 (1989).

24. S. R. Dlouhy, K. Hsiao, M. R. Farlow, T. Foroud, P. M. Conneally, P. Johnson, S. B. Prusiner, M. E. Hodes, and B. Ghetti, Linkage of the Indiana kindred of Gerstmann-Straussler-Scheinker disease to the prion protein gene. *Nat. Genet.* **1**, 64–67 (1992).

25. A. F. Hill, M. Antoniou, and J. Collinge, Protease-resistant prion protein produced *in vitro* lacks detectable infectivity. *J. Gen. Virol.* **80**, 11–14 (1999).

26. G. C. Telling, P. Parchi, S. J. DeArmond, P. Cortelli, P. Montagna, R. Gabizon, J. Mastrianni, E. Lugaresi, P. Gambetti, and S.B. Prusiner. Evidence for the conformation of the pathologic isoform of the prion protein enciphering and propagating prion diversity. *Science* **274**, 2079–2082 (1996).

27. J. A. Mastrianni, S. Capellari, G. C. Telling, D. Han, P. Bosque, S. B. Prusiner, and S. J. DeArmond, Inherited prion disease caused by the V210I mutation: transmission to transgenic mice. *Neurology*, **57**, 2198–2205 (2001).

28. P. Piccardo, J. J. Liepnieks, A. William, S. R. Dlouhy, M. R. Farlow, K. Young, D. Nochlin, T. D. Bird, R. R. Nixon, M. J. Ball, C. DeCarli, O. Bugiani, F. Tagliavini, M. D. Benson, and B. Ghetti, Prion proteins with different conformations accumulate in Gerstmann-Straussler-Scheinker disease caused by A117V and F198S mutations. *Am. J. Pathol.* **158**, 2201–2207 (2001).

29. F. Tagliavini, P. M. Lievens, C. Tranchant, J. M. Warter, M. Mohr, G. Giaccone, F. Perini, G. Rossi, M. Salmona, P. Piccardo, B. Ghetti, R. C. Beavis, O. Bugiani, B. Frangione, and F. Prelli, A 7-kDa prion protein (PrP) fragment, an integral component of the PrP region required for infectivity, is the major amyloid protein in Gerstmann-Straussler-Scheinker disease A117V. *J. Biol. Chem.* **276**, 6009–6015 (2001).

30. P. Parchi, S. G. Chen, P. Brown, W. Zou, S. Capellari, H. Budka, J. Hainfellner, P. F. Reyes, G. T. Golden, J. J. Hauw , D. C. Gajdusek, and P. Gambetti. Different patterns of truncated prion protein fragments correlate with distinct phenotypes in P102L Gerstmann-Straussler-Scheinker disease. *Proc. Natl. Acad. Sci.* **95**, 8322–8327 (1998).

31. M. Yamada, Y. Itoh, A. Inaba, Y. Wada, M. Takashima, S. Satoh, T. Kamata, R. Okeda, T. Kayano, N. Suematsu, T. Kitamoto, E. Otomo, M. Matsushita, and H. Mizusawa, An inherited prion disease with a PrP P105L mutation: clinicopathologic and PrP heterogeneity. *Neurology* **53**, 181–188 (1999).

32. K. K. Hsiao, D. Groth, M. Scott, S. L. Yang, H. Serban, D. Rapp, D. Foster, M. Torchia, S. J. Dearmond, and S. B. Prusiner, Serial transmission in rodents of neurodegeneration from transgenic mice expressing mutant prion protein. *Proc. Natl. Acad. Sci.* **91**, 9126–9130 (1994).

33. R. S. Hegde, J. A. Mastrianni, M. R. Scott, K. A. DeFea, P. Tremblay, M. Torchia, S. J. DeArmond, S. B. Prusiner, and V. R. Lingappa, A transmembrane form of the prion protein in neurodegenerative disease. *Science* **279**, 827–834 (1998).

34. J. Tateishi, T. Kitamoto, K. Doh-ura, Y. Sakaki, G. Steinmetz, C. Tranchant, J. M. Warter, and N. Heldt, Immunochemical, molecular genetic, and transmission studies on a case of Gerstmann-Straussler-Scheinker syndrome. *Neurology* **40**, 1578–1581 (1990).

35. R. S. Hegde, P. Tremblay, D. Groth, S. J. DeArmond, S. B. Prusiner, and V. R. Lingappa, Transmissible and genetic prion diseases share a common pathway of neurodegeneration. *Nature* **402**, 822–826 (1999).

36. J. Tateishi, T. Kitamoto, M. Z. Hoque, and H. Furukawa, Experimental transmission of Creutzfeldt-Jakob disease and related diseases to rodents. *Neurology* **46**, 532–537 (1996).

37. J. Tateishi, P. Brown, T. Kitamoto, Z. M. Hoque, R. Roos, R. Wollman, L. Cervenakova, and D. C. Gajdusek. First experimental transmission of fatal familial insomnia. *Nature* **376**, 434–435 (1995).

38. P. Brown, C. J. Gibbs Jr, P. Rodgers-Johnson, D. M. Asher, M. P. Sulima, A. Bacote, L. G. Goldfarb, and D. C. Gajdusek, Human spongiform encephalopathy: the National Institutes of Health series of 300 cases of experimentally transmitted disease. *Ann Neurol.* **35**, 513–529 (1994).

39. J. Chapman, P. Brown, J. M. Rabey, L. G. Goldfarb, R. Inzelberg, C. J. Gibbs Jr, D. C. Gajdusek, and A. D. Korczyn, Transmission of spongiform encephalopathy from a familial Creutzfeldt-Jakob disease patient of Jewish Libyan origin carrying the PRNP codon 200 mutation. *Neurology* **42**, 1249–1250 (1992).

40. H. Bueler, M. Fischer, Y. Lang, H. Bluethmann, H. P. Lipp, S. J. DeArmond, S. B. Prusiner, M. Aguet, and C. Weissmann, Normal development and behaviour of mice lacking the neuronal cell-surface PrP protein. *Nature* **356**, 577–582 (1992).

41. H. R. Büeler, A. Aguzzi, A. Sailer, R. A. Greiner, P. Autenried, M. Aguet, and C. Weissmann, Mice devoid of PrP are resistant to scrapie. *Cell* **73**, 1339–1347 (1993).

42. A. Sailer, H. Bueler, M. Fischer, A. Aguzzi, and C. Weissmann, No propagation of prions in mice devoid of PrP. *Cell* **77**, 967–968 (1994).

43. G. R. Mallucci, S. Ratte, E. A. Asante, J. Linehan, I. Gowland, J. G. Jefferys, and J. Collinge, Post-natal knockout of prion protein alters hippocampal CA1 properties, but does not result in neurodegeneration. *EMBO J.* **21**, 202–210 (2002).

44. G. Mallucci, A. Dickinson, J. Linehan, P. C. Klohn, S. Brandner, and J. Collinge, Depleting neuronal PrP in prion infection prevents disease and reverses spongiosis. *Science* **302**, 871–874 (2003).

45. S. Brandner, S. Isenmann, A. Raeber, M. Fischer, A. Sailer, Y. Kobayashi, S. Marino, C. Weissmann, and A. Aguzzi, Normal host prion protein necessary for scrapie-induced neurotoxicity. *Nature* **379**, 339–343 (1996).

46. D. R. Borchelt, M. Scott, A. Taraboulos, N. Stahl, and S. B. Prusiner, Scrapie and cellular prion proteins differ in their kinetics of synthesis and topology in cultured cells. *J. Cell Biol.* **110**, 743–752 (1990).

47. A. Taraboulos, A. J. Raeber, D. R. Borchelt, D. Serban, and S. B. Prusiner, Synthesis and trafficking of prion proteins in cultured cells. *Mol. Biol. Cell.* **3**, 851–863 (1992).

48. D. C. Bolton, M. P. McKinley, and S. B. Prusiner, Molecular characteristics of the major scrapie prion protein. *Biochemistry*. **23**, 5898–5906 (1984).

49. R. F. Marsh, B. E. Castle, C. Dees, and W. F. Wade, Equilibrium density gradient centrifugation of the scrapie agent in Nycodenz. *J. Gen. Virol.* **65**, 1963–1968 (1984).

50. C. Korth, B. Stierli, P. Streit, M. Moser, O. Schaller, R. Fischer, W. Schulz-Schaeffer, H. Kretzschmar, A. Raeber, U. Braun, F. Ehrensperger, S. Hornemann, R. Glockshuber, R. Riek, M. Billeter, K. Wuthrich, and B. Oesch, Prion (PrPSc)-specific epitope defined by a monoclonal antibody. *Nature* **390**, 74–77 (1997).

51. E. Paramithiotis, M. Pinard, T. Lawton, S. LaBoissiere, V. L. Leathers, W. Q. Zou, L. A. Estey, J. Lamontagne, M. T. Lehto, K. H. Londejewski, G. P. Francoeur, M. Papadopoulos, A. Haghighat, S. J. Spatz, M. Head, R. Will, J. Ironside, K. O'Rourke, Q. Tonelli, H. C. Ledebur, A. Chakrabartty, and N. R. Cashman, A prion protein epitope selective for the pathologically misfolded conformation. *Nat. Med.* **9**, 893–899 (2003).

52. M. B. Fischer, C. Roeckl, P. Parizek, H. P. Schwarz, and A. Aguzzi, Binding of disease-associated prion protein to plasminogen. *Nature* **408**, 479–483 (2000).

53. J. Safar, H. Wille, V. Itri, D. Groth, H. Serban, M. Torchia, F. E. Cohen, and S. B. Prusiner, Eight prion strains have PrP(Sc) molecules with different conformations. *Nat. Med.* **4**, 1157–1165 (1998).

54. S. B. Prusiner, M. Scott, D. Foster, K. M. Pan, D. Groth, C. Mirenda, M. Torchia, S. L. Yang, D. Serban, and G. A. Carlson, Transgenetic studies implicate interactions between homologous PrP isoforms in scrapie prion replication. *Cell* **63**, 673–686 (1990).

55. G. C. Telling, T. Haga, M. Torchia, P. Tremblay, S. J. DeArmond, and S. B. Prusiner, Interactions between wild-type and mutant prion proteins modulate neurodegeneration in transgenic mice. *Genes Dev.* **10**, 1736–1750 (1996).

56. G. C. Telling, M. Scott, J. Mastrianni, R. Gabizon, M. Torchia, F. E. Cohen, S. J. DeArmond, and S. B. Prusiner, Prion propagation in mice expressing human and chimeric PrP transgenes implicates the interaction of cellular PrP with another protein. *Cell* **83**, 79–90 (1995).

57. G. A. Carlson, P. A. Goodman, M. Lovett, B. A. Taylor, S. T. Marshall, M. Peterson-Torchia, D. Westaway, and S. B. Prusiner, Genetics and polymorphism of the mouse prion gene complex: control of scrapie incubation time. *Mol. Cell Biol.* **8**, 5528–5540 (1988).

58. D. A. Stephenson, K. Chiotti, C. Ebeling, D. Groth, S. J. DeArmond, S. B. Prusiner, and G. A. Carlson, Quantitative trait loci affecting prion incubation time in mice. *Genomics*. **69**, 47–53 (2000).

59. G. G. Kovacs, G. Trabattoni, J. A. Hainfellner, J. W. Ironside, R. S. Knight, and H. Budka, Mutations of the prion protein gene phenotypic spectrum. *J. Neurol.* **249**, 1567–1582 (2002).

60. J. D. Wadsworth, A. F. Hill, J. A. Beck, and J. Collinge, Molecular and clinical classification of human prion disease. *Br. Med. Bull.* **66**, 241–254 (2003).

61. B. Ghetti, P. Piccardo, M. G. Spillantini, Y. Ichimiya, M. Porro, F. Perini, T. Kitamoto, J. Tateishi, C. Seiler, B. Frangione, O. Bugiani, G. Giaccone, F. Prelli, M. Goedert, S. R. Dlouhy, and F. Tagliavini, Vascular variant of prion protein cerebral amyloidosis with tau-positive neurofibrillary tangles: the phenotype of the stop codon 145 mutation in PRNP. *Proc. Natl. Acad. Sci.* **93**, 744–748 (1996).

62. T. Kitamoto, R. Iizuka, and J. Tateishi, An amber mutation of prion protein in Gerstmann-Straussler syndrome with mutant PrP plaques. *Biochem. Biophys. Res. Commun.* **192**, 525–531 (1993).

63. U. Finckh, T. Muller-Thomsen, U. Mann, C. Eggers, J. Marksteiner, W. Meins, G. Binetti, A. Alberici, C. Hock, R. M. Nitsch, and A. Gal, High prevalence of pathogenic mutations in patients with early-onset dementia detected by sequence analyses of four different genes. *Am. J. Hum. Genet.* **66**, 110–117 (2000).

64. R. Chiesa, P. Piccardo, B. Ghetti, and D. A. Harris, Neurological illness in transgenic mice expressing a prion protein with an insertional mutation. *Neuron* **21**, 1339–1351 (1998).

65. R. Chiesa, P. Piccardo, E. Quaglio, B. Drisaldi, S. L. Si-Hoe, M. Takao, B. Ghetti, and D. A. Harris, Molecular distinction between pathogenic and infectious properties of the prion protein. *J. Virol.* **77**, 7611–7622 (2003).

66. D. A. Butler, M. R. Scott, J. M. Bockman, D. R. Borchelt, A. Taraboulos, K. K. Hsiao, D. T. Kingsbury, and S. B. Prusiner, Scrapie-infected murine neuroblastoma cells produce protease-resistant prion proteins. *J. Virol.* **62**, 1558–1564 (1988).

67. H. M. Schatzl, L. Laszlo, D. M. Holtzman, J Tatzelt, S. J. DeArmond, R. I. Weiner, W. C. Mobley, and S. B. Prusiner, A hypothalamic neuronal cell line persistently infected with scrapie prions exhibits apoptosis. *J. Virol.* **71**, 8821–8831 (1997).

68. P. J. Bosque and S. B. Prusiner, Cultured cell sublines highly susceptible to prion infection. *J. Virol.* **74**, 4377–4386 (2000).

69. I. Vorberg, A. Raines, B. Story, and S. A. Priola, Susceptibility of common fibroblast cell lines to transmissible spongiform encephalopathy agents. *J. Infect. Dis.* **189**, 431–439 (2004).

70. H. A. Kretzschmar, Neuropathology of human prion diseases (spongiform encephalopathies). *Dev. Biol. Stand.* **80**, 71–90 (1993).

71. S. B. Prusiner and S. J. DeArmond, Molecular biology and pathology of scrapie and the prion diseases of humans. *Brain Pathol.* **1**, 297–310 (1991).

72. E. A. Asante, I. Gowland, J. M. Linehan, S. P. Mahal, and J.Collinge, Expression pattern of a mini human PrP gene promoter in transgenic mice. *Neurobiol. Dis.* **10**, 1–7 (2002).

73. C. Lemaire-Vieille, T. Schulze, V. Podevin-Dimster, J. Follet, Y. Bailly, F. Blanquet-Grossard, J. P. Decavel, E. Heinen, and J. Y. Cesbron JY, Epithelial and endothelial expression of the green fluorescent protein reporter gene under the control of bovine prion protein (PrP) gene regulatory sequences in transgenic mice. *Proc. Natl. Acad. Sci.* **97**, 5422–5427 (2000).

74. S. J. DeArmond, S. L. Yang, A. Lee, R. Bowler, A. Taraboulos, D. Groth, and S. B. Prusiner, Three scrapie prion isolates exhibit different accumulation patterns of the prion protein scrapie isoform. *Proc. Natl. Acad. Sci.* **90**, 6449–6453 (1993).

75. P. Gambetti, Q. Kong, W. Zou, P. Parchi, and S. G. Chen, Sporadic and familial CJD: classification and characterisation. *Br. Med. Bull.* **66**, 213–239 (2003).

76. J. M. Warrick, H. L. Paulson, G. L. Gray-Board, Q. T. Bui, K. H. Fischbeck, R. N. Pittman, and N. M. Bonini, Expanded polyglutamine protein forms nuclear inclusions and causes neural degeneration in Drosophila. *Cell* **93**, 939–949 (1998).

77. P. W. Faber, J. R. Alter, M. E. MacDonald, and A. C. Hart, Polyglutamine-mediated dysfunction and apoptotic death of a Caenorhabditis elegans sensory neuron. *Proc. Natl. Acad. Sci.* **96**, 179–184 (1999).

78. E. A. Nollen, S. M. Garcia, G. van Haaften, S. Kim, A. Chavez, R. I. Morimoto, and R. H. Plasterk, Genome-wide RNA interference screen identifies previously undescribed regulators of polyglutamine aggregation. *Proc. Natl. Acad. Sci.* **101**, 6403–6408 (2004).

79. J. F. Bazan, R. J. Fletterick, M. P. McKinley, and S. B. Prusiner SB, Predicted secondary structure and membrane topology of the scrapie prion protein. *Protein Eng.* **1**, 125–135 (1987).

80. B. Hay, R. A. Barry, I. Lieberburg, S. B. Prusiner, and V. R. Lingappa VR, Biogenesis and transmembrane orientation of the cellular isoform of the scrapie prion protein. *Mol. Cell. Biol.* **7**, 914–920 (1987).

81. B. Hay, S. B. Prusiner, and V. R. Lingappa, Evidence for a secretory form of the cellular prion protein. *Biochemistry* **26**, 8110–8115 (1987).

82. N. Stahl, D. R. Borchelt, K. Hsiao, and S. B. Prusiner, Scrapie prion protein contains a phosphatidylinositol glycolipid. *Cell* **51**, 229–240 (1987).

83. F. Wopfner, G. Weidenhofer, R. Schneider, A. von Brunn, S. Gilch, T. F. Schwarz, T. Werner, and H. M. Schatzl, Analysis of 27 mammalian and 9 avian PrPs reveals high conservation of flexible regions of the prion protein. *J. Mol. Biol.* **289**, 1163–1178 (1999).

84. D. A. Harris, D. L. Falls, F. A. Johnson, and G. D. Fischbach, A prion-like protein from chicken brain copurifies with an acetylcholine receptor-inducing activity. *Proc. Natl. Acad. Sci.* **88**, 7664–7668 (1991).

85. T. Simonic, S. Duga, B. Strumbo, R. Asselta, F. Ceciliani, and S. Ronchi, cDNA cloning of turtle prion protein. *FEBS Lett.* **469**, 33–38 (2000).

86. C. S. Yost, C. D. Lopez, S. B. Prusiner, R. M. Myers, and V. R. Lingappa, Non-hydrophobic extracytoplasmic determinant of stop transfer in the prion protein. *Nature* **343**, 669–672 (1990).

87. H. Bueler, A. Raeber, A. Sailer, M. Fischer, A. Aguzzi, and C. Weissmann, High prion and PrPSc levels but delayed onset of disease in scrapie-inoculated mice heterozygous for a disrupted PrP gene. *Mol. Med.* **1**, 19–30 (1994).

88. N. Stahl, D. R. Borchelt, and S. B. Prusiner, Differential release of cellular and scrapie prion proteins from cellular membranes by phosphatidylinositol-specific phospholipase C. *Biochemistry* **29**, 5405–5412 (1990).

89. R. S. Stewart, B. Drisaldi, and D. A. Harris, A transmembrane form of the prion protein contains an uncleaved signal peptide and is retained in the endoplasmic reticulum. *Mol. Biol. Cell.* **12**, 881–889 (2001).

90. J. Ma and S. Lindquist, Wild-type PrP and a mutant associated with prion disease are subject to retrograde transport and proteasome degradation. *Proc. Natl. Acad. Sci.* **98**, 14955–14960 (2001).

91. B. Drisaldi, R. S. Stewart, C. Adles, L. R. Stewart, E. Quaglio, E. Biasini, L. Fioriti, R. Chiesa, and D. A. Harris, Mutant PrP is delayed in its exit from the endoplasmic reticulum, but neither wild-type nor mutant PrP undergoes retrotranslocation prior to proteasomal degradation. *J. Biol. Chem.* **278**, 21732–21743 (2003).

92. D. R. Brown, B. Schmidt, and H. A. Kretzschmar, Role of microglia and host prion protein in neurotoxicity of a prion protein fragment. *Nature* **380**, 345–347 (1996).

93. M. Daniels, G. M. Cereghetti, and D. R. Brown, Toxicity of novel C-terminal prion protein fragments and peptides harbouring disease-related C-terminal mutations. *Eur. J. Biochem.* **268**, 6155–6164 (2001).

94. S. Lehmann and D. A. Harris, Two mutant prion proteins expressed in cultured cells acquire biochemical properties reminiscent of the scrapie isoform. *Proc. Natl. Acad. Sci.* **93**, 5610–5614 (1996).

95. L. Solforosi, J. R. Criado, D. B. McGavern, S. Wirz, M. Sanchez-Alavez, S. Sugama, L. A. DeGiorgio, B. T. Volpe, E. Wiseman, G. Abalos, E. Masliah, D. Gilden, M.

B. Oldstone, B. Conti, and R. A. Williamson, Cross-linking cellular prion protein triggers neuronal apoptosis *in vivo*. *Science* **303**, 1514–1516 (2004).

96. L. Ivanova, S. Barmada, T. Kummer, and D. A. Harris, Mutant prion proteins are partially retained in the endoplasmic reticulum. *J. Biol. Chem.* **276**, 42409–42421 (2001).

97. J. Ma, R. Wollmann, and S. Lindquist, Neurotoxicity and neurodegeneration when PrP accumulates in the cytosol. *Science* **298**, 1781–1785 (2002).

98. R. S. Hegde and N. S. Rane, Prion protein trafficking and the development of neurodegeneration. *Trends Neurosci.* **26**, 337–339 (2003).

99. J. Ma and S. Lindquist, De novo generation of a PrPSc-like conformation in living cells. *Nat. Cell Biol.* **1**, 358–361 (1999).

100. Y. Yedidia, L. Horonchik, S. Tzaban, A. Yanai, and A. Taraboulos, Proteasomes and ubiquitin are involved in the turnover of the wild-type prion protein. *EMBO J.* **20**, 5383–5391 (2001).

101. J. Ma and S. Lindquist, Conversion of PrP to a self-perpetuating PrPSc-like conformation in the cytosol. *Science* **298**, 1785–1788 (2002).

102. B. Tsai, Y. Ye, and T. A. Rapoport, Retro-translocation of proteins from the endoplasmic reticulum into the cytosol. *Nat. Rev. Mol. Cell. Biol.* **3**, 246–255 (2002).

103. L. Ellgaard and A. Helenius, Quality control in the endoplasmic reticulum. *Nat. Rev. Mol. Cell. Biol.* **4**, 181–191 (2003).

104. N. F. Bence, R. M. Sampat, and R. R. Kopito RR, Impairment of the ubiquitin-proteasome system by protein aggregation. *Science* **292**, 1552–1555 (2001).

105. X. Roucou, Q. Guo, Y. Zhang, C. G. Goodyer, and A. C. LeBlanc, Cytosolic prion protein is not toxic and protects against Bax-mediated cell death in human primary neurons. *J. Biol. Chem.* **278**, 40877–40881 (2003).

106. Y. Bounhar, Y. Zhang, C. G. Goodyer, and A. LeBlanc, Prion protein protects human neurons against Bax-mediated apoptosis. *J. Biol. Chem.* **276**, 39145–39149 (2001).

107. S. J. Kim, R. Rahbar, and R. S. Hegde, Combinatorial control of prion protein biogenesis by the signal sequence and transmembrane domain. *J. Biol. Chem.* **276**, 26132–26140 (2001).

108. S. J. Kim and R. S. Hegde, Cotranslational partitioning of nascent prion protein into multiple populations at the translocation channel. *Mol. Biol. Cell* **13**, 3775–3786 (2002).

109. P. Walter and A. E. Johnson, Signal sequence recognition and protein targeting to the endoplasmic reticulum membrane. *Annu. Rev. Cell Biol.* **10**, 87–119 (1994).

110. R. S. Stewart and D. A. Harris, Mutational analysis of topological determinants in prion protein (PrP) and measurement of transmembrane and cytosolic PrP during prion infection. *J. Biol. Chem.* **278**, 45960–45968 (2003).

111. P. K. Panegyres, K. Toufexis, B. A. Kakulas, L. Cernevakova, P. Brown, B. Ghetti, P. Piccardo, and S. R. Dlouhy, A new PRNP mutation (G131V) associated with Gerstmann-Straussler-Scheinker disease. *Arch. Neurol.* **58**, 1899–1902 (2001).

112. D. T. Rutkowski, V. R. Lingappa and R. S. Hegde, Substrate-specific regulation of the ribosome- translocon junction by N-terminal signal sequences. *Proc. Natl. Acad. Sci.* **98**, 7823–7828 (2001).

113. S. J. Kim, D. Mitra, J. R. Salerno, and R. S. Hegde, Signal sequences control gating of the protein translocation channel in a substrate-specific manner. *Dev. Cell* **2**, 207–217 (2002).

114. B. Jungnickel and T. A. Rapoport, A posttargeting signal sequence recognition event in the endoplasmic reticulum membrane. *Cell* **82**, 261–270 (1995).

115. D. Gorlich, E. Hartmann, S. Prehn, and T. A. Rapoport, A protein of the endoplasmic reticulum involved early in polypeptide translocation. *Nature* **357**, 47–52 (1992).

116. S. Voigt, B. Jungnickel, E. Hartmann, and T. A. Rapoport, Signal sequence-dependent function of the TRAM protein during early phases of protein transport across the endoplasmic reticulum membrane. *J. Cell Biol.* **134**, 25–35 (1996).

117. R. D. Fons, B. A. Bogert, and R. S. Hegde, Substrate-specific function of the translocon-associated protein complex during translocation across the ER membrane. *J. Cell Biol.* **160**, 529–539 (2003).

118. C. V. Nicchitta and G. Blobel, Lumenal proteins of the mammalian endoplasmic reticulum are required to complete protein translocation. *Cell* **73**, 989–998 (1993).

119. J. Tyedmers, M. Lerne, M. Wiedmann, J. Volkmer, and R. Zimmermann, Polypeptide-binding proteins mediate completion of co-translational protein translocation into the mammalian endoplasmic reticulum. *EMBO Rep.* **4**, 505–510 (2003).

120. R. S. Hegde, S. Voigt, and V. R. Lingappa, Regulation of protein topology by trans-acting factors at the endoplasmic reticulum. *Mol Cell.* **2**, 85–91 (1998).

121. P. Chien, J. S. Weissman, and A. H. DePace, Emerging principles of conformation-based prion inheritance. *Annu. Rev. Biochem.* **73**, 617–656 (2004).

122. K. Si, S. Lindquist, and E. R. Kandel, A neuronal isoform of the aplysia CPEB has prion-like properties. *Cell* **26**, 879–891 (2003).

123. R. S. Hegde and V. R. Lingappa, Regulation of protein biogenesis at the endoplasmic reticulum membrane. *Trends Cell Biol.* **9**, 132–137 (1999).

Chapter 17

CONCLUSION: INTERVENTION, THE FINAL FRONTIER

David R. Brown

Department of Biology and Biochemistry, University of Bath, Bath BA2 7AY, UK

The chapters of this book cover large areas of the research into prion diseases. The focus of this book is specifically on neurodegeneration, the central pathological hallmark of prion diseases. Much of what has been covered does not specifically address a "mechanism" and hence could be considered not directly relevant to the cell death mechanism. This would be an unfortunate thing. Cell death does not occur outside of the cellular environment. Furthermore, in this particular group of diseases, cell death is closely linked with aberrations in protein folding and metabolism. In a sense, neuronal death in the brain of a CJD patient does not occur because of a toxic insult or from failure of blood supply or exposure to cellular contents as in ischemic death in stroke. It is the result of an uncontrollable cascade of events resulting from the misfolding of a single protein. Many similarities have been drawn to Alzheimer's disease because in that disease there is also the deposition of an abnormal protein, the beta-amlyloid protein. Although there are similarities, not all amyloidogenic diseases are the same, as some misinformed casual observers might suggest. First, there is a considerable body of evidence from a variety of sources to suggest that beta-amyloid is not involved in the neuronal death occurring in Alzheimer's disease and the protein is but a "tomb stone" marking the site of abnormal metabolism that might have a causative role. Second, the amyloid protein that is generated is the result of the change in a metabolic breakdown product of the amyloid precursor protein. Finally, the likelihood of getting Alzheimer's disease increases with age. In other words the perturbations that occur resulting in Alzheimer's disease pathology result from part of the aging process and are quite common. Alzheimer's disease claims a million people each year while prion diseases are only the cause of one death in a million. Therefore, the biology of the prion protein, and the progress of the disease and the many changes that result because of the disease in the brain of an animal or insights from *in vitro* models are all essential parts of the bigger puzzle that once solved might answer the question of what causes neuronal death in prion diseases.

As molecular techniques develop perhaps further insights into the disease mechanism will be found. The current standpoint is that when prion protein expression is switched off during disease, cell death is halted. At one level this merely confirms what was found from cell culture studies ten years ago. At another level it is such a powerful confirmation of that finding that the direct role of PrP^c expression in the toxic mechanism cannot be ignored. Similarly, lack of expression of PrP^c as the cause of cell death can be eliminated because of the variety of transgenic models that have shown lack of expression does not cause spontaneous cell death. This is probably because of compensatory mechanisms causing upregaulation of other proteins or molecules that can substitute for PrP^c functionally. In contrast, loss of the function of PrP^c while the protein is still being express is emerging as a strong candidate for an aspect of the toxic mechanism. Inhibition of the the function of PrP^c can come about two ways, either by the conversion of the protein to PrP^{Sc} or by inhibition due to interaction with PrP^{Sc} or another molecule. The consequences of this "direct effect" are in need of further study and are tightly linked with research to develop our understanding of the function of PrP^c. This loss of function is also not sufficient for cell death in prion disease. Mice expressing mutant proteins lacking the octameric repeat region or other elements do not show any significant neuronal loss. Therefore, loss of function of PrP^c does not cause neuronal death in prion disease.

Altering the structure or length of PrP^c does cause neuronal death. Expression of truncated PrP molecules in transgenic mice cause neuronal death. This cell death is not necessarily like that seen in prion disease but these results do highlight the potential for PrP to be directly toxic to neurones. Indeed, deletions are not necessary for this effect as it has also been shown that some point mutation can cause a phenotype in mice that leads to spontaneous neuronal loss. Some researchers have suggest that these point mutation alter the protein to have a transmembrane orientation. However, other evidence, especially from cell culture studies suggest that the protein is not be correctly anchored to the cell membrane and remains trapped in the endoplasmic reticulum. Expression of proteins in the wrong place or "ectopically" can also cause cell death and overexpression systems for PrP are also likely to have similar effects. This has been observed for the homologue of PrP, doppel. Expression of doppel in the brain, especially in the absence of normal PrP^{c} expression, results in the death of neurones. Once again, this in not prion disease and there appears to be no role of doppel in any prion disease cell death mechanism.

Although some researcher still pursue the hypothesis that cell death in prion disease is not linked to direct or indirect effects of, the vast majority of evidence suggest that it is, or at least that a PrP molecule of some form is involved. Therefore, most of the current and future research is likely to focus on exactly how PrP causes neuronal death. The many approaches in the present book show the wide variety of ways this has been pursued. As mention in the introduction, the relevance of a lot of this research is one of inference and background rather direct observation of phenomena or mechanism causally related to the death of neurones. This is a necessary and important thing as it provides a more realistic view of the whole biological event of "prion disease". It is also far superior to a disturbing current trend in much of the literature. This trend involves researchers identify a "change" either in the expression of a particular protein or a signalling pathway and then stating that this is directly related to neuronal death. This is purely wishful thinking and often plainly illogical. Any event can have multiple consequences. In biology the event may cause cell death but the consequences of the multi-step pathway from initiation to final cellular destruction are likely to be vast in number. Additionally, the event of prion disease may have the consequence of neuronal death but that pathway leading to death is only one consequence of prion disease. There are likely to be a multitude of biological consequences that are totally unrelated to the cell death observed. Therefore insistence that altering the expression of a protein has consequences for neuronal death are meaningless until it is shown that inhibiting that change, specifically, decreases cell death.

The emerging picture for many neurodegenerative diseases is that the mechanism of cell death is multi-factorial. This is probably a simplistic way of saying that a lot is going on during cell death and there is no salient characteristic that can be targeted. On the other hand, when a number of factors are involved then, potentially, diminishing one of them might aid neuronal survival. This, unfortunately, is of no potential benefit to the disease sufferer. In prion disease, diagnosis before death is still so poor that neurodegeneration is already highly advanced before a potential treatment could be applied. Although arresting cell death in a patient who is practically already permanently brain damaged might be of clinical interest, it is of little practical use. This has, sadly, been

Figure 17.1. The pathway leading to cell death. Identification of the targets for intervention is necessary to produce a strategy for treatment. The steps in this pathway coincide with the targets for intervention.

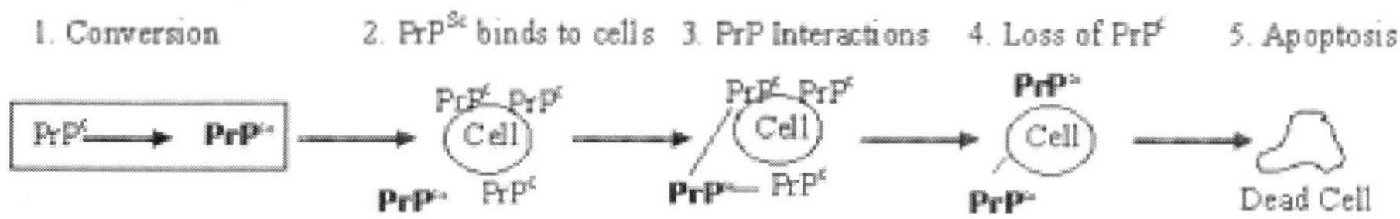

the case for all attempted treatments of CJD to date. Diagnosis is still very difficult and is likely to remain so even when better diagnostic tests are available, for the simple reason that the primary symptoms of the diseases are very similar to a huge range of neurological conditions. This pessimistic scenario does not detract from the need to find an effective treatment for these diseases.

The understanding of neurodegeneration in prion disease is sufficiently advance for consideration of strategies to inhibit cell death to be possible. The most straightforward way to develop such a strategy is to identify cellular targets where intervention could arrest or inhibit cell death. The chapters in this book contain much information that allows these targets to be identified.

These events constitute the direct effects of toxic PrP forms to neurones. The various events that signify these points of intervention can be summarised as follows:

1. Formation of the abnormal PrP protein.
2. Interaction of the abnormal protein with cells.
3. Entry of the protein into the cell (uptake)
4. Possible involvement of the host protein either as a site at which the abnormal protein binds or as a substrate to form the abnormal protein.
5. Loss of functional cellular prion protein.
6. Initiation of cell death mechanisms—signalling cascades

Other factors may also be involved which could be targets for intervention. These include possible oxidative stress generated in the brain or the interaction of the abnormal protein with cells such as microglia and astrocytes that have been suggested to be involved in the cell death mechanism. The latter targets would be those considered to be "indirect" in that they are not a result of the effect of the protein directly on the cell fated to die. However, such a description is rather a misnomer as even the so called "direct" effects must be mediated by proteins or other molecules. As a neurone cannot survive in the brain in the absence of the support cells then this distinction between direct and indirect effects is ludicrous. Indeed, oxidative stress is something that happens to cells all the time. The ability of the cell to respond and deal with the stress is the single factor that determines whether this stress will kill the cell. Oxidative or stressing substances can also be generated by neurones as well as glial cells as a result of prion disease and these substances could then initiate death of neighbouring neurones. In other words the complexity of the system precludes simplistic descriptions of cellular responses to their environment.

In the case of prion disease and probably most forms of neurodegeneration the neurone cannot be considered separately from its cellular environment. The clearest demonstration of this is in the chapter by Liberski and Ironside describing the intricate changes to the pathology of the brain in the many different forms of prion diseases. As these changes are complex and vary between different types of TSEs and even between different strains of the one disease, then it is very likely that the cause of neuronal death may differ as well. Identification of underlying causes common to all the forms is also likely. Despite the insistence of some, it is unlikely that these causes will be easily defined as direct or not or would even need to be.

Strategies that deal with the neurone and changes to it are most likely to deal effectively with the changes to neurones. However, strategies that don't include consideration of the cells environment are only likely to temporarily abate any ongoing neuronal death. Once glial cells in the brain are activated, generating known inflammatory cytokines, oxidative substance or causing a relative increase in the level of toxic glutamate, neuronal death is likely to continue as a result of these changes, independently of any changes specific to prion disease. Targeting neurones as the point of intervention is only likely to have long term preventive consequences if the intervention occurs before the cascade of events involves intercellular changes such as neuronal-glial interactions.

Strategies to be investigated to inhibit cell death include the following:

1. Create agents that remove PrP^{Sc} or abolish its formation
2. Block PrP^{Sc} interaction with cells.
3. Temporarily switch off expression of PrP^{c}. Requires analysis of factors that regulate the levels of PrP expression.
4. Map the intracellular signalling pathways that lead to neuronal death. Then use agents which inhibit cell death through the identified pathway.
5. Prevent changes to cells resulting in loss of PrP^{c} activity

The investigation of these possibilities remains. Developing an effective means of intervention remains the final frontier for prion diseases. Although many of us will cross to the unknown country before this occurs, it is likely that in the generations to come prion disease will meet its own nemesis. In terms of the voyage home to this end, the contents of this book amply demonstrate that we have at least made first contact with this possibility.

Index